Praise for *Breath from Salt*

"A thorough and engrossing saga packed with information, *Breath from Salt* is really about the transformative power of hope. It arrives at just the right moment, when too many of us have forgotten what human faith, ingenuity, and relentless determination can accomplish."

—Dan Fagin, author of the Pulitzer Prize–winning *Toms River*

"Beautifully written, *Breath from Salt* brings to life the amazing story of a passionate community that dared to dream, work together, and take extraordinary risks to bring life-saving treatments to those with cystic fibrosis. Timely and important, the breakthroughs described in this book are likely to accelerate cures for many other genetic diseases."

—Dr. Preston W. Campbell, former president and CEO of the Cystic Fibrosis Foundation

"Timely and inspiring, Bijal Trivedi's *Breath from Salt* performs several rare feats. It captures—through rigorous research, rich scene-setting, and deep empathy—a revolution-in-progress: genetic medicine's ability to conquer new fronts in the war on disease. It chronicles the alchemy of pain, loss, science, and money into forging bold new cures. It heaps 'firsts' one on top of another into a template for innovation in health. It's an expansive, moving, hopeful, multigenerational saga with the intimate tempo, pacing, dynamics, and attention to character of a mystery novel . . . A truly masterful combination of science writing, deft historical insight, narrative dexterity, thoughtfulness, and compassion."

—Barry Werth, author of *The Billion-Dollar Molecule*

Breath
from Salt

Breath
from Salt

A Deadly Genetic Disease,
a New Era in Science, and the Patients and
Families Who Changed Medicine Forever

BIJAL P. TRIVEDI

BenBella Books, Inc.
Dallas, TX

BenBella Books, Inc.
10440 N. Central Expressway
Suite 800
Dallas, TX 75231
www.benbellabooks.com
Send feedback to feedback@benbellabooks.com

BenBella is a federally registered trademark.

Printed in the United States of America
10 9 8 7 6 5 4 3 2 1

Library of Congress Control Number: 2020018303
ISBN 9781948836371 (trade cloth)
ISBN 9781950665501 (electronic)

Editing by Leah Wilson
Copyediting by James Fraleigh
Proofreading by Michael Fedison and Sarah Vostok
Indexing by WordCo Indexing Serivces, Inc.
Text design by Publishers' Design & Production Services, Inc.
Text composition by Katie Hollister
Cover design and photography by Sarah Avinger
Printed by Lake Book Manufacturing

Distributed to the trade by Two Rivers Distribution, an Ingram brand
www.tworiversdistribution.com

Special discounts for bulk sales are available.
Please contact bulkorders@benbellabooks.com.

For my mum and dad, Bhavna and Pravin Trivedi, whose love and encouragement make all things possible.

CONTENTS

Prologue   *xv*

PART 1

1   Hello, Joey   *1974*   3

2   The Treatment Plan   *1974*   11

3   Case 44: Babies Hospital, New York City   *1935*   17

4   A New Disease   *1936–1937*   23

5   An Equal Opportunity Disease   *1938–1942*   28

6   Christmas Homecoming   *1974–1977*   34

7   The Sweat Test   *1943–1960*   45

8   A Tribe of Desperate Parents   *1950–1955*   55

9   Lessons from Polio   *1955–1960*   64

10 The Registry   *1960–1966*   72

11 The Therapist   *1977*   80

12 A Disease in Search of Ideas   *1964–1980*   85

13 The Hitman Cometh   *1979–1982*   97

14 Salt and Water   *1970–1981*   104

15 Salty Boy   *1981–1983*   112

16 Birth of an Advocate   *1984*   123

17 Out of Many, One   *1978–1984*   129

PART 2

18   The Gene Hunters   *1980–1984*   139

19   Lucky Number Seven   *1984–1985*   149

20   Kate   *1986*   160

21   To Screen or Not to Screen Newborns   *1985*   166

22   Michigan   *1985*   172

23   Joey's Long Goodbye   *1986*   185

24   Mad Pursuit   *1987*   191

25   The Gene   *1989*   206

26   Runny, Like Water   *1989–1994*   214

27   Venture Philanthropy—Funding Drug
     Development   *1977–1999*   224

28   The Gene *Is* the Medicine   *1989–1991*   236

29   Transforming the Lungs   *1992–1993*   247

30   First-in-Human Trials   *1993–1995*   253

31   The Beall Curve   *1996–1997*   264

PART 3

32   The Joey Fund   *1989–1999*   275

33   A Network for Developing Therapeutic Drugs   *1997–1998*   283

34   Aurora   *1998–1999*   291

35   The Gates Open   *1987–2000*   307

36   A Tale of Four Families—CF in the New Millennium
     1999–2005   318

37   The Takeover   *2001*   327

38   Getting the Band Together   *2001–2003*   339

39   Pay to Play   *2002–2004*   354

40 Molecular Architects of Vertex West *2004–2005*   363

41 Rat to Man *2003–2004*   371

42 A Christmas Gift *2004–2005*   379

43 The Lucky Four *2005–2007*   387

44 The Doorman Launches
the Era of Genetic Medicine *2007–2008*   395

45 Disruption *Fall 2008*   406

46 Tasting Like Average People *2010–2012*   411

47 Milestones to a Cure *2004–2015*   425

48 Tackling the Common Mutation *2004–2013*   431

49 What Mutation Are You? *2013–2014*   441

50 The Mother of All Deals *2014*   447

51 Very Personal Clinical Trials *2012–2017*   453

52 The Triple *2015–2017*   462

53 The Home Stretch *2018*   471

54 The Leftovers *2015–2020*   477

55 A New Generation *2017–2020*   487

Epilogue *2019–2020*   495

Acknowledgments   500
*History of the Cystic Fibrosis Foundation
    and CF Science and Drug Discovery*   506
Life Expectancy of Cystic Fibrosis Patients, 1940–2017   510
Endnotes   511
Index   541
About the Author   558

Dum spiro spero . . .
While I breathe, I hope.[1]

PROLOGUE

It isn't often that a journalist has the opportunity to chronicle a revolution—not a singular discovery, or incremental advance, but a true sea change in science and medicine—all in real time.

When I was assigned to write a piece for *Discover* magazine in 2012 about what the US Food and Drug Administration described as a "breakthrough therapy"* for cystic fibrosis, I thought the story was simple: a rare and deadly disease, a promising new drug, and a bright future for thousands of suffering children. But over the course of a few months I realized that what I'd seen at first was just the tip of the iceberg. This story was much deeper. Much more special. And had implications that reached far beyond this disease.

Cystic fibrosis, or CF, is a fatal inherited lung disease that once killed nearly every child born with it before their first birthday. The disease destroys the pancreas and fills victims' lungs with a thick, suffocating mucus that starves them of air and allows bacterial infections to thrive and destroy their airways. And it's one of the few conditions that causes extremely salty sweat. People with CF have trouble extracting nutrients from the food they eat, but ultimately it is the lung infections that kill. This newly approved medicine was the first treatment that fixed the disease's underlying cause.

Like most students who have ever taken a biology course, I had heard of cystic fibrosis. It's described in introductory texts as the most common genetic disease among Caucasians, and featured as an example of disease caused by a mutation in a single gene. One copy of a mutated cystic fibrosis gene makes a person a "carrier"—a symptomless vehicle for deadly genetic cargo. If two carriers have a child, there is a one-in-four chance that the child will inherit a bad copy from each parent and

---

* Andrew Pollack, "F.D.A. Approves New Cystic Fibrosis Drug, *New York Times*, February 1, 2012.

be born with the disease. And though the disease itself is rare, affecting only 30,000 in the US, one in twenty-nine Caucasians, and one in thirty-five Americans overall, are carriers.

In 1989, when I was still in college, scientists identified the gene mutation that causes cystic fibrosis. Everyone expected a cure would follow. But it didn't. Efforts to develop therapies withered. Angry, disappointed parents, their hopes crushed, were left to watch their children die. And by the mid-1990s, cystic fibrosis was deemed incurable.

That prognosis seemed to be reversed, at least for some, with the FDA's 2012 approval of this new drug, Kalydeco. What made Kalydeco's development possible, I discovered, was that—at first—it wasn't paid for with stockholders' money or investments from corporate giants. Instead, the nonprofit Cystic Fibrosis Foundation, based in Bethesda, Maryland, had somehow come up with hundreds of millions of dollars to support its earliest development.

One of the heroes responsible for driving that outpouring of financial support was Joe O'Donnell. He spearheaded an ambitious campaign to raise more than a quarter-billion dollars from philanthropic donations, which the foundation then invested in a company called Vertex Pharmaceuticals to develop a drug. This new strategy, dubbed "venture philanthropy," was the first time a health nonprofit had used the approach to drive the creation of a medicine from scratch.

In Boston, Joe O'Donnell is a legend. Half Irish, half Italian, he was born in the hardscrabble working-class town of Everett just five miles from downtown. His father was a Golden Gloves boxing champion turned cop; his mother, a homemaker. He attended Harvard College, where he was a football and baseball star, then the university's business school, and then he became a wealthy, influential businessman—and a philanthropist who could raise huge sums with a Midas touch. That's what he did for the Cystic Fibrosis Foundation—and what he continues to do.

But most importantly for this story, Joe is a father. His first child, a son named Joey, died from cystic fibrosis in 1986, when he was just twelve years old.

IN MAY 2012, TO COMPLETE MY RESEARCH FOR THE MAGAZINE ARTICLE, I flew to Boston to interview Joe O'Donnell and his wife, Kathy. I'd arranged to meet them at the headquarters of one of Joe's many companies. His longtime assistant, Janice, ushered me into a conference room with a stunning view of Boston and the Charles River. The windowsill was lined with brightly colored bowling pins embossed with "The Joey Fund."

Joe was an imposing figure, tall and stocky, filling the doorway as he entered the room; he had the ruddy complexion and pale blue eyes of an Irishman, but his effusive, booming welcome signaled his Italian bloodline. Kathy, all Irish American, was petite and elegant, with a silver-blond bob that framed her high pink cheekbones and blue eyes. She had a bubbly warmth, grasping my hand and smiling as we sat.

Our conversation began with Joey.

As we sat together, Joe and Kathy shared a rich and intimate story of their beloved son and the twelve priceless years they spent with him. They told plenty of warm and funny and joyous anecdotes of a little boy who made the most of every moment of his short life. What I didn't expect, since we'd just met, was for them to open up and share candid, brutal details of their ordeal, how the disease progressively stole their child, and their final moments with Joey. Kathy broke down. Joe handed her tissues, his chin puckering and eyes brimming with tears as he shared the conclusion of the story when Kathy could not. I was unprepared for my own tears; as the mother of two children, I was deeply affected by their descriptions of Joey's suffering. Even some twenty-six years later, I could see losing their child was still an open wound.

And yet, they were not bitter. They were still pouring energy and soul into fundraising for cystic fibrosis. When I asked them about it, they told me why they remained so deeply invested in this cause. I learned the epic saga of a community of parents and patients and volunteers, some of whom have spent more than sixty-five years advocating for this disease and fighting for a cure. The O'Donnells told me stories of bull-headed scientists and CEOs who refused to quit until they delivered a treatment for patients. I heard about how the Cystic Fibrosis Foundation and Vertex, together, launched the personalized-medicine revolution—in which drugs are designed to match patients' individual mutations—and about the entirely new classes of drugs the company's scientists invented. And Joe explained the foundation's innovative model for funding the development of treatments for a disease once believed to be a death sentence.

WHEN I LEFT THE CONFERENCE ROOM, MY BRAIN WAS ON FIRE. IT DIDN'T calm down for the next three days. Cystic fibrosis had given me a new lens through which to view the progress of medicine. It provided a compelling illustration of the power of patient advocacy. And perhaps most exciting was that the new types of medicines developed through the partnership between the foundation and Vertex—and the strategy for doing so—made the process a test case for other diseases, too. In the coming decades, I realized, the lessons of cystic fibrosis could be meaningful to millions—saving individuals suffering both from rare diseases, like Duchenne muscular dystrophy, and common ones, like cancer and Alzheimer's.

After three sleepless nights, I FedExed an impassioned plea to the O'Donnells explaining why I wanted to tell their story in a book and why it was important. They invited me up to Boston a few weeks later to talk about what this would involve and how much time it would take—for both them and me. Then they agreed, Joe sealing the deal the same way he did with most of his business partners—not with lawyers and contracts, but with a handshake.

When I began working on this story, there was one game-changing drug—Kalydeco—that worked for 4 percent of patients with cystic fibrosis. It was a great tale that I figured might take about three years to write. What I didn't anticipate, as I interviewed people and began writing, was how fast Vertex's science would progress.

Kalydeco was just the beginning. As I got to know the biologists and chemists, families, fundraisers, and leaders of the CF Foundation who had committed to this effort for decades, more drugs were developed and approved. Seriously ill patients were suddenly given new hope, as drugs designed to fit their mutations began to rapidly transform their health. As this book goes to press, there are drugs for 95 percent of all patients with cystic fibrosis. And the fight will go on until there is a cure for all.

That 2012 meeting with the O'Donnells launched my seven-year immersion into the world of cystic fibrosis, the lives of those touched by the disease, and those fighting to cure it. *Breath from Salt* is the narrative of not just Joe and Kathy O'Donnell and their son Joey, but of a heroic community—the parents and patients who have been devastated by this disease, the physicians and scientists who have pioneered cures, the philanthropists who paid for them, and the people who had the vision to believe that progress was even possible. I am humbled by their trust in sharing their lives, and moved by their generosity, genius, fortitude, and strength. This is their story—so far.

PART 1

CHAPTER 1

Hello, Joey

1974

A smart mother makes often a better
diagnosis than a poor doctor.

—August Bier, German surgeon

Joe and Kathy O'Donnell stumbled out of the doctor's office and into a broad, bustling hospital stairwell. As a steady stream of nurses and patients rushed past them, they collapsed, sobbing. It was October 30, 1974, and a specialist at Boston's Massachusetts General Hospital had just confirmed what Kathy had known for five months: their infant son, Joey, was severely ill.

On May 1, 1974, he'd been born breech, three weeks premature and weighing six pounds, three ounces; he looked generally healthy, if a little small. But in the four months since they'd brought him home, Joey had "failed to thrive"—doctor's code for a baby who doesn't gain weight. He choked when Kathy nursed him, with violent, racking coughs that burst blood vessels, peppering his pale skin with tiny red spots. His frequent, pungent stools were a sign that food was slipping through his body undigested, depriving him of the nutrients he needed to grow.

Kathy brought Joey to their pediatrician almost weekly, and each time he dismissed her concerns. "Calm down," he said the first time, putting his arm around Kathy and escorting her out of the exam room. "Joey is fine. Sometimes it takes a while for a baby to learn how to latch on and nurse." When she brought him in again the following week, he told her to take him home. "Practice, Mrs. O'Donnell. This is your first baby. He's fine."

Despairing over the lack of any improvement, Kathy took Joey to the emergency room, where the same pediatrician met and admonished her again. This is becoming a habit, he told her, wagging his finger at her as if she were a small child.

But Kathy stood her ground. Ragged from sleep deprivation, she told him in a clipped, terse tone that Joey was perpetually hungry, that she fed him every hour, day and night, and he weighed the same as the day he was born. Obviously, he wasn't getting any food.

The doctor glanced at the baby and then locked eyes with her. Kathy could see that the man thought she was hysterical. "It's clear that Joey doesn't like the taste of your breast milk. Switch to formula." He scribbled out a prescription. Try this brand, he said.

She stared, incredulous. Whoever heard of a baby not liking their mother's milk?

She followed his orders, but it made no difference: Joey coughed up sour-smelling vomit after every feeding. Kathy changed his feeding schedule. Fitted a different nipple on his bottle and shifted the angle to slow the flow of liquid. Spooned soy milk and cereal into his tiny mouth. But everything she poured in came out, either immediately or as he slept, with vomit pooling around the side of his small, delicate face. Then something else impelled Kathy to drive back to the ER.

What's troubling you this time? the pediatrician asked in what seemed to Kathy an increasingly patronizing tone.

She was worried about the dots, she said, pointing to Joey's tiny hands. They covered his body. She pulled up his shirt. Red and brown spots speckled his entire torso.

The doctor stepped close and squinted at him. "Those are freckles, Mrs. O'Donnell!" he said, not bothering to conceal his exasperation. He chided her for getting agitated about the dots and sent her home.

But Kathy did not remain home with Joey. She returned repeatedly. Her frequent office visits and phone calls over the next months plainly irritated the pediatrician, and though he brushed off her concerns, he did sometimes run a blood test or perform a basic checkup on Joey. On one occasion he performed something called a "sweat test." Kathy had no idea what that was but she continued to put her faith in the expertise of the doctor, who assured her the results were normal.

ARMED WITH DIAPERS, FORMULA, AND CEREAL, JOE AND KATHY took their son with them everywhere during those first four months. As the reluctant Massachusetts spring gave way to summer, they showed off their beautiful boy to friends and family, took him to Cape Cod and to his father's semiprofessional baseball games. Though Joey barely ate, the trickle of nutrients he received from Kathy's unceasing efforts was enough to sustain his bright-eyed, cheery disposition.

Still, Kathy's suspicions plagued her. Her sister-in-law, Mary, had a son that was born six weeks after Joey, yet he was already heavier, packing on pounds as Joey struggled to gain ounces. Then, on the morning of September 24, when Joey was almost five months old, a bout of relentless vomiting sent them back to the hospital. This time, a different doctor happened to examine Joey. "These aren't freckles, Mrs. O'Donnell. These are petechiae, broken blood vessels sometimes caused by coughing. Your son is very sick. You must take him directly to Mass General."

Kathy called Joe from a pay phone in the hospital lobby and shared the news. The doctor didn't know what was wrong, she told him, but the seriousness and urgency of his assessment had frightened her—though she was relieved to hear another opinion that echoed her own fears. Perhaps now they could figure out what was afflicting Joey.

It was a quick fifteen-minute drive from Harvard Business School, where Joe worked as a recruiter for the executive education programs, to Massachusetts General Hospital in downtown Boston. The hospital, known locally as Mass General, opened in 1821[1] and was the first New England hospital offering round-the-clock care to the general public. It became the teaching hospital for Harvard Medical School and the site of numerous medical breakthroughs, including the first public demonstration of anesthesia in 1846, the first use of X-rays to diagnose patients in 1896,[2] and the first reattachment of a severed limb in 1962.[3] If anyone could figure out what was wrong with Joey, Joe thought, it would be the doctors there.

Joe and Kathy sat by Joey's crib on the fifth floor of Mass General's Vincent Burnham building, which was filled with other sick babies. Doctors dropped by sporadically to assess Joey, each one ordering a test or two. By 9 PM, it was clear that they wouldn't figure it out that evening. The nurses encouraged the couple to go home and sleep—Joey was in good hands. But Kathy refused to go home, insisting she would not leave her baby. She chose to wrap herself in a blanket and curl up on the floor beside Joey's crib, not wanting to miss a chance to speak to any doctors should they come by.

The next morning, there were still no answers. Joe could see in Kathy's stiff, exhausted frame and red-rimmed eyes that her twenty-four-hour vigil was not sustainable; they immediately agreed that if Joey was going to stay here for several nights while the doctors and nurses put him through an incomprehensible array of tests, then Kathy and Joe would take shifts so that she could go home and sleep. As Joey's stay stretched on, Kathy would arrive at the hospital at 6 AM and stay twelve hours; Joe would come at 6 PM and stay till midnight. Together, they would spend eighteen to twenty hours a day by Joey's side—though they barely saw each other. By the time Joe came home, Kathy was asleep, and she was gone long before he arose.

Days soon turned into weeks as Kathy watched helplessly while nurses prodded Joey, piercing him with needle after needle in their efforts to figure out what was wrong with him. Joey's broad range of symptoms muddied his diagnosis because, in addition to everything else, urine tests revealed Joey had a congenital viral infection called cytomegalovirus. The doctors knew the virus was linked to small size at birth, lung infections, hearing and vision loss, premature birth, muscle weakness, and seizures. But it didn't explain Joey's critical state or persistent rash.

When Joe arrived on the fourth day, he found Kathy distraught. An error had been made and Joey was given the wrong medication at the wrong time. Already short fused and frustrated, Joe grabbed Joey's chart from the end of the crib and noisily flipped through it, trying to make sense of what had been noted by whom and when. When a doctor came through on rounds, Joe pounced and interrogated him for a solid fifteen minutes, pointing out all the things that hadn't been done for his son. As the doctor left, scowling, Kathy grabbed Joe's arm firmly and led him into the hallway.

"Look, this is Joey's home right now, it's *our* home," she said bluntly. "So what we can't do is what you just did. We can't go in and grill these people. Mistakes will be made. These people are human. But they're working on our side—and you can alienate them. I've seen it happen. I'm being nice to them. I need them. And you need them, you dope. So stop it."

Joe realized that Kathy was right. He liked to be in control, to call the shots. Now Kathy was telling him that, when it came to Joey and his hospital, Joe had to stop. The next day he walked in with two giant bags of popcorn for the nurses.

Joey shared a room with an ever-changing and often tragic carousel of other young patients. One of the first was a baby, brought there by a poor single mother who had hitchhiked from Maine. The doctors told her that her child was blind and brain dead. Some children were missing limbs. Others were gaunt and bald—dying from leukemia. By comparison, Joe thought, Joey looked good. Perhaps he wasn't as sick as they thought.

Days dragged as the doctors' list of potential diagnoses hopscotched from one terrible disease to another. Spinal meningitis. Leukemia. After two and a half weeks in a hospital crib, the doctors were still stumped—but Joey was undoubtedly in better health than when he arrived. On October 10, they sent him home without a definitive diagnosis. Antibiotics had treated his urinary tract infection; he was feeding a little and had put on several ounces. There was nothing else the doctors could do for him. He still had the mysterious petechiae, but otherwise he looked like he was on the mend.

Ten days later, on October 20, Joey was readmitted. Nearly six months old now, he still only weighed about eight and a half pounds. He was vomiting again. His body was lethargic and floppy, his skin was covered in an ever-denser coat of speckles, and he had a raging case of pink eye. Dr. David Walton, a pediatric ophthalmologist, examined him and found that Joey's eyes were dry, which left the surface vulnerable to infection. And the clear protective covering of the eye—the cornea—was cloudy. Joey was looking at the world through a fogged windshield.

Dr. Walton suspected a severe lack of vitamin A, the most common cause of blindness worldwide. But after consulting with Kathy, he found there was no shortage of vitamin A–rich foods in Joey's diet. There was, however, another possibility, which Walton knew was linked to such eye problems—an inherited disease particularly common in New Englanders of Irish descent.

Walton suggested that Joey's sweat should be tested. Kathy told him that this had been done already, months earlier, and the results were normal. But Walton knew that the test was often botched by inexperienced doctors at small hospitals, and requested it be redone at Mass General.

The next day, October 21, Kathy was standing by the elevator when a young resident who had been caring for Joey stepped out with the results. The sweat test had confirmed Walton's suspicions. Joey had hyper-salty sweat, the resident explained—the hallmark of an inherited and fatal disease called cystic fibrosis.

Kathy was terrified. She had never heard of cystic fibrosis and had no idea what the resident was talking about.

The hospital's geneticist, the resident said, could tell them more.

It was just after 2 pm when Kathy called Joe, sobbing, her trademark Irish stoicism cracking. Joe knew immediately that whatever Kathy was about to say was dire. His wife was congenitally calm; it took a lot to frazzle her. In fact, in the four years they had been married, Joe had rarely seen her shed tears. Now she was incoherent. He felt his pulse quicken as he pressed his ear to the receiver, straining to decipher her words through the sobs and the din of the hospital. She was hysterical and her fear filled him with panic. "Cystic what?" he asked.

He had no idea what she was saying. But he seized on the one word that he understood: *fatal*.

He rushed straight out of his office and sped to the hospital, where Kathy sat, her face flushed and tear streaked. Joe held her until they were led to the office of a medical geneticist, where they learned Joey's fate. The geneticist, short with a goatee and dark hair, shook Joe's hand but barely nodded at Kathy. He reviewed Joey's records and, without making eye contact with either parent, delivered the facts with speed and clinical precision.

This is a genetic disease, the doctor said. Everyone carries two copies of the gene that, when mutated, causes cystic fibrosis. Each of you carries one normal and one mutated version of this gene, and you each gave your son a bad copy. He inherited this from both of you, adding, "Receiving two bad copies is lethal."

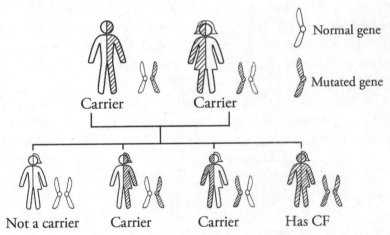

When two carriers have a child, in every pregnancy there
is a one in four chance that the child will have CF.

The words sounded garbled and foreign to the O'Donnells, as if the conversation were happening underwater in a distant ocean. Inherited? Neither Joe nor Kathy could remember anyone in either family who had died young—and both of them were in great health. They'd never crossed the threshold of a hospital before Joey's birth.

The geneticist continued his rapid-fire, monotone primer. Children with the disease have salty skin. They produce thick mucus that clogs the airways and promotes severe bacterial infections that destroy the lungs and lead to death. The life expectancy for children with the disease is five years. "But your son's lungs are badly damaged. He probably won't live a year," he said, then paused. He probably wouldn't get out of the hospital. If he did, it wouldn't be for long. "Your son's death is going to be slow and painful—and you ought to know that." Joe felt as though he had been shot.

The message was clear: their baby was dying, and further investment by Joe and Kathy would be pointless. The geneticist's assessment left no space for hope.

That was it. Just fifteen minutes after he'd shown them in, the doctor ushered Joe and Kathy, shell-shocked and shaking, out of his office, and shut the door. The world slid out from under them. Their dreams for Joey, their life with him, shattered.

The doctor's blow-by-blow description of Joey's future reduced Joe—a tough, six-foot, 195-pound former Harvard baseball and football star—to an inconsolable, heaving mass of grief. He and Kathy collapsed in a nearby stairwell, stunned and oblivious to those hurrying past them as they held each other and cried.

CONFUSION AND BEWILDERMENT WERE COMMON IN PARENTS WHOSE children received a cystic fibrosis diagnosis. In 1974, CF was an enigma even to most physicians, who, if they knew anything, knew little more than it was a fatal genetic disease.

Doctors couldn't explain why some cases of cystic fibrosis were more severe than others. Why patients tasted salty when kissed. Why the illness cruelly snatched the lives of some children before their fifth birthday, while others lived longer. Genetic medicine was embryonic. No one had yet discovered a genetic mutation responsible for causing cystic fibrosis, or any other disease. The tools for reading and studying DNA, and cutting and pasting genes together, were just being developed. And there was no genetic testing, prenatal or otherwise, for diseases like this one. Even if parents knew a disease ran in their family, there was no way to control whether they would pass the genes to their children. Not only would Joe and Kathy's firstborn die, but the couple had no way to guarantee future children wouldn't suffer the same fate.

AFTER MORE THAN AN HOUR IN THE STAIRWELL, JOE AND KATHY gathered their strength and walked to a waiting area. There, Kathy sat, head in hands, as Joe left to make a call on a pay phone a few feet away. He dialed his friends at Harvard.

Who was the top doctor for cystic fibrosis? he asked quietly, his voice hoarse. They would find out and call him back. Joe leaned against the wall, eyes closed, for about twenty minutes, waiting for an answer.

Dr. Henry Shwachman was the doc they wanted, he came back and told Kathy. He was just a couple of miles away at Boston Children's Hospital and Joe had made an appointment for the next day.

"No," Kathy told him. "Cancel it. We are staying here."

Joe just stared at her. Was she giving up? Didn't she understand what he was saying? He had found the expert, the best guy in the world, the big name who had been working on this disease for over forty years—this guy would give their baby the best shot at life.

But Kathy was adamant. And practical. Shwachman was too old, she told Joe bluntly. He'd be dead, or retired, by the time Joey was five. There was a young doctor right there at Mass General, Dr. Allen Lapey, who had trained down at the National Institutes of Health and knew a lot about this disease. He'd examined Joey after the second sweat-test results came in, and said he was a textbook case and should have been diagnosed immediately.

"I like him," Kathy said. "I want him to be Joey's doctor."

CHAPTER 2

The Treatment Plan

1974

Diagnosis is not the end, but the beginning of practice.

—Martin H. Fischer, physician

A couple of hours had passed since the O'Donnells' meeting with the geneticist when one of Joey's nurses found them in the waiting area. She'd been searching for them so they could meet with Allen Lapey.

Mascara had streaked Kathy's cheeks. Wisps of wet brown hair matted her face. Joe's eyes were red and puffy. The anguish of Joey's diagnosis had engulfed them, making them feel small, unsteady, and wilted under the weight of physical and mental exhaustion. The nurse did her best to assure them that Dr. Lapey would take good care of Joey as she led them to his office in Burnham 4, the children's ward, a floor below Joey and the rest of the infants in Burnham 5.

Dr. Lapey greeted the O'Donnells and sat them down. After asking the nurse to bring them water, Dr. Lapey held their hands and provided, for the first time, some compassion.

"We are going to do everything possible to help little Joey," he told them.

Dr. Lapey hadn't seen Joey when he first came to the hospital because Joey had been brought to the infants' floor, not the children's where his office was located. When he did examine Joey, he couldn't understand why no one—from Joey's pediatrician to the doctors at Mass General—had diagnosed him sooner. Joey's symptoms—vomiting, foul runny stool, lung infections, salty skin, malnutrition—were hallmarks of cystic fibrosis. Dr. Lapey was frustrated by this late diagnosis because it meant Joey's lung disease was now more advanced than it needed to be; his body on a tightrope between life and death. But Dr.

Lapey assured the O'Donnells that Joey was his responsibility now and he knew just what to do.

Dr. Lapey dreaded first conversations with the parents; they were never easy. But the one with the O'Donnells would be particularly tough because Joey's prognosis was bleak. The critical issue was the horrific state of Joey's lungs. When Dr. Lapey pressed his stethoscope on the six-month-old's tiny back he heard a series of crackles, like Velcro being ripped apart, and knew the situation was dire. The sound was evidence that even the tiniest airways in the outermost branches of the lungs were filled with thick, tenacious mucus. Chemical warfare between Joey's immune cells and invading bacteria had permanently destroyed sections of lung, transforming the tissue into pockets of pus. Those areas were now incapable of ever capturing oxygen from a breath.

The initial challenge, he explained, was going after the bacteria. First, he had to quash the infections in Joey's lungs with intravenous antibiotics. He'd then knock out the mucus that was allowing bacteria to multiply and destroy his lungs—a process that included physical therapy. Dr. Lapey could see that his explanation was overwhelming the couple; he paused for a minute to let things sink in.

Once the drugs kicked in, Dr. Lapey continued, the next step would be to get food into Joey so his body could get the nutrition it needed to fight. This would involve pediatric surgeons inserting a G-tube into Joey's abdomen. Joe and Kathy couldn't imagine yet another tube puncturing Joey's fragile body, but Dr. Lapey told them it was painless, and that it meant Joey wouldn't need to swallow formula and other food anymore. They'd use the tube to pour it straight into his stomach.

If all of these interventions went well, then Joey would start to put on weight, get stronger, and hopefully leave the hospital.

ALLEN LAPEY HAD BEGUN STUDYING CYSTIC FIBROSIS SHORTLY AFTER completing his residency in pediatric pulmonology at Boston Children's Hospital in 1968. At the time, the US was embroiled in the Vietnam War and was drafting young men to fight. Lapey was a pacifist and desperate for a spot in the US Public Health Service that would fulfill his draft requirement. His mentors in Boston recommended him for a clinical researcher job at the National Institutes of Health (NIH) in Bethesda, Maryland, in lieu of him serving in Vietnam. With few posts outside of field hospitals for doctors, and even scantier options for pediatricians, researcher positions were coveted opportunities that drew the country's best young physician scientists. Among Lapey's peers were Mike Brown,[1] Harold Varmus,[2] and Robert Lefkowitz,[3] all of whom eventually won Nobel Prizes.

When Lapey arrived in Bethesda, he began working in the lab of Dr. Paul di Sant'Agnese, a pioneering researcher who, Dr. Lapey knew, had devised the first version of the sweat test that would finally diagnose Joey years later. It was a lucky match. This was one of only two labs at the NIH where a pediatrician could work as a clinical researcher, and di Sant'Agnese had a reputation as a compassionate and supportive advisor who made sure that his mentees found good research projects that fast-tracked their careers. The appointment felt right to Lapey for another reason: Dr. di Sant'Agnese had cared for a childhood friend of Lapey's who had lost two siblings to cystic fibrosis and was battling it herself.

Lapey's arrival at the NIH was timely, happening just prior to an impending medical breakthrough that would extend the lives of CF patients. Up until the late 1960s, there were few antibiotics that effectively cured lung infections. This had begun to shift in 1963 when pharmaceutical researchers at Schering[4] discovered a powerful new class of antibiotics, aminoglycosides, that killed bacteria by blocking their ability to make proteins. One of these new drugs was gentamicin. It hadn't been approved by Lapey's arrival in 1968,[5] but di Sant'Agnese was able to prescribe it for "compassionate use," providing them with a powerful new weapon against CF. They gave the drug intravenously to patients infected with the *Pseudomonas aeruginosa,* a ubiquitous and hardy microbe that was particularly dangerous for cystic fibrosis patients and burn victims. It worked like magic; even patients so sick they were on oxygen began to improve. The number of hospital stays dropped, and the duration of the remaining ones were shorter. The impact was extraordinary. During Lapey's year of clinical work with this drug, not a single patient died. Lapey gained confidence, and became inspired by his patients. Many were sickly and had little to look forward to, yet had incredible resilience.

Working with the renowned di Sant'Agnese catapulted Lapey to the top of his field, and after two years at the NIH he had become one of the nation's leading CF experts. When his time there was done, Lapey had a driving desire to find a cause and cure for cystic fibrosis, but di Sant'Agnese, recognizing the young man's talent as a physician, encouraged him to take a position at Massachusetts General Hospital, where another of his mentees was leading a new cystic fibrosis unit. Going to

Mass General was a daring choice when the more prestigious place to study CF was Children's Hospital in Boston, under Harry Shwachman. But the man had a reputation for being overbearing—a contrast to the easygoing di Sant'Agnese. Lapey wanted to join a young team where he wouldn't be corralled by old ways and ideas, in a place where he could develop new therapies to tame this savage disease.

Now, he would use his knowledge to keep Joey alive.

BARELY A WEEK AFTER DR. LAPEY ADMITTED JOEY, HE PUT HIM ON A ventilator—a machine that kept Joey alive by forcing air in and out of his weak, mucus-filled lungs—and prescribed him a continuous stream of antibiotics. Joey's tiny veins were ravaged after five weeks in the hospital, forcing Dr. Lapey to insert an IV tube into a vein on his head. He covered the entry site with a Styrofoam cup, giving Joey the appearance, thought Kathy, of a jaunty 1940s sailor.

Then came the chest therapy. Killing the bacteria wasn't enough; the sticky mucus that had created cozy conditions for microbes to multiply also had to go. And getting rid of it wasn't an easy process. To loosen it, a physical therapist had to tap vigorously on Joey's chest and back in some fourteen locations, like thumping ketchup out of a bottle. It was an hour-long ritual that needed to be repeated three times a day. But it worked. After just forty-eight hours, Joey started to cough up the noxious gunk.

Joe and Kathy continued their shifts at the hospital. On the dawn-to-dusk shift, Kathy watched, hawk-eyed, as the nurses fed Joey and administered physical therapy. She would eventually need to do this herself—and she was determined to learn how to do it right. Joe arrived directly after work and the couple would spend an hour together talking over a meal or a cup of coffee in the hospital cafeteria. The schedule tempered their conversation, limiting it to the essentials. She told him what happened to Joey that day. He told her about work. They bid each other good night. Then Kathy would leave while Joe sat at Joey's bedside until midnight.

Kathy rarely left the hospital except when Joe came to relieve her. When she absolutely had to leave Joey during the day, she checked in every couple of hours, calling the nurses' station and letting them know how and where they could contact her. Though Joey was improving, he was still incredibly sick, and they couldn't assume his condition was stable. In the pre–cellphone, pre-internet era, leaving even for a few minutes was nerve-racking.

On November 5, 1974, with Joey steadily improving, the couple left the hospital to see friends—something they had largely stopped doing since Joey had been born six months earlier. They were awaiting election results in the Massachusetts governor's race between incumbent Francis Sargent and Michael Dukakis[6] when a nurse called them at 12:15 AM. Joey had taken a bad turn and was tinged blue, a sign of low oxygen. His lung infection had surged, and just after midnight doctors had rushed him into the intensive care unit. His damaged lungs couldn't harvest enough oxygen to fuel his body and his tiny heart was sputtering, too weak to pump and threatening to stop. Blood was backing up into his lungs and pooling in his legs, ankles, belly, and liver. The fluids in his lungs were drowning him from within.

When the couple arrived, Joey was still in the ICU, surrounded by a team of surgeons. Desperate to see his son, Joe barged in to see at least five tubes erupting from Joey's frame—three more than when Joe had seen him just a few hours earlier. A tube from the ventilator into Joey's trachea forced air into his wet lungs to keep him breathing. A catheter in his neck monitored the soaring pressure in the right side of his heart. An intravenous line in his wrist delivered antibiotics. A catheter removed urine. A tube in his nose suctioned out mucus. Joe swiftly backed out and into the waiting room with Kathy, apprehensive about whether Joey could possibly survive.

The doctors worked hard, administering antibiotics for the infection and diuretics to mobilize the pooling fluids. The ventilator continued to force air into Joey's lungs, and as the fluids normalized, his weary, walnut-sized heart began to pump again. A surgeon came to speak with them in the waiting room: "If he makes it through the night, he's got a pretty good chance of recovering and leaving the hospital."

By midmorning, Joey's blue eyes were open and he was peering around. The infection had ebbed and a hint of color had seeped back into his face. The doctors told Joe and Kathy to go home and sleep. Joey would recover.

And he did. Four days later doctors removed the breathing tube. Over the next four weeks, Joey fought gallantly. Weak from heart failure, lung infection, malnutrition, and a cluster of other complications that cystic fibrosis had inflicted, he began to make slow, steady improvements.

After five weeks under Dr. Lapey's care, on November 27, Joey's lungs began clearing, and Dr. Lapey arranged for surgeons to insert the

G-tube into the infant's stomach. The tube was a donut-shaped valve with a flip-top button like those on an inflatable raft. When the button was popped open, nutritious infant formula could be funneled directly into Joey's stomach, bypassing his mouth and the coughing reflex. A clamp on the outside kept food and stomach acid from oozing out.

Between the therapy and antibiotics, and the nutrients pouring in through his feeding tube, a miracle happened: Joey began to grow.

On December 2, just after Thanksgiving, another lung infection erupted. But Dr. Lapey reacted quickly and the next round of powerful antibiotics wiped it out.

Kathy remained at Joey's side. She wanted to hold him and feed him and love him every moment. Sitting next to Joey's crib, she learned how to feed him through the G-tube, give him therapy, and administer his half-dozen medications. And every time another child with CF came onto the ward, she'd befriend and engage the parents, learning from them, sleuthing for any clues on how to deal with this disease.

On December 9, Lapey decided that Joey was well enough to go home, which was both joyful and terrifying for the O'Donnells. Kathy would be taking care of Joey without the friendship and camaraderie of the nurses to keep her upbeat and without Dr. Lapey's reassuring presence nearby. But for the first time in almost three straight exhausting months, except for a brief window in October, the O'Donnell family all slept under the same roof again.

Joe and Kathy were relieved. But the geneticist's predictions of Joey's slow and painful death haunted them. Although the disease had been characterized nearly forty years before, its cause remained a mystery and effective treatments were elusive.

CHAPTER 3

Case 44: Babies Hospital, New York City

1935

If you know you are on the right track, if you
have this inner knowledge, then nobody can
turn you off . . . no matter what they say.

—Barbara McClintock, Nobel Prize winner in physiology

Cigarette hanging loosely from her lips, pathologist Dorothy Hansine Andersen peered through a veil of smoke at the little girl lying on the metal table in front of her, the barest warmth long absent from her translucent skin—a reminder of her untimely, unnatural death.

Andersen had seen the child, whom she labeled "MD" in her notes, when she entered the hospital for the first time a year earlier, alive but miserable. Just shy of her second birthday, she had skeletal limbs and a potbelly, a sign she was malnourished, like the African children Andersen had seen in the medical journals—although in this case it was not due to a lack of food. When her mother's milk failed to come in, she'd fed her daughter lactic acid, cow's milk, corn syrup, and water—the prescribed diet for newborns in 1935—for the first three months. By four months she'd added orange juice and cereals. Vegetables at six. And liver, lamb chops, and bacon beginning at seven months. Cod liver oil and a preparation of vitamin D[1] in winter. Cereals gave the girl persistent foul green diarrhea, up to eight times a day. But at thirteen months she refused to eat it and her gut calmed down. The girl's mother told doctors MD stuck to the diet and ate voraciously, but still failed to grow and develop like other children. When she was admitted to the hospital, she was too weak to walk.

After a few weeks, the doctors diagnosed MD with celiac disease, a condition recognized in the first century AD by the Greek physician Aretaeus of Cappadocia.[2] He called the disease *koiliakos* after the Greek word *koelia*, which means abdomen. "If the stomach be irretentive of the food and if it pass through undigested and crude," Aretaeus wrote, "and nothing ascends into the body, we call such persons coeliacs."

In 1888, British physician Samuel Gee described celiac as "a kind of chronic indigestion which is met with in persons of all ages, yet is especially apt to affect children between one and five years old. Signs of the disease are yielded by the fæces; being loose, not formed, but not watery; more bulky than the food taken would seem to account for; pale in color, as if devoid of bile; yeasty, frothy, an appearance probably due to fermentation; stinking, stench often very great, the food having undergone putrefaction rather than concoction."[3] Gee noted that in those diagnosed with the disease the belly protruded and weight loss was particularly visible on the limbs—a description clearly matching MD.

Today we know that celiac disease is a digestive malady triggered by the body's intolerance of gluten, a protein present in wheat, barley, and rye that gives bread its springiness and pasta its appealing gumminess.[4] To treat the disease in the 1920s, pediatrician Sidney Haas developed the "banana diet," which consisted of only banana powder, milk protein, water, eggs, scraped beef, vegetables, cheese, and bananas.[5] It was effective, we now know, because it cut out bread, crackers, potatoes, and all cereal grains. MD's new diet agreed with her delicate digestion, and after about two months in the hospital, she finally went home.

A year passed and, despite the digestive malady that initially led to her hospitalization, MD grew taller, gaining weight and strength. However, she had to return to the hospital for surgery to correct her torticollis, a birth defect that permanently torqued her neck, thrusting her chin in the air and her gaze skyward. Forty-eight hours after the operation her temperature spiked to 103.5 degrees Fahrenheit. Pneumonia set in. A throbbing infection erupted in both ears. Three weeks later, she was dead and lying on Dorothy Andersen's table for autopsy.

IN 1935, ANDERSEN WAS A RARITY IN AMERICA: FEWER THAN 5 PERcent[6] of practicing physicians were female, and they were strongly encouraged to focus on just pediatrics, obstetrics, gynecology, and housekeeping elements of public health and preventive medicine[7]—though social pressures at the time often dissuaded patients from seeing women physicians altogether.[8] Andersen was one of an even smaller group of women who held both a medical and a doctoral degree. She was pragmatic and worked tirelessly, married only to her work. She shunned makeup and wore her hair in a tight Victorian bun. She chain-smoked and, after hours, drank hard.

Born on May 15, 1901, in Asheville, North Carolina, she was the only child of Hans Peter Andersen, who hailed from the Danish island of Bornholm, and Mary Louise Mason, who was descended from a prominent New England family.[9] Andersen's father died in 1914, when she was just thirteen, and she and her mother resettled in St. Johnsbury, Vermont. Six years later, her mother died while Andersen was studying at Mount Holyoke College. Without money or family, she still managed to attend medical school at Johns Hopkins University, simultaneously working as a researcher[10] for Florence Rena Sabin, the first female faculty member at Johns Hopkins,[11] to help fund her studies. She completed her medical degree in 1926, and then taught anatomy for a year at the University of Rochester before applying for a surgical residency at Strong Hospital—which she was denied, historians believe, because she was a woman.[12]

Refusing to be swayed by social mores of her time and determined to have a career in medicine, Andersen accepted a position as a research assistant in a pathology lab at Columbia University's College of Physicians and Surgeons. There, she focused on endocrinology—the study of hormone-secreting glands that orchestrate growth, reproduction, mood, digestion, and sleep. In 1935 she received her doctorate in medical science and accepted a job as a pathologist at Babies Hospital.[13]

As a pathologist, Andersen autopsied bodies of patients whose deaths were inexplicable or of academic interest. It was a profession suited to Andersen's solitary nature, and she took pride in the methodical work it required. Once the patients' bodies were delivered by gurney to the hospital's dank bowels, where Andersen presided, the highly choreographed autopsy procedures began. There was a specific order for weighing, measuring, cutting, and sectioning each organ. Andersen would take small slivers of suspicious tissues and send them to a histology lab, where the tissue would be dyed, embedded in paraffin wax, and sliced paper thin so she, and others, could examine them under the microscope.

One of her first autopsies was ten-year-old Katherine Woglom,[14] whose illness began with a sore throat. She was first brought into the hospital on May 16, 1935, half-comatose after falling ill with a raging infection of the epiglottis—the flap of cartilage that covers the trachea and prevents food from sliding into the lungs. The swollen cartilage was blocking her windpipe, laboring her breathing. In the middle of the night on May 17, the surgeons performed an emergency tracheostomy.[15]

After surgery, with airflow restored, Katherine looked well. But over the next three months, she swung like a pendulum between health and sickness. Pneumonia sprang up in both lungs. She bounced back. Then a fever spiked, her neck stiffened, and her knee-jerk reflexes slowed. Her neurological symptoms suggested an infection, but a spinal tap revealed nothing. Katherine's physician at Babies Hospital tracked down a shipment of a new antibiotic—Prontosil, discovered in Germany—that had just arrived in America and seemed to destroy some forms of bacteria.[16] He injected her with it two to three times a day for almost two weeks. Her fever dropped almost immediately, but then crept up again, and the strange neurological symptoms remained. After Katherine died, four months after entering the hospital, Andersen performed the autopsy. She successfully identified the *Haemophilus influenza* bacterium as the cause of Katherine's meningitis and neurological symptoms—a feat that earned her respect among male colleagues for her astute and meticulous observations.

And now, some eighteen months or so later, Andersen had a new mystery: MD, Case 44.

WITH THE CHILD LYING ON THE TABLE, ANDERSEN OPENED HER NOTE-book, arranged her instruments, and began the autopsy. Age at death: three years and one month. Body pale, but well nourished. A bright-red slash on her right clavicle where surgeons had corrected her twisted neck. The contour of her chest normal, with a plateau at the chest bone. The belly slightly distended, but greatly improved from a year ago. With scalpel in hand, she scored the girl's chest and inserted a surgical buzz saw, grinding her way through the sternum. She cracked the breastbone in half and stretched the ribs apart like two halves of a walnut. The heart and aorta—the hose-like vessel that carries blood to the rest of the body—looked normal. She removed the heart, weighed it, and put it aside.

She removed the right lung and weighed and filleted it, opening it like a book. As a pathologist, she'd seen many healthy lungs—spongy, soft, and pink. This was not one. A healthy lung could be squeezed, chasing pockets of air into other chambers of the organ that then expanded, balloon-like. This one was heavy and dense, pocked with dime-sized abscesses that were filled with plugs the color of dried blood or the hue of yellow cheese. Some of the mucus plugs were green, instead; the shade, Andersen knew, was dependent on the bacteria living there. The tissue between the abscesses was mottled and tough, incapable of the stretching and relaxing that breath inspired. The bronchial airways were cemented with thick, sticky mucus. The left lung was a little lighter than the right, but when split open it looked exactly the same.

Next, Andersen removed and weighed the soft, bright-red spleen. It looked normal. The liver was a pale yellow rather than reddish brown, but otherwise normal as well. The stomach was small, but also normal. She then lifted it to examine the pancreas nestled below.

Something wasn't right. Andersen leaned in for a closer look, and then removed the pancreas to place it on the scale. It was just half the expected size. As she inserted her scalpel into the banana pepper–shaped organ, she heard a scraping sound, as if the pancreas were filled with grit.

The healthy pancreas is an unsung gland sitting discreetly in the abdomen, shielded by the liver and stomach, quietly secreting various juices. Some of its cells make insulin, a hormone vital for lowering sugar levels in the bloodstream, while others make a hormone called glucagon, which increases blood sugar when the body needs energy. Still others churn out acidic digestive juices, without which the intestine cannot break down food and extract nutrients.

Most of MD's pancreas, in contrast, was filled with fibrous cysts—something Andersen had never seen before—and had only a few remaining insulin-producing cells. In a healthy pancreas, a long central tube called the pancreatic duct runs the length of the organ, collecting digestive fluids and passing them to the bile duct, which empties into the intestine. But when Andersen searched for MD's pancreatic duct, she could only find a short half-inch section. The rest of the pipe, if it existed, was lost in a mass of tough scar tissue.

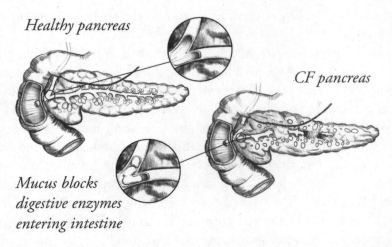

Healthy pancreas

CF pancreas

Mucus blocks
digestive enzymes
entering intestine

In a healthy pancreas, acinar cells produce digestive juices that flow into the pancreatic duct, which empties into the small intestine. In patients with cystic fibrosis, the pancreatic duct is blocked with mucus, causing acidic juices to back up and destroy both acinar and insulin-producing cells.

It was clear that this organ couldn't possibly have been doing its job. That explained the patient's malnutrition. But Andersen had never seen a celiac patient with a pancreas that was so damaged. She began combing the medical journals for clues, spending all her free time hunting through stacks of books and journals in nearby Columbia University's library for mentions of similar cases, and exploring autopsy files of other children who had been diagnosed with celiac disease. She soon discovered reports from Boston to Europe to Australia describing similar fibrous pancreatic cysts in other children who had been classified as celiac patients. But she knew that kids with celiac disease, once they were prescribed the right diet, didn't usually die; they grew quite normally. Perhaps MD didn't have celiac disease at all. In fact, based on the state of the cystic, fibrotic pancreas, she was fairly sure she didn't.

Surrounded by an ocean of medical literature in Columbia's library, Andersen realized this could be an entirely new disease—one that, unlike celiac, was incredibly deadly.

CHAPTER 4

A New Disease

1936–1937

The only true voyage of discovery . . . would be not
to visit strange lands but to possess other eyes, to
behold the universe through the eyes of another.

—Marcel Proust

The first evidence Dorothy Andersen found that others may have
shared MD's condition was in the writings of an Australian physician named Margaret Hilda Harper.

Harper was born April 4, 1879,[1] on a street lit by gas lamps and
clacking with horse-drawn buses and hansom cabs. She was raised in
a home that was a hotbed of culture and intellect, with her father and
stepmother encouraging learned, lively conversation and a healthy irreverence for "polite society." In this stimulating environment, she grew
into a stellar student, attending Melbourne University, studying in Italy,
and earning her medical degree from Sydney University in 1906. She
was a striking woman, tall and elegant, and unfailingly practical; she
kept her long, brown, wavy hair braided and tidily wrapped around her
head, and she was uninterested in clothes, except that her dresses be of
good quality and have at least one pocket. Despite her many suitors, she
had little interest in the conventional life of marriage.

By 1935, a year or so before Dorothy Andersen would begin her
autopsy of MD, Harper was already a household name in Australia. In
the 1920s, she launched the new science of "mothercraft," which promoted the benefits of breastfeeding and maintaining good hygiene in
the home to improve the odds of survival for premature babies and newborns. During that same decade she was also advocating for immunizing children with smallpox, diphtheria, and pertussis vaccines.[2] (The

tetanus vaccine only became available for civilians after World War II.) Thanks to her work, infant mortality had plummeted from a peak of 75 per 1,000 infants in 1914 to just 19 per 1,000 by 1930.

As busy as she was, Harper enjoyed keeping up with the latest medical findings. While reading the *Quarterly Journal of Medicine*, she noticed several physicians writing about pancreatic steatorrhea—the medical term for greasy, foul-smelling stool.[3] What was striking to her was that these cases were not identical. They could be stratified depending on whether there were issues with the pancreas or the lungs. In the two cases that piqued her interest,[4] pneumonia was the cause of death, and the two children looked half their ages and had a deformed pancreas and liver.

The first child, a girl known as "Baby A," was the fifth child in a family with no other notable medical history. She had digestive issues and couldn't gain weight, despite her doctors trying every possible dietary permutation, including the elimination of gluten—a sign that this wasn't celiac disease. The girl succumbed to pneumonia before she could celebrate her first birthday and no autopsy was performed. The second case, "Baby B," was the second child in her family. The first had died shortly after birth from unknown causes. Baby B's autopsy revealed a grossly deformed pancreas: a mishmash of fibrotic scar tissue where fat, muscle, and nerves were incoherently squished together with random clusters of insulin-producing cells. This pancreas couldn't make the juices critical for digesting food and fats, which was why Baby B's malnutrition was severe and ultimately fatal.

Harper wrote up her observations and published them in the November 15, 1930, issue of the *Medical Journal of Australia.*[5] She didn't say explicitly that oily stool, a deformed pancreas, and stunted growth represented a new disease, but she made clear that all conditions with fatty feces were not alike.

Reading Harper's report some six or so years later, Andersen recognized the condition she'd described as mirroring MD's. And as she dug deeper, she saw that Harper wasn't alone in her observations. Andersen found other cases much closer to home. Pioneering Chicago neonatologist Arthur Parmelee,[6] who specialized in congenital diseases afflicting newborns, had written in the literature about celiac cases that didn't respond to the usual dietary hacks.

Parmelee was inspired to probe further after the passing of a couple of his patients, including Barbara W. and Jane C., whom he suspected did not die from celiac disease despite their official diagnoses.[7] Barbara died at eleven years old with infections raging in her sinuses and lungs, and a protuberant abdomen; she had become increasingly emaciated toward the end. Superficially, her pancreas looked normal, but under the microscope it appeared as if a tornado had blown through the organ, scrambling the grape-like bunches of cells that produced the digestive juices.

Jane C. had died at just four and a half years old. When Parmelee examined her tests, he saw that, two weeks before her death, sugar levels in her urine rose—a sign that her pancreas was not churning out the insulin she needed. The autopsy report also called into question a simple celiac diagnosis when it described Jane's pancreas—"There was no typical pancreatic tissue except at the head [of the organ]."[8]

Scouring the literature, Parmelee homed in on the same examples[9] that Harper described in her 1930 publication, as well as other local cases.[10] Parmelee noticed that many sick children were grouped under the banner of celiac disease because they suffered from steatorrhea, celiac's chief symptom—but other signs like a diseased pancreas and pneumonia didn't fit. As Parmelee wrote in a December 1935 journal article, "A study of the literature plainly shows the confusion regarding the clinical conditions classified as celiac disease."[11]

Others studying celiac had clearly also noticed the peculiarities of certain cases, just as Andersen had. But no one else was putting the clues together to reach the same conclusion: that this was a new disease.

A team just to the north of Andersen, in Boston, had been drawn into the study of celiac disease from a seemingly unrelated inquiry: vitamin A deficiency. Kenneth Blackfan and Simeon Burt Wolbach, physicians at Harvard Medical School, had wondered in the early 1930s how common vitamin A deficiency was, and decided to dig through a decade's worth of autopsy reports to investigate. They discovered thirteen infants and children with vitamin A deficiency, almost all of whom died from pneumonia. This wasn't too surprising to Wolbach, an expert on vitamin deficiencies. It had been established that lack of vitamin A was linked to eye infections. But Blackfan and Wolbach suggested that vitamin A did more than just protect the eyes; it had effects throughout the body. They knew that when vitamin A was eliminated from the diets of rats and guinea pigs, the mucous lining of their respiratory passages degraded, destroying the cilia—the microscopic hairs on the airway cells that sway back and forth, sweeping the passages clear of infectious agents and debris. Without the cilia's tireless sweeping, particles and microbes lodged in the lungs could cause infection. When they published their findings in 1933, Blackfan and Wolbach hypothesized that the reason people with a vitamin A deficiency were susceptible to lung infections was that the tissue lining their airways didn't secrete enough mucus to keep the cilia lubricated, preventing their sweeping action that cleared breathing passages of microbes.[12] However, what intrigued Andersen was that more than half of the thirteen children Blackfan and Wolbach described had a decrepit, ravaged pancreas.

Leapfrogging off his work with Wolbach, in 1934 Blackfan teamed up with another Harvard Medical School pediatrician, Charles D. May, the world's reigning expert on childhood nutritional disorders.[13] For the next few years, the duo slogged through 2,800 autopsy reports of children in Boston from the previous fifteen years. Thirty-five of the children had a dysfunctional pancreas—something that no one appeared to have suspected when they were alive. Most of these kids were wasting away from malnutrition and suffering from fatty stools and chronic pneumonia. In every case, the pancreas had been annihilated.

It was clear to Andersen from the clues scattered across the literature that this celiac-esque syndrome was a distinct new disease. But she needed to prove it. She decided to write to the authors of journal articles describing fibrous, damaged pancreases and request samples of tissue collected during the autopsies of these so-called celiac cases, which she could then examine in context of the deceased patients' clinical histories. Her research turned up twenty celiac cases from the literature—including Harper's and Parmelee's—and seven more from Blackfan and Wolbach's reports on vitamin A deficiency. In addition, she used donated pancreatic tissue samples to discover twenty-two new cases, which came from just three sources: one from the Children's Hospital of Winnipeg, Manitoba; one from a nearby colleague at Mount Sinai Hospital, New York; and twenty from the pathology files of her own Babies Hospital in New York.

In an article she prepared for the *American Journal of Diseases of Children* in 1938, she meticulously described[14] the forty-nine cases that showed pancreatic damage, noting that they fell easily into distinct groups. Five of the children didn't even survive a week, dying shortly after birth from tar-like feces that clogged the bowel. All five had a malformed pancreas, implying that the condition began before birth.

A second group, of nineteen children, lived a little longer, but all died by the age of six months from lung infections and exhibited the same organ damage she'd seen in Case 44: heavy lungs packed with dense multicolored plugs of mucus blocking the airways and a shrunken, fibrous pancreas.

The third group, the remaining twenty-five cases, survived anywhere from six months until their early teens. Most had been diagnosed with celiac disease and their growth was stunted, their muscles atrophied. The trajectory of disease in this group mirrored that of group two. But the longer these children lived, the more damage their lungs and pancreas sustained.

With so many of these cases being misclassified as celiac disease, it was tough to figure out how common this newly identified disease was in the general population. But with four suspicious fatalities at her hospital in just the last year, Andersen believed the condition couldn't be that rare. So, based on 1,000 autopsies done at Babies Hospital, she did some math. Of the 605 that included an analysis of the pancreas, twenty described an organ so horribly damaged it was useless—a staggering 3.3 percent.

The number suggested this previously unrecognized fatal condition was shockingly common.

CHAPTER 5

An Equal Opportunity Disease

1938–1942

It is more important to know what sort of person has a
disease than to know what sort of disease a person has.

—An old proverb

Poring through the literature and over slides of pancreases in her lab still left Dorothy Hansine Andersen with many questions. If this disease was so common, she wondered, why was it so underreported and understudied? Perhaps, when children died of pneumonia, no one looked past their lungs? Two of three of the cases she had personally investigated were girls; were boys and girls equally affected? How did children get this disease? Was it infectious, or inherited?

In thirty-six of the forty-nine cases, Andersen had the family's medical history. In three instances, two siblings from the same family were discovered to have a fibrous pancreas when they were examined postmortem. In five cases, the patient had an older sibling who had also died in early childhood, but no postmortem had been done. Of these cases, a couple of the patients also had healthy siblings. Fifteen patients only had healthy siblings; ten were the couple's first child. Could parents transmit this disease to only some of their children?

Most of the young victims were of European descent, from Germany, England, Holland, Italy, and Ireland. There were cases in Australia and Canada and the United States, all former British colonies; and Puerto Rico, which had endured an influx of Spanish blood for four hundred years. But the disease wasn't restricted to white people; as Andersen noted in her journal, "One of the patients of the Babies Hospital was a Negro." The disease also didn't discriminate between rich and poor, hitting families at every rung of the economic ladder.[1]

As a pathologist, Andersen was particularly curious about what caused the deadly blockage in the pancreatic duct that trapped digestive acids. Even newborns had a damaged pancreas, so the evidence suggested that the problems began in the womb. But what caused them? Was there a congenital deformity that blocked the pancreatic duct, causing the digestive juices to corrode the gland from within? Or was something else the cause? The autopsies revealed that in all three groups of children the immune system was in overdrive, triggering inflammation. Was the damage to the pancreas a secondary result of inflammation? Was vitamin A deficiency responsible? That could explain the multiple cases of the disease in one family, if the mother consumed the same foods during both pregnancies.

As Andersen mulled over these questions, she learned of a young researcher working on his doctorate at Yale University who shared her interest in the pancreas and its link to celiac diseases. Jacob "Jack" Greenberg had a hunch that a birth defect caused celiac disease by blocking the pancreas, trapping the acidic juices inside and destroying the gland from the inside out. To test his hypothesis, he tried to induce celiac in kittens, a common research animal at the time, by suturing the pancreatic duct to trap the gland's digestive brew inside. To his surprise, the kittens did not develop celiac disease; instead, they became emaciated and starved to death.[2] When he examined leaf-thin slices of the animals' pancreases under the microscope, he saw normal tissue had been replaced with fibrous connective tissue. Greenberg concluded that preventing pancreatic fluids from entering the intestine did not cause celiac disease but rather some other, more severe condition. But more interesting to him was that the cats' deformed pancreases—also exhibited by Andersen's Case 44—mirrored those of other patients at Yale's hospital. One family had lost three children there with this same constellation of symptoms.

As Greenberg delved into the literature, he discovered that similar experiments[3] had been done in France nearly eighty years earlier by the prominent French physician Claude Bernard. Bernard was intrigued by the differences in digestion between carnivores like dogs and herbivores like rabbits and began experimenting to figure out how the pancreas affected digestion. Contrary to other French scientists, he postulated that the pancreas—not the stomach—was the most critical organ for digestion, due to its potent mix of digestive juices, and proceeded to prove it in dogs. When he plugged their pancreatic ducts, the dogs, like Greenberg's cats, couldn't digest their food. They became emaciated, produced stool heavy with oil and undigested food, and ultimately starved despite insatiable appetites. He published the work in 1856, earning a prize in experimental physiology from the French Academy of Sciences.[4] Bernard also connected what he observed in dogs to a case that had been described by another French physician, Dr. de la Tremblaye,[5] who in 1826 wrote about a child who died from "fatty diarrhea" and whose autopsy revealed a diseased pancreas. These studies supported Andersen's hypothesis that blocking the pancreatic duct caused a new condition similar to the one she was now calling "cystic fibrosis of the pancreas."

By 1938, Andersen finally had a critical mass of research, with examples of this new disease dating back more than a century, and dozens of case studies from the last decade alone. Her own research and astute synthesis of the medical literature led to her publication of the first complete description of this new disease—"Cystic Fibrosis of the Pancreas and Its Relation to Celiac Disease"—in the *American Journal of Diseases of Children* in 1938.

In her report, she connected the malfunctioning pancreas to the disease in the lung and gut—something no one else had done. When the pancreas was destroyed and digestion curbed, she wrote, children with cystic fibrosis couldn't break down food and absorb the fats that contained vitamin A. This led to a vitamin A deficiency, eye problems, and lung infections. She also noted that children with this disease who survived their first year had symptoms that were "clinically indistinguishable" from celiac disease and even proposed a hypothetical test to distinguish the two.[6]

Andersen's study was well received. In the discussion section of her paper, other pediatricians described patients they had examined; their comments supported her conclusions and they added a few observations of their own. Dr. Frederic H. Bartlett of New York described an infant who died at three months old, whose mother told him that another of her babies had died at eight weeks. "Whatever my first baby had, this baby has, too," she said. Bartlett noted, "It is interesting that this condition has been found in members of the same family," adding later, "the condition should be considered a new clinical entity, and I feel that Dr. Andersen has contributed a great piece of work" by deciphering the entire constellation of symptoms connected to cystic fibrosis of the pancreas.

After the paper's publication in 1938, word of the new disease spread and cases started to trickle in—not only to Babies Hospital, but specifically to Andersen herself. Because she was a pathologist, and a woman, she had never expected to assess living patients—but the hospital allowed her to begin examining small children brought to her office in the arms of desperate parents. She sensed that early diagnosis was key to helping these children. If the disease was diagnosed soon after birth, the child could be given nutritional and vitamin A therapy that might reduce the infections in the eyes and lungs. But now she faced a new dilemma: how to separate the children with celiac disease from those with cystic fibrosis. In her paper describing the disease, she suggested that because the pancreas was destroyed in children with CF, the organ didn't produce the critical enzymes essential for digesting food. She reasoned that children with cystic fibrosis would lack pancreatic enzymes, whereas children with celiac disease would secrete them. If so, a measure of the varying levels of digestive enzymes in the upper section of the intestine—the duodenum—could provide a valuable test for diagnosing CF. But she had to prove it.

First, she needed to collect and measure pancreatic enzymes from both healthy children and those suffering from a range of maladies—including those with symptoms similar but unrelated to cystic fibrosis—to establish normal baseline levels. She knew from earlier studies[7] that the enzymes that digested protein, fat, and starch—trypsin, lipase, and amylase, respectively—were present in the gut of healthy infants. But what about sick ones? To find out, she extracted and tested the intestinal juices of ninety-eight healthy and sick children.

Collecting these pancreatic juices was not a trivial procedure. She had to thread a tube through the subject's nose or mouth, using a fine silk thread with a tiny half-inch metal dumbbell at the end to coax the tube down the throat and through the stomach until it reached the duodenum, where the stomach emptied food and the pancreas dumped its digestive juices. The natural squeezing action of the esophagus that carried food from the mouth to the stomach also pulled down the tube. A nurse then would lift the patient, with the tube inserted, onto a fluoroscopic table—a device invented in 1895 that used X-rays to image the organs, skeleton, and other tissues moving and working in real time—so Andersen could see where the mouth of the tube ended up. Fluoroscopy was a dangerous procedure because, even though the risks of radiation were becoming known, neither the patients nor the doctors wore protective gear. But Andersen limited the children's exposure by working as quickly as possible. When she was sure the tube was in place, she taped the top end to the side of the patient's face so it wouldn't slide further down the gut. Then she attached a syringe to the tube and began suctioning out the digestive juices.

For Andersen, the first few efforts at collecting the duodenal liquids were harrowing—the tube tripped the children's gag reflexes and several vomited, though their bellies were empty. For other children, the procedure was too traumatic, causing them to scream and flail. But with the aid of a couple of experienced pediatric nurses, Andersen soon got the hang of it. In some children, the procedure was quick—just fifteen minutes to gather the necessary one and a half teaspoons of the vital liquid. But with babies it could take a full hour. Once Andersen had a sample, she went straight to her lab to analyze the new material.

Research for the study took almost four years, but it was worth the effort: her hypothesis was correct. After testing ninety-eight sick children—suspected celiac cases, the undernourished, some with chronic diarrhea, and fifteen with symptoms consistent with cystic fibrosis—she found that the digestive juices of the children with cystic fibrosis were all lacking one key ingredient present in healthy children and those with celiac disease: the digestive enzyme trypsin. Her research method soon became the definitive diagnostic test for the patients who were now lining up to see her at Columbia. Andersen published a description of her diagnostic test for cystic fibrosis, the first developed, and then shared it with her colleagues at the annual meeting of the American Pediatric Society in Hot Springs, Virginia, on May 23, 1941.[8]

With this second publication, Andersen was becoming a well-known scientist. Her colleagues saw her as determined, ambitious, and highly intellectual, but her growing prominence in this new field wasn't all that mattered to her. The families and sick children who came to see her experienced a different side of Andersen, one that was tender and compassionate. She cherished her relationships with the mothers and patients and stayed in close contact with them long after she'd finished their treatments, in some cases all the way until her own death in 1963. In their letters to her, the mothers shared intimate portraits of their children, with detailed descriptions of their physical appearances, weights, diets, and dispositions, and the medications they were taking. Many also sent Andersen recipes that their children tolerated, even enjoyed, that might help other sick children gain weight. In a letter dated 1944, one mother told Andersen that Knox Gelatin's cookbooks were a particularly rich source of recipes and also included a selection of her own trifles and puddings, cookies, and creams. All the letters included photographs of slender, smiling children that Andersen had personally diagnosed with CF.

Andersen always responded to these effusive letters with the utmost care—and delight—from her office at Babies or from her rustic cabin in the Kittatinny Mountains in northern New Jersey, where she would spend time away from the city. She was acutely aware of the challenges these children faced and was always generous with advice and time. As World War II was drawing to a close and food rationing was still mandated, Andersen wrote a letter to the Office of Price Administration to support one mother's request for higher sugar rations—a substitute for starch, which was difficult for children with cystic fibrosis to digest.

Even when families moved away, they stayed in touch with Andersen, whom they trusted would provide the best counsel and latest recommendations for therapies and medicines. In a letter to a Mrs. V. W., Andersen enclosed a diet that she recommended for pregnant mothers who already had a child with cystic fibrosis. She added, "I would be grateful if you would keep me informed about how you feel and be sure to have the baby tested within a month or six-weeks of birth"—a reminder that the disease seemed to run in families.

Her continued interest and attention was appreciated. One mother named J. O. wrote to her in February 1946, telling her, "You are an extra parent to these children and you give them love and sympathy that is quite separate from your intellectual gifts."[9] It was just one of many affirmations of Andersen's warm bedside manner, ironic especially because she could never be a full-time clinician.

But that was, perhaps, less important to her now than it once had been. While no one knew the cause of this disease, thanks to Andersen's work it was at least possible to accurately diagnose children with it. Now physicians could give them the nutritional support they needed—a small but critical step toward keeping these children in good health for as long as possible.

CHAPTER 6

Christmas Homecoming

1974–1977

> To raise a child with additional needs is to inhabit a
> different country from those around you. You will
> have your own customs, rules, rituals, habits, mores
> and vocabulary. People may visit, but they will never
> truly know what it is to live within the border.
>
> —Maggie O'Farrell

For Kathy O'Donnell, the outside world in 1974 was a blurred back-
drop to the personal chaos of the last eight months, four spent at
the hospital. The Watergate scandal climaxed and Nixon resigned;[1]
Muhammad Ali beat George Foreman in the Rumble in the Jungle;
and on September 12, 1974, just a few miles from Massachusetts General
Hospital, street fights and riots erupted as black children were bused to
white schools in South Boston and Hyde Park and white children were
bused to black schools in Roxbury and Mattapan.[2]

The only date that mattered to Kathy was December 9, the day Joey
came home. Kathy felt optimistic. Dr. Lapey had saved him. And he
could do it again. Over and over if necessary. "We can take care of this,"
he'd told her as she left the hospital. It calmed and reassured her. But
being home was a big adjustment. Kathy had grown used to the com-
motion of the hospital: the crowded hallways where the nurses wheeled
meal carts; the beeping of lifesaving monitors and the rattle of IV stands
moving from room to room; the hourly conversations with the cheerful
young nurses who had become her friends; the quick chats with doctors
bringing fresh-faced residents on rounds; and the soothing daily visits
from the sprightly Dr. Lapey, with his colorful ties and socks with purple cows. Without the support of this army of nurses and doctors, she

felt isolated. Though prepared to perform therapy and the other daily medical procedures Joey needed—she had watched the nurses at the hospital for months—her confidence thinned. Once Joe left for work every morning, it was just Joey, cystic fibrosis, and her.

The fact that Joey had been discharged from the hospital signaled to Kathy that his health was stable, at least momentarily. Joe was more pragmatic, always cognizant of his son's fragility and aware that the boy, now just over six months old, might not live to see his first birthday. So he and Kathy navigated life slowly, carefully, day by day. They avoided making any grand plans—but Christmas was around the corner.

For Joe and Kathy's Irish Catholic family, Christmas was a big deal. And even with the new responsibility of caring for Joey at home, the couple was excited to celebrate the holiday with Joe's family in Everett, as per tradition. When they arrived at Joe's parents' home on Christmas morning, the house was already packed. Joe's mother, Teresa, and father, Joe—Nonni and Poppy to their grandkids—were in the kitchen and living room. Nonni's sister, Aunt Lucy, who had no children of her own and had always spoiled Joe and his brothers with birthday cakes and sweets, was cooking. Joe's younger brother, Dennis, and his wife, Mary, sat on the couch with their baby, Denny, who was six weeks younger than Joey but now a whopping fifteen pounds heavier. Joe's youngest brother, Neil, sat talking to Poppy's brother, Uncle Ted, and Ted's wife, Aunt Ruth. In the corner of the living room, Poppy had placed a Christmas tree that Nonni had covered with decorations Joe and his brothers had made when they were children. Underneath the tree were brightly wrapped presents for everyone.

As Joe and Kathy opened the door, the house exhaled the mouthwatering aromas of Nonni's northern Italian cooking. Poppy reached out to grab his grandson and bring him into the living room, where he was immediately surrounded by family cooing delightedly and tickling him gently under the chin. Joe headed into the kitchen to kiss Nonni and hover hungrily over the food—traditional dishes featuring paper-thin slices of prosciutto di Parma, crispy pancetta, sharp nutty Parmigiano Reggiano, sweet-acidic balsamic vinegar, and the famous handmade stuffed pasta, tortellini, which she called "tutline," that originated from the rolling hills of Emilia-Romagna where Nonni's mother had grown up.

Every year at Christmas, Nonni would take her mother's rolling pin—one of the few possessions she brought from her hometown of Modena, Italy, when she came to America—flatten the pasta dough, and cut it into squares. Then, under Nonni's supervision, Kathy, Mary, and Joe's aunt Lucy and aunt Ruth would construct the tutline: a dollop of ground meat, a little salami, and a pinch of cheese, all placed on a square of dough, capped with another square, and pinched at the edges to make plump, pale-yellow pasta pillows. They would keep going until they had a couple hundred. When the tutline were ready, Nonni dropped them into the simmering broth for a couple of minutes before ladling them into bowls. As soon as Joe saw the pasta submerged he'd reach in and try to snag the tutline directly out of the pot, while Nonni feigned annoyance. The same scene played out every year.

Just before the family sat down for the holiday meal at 2 PM, Kathy unclamped Joey's G-tube and poured in his formula as he smiled contentedly. At the table, Joey was passed from one doting relative to another.

That evening, Kathy and Joe, stuffed from a scrumptious meal, and Joey, asleep after hours of cuddling, left Everett and went to Kathy's brother, Dick's, home in Melrose, where Dick's wife, Joan, their two kids, and Kathy's parents, Margaret and Albert Kelliher, were waiting for them. There they sat again, ate more dessert, and exchanged gifts as Dick and Joan fussed over Joey adoringly.

It was a joyous beginning to what they hoped would be a better year.

FOR KATHY, KEEPING JOEY HEALTHY WAS A FULL-TIME JOB. HE NEEDED three one-hour sessions of physical therapy every day to clear out his lungs and keep them infection-free. Kathy had to repeatedly clap her cupped hand on fourteen locations on his back and sides with the rhythm of a galloping horse to dislodge viscous phlegm that Joey expelled in large globs. She did his therapy alone except when Joe stepped in on the weekends. Then there were the powdered medicines she had to feed him before every meal, as well as the pancreatic enzymes that Kathy mixed up with applesauce and spooned into his mouth to help him digest his food. Joey had lost his sucking reflex early and couldn't drink from a bottle or cup, so at first Kathy fed him his predigested formula through his G-tube—sometimes even when he was asleep. She'd unclamp the tube, pour in the liquid, then reclamp without waking him. Within a few weeks of coming home, however, Kathy also tried solids like cereal, introducing new foods slowly as Joey began learning to eat and swallow without coughing or gagging. As his appetite grew and he got the hang of eating, Kathy spent more and more time feeding him.

The family soon fell into a routine. Kathy cared for the baby during the day and Joe, who was a night owl by nature, took the 2:00 AM feeding. The learning curve was steep as the couple struggled to apply their growing knowledge and experience to Joey's life. There were hundreds of tiny decisions to make each day. But all the care paid off. Joey began to grow.

Once Joey was able to hold up his head, Kathy secured a jumper seat in the kitchen doorway so that Joey could bounce around. It wasn't just a way to entertain him; it was therapeutic, as it helped dislodge the gunk in his lungs. When it came time for Joey's inhaled medications, Kathy put a nebulizer mask over his nose and mouth and then played games with him as he breathed the medicines in. She was amazed he was so accommodating, accepting these procedures without protest. But for Joey, they were just part of his life; he didn't know anything different.

Despite Kathy and Joe's vigilance, sickness was unavoidable. The first infection came in early February 1975, barely two months after they'd brought him home. The illness began stealthily with a cough—tiny barks that triggered vomiting. The next day, Joey developed a fever and became limp and listless. It was a pattern that Kathy would soon learn to recognize: first, the coughing, and then, within a couple hours, the fever, as the latest infection in his lungs began to spread, making him languid and unresponsive. She put Joey in the car and calmly drove to Mass General, where she met Dr. Lapey and hugged the nurses she knew so well. They quickly inserted a drip line of antibiotics in Joey's tiny wrist and pumped his fragile body full of antibiotics as he lay lethargic in the hospital crib.

Kathy had faith that Dr. Lapey would make him well again. And he did. Two weeks later, Joey was back home. But that first year out of the hospital, particularly the icy winter and chilly spring, was a lonely time for Kathy, who spent a great deal of time inside caring for Joey. Though she and Joe had family close by and a circle of supportive friends, they could do little other than offer moral support. Both Kathy's mother and Nonni would drop by to keep Kathy company; her sister-in-law Mary visited with her nephew, Denny; and her close girlfriends, Connie and her younger sister, Florence Martin, would come armed with lunch and news and gossip about their friends and family. Things were a little different for Joe; he wouldn't talk about Joey with anyone other than Kathy. Joe, as the oldest of three, was used to shepherding and guiding

his two younger brothers and preferred to solve problems on his own
rather than ask for help. Joe and Kathy handled stress differently. Joe's
strategy was compartmentalization. It was the only way he could cope
with the anxiety of Joey's disease. For him, hard work and doing something he could control—because he had so little of it, when it came to
Joey's sickness—was what he needed. Kathy understood this and gave
Joe the space to do so, enduring long days without him. But in the back
of his mind, whether he was sitting by Joey's side in the evenings at the
hospital or at home late at night, Joe was always thinking about ways he
could fight this disease.

Early on, Dr. Lapey had given the O'Donnells the phone number for the local Massachusetts chapter of the National Cystic Fibrosis
Research Foundation and encouraged them to reach out. The foundation was a group started by the parents of children with CF, and the
O'Donnells were hopeful it could provide tips about caring for Joey,
companionship with other parents, and perhaps news of promising
drugs and treatments in development. Most critically, they were hoping for a more optimistic outlook than the one the geneticist had given
them some three months earlier. Joe called, but the chapter had little to
offer—just meetings in church basements and sporadic fundraisers by
motivated parents. The fundraisers' main purpose was to support medical care for children at the local CF care center at Boston Children's
Hospital, sending whatever meager dollars left over to national headquarters to support research.

Research was important—after all, it was the path to treatments,
the way forward. But nobody knew what caused the disease, and without that information, researchers didn't have a target for a potential
drug. Any progress in quality and length of life could only stem from
better nutrition and physical therapy. The best hope from medicine at
the time was more powerful antibiotics to target lung bacteria, before
they destroyed the lung tissue and killed the child. Joe and Kathy
attended several bleak fundraising dinners in Boston where the doctors who spoke bore no good news. There were no new treatments in
the pipeline.

The Massachusetts chapter meetings also fell short of what the
O'Donnells were hoping for. The support group consisted mostly of
women, who sat in a circle in a church basement to talk about the problems and hardships that cystic fibrosis had brought to their lives and the
challenges of having sick children. Some of the mothers were physically

and emotionally exhausted, overwhelmed with caring for their children. When given the chance to talk, they cried uncontrollably as they poured out their stories. Others ranted against their doctor's advice—what was the point? Some described just appeasing their sick child, whom they were convinced would soon be dead. A few were extremely poor—desperate for help, money, and childcare. These parents had lost all hope, having written off their sick children and discarded their dreams of a future. Joe and Kathy went a few times, but quickly decided—no more.

That first year after Joey came home, the couple didn't have much of a life outside the house. They worried about leaving him because no one other than Kathy knew how to do his chest therapy or care for him. Occasionally, they went to dinner or a movie, and Nonni and Poppy or Kathy's parents would stay with Joey. Poppy would sit nervously on a chair immediately outside Joey's bedroom, reading a book, ready to come to his grandson's aid if he heard a cough or cry. But Kathy could see their parents sigh with relief whenever she and Joe returned. Hiring a babysitter was impossible; no one would look after a kid with a tube in his stomach.

As word of Joey's illness spread, Kathy and Joe began to hear from others with the same disease. Kathy's friend Florence Martin had a close friend, Denise, with cystic fibrosis. Kathy invited Florence, who everyone called Bubs, and Denise to visit. Denise was twenty-two years old and a senior in college, an excellent math student who planned on teaching elementary school children after she graduated. Kathy drank in the image of this young woman. She had so far thwarted the life expectancy for CF by six years,[3] buoying Kathy's hopes that Joey might be as lucky.

Denise told Kathy that her younger brother, who also had cystic fibrosis, was much healthier than she was because he was really active; his jumping, running, and incessant movement helped to keep congestion from settling in his lungs. Denise, by contrast, had to go to the hospital up to four times a year for regular "cleanouts"—the term used to describe the two-week-long hospital stays during which the doctors gave cystic fibrosis patients powerful antibiotics to wipe out lung infections. And Denise shared an important tip: as she and her brother grew larger, and required more vigorous chest therapy, her parents had hired a physical therapist, who had been critical to keeping both of them healthy. Kathy filed the information away for later. In the meantime, she had to keep Joey alive long enough to need it.

By the middle of 1975, when Joey was just over a year old, it was becoming evident that Joe needed to make a lot more money to cover Joey's medical bills. After working as the associate dean of MBA students at Harvard Business School for three years, he had spent the past two years in the executive education program, recruiting and admitting promising young businesspeople. It was a prestigious position, but he made only $13,000 a year (the starting salary for an MBA in the early '70s)—the equivalent of around $71,000 in 2019. He cared about the job and the people, who, once they heard about Joey's diagnosis, had rallied behind the family. The dean of the business school, Joe's boss, had come to talk with Joe and left him a check for $10,000, explaining he had seen the impact a disease like this could have on a family. It was a kindness that Joe never forgot. And the money was welcome. Joey's first year of hospital bills totaled $32,000 (or $160,000 in 2019), some $6,000 of which had to be paid out of pocket. Before Joey was born, Kathy had worked full-time. She'd run the personnel department at a poverty-fighting nonprofit, Action for Boston Community Development.[4] She'd loved the job and had pulled in a decent salary—more than Joe—and had planned to go back to work after maternity leave. But that just wasn't possible now. Joey needed her.

Joe had always been an entrepreneurial guy. He'd started a housing service for first- and second-year students, many of whom were minorities, while at business school, and after graduating he stayed on to run it while also working as an associate dean. Racial conflicts had peaked in Boston a few years earlier, but it was still really tough for African American students to find apartments. Joe approached landlords who advertised vacant units, and if they refused to rent to minorities, he challenged them in court.

Now Joe again needed to blend the skills he learned in business school with his instinctive street smarts to find a more lucrative job. He had always known that he wanted to start his own business someday, but Joey's health added pressure that drained some of the thrill. With Joey's medical expenses, it was imperative for Joe not only to stretch his entrepreneurial muscles but to succeed.

Joe was competitive from an early age; whether it was cards or football or baseball, he played to win. He was also a born businessman and dealmaker. As a kid, Joe would watch his town's legendary football and baseball teams practice at the stadium next door to his house. During the games, Joe and his buddies would hop the fence and sneak in. It was here that Joe spotted his first business opportunity. At age nine, with no money to buy candy or ice cream, Joe approached the ice cream truck outside the stadium.

How about I help you out and take ice creams into the stands? Joe asked the ice cream vendor.

The guy looked at him suspiciously, trying to gauge Joe's age.

Joe told the vendor he could sell to the 7,000 or so people watching the game by working his way from section A to G, going all the way up and all the way down. The vendor agreed and told Joe he would give him one penny for every ice cream he sold.

At the end of that first triumphant day, he walked home with a handsome $1.50 in his pocket—more than enough to buy ice cream for him and all of his friends.

It was a great job until one day, during one of the last games of the season, another kid a foot or so taller and a solid thirty pounds heavier stole a fistful of cash from Joe's pocket. The vendor screamed at Joe for losing the money and swiftly fired him.

The experience left him shamed and in tears—but it didn't dampen his enthusiasm for working. The ice cream gig had given nine-year-old Joe a feeling of control over his own life and an intoxicating taste of independence. With the money he'd earned selling ice cream, he'd been able to buy candy and toys for himself and his friends and even put the extra in the bank.

His next job was sitting in the back of a bowling alley, where he reset the pins and returned the balls; lanes were not yet automated. Once he proved he could handle that simple task and was a responsible employee, the bowling alley owner promoted him to manager, and he began hiring his friends. That was a turning point for Joe. He didn't just enjoy earning money; he liked getting promoted and felt good hiring other employees.

By junior year of high school, Joe had ratcheted up the negotiating skills he'd used to get his first job selling ice cream. It was prom season, and Joe approached the owner of the town's tuxedo store with an aggressive proposal. I'm going to bring in fifty guys to rent tuxes, Joe told him. He knew that at least thirty or thirty-five would have come anyway.

The owner started shooing him out of the store.

Joe pushed. He could bring in fifty guys. What he wanted in return was a free tux and a hundred bucks, he told the store owner, recalling business advice an uncle had once shared: Don't be afraid to act stupid. If you want one, ask for six.

The owner stared at Joe as if he were nuts. Joe said, I'm president of my class, captain of both the football and baseball teams. I could bring a lot of people into this store.

They'll all come here anyway, countered the owner.

Joe nodded politely—and told him he was going to see Max Ross, the competition in the neighboring town of Malden, instead.

The guy gulped, and for two years, Joe had a free tux and a hundred bucks.

IN 1976, JOE LEFT HARVARD AND HIS JOB AS ASSOCIATE DEAN OF EXECU-tive education. Through a colleague, Joe landed a position as president of a small company called Drive-In Concessions of Mass., Inc., that managed concessions stands at forty drive-in theaters. It had fifty employees and $1.5 million in gross revenues, and the job paid $60,000, more than four times Joe's Harvard salary. But after just six months, Joe felt hemmed in and claustrophobic. He was ready to grow the business and wanted to take on more theaters, expand the stands at existing ones, and dive into other markets. He wanted to partner with local business-men, to provide financing and help them expand their own businesses in exchange for a portion of their companies. But every decision had to be cleared with the head of the company at headquarters in Buf-falo, New York. Joe frequently flew back and forth for meetings, which exhausted him and took a toll on Kathy, too.

In December 1976, after just six months, and increasingly frustrated by the fact that he couldn't change the way this century-old company functioned, he gave notice, offering to stay for three more months until the company had hired a replacement. He told his boss that he planned to launch his own company and promised not to poach any of the cur-rent customers.

His boss, not wanting to let Joe go, countered: How about we become partners? Fly to Buffalo and we'll draw up the papers.

It was a possibility Joe hadn't considered. Buying a share in Drive-In Concessions was an exciting proposition—it wouldn't eliminate all the travel, but he'd have more freedom to innovate, expand the busi-ness, and make more money. Joe accepted the offer and flew to Buffalo, hoping to leave with a 50 percent share in the company. Concessions was the type of business he liked and understood: it was dynamic and personal, with one-on-one negotiations, and focused on food and entertainment.

After paying all of Joey's hospital bills, the O'Donnells didn't have a cent to buy into the partnership. So Joe signed papers agreeing to pay $1.5 million for half the company over the next ten years. And after eight hours of promising negotiations, Joe's boss called Joe into his office.

A weathered seventy-year-old Italian man with wrinkles gouged into his face like canyons, the man sat in his black leather chair carving an apple with a jackknife. Joe reached out his hand, expecting congratulations.

Joe, he said instead, continuing to focus on the fruit, this is a whole lot of bullshit. I don't want anyone leaving the company; it would set a bad precedent.

Joe was stunned. He had flown all the way to Buffalo and spent the entire day signing papers. What's the problem? he asked.

The deal's off, his boss said, still not looking up. This isn't going to work. Look, you can have your old job back. But if you leave, and start your own business, we'll sue you. You're not going to steal our clients.

Joe walked out of the office, slammed the door, and caught a cab to the airport. The veins in his temples throbbed, and he felt as if his head might explode. Joey was sick, but he had no intention of returning to his job, and now he had the pending threat of a lawsuit if he struck out on his own. What was he thinking? He had no safety net. He took a 9:00 PM flight back to Boston, drove himself home, and went directly to Joey's room to feed him, as he did every night. But just as he'd finished pouring the formula into the tube, before he could replace the clamp, Joey coughed in his sleep, launching the liquid—now mixed with stomach acid—back out of the tube and right in Joe's face.

Cursing under his breath and dripping with the sour-smelling vomity mess, Joe wiped himself off and then walked downstairs, collapsing in an armchair in the basement. He had a sick child gasping for breath, a mortgage, no job, no health insurance, and no backup plan. If this were a movie, he thought, he'd be lighting up a cigarette and downing whiskey. He didn't have cigarettes because he didn't smoke. He didn't really drink, either, but he found an old bottle of whiskey, poured it straight, and took a sip. His throat burned. John Wayne he was not, he thought.

As he sat in the chair, trying to stave off stomach-clenching panic, the phone rang. It was 1:00 AM. No one called at that time unless someone was hurt or dead. He picked up.

It was the boss. I'm sending over our guys with papers to sign in the morning.

Joe was silent, so confused and tired he couldn't even respond.

Joe, you still there? the man barked when Joe didn't answer. We had to see that you had the stones for this job. We needed proof you wouldn't cave under pressure. It's a big step you're taking and this is a rough business. You passed the test.

The next morning at 9 AM, a colleague from the company knocked on his door and handed him an envelope, just as his boss had promised. Joe invited him in, and after some thirty minutes of signing papers, he owned a portion of the company; in five years he would own half.

The family was now on firm financial footing, but Joey's prospects remained grim. In 1977, all doctors could do was treat his symptoms: antibiotics to kill the bacteria that caused infections, physical therapy to clear the muck from his lungs, and oral pancreatic enzymes to break down his food. For a while, his crib was engulfed by a giant "mist tent"—a type of humidifier that resembled a giant teepee—which physicians believed moistened the sticky gunk in the lungs, helping kids cough it out. It was a technique that helped a few children, but left most of them uncomfortable, frightened, and dripping with sweat—barely able to see their parents through a gray fog when they lifted a flap to check on them.

Almost forty years since the disease's discovery, very little had changed. And its cause remained a mystery.

CHAPTER 7

The Sweat Test

1943–1960

Both tears and sweat are salty, but they render a different result. Tears will get you sympathy; sweat will get you change.

—Jesse Jackson

By 1943, Dorothy Andersen was busy with speaking engagements and diagnosing patients, and had barely any time for her own research. Her pancreatic enzyme test was accurate but time consuming, invasive, and unpleasant for children. There had to be an easier way to diagnose these children, but she was stretched too thin to investigate further. She needed help.

Andersen put out word that she was looking for a pediatrician to care for children with this new disease, and quickly came across Paul di Sant'Agnese, an Italian immigrant who had just completed a four-year residency and internship at New York Postgraduate Hospital, which was affiliated with Columbia University. Di Sant'Agnese was a charming, slim, soft-spoken man keen to work with Andersen. He'd followed her work and told her about several young patients he'd lost to lung disease who he believed had suffered from cystic fibrosis.[1] To Andersen, he seemed to be the ideal colleague and partner.

Born in Rome in 1914, di Sant'Agnese came from a distinguished family. Various Italian monarchs had bestowed titles of nobility on both his great-grandfather and grandfather for meritorious service: the former was a legislator and an engineer whose work had helped connect the country with a network of railways, and the latter was considered the finest obstetrician-gynecologist in Europe during the 1920s and '30s, whose clientele included Italian and European royalty and some of the continent's richest and most influential women.[2]

Di Sant'Agnese's father had continued the medical tradition and also became a distinguished obstetrician-gynecologist. His mother was an accomplished pianist. He was raised bilingual, learning to speak English from his nanny. When he completed college, he, too, studied to be a doctor, at the University of Rome Medical School, where he graduated with honors in 1938. A year later, before Italy joined World War II, he boarded a ship to America to continue his education.

At first, di Sant'Agnese found the sights and sounds of New York jarring and Americans hard to understand. He found both the accent and the expressions baffling. As he made his way from Lower Manhattan uptown, he figured out that "Toidy Toid Street" was Thirty-third Street. And when an impertinent New Yorker asked him "what neck of the woods" he hailed from, he explained that he didn't come from the woods, "but from Rome, Italy, the eternal city and the hub of the world."

But di Sant'Agnese was adaptable and affable, and quickly began to appreciate and embrace his new home. He sealed his future in America on October 1, 1940, the first day of his internship, when he met and fell in love with the night nursing supervisor at New York Postgraduate Hospital. Bonded by their love of medicine and many other common interests, the two married, and she helped di Sant'Agnese master reading and writing in English, providing critical support for him as he wrote scientific papers and corresponded with other scientists. And in 1943, di Sant'Agnese joined Andersen in her work.

At the beginning of their partnership, there were only a handful of children with CF for di Sant'Agnese to treat, but as Andersen traveled up and down the East Coast lecturing as the leading expert on the disease, fearful parents poured into Babies Hospital with their kids.

Di Sant'Agnese loved his job as a pediatrician but increasingly hated testing these children. When the test was positive, he had to deliver the diagnosis of cystic fibrosis—a death sentence—to desperate, terrified parents. Over and over, he would stand by, gut wrenched, as the children got sicker and sicker, many of them dying before their first birthday.[3] The only treatments available were first-generation antibiotics—sulfa drugs like the one given to Katherine Woglom,[4] whose autopsy Andersen had performed some eight years earlier. But against the microbes that infected the lungs of CF patients, those drugs were largely useless.

Scientists developing medicines for wounded soldiers in World War II, however, delivered an unexpected treatment: penicillin. Though discovered back in 1928 by Scotsman Alexander Fleming, it wasn't turned into a drug until 1943, when scientists at Northern Regional Research Laboratory in Peoria, Illinois, figured out how to scale up its production.[5] The drug was primarily for use in the war and supplies were rationed to ensure an adequate supply for D-Day, but a small quantity was spared for research. The War Production Board gave Boston physician Chester Keefer the unenviable role of "penicillin czar," in charge of rationing and distributing small quantities of the drug for civilian use and monitoring the outcomes. An article in the *New York Herald Tribune* contributed to the demand for the drug: "Many laymen—husbands, wives, parents, brothers, sisters, friends—beg Dr. Keefer for penicillin. In every case the petitioner is told to arrange that a full dossier on the patient's condition be sent by the doctor in charge. When this is received, the decision is made on a medical, not an emotional basis."

Keen to try the new drug on the sick kids in his care, in 1943 di Sant'Agnese petitioned Keefer on behalf of his patients. With a dossier containing notes on several patients in hand, he stood in line for hours at Columbia University, where a uniformed colonel distributed samples to approved recipients. Di Sant'Agnese's patience paid off, and he walked away with several vials of the precious drug.

The impact was dramatic. Children on the brink of death, their lungs racked with disease, were snatched back into the realm of the living. Their fevers declined, their appetites increased, and they began to gain weight. A year later, when the penicillin supply became more plentiful, Andersen and di Sant'Agnese incorporated it into a treatment regimen that, while not a cure, extended children's lives. By March 1945, penicillin production had skyrocketed and the government lifted all restrictions on its use, making it available to everyone at their corner pharmacy. If children were diagnosed before bacteria colonized their tiny lungs, then there was a chance that diet and regular antibiotics could stave off ill health—making it all the more important to find an easier method of diagnosis.[6]

BY THE MID-1940S, DI SANT'AGNESE WAS BUSY ATTENDING TO THE more than six hundred patients who had been referred to Andersen, as she continued her research and lectured around the country.[7] He ran Babies Hospital's celiac clinic, which catered to cystic fibrosis patients and saw at least six patients every afternoon. He also taught students and served as chief of pediatric clinics. All was going well for the energetic

doctor—until May 2, 1946, when he started to feel weary and fever-
ish. He developed a splitting headache and started vomiting. His colleagues diagnosed him with acute disseminated encephalomyelitis[8]—a
bacterial- or viral-triggered inflammation that had gripped his brain
and spinal cord. He was racked with seizures and fell into a coma. His
doctor friends speculated that even if he survived, he would be blind,
weak, partially paralyzed, and incapable of caring for himself; he might
require institutionalization.

He woke from the coma after two months and, much to his doctors'
and Andersen's amazement, gradually recovered, though the sickness
had taken a huge toll physically and psychologically. Damaged nerves
paralyzed his right hand, forcing him to learn to write with his left.
His balance was poor. He had trouble controlling his eye movements.
He tired easily[9] and was from then on plagued with worries about his
health, fearing another disease would strike without warning. Yet, as
soon as he was able, he went back to work, focused more heavily on his
research.

As the 1940s wore on, somewhere between fifty and sixty new cystic fibrosis patients were arriving at Babies Hospital for treatment every
year. Each new case required the same invasive procedure of inserting
a tube into the intestine. But then, during a scorching heat wave in the
summer of 1948, di Sant'Agnese noticed a strange symptom of CF that
would change the way the disease was diagnosed.

When he entered the hospital one morning that summer, having
passed dozens of men in sweat-wilted collars and women with melted
makeup, a nurse from the pediatric ward told him ten children had
been admitted with heat exhaustion. [10] In the clinic, he was surprised
to see five familiar faces, children he'd recently diagnosed with cystic
fibrosis. A sixth child looked like he, too, might have the disease, but
di Sant'Agnese hadn't tested him. They were sweaty and pale, severely
dehydrated, and on the verge of heart failure. Di Sant'Agnese quickly
tended to them one by one with a team of nurses, inserting IVs into
their limp veins to replenish their salts and water. One child died within
an hour of reaching the hospital, but the others recovered once they
were rehydrated.

It seemed like a strange coincidence that so many of the kids suffering from heat stroke on that particular day were also his patients. But he
forgot about the incident—until the same thing happened during a hot
spell the following summer, in 1949. Again, he saw children he knew.

His cystic fibrosis patients were clearly more vulnerable to heat stroke. Sitting in a room with one patient, di Sant'Agnese noticed that when the little boy put down his glass after gulping down water, the surface was decorated with ghostly white, salty fingerprints. The other children left similar prints, which inspired the physician to begin investigating the physiology of sweat.

While hunting through the musty stacks at Columbia's library, he pulled out a hefty tome by the famous Japanese physiologist Yas Kuno. His 1934 book, *The Physiology of Human Perspiration*, was widely regarded as the definitive study on the science of sweat.[11] In it, Kuno estimated that two million sweat glands covered the skin's surface, modulating body temperature by releasing water through the pores, which carried away excess heat and cooled the body. He described this mechanism and the different types of sweat glands.

But even in Kuno's comprehensive text, there was little about the chemistry of sweat or the quantities of electrolytes—sodium, chloride, and potassium—that gave sweat its trademark saltiness. So di Sant'Agnese turned to medical journals, where he discovered other scientists had calculated these values in healthy subjects by placing volunteers in a heated room to induce sweating. He decided to replicate these experiments at Columbia Presbyterian Hospital to see whether children with cystic fibrosis sweated excessively, or if their sweat was chemically different from others.

Early in the spring of 1952, he tested four teenagers—two with cystic fibrosis and two without. One by one, each teen lay on a bed in a steamy room, where di Sant'Agnese had cranked the temperature to a balmy 90 degrees Fahrenheit, with 50 percent humidity. He placed a square swatch of gauze on each subject's belly, covered it with plastic, and taped it in place, then covered them with a sheet and blanket and left them to sweat for up to two hours, in order to collect enough sweat to test. Then he ran the sweat-soaked gauze to his lab to analyze the liquid.

The results were electrifying. All four children sweated out roughly the same volume of liquid, but the sweat from those with cystic fibrosis was chemically unique, with higher salt. To prove that this wasn't a fluke, over the next few months di Sant'Agnese tested forty-three children with the disease and fifty without it. Sweat from the children with CF had dramatically higher levels of sodium and chloride (the two halves of table salt) and potassium—but it was the chloride that really jumped out. In patients with cystic fibrosis, chloride levels were between 60 and 160 mEq/L (milliequivalents per liter), or between three and five times higher than in people without the disease.[12]

Perhaps a high sweat chloride level was a hallmark of this disease, di Sant'Agnese thought. If so, maybe sweat could be used to diagnose it—and replace Andersen's cumbersome, intrusive enzyme test. It was an exciting result and he was keen to share it with fellow physicians and researchers.

He first shared the results in May 1953,[13] when he presented his research at the annual meeting of the American Pediatric Society in Atlantic City. He was barely able to contain his excitement as he told a room packed with scientists and physicians that children with this lethal genetic lung disease had a chemically unique sweat extremely high in chloride. But at the end of his presentation, as di Sant'Agnese eagerly awaited questions, the room was silent. None of the physicians assembled could see a connection between salt, or chloride, and the diseased pancreas or infection-prone lungs.

It was a disappointing reception, but di Sant'Agnese wasn't deterred. He presented his work to a small cluster of colleagues when the distinguished Japanese sweat physiologist, Yas Kuno, visited Columbia University. Kuno listened attentively—then stood up and spat out a single word, "Impossible," before marching out of the room.

Not even Dorothy Andersen, now a nationally recognized expert on CF, was entirely convinced. She sat in the audience along with Kuno, smoking and looking skeptically at her colleague. He was a talented physician, but she was dubious about his conclusions.

IT WASN'T UNTIL AFTER DI SANT'AGNESE PUBLISHED HIS FINDINGS LATER that year, in November 1953, that someone finally appreciated the implications of his discovery: an opinionated pediatrician named Harry Shwachman. After learning about cystic fibrosis in the late 1940s from Dorothy Andersen's publications,[14] Dr. Shwachman had begun using her diagnostic test to distinguish CF from celiac disease at Boston Children's Hospital. So Shwachman—the physician first recommended to Joe and Kathy O'Donnell—had a vested interest in di Sant'Agnese's new discovery.

Just a year after passing his pediatric specialty exam in 1941, Shwachman was drafted and immediately sent to Fort Monmouth in New Jersey.[15] On a weekend trip to New York City, he met his future wife, Irene, but held off making plans because he expected to be stationed in Africa.

When he learned he would be sent to a base in Puerto Rico instead, Irene and Harry were married and she went with him to the island, where he ran a large army laboratory until 1946.[16] When he returned to Boston he began reestablishing his pediatric practice, and in 1947 he took over the Children's Hospital's Nutrition Clinic, which at the time handled CF cases. He was the logical choice because, before Shwachman was drafted, he had worked for the clinic's previous leader, Dr. Charles May—the world expert on nutritional disorders, who was one of the first to link pancreatic damage with lung disease and whose work Dorothy Andersen had referenced in her seminal paper.

When Shwachman took over the clinic, Dr. Sidney Farber, the head of pathology at Boston Children's Hospital (who would later revolutionize the treatment of childhood cancer), tasked him with evaluating every child suspected of having cystic fibrosis. Since then, Shwachman alone had tested about 3,000 children—so many that, he boasted, he could handle five "duodenal procedures" in a single morning.

Though he could be arrogant, Shwachman was a dedicated physician, working weekends to make sure he gave his patients the attention they needed. He was fiercely devoted to these sick children, gentle and animated when speaking to them. He was also deeply protective of them, and though he knew how much effort it took to keep them healthy, he had no patience or sympathy for overwhelmed parents. They had to adhere to all the treatments and therapies; otherwise, their child's health would slide fast.

When he read about di Sant'Agnese's work, he wasted no time, traveling to New York in early 1954 to meet him and learn about his new sweat test. Di Sant'Agnese was eager to share his discovery with another interested physician and took him to the special rooms where he had done the work. Shwachman came away impressed but frustrated: di Sant'Agnese's test required the use of specially designed rooms with a constant temperature and high humidity (which prevented the sweat from evaporating), and Shwachman didn't have access to a similar facility.

When he returned to Boston, a colleague suggested a less refined but equally effective approach: inducing sweat production by wrapping patients in plastic. Inspired, Shwachman began having children and teenagers step into a large plastic bag—which he secured with tape just below their chins, leaving only their faces peeking out—and stay there for forty-five minutes while he collected their sweat. It was a weird and uncomfortable procedure, but the children and their parents trusted Shwachman and dutifully agreed. The strange-looking test turned out to be both quick and reliable, and Shwachman verified that children with cystic fibrosis did indeed produce saltier sweat than healthy children.[17] Di Sant'Agnese was thrilled by the speed and precision of Shwachman's work—and relieved that his findings had been replicated. But sweat bags were awkward and risky; patients could overheat and die.[18] They needed something better.

Fortunately, a more precise approach was just a few years away, thanks to the work of a former intern of Shwachman's named Lewis Gibson. Gibson had trained at Johns Hopkins University from 1949 to 1953, completed his residency and internship at Boston Children's Hospital, and then moved to the NIH in 1955 to become part of the US Public Health Service. Gibson wanted to figure out why CF patients had such salty sweat and began studying the science of sweating.[19]

He learned that the autonomic nervous system—which controlled heart rate, digestion, and breathing—also controlled sweating. Gibson hypothesized that CF patients' hyper-salty sweat was caused by an imbalance in the autonomic nervous system, and he wanted to test his hypothesis. He knew he could induce sweat production with adrenaline or acetylcholine, but these were powerful drugs. Adrenaline quickened the heart rate and too much could trigger a heart attack. Meddling with acetylcholine could wreak similar havoc on the nervous system. It was reckless to inject a child with a potent hormone or neurotransmitter just to induce sweating.

Then, one day, Gibson recalled a trick he had seen in medical school—a technique called iontophoresis. During a lecture, a physiology professor used a mild electric current to drive histamine—a substance the body releases during allergic reactions—through the surface of his skin to prove that the chemical triggered hives. Perhaps, Gibson mused, he could use an electrical current to drive a minuscule quantity of a sweat-inducing drug directly through the skin to induce a small zone of perspiration. He began testing several drugs he knew would trigger sweating, including adrenaline and acetylcholine, by zapping the substances into the skin of some CF patients and healthy volunteers. While the drugs did cause some local sweating on the arm where the electrode was, Gibson didn't gain any insights into the difference between the two groups' sweat. Uncertain why the work wasn't panning out, he essentially abandoned this line of research.

After a few years at the NIH, however, in 1957 Gibson returned to Johns Hopkins University to complete his residency, where he happened to meet Dr. Robert E. Cooke, a pediatrician and chairman of pediatrics. When Gibson learned that Cooke had expertise with iontophoresis and CF patients while at Yale University, he suggested that they use iontophoresis to diagnose CF.

The new diagnostic test Gibson and Cooke came up with involved placing a small circle of filter paper saturated with the safest of all the sweat-inducing drugs—pilocarpine—under a small positive electrode taped to a child's forearm. The negative electrode was taped elsewhere on the body. When the current was switched on, it pushed the drug into the child's skin for five minutes, creating a warm, tingling sensation. The pilocarpine-saturated paper was then removed and replaced with clean filter paper, which was left in place for thirty minutes to collect sweat. The sweat on the paper was then analyzed for chloride content.

The test's results were consistent with the sweat-bag test, revealing stark differences between patients, the patients' relatives (including parents, who were carriers, and other relatives, whose statuses were unknown), and healthy volunteers. Gibson tested twenty-five patients with cystic fibrosis and all had chloride levels above 80 mEq/L; neither of the other groups had levels higher than 60.[20]

Gibson and Cooke published their new test in 1959. Physicians liked it; it was safe, fast, painless—and, most importantly, accurate and reproducible. Though the new test also did not work for newborns, it worked for any patient old enough to produce sweat, and over the next decade, it became the gold standard for CF diagnosis.

As GIBSON AND COOKE WERE FINESSING THEIR SWEAT TEST IN THE late 1950s, Paul di Sant'Agnese was continuing to untangle cystic fibrosis's complexity, even as Andersen's research interests broadened. Andersen was as productive as ever, discovering another metabolic disease and pioneering the study of congenital heart defects. Her collection of infant hearts with cardiac problems, which she began gathering in 1935 from her autopsies, formed the basis for a surgical program in open heart surgery.[21] But in the late 1950s, Dorothy Andersen became ill and was in her laboratory less and less. And, in 1962, she was diagnosed with lung cancer after decades of chain smoking.[22]

Di Sant'Agnese missed her cantankerous presence; she had always been a good mentor and friend. On many occasions he had brought his family to her Kittatinny Mountain cabin, where they shared a meal while his children played outside. But even without Andersen, he continued to probe the disease. Most importantly, he showed that *cystic fibrosis of the pancreas*, as Andersen had described the disease in her 1938 paper, was a misnomer. This disease affected the sweat glands, the gut, the liver, and the lungs—and most children and teenagers died from bronchial infections caused by the murderous microbes *Staphylococcus aureus* and *Pseudomonas aeruginosa*.[23]

In 1958, di Sant'Agnese was invited to speak at the NIH about "his favorite disease," and soon afterward he was asked to lead the new Pediatric Metabolism Branch at the National Institute of Arthritis and Metabolic Diseases. The offer was more than a promotion: it was recognition of a new strange disease, acknowledgment that it deserved to be understood, and a commitment to discovering its roots and finding a cure.

The timing of the NIH's recognition of this new disease wasn't a coincidence. This newfound focus on cystic fibrosis was a direct result of parents of children with CF joining together to raise awareness of this obscure but deadly disease and to secure congressional funding to discover the cause and find a treatment.

CHAPTER 8

A Tribe of Desperate Parents

1950–1955

> Never doubt that a small group of thoughtful
> committed citizens can change the world.
> Indeed, it is the only thing that ever has.
>
> —Margaret Mead, cultural anthropologist

One Sunday morning back in 1950, while Dr. Andersen was working in her lab, she received a call from a distraught parent and newly minted pediatrician named Milton Graub.[1] He had just been told that his two-year-old son, who suffered from a racking cough and abnormal stools, had a syndrome called exudative diathesis[2] that was common in poultry and caused by a selenium deficiency. Graub thought the diagnosis was absurd. He had been frantically reading medical reports to find a more plausible explanation when he stumbled across Dr. Andersen's paper introducing cystic fibrosis of the pancreas. The symptoms seemed to match his son's. Anxious and desperate, he immediately called Dr. Andersen to talk with her.

Dr. Andersen had sighed when she heard the phone ring. Though she treasured her quiet Sundays, the only day she could focus exclusively on her research without interruption, she answered the phone. She listened carefully as Graub shared his son's medical history, and then asked if they could see her the next day.

Andersen administered her duodenal enzyme test on the child, and the results came back three days later: the little boy was the latest victim of cystic fibrosis. And when Graub and his wife, Evelyn, had a second child, a girl, about a year later, she also had the disease.

The Graubs felt helpless. Andersen had told them there was nothing that could be done for their son except make him comfortable—that his early death was inevitable. And now their daughter had a similar prognosis. They were frustrated that so little was known. There was no treatment. There was nowhere to go for support beyond the few doctors who specialized in the disease. No one had any idea what caused CF—and no one, as far as the Graubs knew, was even trying to find out. They wanted to change all of that.

As word of the Graubs' misfortune spread through their circle of friends and neighbors in Philadelphia, a local pharmacist named Dr. Lesnick called the Graubs to share the story of a relative of his who lived in Los Angeles and had recently lost her granddaughter to cystic fibrosis. After her death, a small group of friends and family established a fund named in her memory—the Jenny Lesnick Fund. As far as Milton Graub could tell, this was the first fund organized to raise dollars for this rare disease, and after a telephone conversation with Jenny's grandmother, Milton and Evelyn Graub were inspired to unite locals whose children or friends' children were sick with cystic fibrosis.

In 1952, they gathered a small group of Philadelphia parents to learn about the problems they all faced and find a way to do something about CF—although they had no clue what, or how. But this new organization was a start. Among the monthly meeting's attendees were three couples who had at least one child with CF: the Graubs, a truck driver and his wife, and a builder and his wife. A fourth couple, a bank vice president and his wife, were not parents of a sick child, but were interested in the cause.

The Graubs were keen to expand the group and spent much time brainstorming ways to identify and meet other families. But thirty years before personal computers and half a century before widespread internet access, it was a challenge to find families touched by CF. One strategic hot spot for connecting with new patients was di Sant'Agnese's clinic in New York City. Dorothy Andersen was recognized as the leader in the field of CF, and most suspected cases on the East Coast were brought to her. After diagnosing children, she referred them to di Sant'Agnese for regular medical care. So, with the doctor's permission, Evelyn Graub stationed herself in di Sant'Agnese's waiting room during her kids' visits, enthusiastically introducing herself to everyone and telling them about the group and their mission to raise funds. The Graubs also devoted much of their weekends to visiting every family with a newly diagnosed child within a 100-mile radius of their Philadelphia home to offer support—and to ask for help spreading the word about the disease and their group.

It was an enormous commitment. Milton Graub had a booming pediatric practice, and their sick children needed constant care and frequent trips to New York to see di Sant'Agnese. But the couple believed that building this new group was the best possible use of their time and energy. It was the only chance they had to save the lives of not just their children, but of all those with this disease.

Because the Graubs wanted to raise the profile of cystic fibrosis beyond the immediate circle of affected families, they hired a professional publicist in Philadelphia—Sy Shaltz, who had lost his twelve-year-old son to leukemia—to launch an awareness campaign. Shaltz empathized with their drive and goals, and not only offered up services for free but also donated money raised for leukemia research in his son's memory to the CF group instead.

The Graubs' tactics were successful; momentum built. By 1954, thirteen other local volunteer branches dedicated to CF had sprouted up within that 100-mile radius, they had an official office, and they'd hired an executive director to coordinate. But each of these little groups was a ship unto itself. There were no rules and no overarching missions on which to combine efforts.

That year, the Graubs hosted their first big fundraiser, "Friends of Cystic Fibrosis," which brought in a whopping $14,000—a windfall for that era, especially considering that most of the 250 guests had never heard of the strange disease. The guest speaker was Dr. Wynne Sharples, a nonpracticing pediatrician and mother of two children with CF. They'd chosen her because she was a socialite with the means, connections, visibility, and time to dedicate to the cause—a potentially powerful ally in the fight to treat the disease. With two sick children of her own, Sharples understood the urgency of finding the cause and cure.

Just a year later, in 1955, the Graubs learned from di Sant'Agnese during a visit to the doctor's office that Sharples was working on the legal framework to launch a national CF foundation of her own. Sharples had mentioned nothing about it when they asked her to speak at their event. When Sharples did eventually reach out, later that year, she told them that she had already secured a charter for what she called the Children's Exocrine Research Foundation and asked Dorothy Andersen to join the Medical Education Committee.[3] As the new foundation's self-proclaimed president, she wanted Milton Graub to wrap all his groups into her foundation and to represent the Philadelphia chapter at the founders meeting in New York City, where they would decide on a board. Graub and his wife had spent the last three years working to bring more recognition to cystic fibrosis, and he quickly appreciated that with Sharples, they'd be able to bring greater numbers of CF families together and raise more money faster.

Wynne "Didi" Sharples was a smart, well-educated woman.[4] She had graduated from Foxcroft School and Radcliffe College and then had trained as a pediatrician at Columbia University's College of Physicians and Surgeons in New York.[5] She was also a member of the National Society of Colonial Dames, an organization founded in 1891 for women descended from leaders in colonial America.[6] In her final year of medical school, she married a Russian-born petroleum engineer, George de Mohrenschildt. After her graduation, the couple moved to Dallas for her residency at Baylor Hospital. In 1953, they had a son, Sergei, and in 1954, a daughter, Nadejeda. Both were born with cystic fibrosis.

Shortly after Sharples's children were born, her wealthy industrialist father, Philip, and his wife arranged a meeting with Philip's friend, the director of Boston Children's Hospital, to ask for advice. The director summoned Harry Shwachman, by then the leading physician for cystic fibrosis, who agreed to care for Philip's grandchildren.

Wynne Sharples was impressed by Dr. Shwachman. When she met the doctor and saw how hard he worked, she even suggested to her father that the doctor might benefit from a grant. He agreed, and wrote Shwachman a check for $10,000, which he used to fund a room at the hospital dedicated to testing the sweat of children with suspected CF.

Dr. Shwachman was equally enamored with Sharples, treating her like royalty, giving her free drugs for her children, and never charging her for visits, even though she easily could have paid.[7] It was during one of those visits that she suggested to Shwachman that there ought to be a national organization to promote cystic fibrosis research. He agreed.

Wynne Sharples knew about the Graubs' efforts, and while she recognized that they were worthwhile, she also found them amateurish. Advancing research on this disease would require bringing together top scientists to find a cure, wealthy society types to bring awareness and funds, and politicians to wield influence. She needed to create a national foundation with a high profile even as grassroots efforts continued to bring in money.

Together with her husband,[8] Sharples launched such a foundation, which a year later had changed its name from the Children's Exocrine Research Foundation to the National Cystic Fibrosis Research Foundation. Though Sharples was based in Dallas, Texas, the foundation began with five chapters all in other locations—New York, Los Angeles, Philadelphia, Hartford, and Boston. The thirteen volunteer groups that the Graubs had pulled together fell under the Philadelphia umbrella, as Sharples had wanted, and the inaugural meeting was held there in 1955.

Sharples proclaimed herself chair of the Board of Trustees. The members of the scientific advisory board included Drs. Sidney Farber and Charles May from Boston Children's Hospital, along with a couple of other distinguished physicians. And on the medical education committee were Andersen, di Sant'Agnese, Shwachman, and two other physicians familiar with CF.

Sharples was opinionated, set on running the foundation her way from the beginning. She wanted the chapters of the national organization to raise money, but she didn't want them represented on her board of trustees. Rather, she handpicked people she considered to be "exemplary citizens" to fill the positions. This elitist attitude sparked an outcry from chapter leaders, who were parents of sick children and determined to have a say in how the foundation was run. Milton Graub was certain that a board needed parents who had the drive and the commitment to fight for these children. In the end,[9] Sharples compromised by including two representatives from each of the five chapters, along with the ten individuals she had already selected, to form a twenty-person board.

Several of Sharples's chosen trustees, however, had been personally affected by the disease. Included in her selections were her father, Philip Sharples; and George Frankel, a wealthy, hard-nosed oilman with a soft spot for children, who also had skin in the game—his daughter, Doris Tulcin, had recently given birth to a girl with CF.

WHEN BABY ANN WAS BORN IN JUNE 1953, DORIS TULCIN KNEW HER child was sick. Her daughter looked very different from her son, Roger, who had been born healthy a few years earlier. Ann was small, her skin gray and wrinkled with a texture akin to elephant hide, and she had severe intestinal issues that kept her in the hospital for an extra week after birth. She ate, but she didn't gain weight. After three months and a slew of frustrating pediatrician appointments, a nurse friend of Tulcin's read an article by Dorothy Andersen in a nursing journal and recognized that Ann's symptoms matched the ones Andersen described. Tulcin took Ann to Dorothy Andersen in mid-September, and Andersen's signature test confirmed that the baby lacked critical pancreatic enzymes. Putting her arm around Tulcin's shoulders, Andersen shared the bad news: her daughter would likely die within the year.

Tulcin was stunned. She stared at her tiny, perfect Ann, shocked that this odd collection of symptoms could yield such an unforgiving outcome. Andersen told Tulcin, as she had so many parents before her, to take little Ann home, hold her, love her, and try to keep her comfortable. She recommended a diet that might help Ann gain weight, but admitted there was little else to do or say. "We just know so little about this disease," Andersen explained apologetically.

Three weeks after the diagnosis, Ann was feeding well when she became sick with a nasty virus, forcing Tulcin to take her back to the hospital. Still, Tulcin refused to accept Andersen's bleak prognosis. She would fight, though it was a tumultuous time for her, with other family members experiencing crises of their own. Her alcoholic mother was in the hospital. Her husband, Bob, was engulfed in the first of what would be many episodes of major depression, and both Bob and Ann were in the hospital for weeks. But Doris could always rely on her father to fight for her.

George Frankel had a passion for politics and medicine and knew how to get things done. He was active in the Democratic Party in Connecticut.[10] He was also a friend of both the late president Franklin D. Roosevelt and Roosevelt's former law partner, Basil O'Connor, who later became the influential president of the National Foundation for Infantile Paralysis, which funded and drove the development of the polio vaccine. Frankel knew President Harry Truman, and the governor of Connecticut, Abraham Ribicoff, whom he had convinced to run for the office. He also had a long-standing commitment to medical philanthropies, particularly when they involved children. He'd established the children's ward at the New Rochelle Hospital in 1948 and was one of a handful of cofounders of the Albert Einstein College of Medicine.[11]

When Tulcin told Frankel how little was known about the fatal condition afflicting his granddaughter, he reacted with his standard pragmatism. He suggested that they find other families stricken with the same condition and began by asking the chairman of pediatrics at Babies Hospital, where Andersen was based. The hospital chairman connected Frankel and Tulcin with the Graubs' network of volunteer groups in Philadelphia. Then father and daughter reached out to other families in major cities across the country, from Washington, DC, and Boston, to Cleveland and Los Angeles—cities that boasted elite hospitals and universities and seemed like promising places to network. And Frankel promised his daughter he would help her launch her own fundraising group in Scarsdale, New York, like the Graubs' in Philadelphia.

When Ann finally came home from the hospital, she began a slow and steady recovery with help from a home nurse. By mid-1954, Ann's health had improved dramatically, against Andersen's predictions. Tulcin had followed Andersen's advice to feed Ann a banana-rich formula, and her daughter began to grow, becoming deliciously fat and chunky by the time she turned one and reaching all the expected physical and behavioral milestones. Still, Tulcin was acutely aware that Ann's health could nosedive at any moment. And later that same year, Tulcin's sister-in-law gave birth to a girl who also had cystic fibrosis—a bludgeoning reminder that the disease was genetic, and had now struck her family twice. In her Scarsdale basement, Tulcin began brainstorming with friends about possible fundraisers.

Frankel and Tulcin were completely in the dark about the national foundation Sharples was launching until Sharples learned of their networking efforts and invited Frankel to serve on her board of trustees. Frankel knew of the wealthy and influential Philip T. Sharples and his pediatrician daughter Wynne, and thought they were strong advocates for the work that needed to be done; they were well connected and knew how to raise money, and Wynne brought medical insights. He signed on and used his position to help Tulcin launch her own hometown chapter of the National Cystic Fibrosis Research Foundation.

ONCE WYNNE SHARPLES OFFICIALLY LAUNCHED THE FOUNDATION IN 1955—more than fifteen years after Dorothy Andersen's seminal work describing cystic fibrosis—she had her work cut out for her. The public and even most physicians had never heard of the disease. When *Time* magazine[12] wrote about it in 1954, they noted that even for "most doctors, pancreatic fibrosis (also known as mucoviscidosis) is a 'new' disease." In 1956, Paul di Sant'Agnese wrote a letter to the army explaining that his patient CJ, who had survived to the age of eighteen, wasn't healthy enough to be drafted—and an army health officer wrote back declaring that he had searched for cystic fibrosis among the list of official diseases and it didn't exist. (Fortunately, when an army doctor X-rayed

CJ's chest and saw his damaged lungs, he diagnosed him with TB and excused him from duty.) Two years later, in 1958, when di Sant'Agnese penned a chapter about the disease for the *Cecil Textbook of Medicine*, it was the first time cystic fibrosis appeared as a distinct entity in a medical textbook. Perhaps this, he thought, would finally put this disease on the map.

Sharples knew that her first priority as National Cystic Fibrosis Research Foundation president was education. She gathered key insights about the disease from the top CF experts—Andersen, di Sant'Agnese, and Shwachman—and convinced all three to weigh in on her draft booklets for physicians, parents, researchers, and the general public. Another key move Sharples made was producing a quarterly newsletter for foundation members. In it she shared new discoveries and diet recommendations, listed new chapters and their leaders, mentioned upcoming conferences, and included articles on the disease published in national magazines and newspapers along with their estimated audience.

In January 1957, Sharples used the newsletter to alert foundation members that national headquarters would be kick-starting the hunt for a research director. With growing funds to spend thanks to the work of the individual chapters, and only a paltry trickle of research applications coming in, the trustees and the scientific advisory board concluded that they needed a "capable and devoted person to review the problems presented by cystic fibrosis." The research director would be a foundation employee tasked with knowing everything possible about the disease—including which researchers would be most likely to study it. That same individual would also be in charge of encouraging those researchers' interest in this rare condition.

Money, Sharples knew, was the only way to fund research and develop a cure. She hoped that as the public learned more about the disease they would be generous and provide the funds. And she never missed an opportunity to make that clear to reporters reaching out for stories about the disease. On April 12, 1955, the *Bergen Evening Record*[13] ran an article about the grave future of CF kids that quoted Sharples: "Intensive medical research alone will enable us to solve the problem of cystic fibrosis. We know that we can count on the generous support of the American parents toward this end."

Growing media coverage proved Sharples's education efforts were having an impact. A 1957 headline in the *New York Times* proclaimed the foundation's "War on Cystic Fibrosis."[14] The story described the plight of an unnamed family with one child suffering from the disease and another on the way. It was a sizable story, describing the disease's symptoms, the new diagnostic test that Shwachman was using, and how it was based on di Sant'Agnese's discovery of salty sweat. It described how in the 1940s no children with CF made it to adolescence but now—with the right diet and antibiotics—it was possible to extend life into the early teens. It also mentioned how the Philadelphia-based National Cystic Fibrosis Research Foundation was raising funds. The story noted as well that the birth of the foundation was part of a broader trend of national volunteer-driven health agencies, dedicated to specific diseases, popping up all over the country.

Shortly after the article came out, Tulcin's father, George Frankel, encouraged Sharples to move the foundation to New York to give the fledgling organization more gravitas and visibility. He generously offered to house the headquarters in his elite office space on the fourth floor of 521 Fifth Avenue, a ritzy location that shifted the foundation to the center of the corporate world—a premier location for fundraising. Sharples liked the idea, and agreed to discuss it with the board members. She was ready to dive deeper into fundraising and awareness campaigns to catapult the foundation into the big leagues and make this disease known to every household in America.

CHAPTER 9

Lessons from Polio

1955–1960

But if you think of global public goods like polio
eradication, that kind of risk-taking new approach,
philanthropy really does have a role to play there,
because government doesn't do R&D about new
things naturally as much as it probably should.

—Bill Gates

While Wynne Sharples worked hard to raise both funds and the profile of the National Cystic Fibrosis Research Foundation, Doris
Tulcin was busy organizing small local fundraisers with her Scarsdale
chapter. But by 1957, she decided to expand her Scarsdale group into a
Westchester County chapter, moving it from her basement[1] to an office
in White Plains, and hiring a professional named Katherine Earnshaw,
a former director of the Arthritis Foundation, to run it. Earnshaw was
a masterful organizer and taught Tulcin and her friends how to run
their first door-to-door campaign, draft a budget, and become successful event planners and fundraisers.

After a series of small events—intimate lunches and afternoon teas
at the homes of various friends—Tulcin organized her first high-profile
fundraiser: a glamorous ladies' luncheon for three hundred in a majestic
white tent pitched on the sprawling grounds of her parents' estate. It was
one of the largest cystic fibrosis fundraisers to date, and she'd arranged
an impressive lineup of speakers, including Paul di Sant'Agnese and her
father's friend, Connecticut governor Abraham Ribicoff.

The elegant event didn't go quite as she planned. Governor Ribicoff
pulled Tulcin aside when he first arrived and whispered, "What's cystic
fibrosis?"[2] Few could understand Dr. di Sant'Agnese's mellifluous Italian

accent. Tulcin had planned to screen a film about the grim outlook for children with CF, but the projector broke. Then she watched, dismayed, as the audience's attention slipped, and the volume of chatter rose. The governor glanced at Tulcin with an expression that she interpreted as *You need to take charge.*

Her fear of public speaking dwarfed by her fear of the fundraiser failing, Tulcin stepped behind the podium and began sharing the heartbreaking details of her daughter Ann's disease—and the cruel fact that her child only had a few more years to live. With the crowd hushed, Governor Ribicoff leveraged her speech with an impassioned plea for donations, and the inaugural event was a fundraising success. But Tulcin knew it would take hundreds of showy events to bring in enough money to determine the cause of the disease and conquer it, and she was determined to be a part of that effort. She didn't have any scientific training. But she did have connections.

The most influential of these, and a person steeped in the business of curing diseases, was Basil O'Connor, president of the National Foundation for Infantile Paralysis (NFIP, which later became the March of Dimes) and a close friend of her father. Tulcin knew the legendary role O'Connor and the foundation had played in orchestrating polio research and vaccine development, and she wanted to learn from him. She was aware that raising public awareness of CF would be a much tougher battle—polio had had a president in the White House whose suffering had drummed the cause into the American consciousness. But O'Connor's advice and guidance would, at least, provide a place to start.

First, Tulcin shrewdly strengthened the ties between the foundation and the NFIP by holding an event honoring O'Connor in November 1957. Shortly after, she and her father met with him in his New York office, where O'Connor distilled the lessons he'd learned from twenty years in philanthropy and medicine after his chance encounter with a future president.

Franklin Delano Roosevelt had contracted polio in 1921. He and O'Connor had met in 1922 when, as Roosevelt was returning to his Manhattan office, he slipped and fell walking through the marbled lobby, and the ambitious young lawyer came to his aid. They struck up a friendship, and ultimately launched a law firm together, Roosevelt & O'Connor.[3]

O'Connor told Tulcin and Frankel that, in 1924, Roosevelt had traveled to a dilapidated resort called Warm Springs, Georgia, lured by rumors that its hot, mineral-rich waters had healed a fellow polio victim. The magic of Warm Springs filled him with optimism that he would walk again, and in 1926, he purchased the property and built a home there. A year later, O'Connor convinced Roosevelt to convert it into the Georgia Warm Springs Foundation, a move that provided tax benefits and made it eligible for grants that could help spruce up the place. Soon, polio survivors began filling the resort, traveling from all over the country to soothe their muscles and heal, away from the prying eyes and judgment of the healthy. Roosevelt himself spent more than half his time there between 1925 and 1928.

Eventually, however, Roosevelt returned to Manhattan and politics. And after winning the New York governorship in 1928, he passed managerial responsibilities of the foundation to O'Connor, who hired an innovative and aggressive publicist[4] to raise money for the foundation and make Warm Springs a popular destination for the rich. Yet when the stock market crashed in 1929, even donations from the wealthy dried up, pushing Roosevelt's beloved foundation into the red again. Only after Roosevelt became president in 1933 did polio become a national topic of conversation.

Around that time, O'Connor met another talented public relations specialist who suggested fundraisers be held throughout the United States to celebrate Roosevelt's birthday, January 30, 1934, with the profits going to Warm Springs. The fundraising parties held the night before the president's birthday, on January 29, ranged from exquisite black-tie events in glitzy ballrooms in big cities to small-town church suppers and square dances. In a single night, they raised more than $1 million,[5] creating a hefty endowment for Warm Springs. So, the following year, O'Connor earmarked 70 percent of new donations for the care of polio patients and most of the rest for Warm Springs, with only a little going to research.

This, O'Connor emphasized to Frankel and Tulcin, had been a mistake—research was the only way to understand the cause of a disease and cure it.

What put the foundation on the right path was a former scientist who convinced O'Connor to divorce Warm Springs from the president and rebrand the foundation—as was done on January 3, 1938—as the National Foundation for Infantile Paralysis, with a mission of curing polio.

Under constant pressure to raise funds, O'Connor enlisted Roosevelt's close friend, the beloved vaudeville performer and radio show host Eddie Cantor,[6] to join the fight as the host of a new fundraising campaign that Cantor nicknamed the "March of Dimes." During radio broadcasts, Cantor appealed to the American public: "The March of Dimes will enable all persons, even the children, to show our president that they are with him in this battle against this disease. Nearly everyone can send in a dime, or several dimes."

The White House received a flood of mail with more than two million dimes, plus other cash and checks, totaling $1.8 million. The campaign had democratized traditional philanthropy—replacing a few hefty donations from the wealthy with millions of tiny ones from the masses. But when Roosevelt died suddenly of a cerebral hemorrhage on April 12, 1945, just eleven weeks into his fourth term in office, the donations dwindled.

To keep the foundation going, O'Connor turned to celebrity-studded events, including high-end fashion shows with sets designed by Salvador Dalí and clothes modeled by Marilyn Monroe. However, one of the most inspired and enduring attempts to reignite interest in eliminating the disease was a simple grassroots campaign in 1950 called the "Mother's March on Polio," an hour-long house-to-house solicitation on a single January night in Phoenix, Arizona. Those willing to donate left their porch light on. After its success, a national version was organized for the following year.

But even with all the fundraising, O'Connor said, the status of the science was dismal: two useless attempts at vaccines and no answers to the most basic questions about the disease. Realizing that good science was the only way to stop polio, O'Connor hired a new director of research, Harry Weaver, a gifted administrator with a pragmatic and ruthless approach to research funding.

Weaver single-handedly catalyzed the pace of polio research. He favored short-term grants to answer specific questions formulated to move them closer to developing a polio vaccine: How many strains of poliovirus existed? Could large quantities of virus be cultivated to provide the raw material needed for a vaccine? How did the virus invade the nervous system? If the science progressed quickly, researchers received additional funding; if not, the grant was terminated. Weaver also believed in funding conflicting approaches to vaccine design (some used live virus; others, dead), expecting that fierce competition between ambitious scientists would drive progress and eventually identify the best approach. To that end, NFIP fully funded the development of

Jonas Salk's vaccine. Weaver also cultivated goodwill with the scientists' host universities by paying for new facilities, extra research and administrative staff, and more overhead expenses.[7]

The strategy worked. After two years and $1.2 million in research grants, the foundation's scientists were able to answer the key questions, laying the groundwork for the first large-scale vaccine trial in 1954. More than 1.3 million children[8] participated in the trial, which the NFIP organized and fully funded for $55 million. Another $20 million was needed to cover related expenses, even with support and the tireless efforts of unpaid local volunteers. Once the trial was complete, the foundation also funded an independent expert analysis of the results, in which every polio death during the trial was investigated. After a year of analysis, the results were released with much fanfare on April 12, 1955—the tenth anniversary of Roosevelt's death.

Research, O'Connor repeated to Frankel and Tulcin, was the key to a cure.

Tulcin agreed that the fledgling National Cystic Fibrosis Research Foundation needed its own Harry Weaver to find and fund the scientists and institutions that would put them on the path to a cure. But Tulcin was deeply concerned about all the sick children. Wouldn't it be more compassionate, she asked O'Connor, to raise funds to help them?

O'Connor agreed that helping the families was more compassionate—in the short term. But compassion wouldn't save those children's lives. If they wanted a cure, they needed to figure out the cause. And scientific research cost money. *Lots* of money. If you raise money, get scientists to study the patients, and find the cause, he said, then you have a chance at preventing the suffering of all future children born with the disease, and possibly even preventing the very disease itself.

As the meeting concluded, O'Connor offered to fold the National Cystic Fibrosis Research Foundation into the March of Dimes, which, having achieved its goal of curing polio, was looking for new causes. It was a tempting offer; March of Dimes had money, visibility, and a track record of success, having cured polio in only fifteen years. But Frankel, with the support of the other foundation trustees, refused the merger, afraid that the little foundation would drown in a pool of bigger, more visible diseases. He was, however, interested in mining the March of Dimes for human capital. With O'Connor's consent, Frankel suggested to Sharples and the trustees that they hire key March of Dimes staff and enlist their energies in the battle against cystic fibrosis.

THE LATE 1950S WAS A BUSY TIME FOR TULCIN AND HER FATHER. IN 1958, Ann was doing well. She was healthy and playful and full of energy, attending kindergarten and growing like any other five-year-old. For the past three years Tulcin had been taking her to Shwachman for an annual checkup. He was always thrilled to see Ann's progress, and the fact that he never asked Tulcin to change what she was doing for her daughter gave her an enormous boost of confidence. Fortunately for the national foundation, Ann's good health didn't lull Tulcin into complacency or diminish her commitment to the disease. Quite the opposite— it gave her the freedom to ramp up her fundraising.

In 1959,[9] Tulcin's chapter had begun organizing tours of fabulous homes in affluent Westchester. But the event that put Tulcin's chapter on the map was a luncheon and fashion show at New York's Plaza Hotel, where they honored the famous medical philanthropist and healthcare activist Mary Lasker. Lasker was a role model for Tulcin. She was responsible for getting funding to support medical research at the NIH, and associating her with the foundation, Tulcin thought, could only bring more goodwill to her cause.

At the national level, things were also going well. In the late 1950s, Sharples continued to bring attention to the foundation. In 1958, the Greater New York and New Jersey chapters hosted a luncheon in NYC to honor Dorothy Andersen on the twentieth anniversary of her discovery of cystic fibrosis. There was a noticeable bump in newspaper coverage of the disease. Governor Averill Harriman declared CF Week in New York State in May. And Sharples had leveraged her social and political connections to get members of Congress, as well as the NIH, interested in the disease.[10]

But within the foundation itself, the relationship between Sharples and the rest of the leadership was strained. Sharples was strategic and hardworking, and understood the critical need for funding research. Her approach had been effective. But she was also a tough personality; she was uncompromising and made enemies. As part of her awareness efforts, she had commissioned a small film to be made about the disease, based on a script scrutinized and approved by the medical education committee—which at the time included Andersen, Harry Shwachman, Paul di Sant'Agnese, Dr. Robert Denton (Sharples's second husband), and two other physicians. But the film was something of a last straw for Dorothy Andersen regarding the foundation's publicity efforts, and she resigned from the committee on November 4, 1959, writing that she wasn't comfortable with the way in which Sharples publicized the disease.[11]

While the letter was firm, it wasn't unfriendly, and Andersen offered her help in the future should Sharples need it. And perhaps Sharples would have tried harder to persuade Andersen to stay on had she not been on the verge of resigning herself. In a letter written on December 4, 1959, just a month after Andersen's, Sharples explained why she was choosing to leave the foundation that she launched and still cared so much about.

The first reason that had compelled her, she explained, was a basic philosophical difference between her strategy for the foundation and the motives of the two scientific boards. Though she had been trying since 1957 to hire a director of research, the trustees had instead decided to hire a medical director, who would orchestrate both research and medical education. Sharples saw this as a clear conflict of interest. She didn't see how one person could think about both sick, dying children and a research agenda. She wanted someone who would focus exclusively on luring bright minds to study the disease.

"The second major reason for my resignation," she wrote, "is that I cannot approve the new budget, acted upon this October, allocating less than 50 percent of the funds raised by the foundation to research."[12]

Sharples knew in her gut what Basil O'Connor had told Tulcin and Frankel: all that mattered was raising money to fund science. But that wasn't a popular view. Although Tulcin and Frankel were on the same page, Sharples couldn't convince the rest of the trustees. A quiet mutiny had been brewing at the foundation she had launched, and behind her back it sent press releases to the newspapers and telegrams to the chapters explaining that Sharples resigned because the trustees opposed her expansion of research programs. But Sharples's reasons were broader than that. She argued that the trustees were overspending on administration and publicity. Those were necessary, "but they should be a means to an end of more research and medical education rather than an end in themselves."[13]

Sharples went on to found the Cystic Fibrosis Research Institute of Pennsylvania in 1960,[14] and continued to focus on caring for her sick children with her pediatrician husband, Robert Denton. But she also left the National Cystic Fibrosis Research Foundation on solid footing. The number of chapters had swelled from just the original five to seventy. The chapters were getting the hang of fundraising (with Tulcin's doing particularly well), and did better every year. And while Sharples's publicity campaign had not made CF a household name, the foundation was now well respected, with strong supporters at the NIH and in Congress. Less than five years after its launch, the foundation had gathered significant strength, power, and visibility, and they were ready to launch what they saw as the next stage in the battle against the disease: the creation of centers dedicated to treating and caring for children with cystic fibrosis.

CHAPTER 10

The Registry

1960–1966

It is a capital mistake to theorise before one has
data. Insensibly one begins to twist facts to suit
theories, instead of theories to suit facts.

—Sherlock Holmes, "A Scandal in Bohemia"

The first act of the foundation's new president, Robert Natal—a
trucking executive and close friend of Philadelphia board member
Milton Graub—was to fulfill one of his predecessor's goals. He took
George Frankel up on his offer of office space and moved the foundation's headquarters from Philadelphia to New York. His second act was
to follow the board's recommendation to hire Kenneth Landauer, medical director of the NFIP, as the foundation's new director of research
and medical education.

For the NFIP, Landauer had established a network of medical centers across the country just for polio patients, where physicians could
not only provide specialized care, but also learn more about the disease.
By the late 1950s, however, as the vaccine became widely adopted and
the number of new cases dropped, the polio care centers were closing,
putting Landauer out of work. Now Landauer would create similar centers for cystic fibrosis patients.

The centers, Landauer hoped, would solve a couple of serious problems plaguing the treatment of children with CF. The first was that
there were no standard protocols that physicians throughout the US
were required to follow. Remedies varied wildly from doctor to doctor,
hospital to hospital, and most were not very good. Worse, there was no
incentive for anyone to establish a set of best practices that would ensure
every child got the care they needed to maximize their health.

The second problem Landauer hoped the centers would solve was the limited understanding most physicians had of the disease. Cystic fibrosis was a complicated malady that impacted different patients in very different ways. Some were entirely possessed by the disease. Others suffered lung infections but no digestive issues and appeared healthy and well nourished. Some, like Tulcin's daughter Ann, suffered a life-threatening intestinal blockage at birth, but afterward seemed to grow quite normally. There was such a wide assortment of possible symptoms that it was impossible to predict how each child would progress. CF could destroy the lungs with pneumonia and bronchitis, shake the body with a chronic cough, potbelly the abdomen, barrel the chest, and club the fingers—or not.

With physicians at care centers focusing exclusively on CF patients, they would become familiar with all the disease's idiosyncrasies and therefore better able to deliver expert care. They would also, Landauer envisioned, be able to compile clinical observations that would lead to evidence-based best practices that would guide physicians everywhere.

Landauer's first two care centers opened in 1961[1] at Children's Hospital in Boston and New York's Babies Hospital; soon afterward, additional centers were opened in Albany and Cleveland. When Landauer launched the centers, he established two rules based on lessons learned from polio. First, he wanted all of the CF centers, in addition to seeing patients, to engage in research. And although money raised through the local chapters could be used for funding physicians, nurses, and equipment for patient care, grants from the national foundation were strictly for research—just as O'Connor had recommended to Tulcin and Frankel. Second, there would be no limit to the amount of funding directors could request for research. Together, these rules fostered a research boom, with increasing numbers of PhDs studying CF at the centers. Landauer was cleverly initiating a research agenda in a way that Sharples hadn't imagined.

By the mid-1960s, more than thirty centers had been established nationwide. But although these centers provided specialty care, there was no medical consensus on how to fight the disease. Physicians at the centers were still winging it, mashing together treatments based on their personal experience rather than scientific evidence. In general, CF doctors agreed that a low-fat diet that curbed the volume of offensive stools and gas was good, and that when lung infections set in, patients should receive antibiotics. Most doctors were also aware that physical therapy was essential to clearing the lungs, but many didn't stress the need for frequency. But by now, Landauer thought, with so many patients being evaluated at so many centers, physicians should have been spotting patterns and trends, identifying the most effective medicines, and developing evidence-based treatments. And they weren't.

The reason this wasn't happening, Landauer soon realized, was because there was no way to compare one center with another. There was no central database with statistics from patients at all centers, and there were no consequences if a center performed poorly. The leaders at the national foundation had no way of knowing whether a center was providing particularly dismal care, or making breakthroughs that were revolutionizing treatments. Fortunately, the rumor mill was an effective mechanism for sharing good news.

In 1964, the children at most centers were dying at about three years old.[2] But one care center, funded by the Cleveland, Ohio, chapter, was bucking that trend. Patients there died, on average, at age twenty-one—as young adults, rather than children.

The Cleveland chapter of the foundation was launched in 1955 by a small group of families known as the Cousin's Club. Driven by the death of one of their small children and the impending death of another, these families contacted Dr. William Wallace, the chairman of pediatrics at Babies and Children's Hospital in Cleveland, and asked him to launch a research-based treatment program at the hospital. Dr. Wallace agreed, and in 1957 tapped a twenty-nine-year-old doctor named LeRoy Matthews to run what Wallace described to him as a comprehensive treatment and research program for CF patients. Matthews was a Harvard-educated physician[3] who had previously studied under Harry Shwachman and Sidney Farber at Boston Children's Hospital. An entertaining storyteller, he bore an uncanny resemblance to Groucho Marx—short, dark hair, bushy eyebrows, small mustache, and round glasses.

The position of running the CF program in Cleveland was a phenomenal opportunity for a young faculty physician because he would have the autonomy to develop the entire treatment plan from the beginning. At the same time, it was terrifying. Cystic fibrosis was often fatal, and Matthews knew that even though there was now a national foundation, there was no consensus on the best treatment strategies. He was an analytical type, so he embraced the challenge by first methodically identifying the problems and then going on a targeted fact-finding mission. Visiting cities with large populations of CF patients—Boston,

New York, and Philadelphia—he met with the expert physicians to figure out what actions to take and when, and the best procedures and
strategies to incorporate into his own system in Cleveland.

He began in Boston with Harry Shwachman, who taught Matthews about chest physical therapy, with its clapping and vibrating to
loosen the mucus from the airways and make it easier to cough up. He
spent time in Shwachman's Nutrition Clinic,[4] where he witnessed the
benefits of a low-fat diet and vitamin supplements and the importance
of pancreatic enzymes. He also learned Shwachman's strategy of using
multiple antibiotics—sometimes three at a time—to pummel the tenacious bacteria infecting the lungs.

From Boston, Matthews traveled to Babies Hospital in New York
where he met with the field's matriarch, Dorothy Andersen, and Paul di
Sant'Agnese to better grasp the nuances of the disease and the different
methods of diagnostic testing. Di Sant'Agnese's sweat test was considered the most accurate method to diagnose the disease at the time, but
Andersen's test was still the only one that worked for newborns, because
they didn't sweat. While he was in the city, he also met Andersen's colleague at Babies Hospital, Dr. William Blanc, a pioneering pathologist
who shared sophisticated techniques he'd developed to detect and identify bacteria in individual patients' airways so he could determine which
antibiotic therapy would be the most effective.

In Philadelphia, Matthews met with Wynne Sharples and her husband, Dr. Robert Denton, who studied the flow of mucus in the airways
and was experimenting with ways to thin it. One of his innovations was
using a nebulizer to create a super-fine spray of either salt water or a
solution of a chemical called propylene glycol to loosen the hard mucus.
Patients would sleep in "mist tents" that covered their beds, and inhale
the moist air overnight.[5] Shwachman had found that, for some children with particularly viscous mucus, the fine salty mist the nebulizer
sprayed out to moisten their airways reduced their coughing and helped
them sleep through the night.

By the time the Cleveland CF clinic opened in 1957, Matthews combined his newly acquired knowledge to design what he called a "comprehensive and prophylactic treatment program," a pioneering strategy
that would prevent the progression of cystic fibrosis for as long as possible.[6] Early diagnosis was key to preventing lung infections, so he
made sure that his technicians became experts at administering both
Andersen's diagnostic test for newborns and the new Gibson-Cooke

iontophoresis sweat test. Designed for older babies and children, this test would soon surpass di Sant'Agnese's as the gold standard, and Matthews had learned about it while visiting Johns Hopkins University. Once a child was known to have the disease, aggressive treatment would begin.

For each child admitted to his new center, Matthews began with at least a week in the hospital. After confirming new patients' suspected CF diagnosis (if it was done at another hospital) with the sweat test, Matthews blasted their damaged lungs with antibiotics to kill off the infections lurking inside. He prescribed them a low-fat diet and pumped them full of vitamins and digestive enzymes to nourish their fragile, emaciated bodies. Understanding the importance of data collection to measure the impact his treatments had on the health and long-term survival of these children, Matthews used elaborate devices to measure their ability to breathe, calculating how much air they could inhale into their lungs—an important gauge of lung health that served as a baseline for charting progress.

Matthews sensed that the secret to keeping these kids healthy was his personal connection with both parents and patients. He treated the children's hospital stays as a precious opportunity to bond with and nurture the devastated parents, as well as teach them about their child's unfamiliar and daunting disease. It gave him time to fill them with hope and show them how to care for their children at home. He gave them hands-on training in the percussive physiotherapy essential to giving their child a shot at good health: sitting bedside, parents watched and asked questions as Matthews pounded their child's back and chest. He set a strict schedule of oral antibiotics, specially chosen for their efficacy against the unique brew of microbes in each child's lungs. He showed the parents how to use a nebulizer to give their child deeply penetrating aerosol lung medications—which, in the worst cases, had to be administered four times a day. And after the child was discharged, he insisted the family return once a month so he could chart the child's progress, adjust therapies, take X-rays—and ensure the parents were providing their child with the best care possible.

Before Matthews arrived, dozens of children with CF had died each year at Cleveland's Babies Hospital. But just four years after Matthews launched his Comprehensive Treatment Program, he had lost only three patients. The Cleveland CF center had worked magic, lowering its mortality rate to a jaw-dropping 2 percent. Instead of losing young patients before they entered kindergarten, the center was keeping them healthy long enough to graduate high school. Life expectancy for these kids was seven times the national average.

Matthews wasn't shy about telling his colleagues at other centers and at pediatric and CF conferences about his success—eagerly sharing his latest data and results at each annual meeting beginning in 1958. But as news spread about Matthews's miraculous mortality rates, the stories bred jealousy, skepticism, and distrust among these colleagues. What was he doing for these sick kids that they were not? Was he fabricating his data? At the time, there was no independent way to verify Matthews's claim, because there was no central repository that banked the statistics from each care center.

By 1964, the ill will building among doctors within the CF community led a young pediatrician from Minnesota named Warren Warwick to suggest, at the annual meeting of the Academic Pediatric Society, that the foundation create a questionnaire for the more than thirty centers to settle the debate and either confirm or deny the rumors about Matthews.[7] The foundation immediately assigned Warwick the task.

Warwick, a tall, thin Midwesterner, was a perfectionist, detail-oriented, and entirely driven by data. He developed a one-page questionnaire that he asked every center physician to complete for each of the patients they had treated at their center. Each center was labeled with a code known only to the director of that CF center; Warwick's dedicated secretary, Marjorie Stepek; and another employee, Richard Pogue, who created and ran the database. Warwick then assigned each child a unique code and a three-initial monogram, and charted their birth date, height, weight, complications at birth, date of diagnosis, sweat test results, date and length of hospitalizations, lung function—and, when applicable, the date and cause of their death. Stepek helped in completing this Herculean task by relentlessly tracking down any missing data.

When Warwick crunched the numbers, he confirmed that LeRoy Matthews's center did indeed have a 2 percent mortality rate, and that the average age of patients at his center was in fact twenty-one, compared to the national average of three. There had been no deaths at the center in children younger than six since his program began.

Warwick was convinced, but many at the foundation were not, declaring that Warwick and his Minnesota colleagues had botched the statistical analysis. To settle the dispute, the foundation orchestrated a meeting in Philadelphia and hired Cecil J. Nesbitt,[8] an internationally renowned mathematician and expert in actuarial science, to analyze the data again. But Nesbitt's analysis confirmed Warwick's results. Matthews was extending his patients' lives and amplifying their quality of life.

Matthews's success was proof that the other centers were failing, and it convinced center directors that they, too, needed to adopt his Comprehensive Treatment Program. The onus was now on foundation president Natal and research and medical education director Landauer to establish treatment guidelines. Matthews was named an advisor on the foundation's medical council and the first chairman of the Center Committee.[9] He and another center leader from Houston, Gunyon Harrison, then spent months establishing protocols and standards for all centers, accrediting new ones, and recruiting new directors to lead them. From then on, each center director received a confidential yearly report based on annual statistics mined from the center's submitted questionnaires—including that center's rank among all the others. This ranking was kept confidential, both to encourage honesty on the annual questionnaires and to avoid a mass patient exodus from low-ranking centers. The yearly reports were designed to push the centers to improve. Though no official incentives were dangled in front of the directors, the opportunity to brag about the longevity and health of their patients to CF Center Committee inspectors, who made annual visits and recommendations for future funding, was motivation enough.

The foundation recognized that the vast quantity of clinical data Warwick had collected did more than just vindicate Matthews's approach. It was also a way to identify the best and worst practices and establish a standard metric to assess each center's progress—which ultimately meant better care for patients and families. Beginning in 1966, they gave Warwick an annual $10,000 grant to manage the data and—as researchers learned more about the disease—expand the questionnaire and maintain a database of responses, until the foundation assumed control of it ten years later, in 1976.[10]

Warwick's analysis was so well respected that the Canadian CF Foundation asked him to do the same for all of their centers. Individual centers in Australia, England, and Sweden also asked Warwick to provide a report based on their own questionnaires to see how well their centers were performing. The patient registry Warwick originally created for the National Cystic Fibrosis Research Foundation would eventually be replicated by other disease nonprofits and biomedical institutions across the world.

Though neither Warwick nor the foundation's leaders could have known it at the time, the patient registry would eventually prove to be the foundation's most valuable asset—key to understanding the progression and nuances of the disease—and, decades later, it would further revolutionize patient treatment.

CHAPTER 11

The Therapist

1977

Every [physical therapist] has seen remarkable courage in
patients—bravery that sometimes hardly seems possible.
Witnessing it tends to strengthen our own courage and
gives us insights into how we best can serve others.

—Nancy Johnson, physical therapist

The first time John Nadeau met Joe O'Donnell, Joe was sprinkling seed to fill in the yellow and brown patches of lawn of the O'Donnells' Medford, Massachusetts, home. Joe walked up to him, shook his hand, and thanked him for what he would be doing for Joey.

Nadeau was a newly minted physical therapist who had trained at Northeastern University in Boston. He'd graduated in 1974, just a few months before Joey was born, and had been working at Glover Memorial Hospital doing physical therapy when he was approached by a mother who asked him if he could give chest therapy to her son who had cystic fibrosis. She had never met a male physical therapist—they weren't that common then—and she liked that he was closer in age to her seventeen-year-old son than other therapists.

Nadeau was hesitant to take the boy on; though he'd trained in chest therapy, he didn't have much experience with it, and he'd never heard of cystic fibrosis. The woman scribbled down the name and number of the boy's physician, Dr. Harry Shwachman, and pushed the scrap of paper into his hand. "Talk to him, then decide," she said.

Nadeau called Shwachman up and made an appointment to meet him and talk about chest physical therapy. Waiting outside Shwachman's office at Boston Children's Hospital a week later, Nadeau could hear yelling from behind a closed door. He heard a woman's voice from inside asking what she should do with her sick son. The man snapped, "If this boy is going to have a chance of living, it's going to be up to you. And stop being sorry for yourself. You have work to do." The door opened, and a woman emerged, eyes full of tears, as she carried her scrawny son out of the office.

From inside, Shwachman motioned Nadeau to come in, then settled behind his desk. Wasting no time on niceties, he launched into an explanation of what cystic fibrosis did to the lungs and why these kids needed physical therapy—in the most serious cases, three times a day. He told Nadeau about the shocking quantities of mucus that gunked up these kids' lungs. The only way to dislodge it, he said, was to thump it out. He stood up, a little hunched, and grabbed a nub of chalk to begin drawing as he spoke.

As you and I sit here breathing with normal lungs, the air is swirling around our airways as it travels down to our air sacs, he said, drawing what looked like mini tornadoes. All these airways have a mucus coating—a defensive mechanism that acts like flypaper, snagging bacteria and debris and sterilizing the air before it reaches the air sacs. In our lungs, that mucus is constantly flowing, like a river, up the airways and out toward our mouth and our throat. When we clear our throats during the day and swallow the mucus, we kill all the bacteria with stomach acid.

Shwachman turned and looked Nadeau in the eye in a way that made Nadeau feel like the doctor was boring a hole through him—trying, Nadeau assumed, to figure whether he was worth his time.

The problem in these sick kids, he continued, was that the mucus was thick and dry. It didn't have enough water in it, so it didn't flow. Instead it plugged up the lungs, creating a home for bacteria. So they needed therapy to break up this clotted mess. "Understand?"

Nadeau nodded.

Shwachman told Nadeau how he had learned about chest physical therapy from nurses in England, who used it for many pulmonary disorders. He had encouraged one nurse visiting his practice in Boston to develop a version of the clapping-vibrating therapy for all regions of the lungs, for use in CF patients.

Shwachman continued to talk for another twenty minutes as Nadeau kept nodding, only occasionally daring to let his eyes stray toward a nearby corkboard collaged with photos of Shwachman's young patients, notes, and crayon drawings.

When Shwachman finished, he turned to Nadeau. Any questions? He had plenty. Shwachman sat and answered them patiently before handing Nadeau a fat folder of reading material—Shwachman's papers and book chapters. Nadeau thanked him, and as he got up to leave, Shwachman softened, patting him on the shoulder. "You're going to get very close to your patients. You just don't realize what you're getting into."

Nadeau smiled politely, dismissing the enormity of Shwachman's closing pronouncement. After all, it was common for him to become a bit attached to his patients. He liked helping all of them.

That night, as Nadeau began leafing through the contents of the folder, he realized for the first time that he had just met an eminent scientist, a pioneer in the cystic fibrosis field, and was shocked that the man had given him so much time.

A few days later, he received a card in the mail. It was from Dr. Shwachman. "Thank you for what you are doing for my patients," he wrote. "Call me anytime with questions."

As he began working with CF patients, he called Shwachman whenever he was unsure of what he was seeing and hearing. It was standard practice to listen to patients' lungs with a stethoscope before and after therapy—to listen for any changes. But to Nadeau's inexperienced ears, the patients sounded worse afterward, which made him panic. He called to find out why. "Before therapy, I hear air going in and out pretty well," he told Shwachman, "and afterward I'm just not hearing as much air. I don't know what's going on."

"Son," said Shwachman, "you're not listening to enough normal lungs." You're supposed to hear nothing, he explained. If you hear that much air, that's a sick lung.

When Nadeau hung up the phone, he called his wife, who was also a physical therapist, and listened to her lungs by pressing a stethoscope to her back. Then he did the same with his young daughter. The air in her healthy lungs made barely a whisper, whereas the soundtrack of his patients' lungs was a mix of squeaks and wheezes, pops and crackles.

In 1977, the hospital where Nadeau worked held scheduled CF nights once a month at a nearby grammar school so parents could ask questions and mingle with their kids' physicians. This time Nadeau and Jerry Mulcahy, a friend and colleague at Nadeau's hospital who also did chest physical therapy, were asked to go. The room was packed with anxious, exhausted young parents. At the end of that night's presentation, the organizers opened the floor for a Q&A. The mother of one of Mulcahy's patients got up immediately and raised her hand. She said, I just want to let everybody know that I have a therapist that comes to my house every day and he's wonderful, and he's got a partner, and Blue Cross pays for it, and they're sitting—she pointed—over there. All heads turned. The announcement ended the Q&A as parents beelined to Nadeau and Mulcahy.

Before long, Shwachman was sending Nadeau and Mulcahy all of his patients. The need for their skills was enormous, and eventually both men quit their jobs at the hospital to launch a business focusing on chest therapy, driving like madmen to treat children in their homes all over eastern Massachusetts.

Word spread to other physicians, and in late 1977, Dr. Allen Lapey sent a new patient their way: Joey O'Donnell.

Joey was three, and thin, with CF's classic barrel chest and hall-mark cough, when Nadeau first came to the O'Donnells' home. But the disease hadn't diminished Joey's curiosity, and he was eager to meet Nadeau, peeking at him from behind Kathy as she discussed his therapy.

Kathy had the telltale sign of a CF mom: arms lean and sculpted from the three hours a day she spent working on her child's chest. Her skills had been honed from necessity and fear after almost losing Joey as an infant, and by that point she was an expert—so good that if she'd had a license, Nadeau would have hired her for his growing practice. Kathy was masterful at getting Joey to adhere to all parts of his therapy, from wearing the mask, to inhaling his medication, to lying still during treatment. But now that he was bigger, that treatment was becoming an ever-more-strenuous activity, and Kathy needed help for one of his three daily sessions. Nadeau would be that help, and also another set of ears to help monitor the state of Joey's lungs.

By 1977, John was used to three-year-olds who resisted and refused therapy, but Joey was easy. He'd grown so sick so early and had had so much treatment that he was used to being handled; he'd never known life without it. He pretty much ignored Nadeau during his treatments, much more interested in watching the cat-and-mouse antics of Tom and Jerry on TV.

Still, the round-faced boy was a typical mischievous child, and it wasn't long before Joey became adept at delaying his therapy by ini-tiating a daily game of tag upon Nadeau's arrival. Joey would peep at him from behind a wall or under a table, eyes twinkling, and dash off, laughing softly—unable to inhale enough air for a full-bellied laugh—as Nadeau went after him. Joey had big, floppy feet, in contrast to his small, skinny body; it was like watching a rabbit run. Exercise was great for children with CF, so Nadeau was happy to chase him and get his lungs working. The O'Donnells' house had a circular layout to it that made it easy for the two of them to run two or three laps before Joey settled down. Unless, of course, Joey started an impromptu game of hide-and-seek. Nadeau enjoyed their games and began allotting more time for Joey's appointment.

Nadeau didn't follow research or the inner politics of the National Cystic Fibrosis Research Foundation, which had been renamed in 1976 to simply the Cystic Fibrosis Foundation, so he had no sense of whether there would soon be a treatment or cure on the horizon. What he did know and care about was that he and Mulcahy had their hands full providing physical therapy to dozens of sick children. Each of these kids was spirited, even with their hellish schedule of chest therapy and meds and nebulizer treatments. And although these were early days in his chest-therapy practice, Nadeau was beginning to understand what Shwachman told him. Chest physical therapy was tough on the children and he had to perform it on them every day. It was an intimate practice, and as he got to know these children and their families, he became, as Shwachman predicted, very close to his patients.

CHAPTER 12

A Disease in Search of Ideas

1964–1980

Do not hire a man who does your work for money,
but him who does it for the love of it.

—Henry David Thoreau

Bob Natal, the trucking executive turned National Cystic Fibrosis Research Foundation president, had been a steadfast, dedicated, and diplomatic presence during a transitional—though exciting and productive—time for the foundation. Patients were now pouring into Landauer's cystic fibrosis care centers, helping drive patient survival rates slowly upward.

The mandate for these centers was care, teaching, and research. But finding enough financial support for the centers' continued existence was difficult enough, leaving essentially nothing for research. The research taking place there focused on how the disease progressed in patients, but no one was investigating its roots. Natal knew this problem needed to be addressed, but just a few years after becoming president, he developed colon cancer and had to resign. When Milton Graub, Natal's vice president, took the reins in 1964, the first item on his agenda was to hit up longtime NCFRF trustee George Frankel for $87,500—the equivalent of approximately $730,000 today—to support a research fellowship for five years, a rate on par with what the NIH paid. Graub explained that it took most research projects between three and five years to answer even a basic scientific question. Not only did Frankel grant Graub's request, he promised to provide the researcher's annual salary of $17,500 in perpetuity.

But one research fellowship for one scientist wasn't going to cure this disease. So Graub also reached out to the president of Merck and

Companies' Research Division, who had the gravitas to summon nearly any researcher in the biological sciences. Merck's president suggested a meeting at the Harvard Club in New York City and brought along Dr. James Watson, who had just shared the 1952 Nobel Prize in Physiology or Medicine for his codiscovery of the structure of DNA. The two agreed that what the foundation needed was a cohort of outstanding scientists to advise them about the key questions necessary to crack CF. Watson agreed to join the foundation's new scientist advisory group, as did two other Nobel Prize winners—molecular biologist Joshua Lederberg and biochemist Arthur Kornberg—and six other preeminent scientists.

Just as the NFIP's Harry Weaver had set out a series of vital questions that had to be solved to develop a vaccine for polio, the foundation had to formulate a list of critical mysteries regarding the origin of CF that the researchers it funded could investigate. To determine these questions, Graub and his friend and colleague Giulio Barbero, a physician and researcher at Children's Hospital of Philadelphia, hosted a new event called the GAP Conference, short for Guidance, Action, and Projection. Every year, the foundation, with the help of their Nobel-studded team of scientist advisors, would identify a gap in knowledge about cystic fibrosis, and use that as a theme for an annual meeting. They'd invite the brightest young scientists working in that year's theme of interest, with the hope of enticing them to apply their expertise toward cystic fibrosis.

Once again, George Frankel came to the rescue with more funding, specifically to pay the cost of the GAP conferences themselves. But naturally he didn't have the resources to solve all of the foundation's money shortfalls. Graub recognized that the smartest long-term financial solution was for the foundation to convince the NIH fund to support clinical research as well as patient care at the centers that Landauer had launched.

Through a lobbyist connection, Graub petitioned the powerful and compassionate Pennsylvania congressman Daniel Flood for a meeting.[1] Flood was chairman of the Subcommittee on Labor, Health, Education, and Welfare of the House Appropriations Committee, and had a passion for healthcare and medicine.

When Flood accepted the meeting, he had never heard of cystic fibrosis, so Graub, LeRoy Matthews, and Giulio Barbero spent the first half hour educating him on what it was, how it was passed from parents to children, the gruesome symptoms, and the pitiful life expectancy. Graub's ask was two million dollars as a line item in the budget for basic research in CF from the National Institute for Arthritis and Metabolic Diseases (NIAMD). Flood invited them to share their request and testify before an appropriations subcommittee.

In late April 1967, Dr. Paul Patterson, one of the founding members of the foundation's medical team, and Drs. Barbero and Matthews came to testify. Patterson explained the CF community's dire need for research and training. To highlight to the subcommittee the more general relevance of this disease to the US population, he explained that, though CF was rare, between ten and twelve million Americans were carriers of the cystic fibrosis gene and might also suffer from the same metabolic defect as these children, though to a lesser degree. Congressman Flood, who had a flamboyant theatrical style drawn from a post-college stint as a Shakespearean actor, and enjoyed bringing drama to every conversation, ended Patterson's presentation with his typical flair: "You have a real killer, do you not?"

"I am afraid we do, sir," Patterson replied.

Dr. Barbero advocated for funding pulmonary centers more generally, and Dr. Matthews backed him up, describing the stellar results he had achieved with his own program, and sharing powerful pictures of his patients before and after his treatment. Flood explained to the committee that pulmonary disease centers would benefit not just CF but also bronchitis and asthma; there was "a large number of children with crippling and often fatal pulmonary diseases." Everyone would benefit, he argued, if Congress funded centers like the one Dr. Matthews had established.

When Flood took the floor a month later to haggle over cuts he was being asked to make to the NIH budget, he boldly defended his existing budget—including the late addition of Graub's requested $2 million.

It was a huge victory for Graub. He'd helped the foundation get the attention of Congress, secured more funding for the care centers and research, and convinced Congress to mandate that the NIH study this disease. But it had been a grueling journey. As he fought for the foundation, his son Lee's health had been deteriorating rapidly, and in 1968, at just ten years old, Lee died. Physically and emotionally exhausted, Graub had to bow out. His daughter was getting sicker and his practice was getting busier; he needed to focus on his family. So in 1969, after nearly six years, he stepped down.

IN 1964, WHEN MILTON GRAUB FIRST BECAME THE FOUNDATION'S president, Doris Tulcin was recovering from a heartbreaking few years. Her husband was still suffering from depression, and her mother had died the year before, a devastating blow. The one unexpected joy in her life came from Ann's continued good health; at eleven years old, she was still thriving, and had been since that first year of life. Tulcin had told friends and family about Ann's illness so they could keep their children away from Ann if they were sick, but she never made Ann feel fragile or vulnerable or afraid. In fact, it wasn't until Ann was about nine years old that Tulcin told her she had CF.

Over the eight years since Tulcin launched her first chapter in Scarsdale in 1956, she'd become one of the top fundraisers for the foundation. Her Westchester chapter regularly hauled in $80,000 to $100,000 per event—money that went to support her local care center, and to the national foundation to support research. Inspired by the fundraisers Basil O'Connor described, she'd organized dozens of dinners and theater benefits in New York, tours of exquisite and historic homes, and auctions. She marked the foundation's ten-year anniversary in 1965 with an exclusive tour of Joseph Hirshhorn's Greenwich, Connecticut, sculpture garden and home, which was decorated with rarely seen works of Picasso, de Kooning, and Miró, and in 1966 she added movie premieres to her repertoire, beginning with *My Fair Lady*.

Tulcin had long been toying with the idea of taking on a larger role at the foundation, and that feeling grew when Milton Graub's successor, Robert McCreery, took the reins.

McCreery's son didn't have CF, but he'd become entwined with the Cleveland chapter after LeRoy Matthews figured out that McCreery's son's breathing problems were induced by food allergies. His friendships with CF parents, chapter activities, and fundraising efforts led the Ohio and West Virginia chapters to elect him as their regional trustee. A year later, with Graub as president, he was elected a national trustee. When Graub was ready to leave, he convinced McCreery to take the job. McCreery reluctantly accepted and then began his tenure in 1970 with a series of bad decisions, including moving the foundation's New York headquarters to Atlanta, Georgia, for no good reason other than cheaper office space. There, fifty full-time employees ran education programs, supported the care centers, and conducted government outreach and public relations.

Doris Tulcin was perplexed and irritated by the move to Georgia, and her conscience nagged at her for not contesting the decision. That prick of conscience only grew worse when her beloved father, George Frankel, died suddenly from a stroke in the early spring of 1971. Tulcin was heartbroken and set out to honor his legacy by protecting the foundation. She was well known and well liked, and in the late 1960s had served one term as a regional trustee of the New York CF chapter.[2] Now the board of trustees had elected her to serve as a national trustee. In May 1971, her first assignment in that role was to travel the country to meet with the individual chapters and ask them to pony up more money for the perennially cash-starved national foundation.

Tulcin had no idea how contentious the meetings would be. The priorities of the national foundation and the local chapters were very different. National wanted chapter funds for research. The chapters didn't want to jeopardize the health of their children for research that seemed so nebulous and far from a cure. Parents felt strongly that the money they raised should support the local care centers that kept their children alive.

Tulcin began her quest in Chicago. Then, she traveled to Ohio to appeal to the Cleveland chapter—where LeRoy Matthews's exemplary center was based—asking them to give more of their funds to national to support research into the disease's cause. Robert Dresing, a parent sitting in the second row of the audience, stood up the moment she finished and bluntly accused her of shamelessly attempting to steal funds from their children and the local doctors and specialists who cared for them.

Dresing, a tall and handsome businessman with dark, slicked-back hair who had cofounded the Ethan Allen retail franchise in 1963, was a charismatic figure with a commanding presence. His four-year-old son had been diagnosed with cystic fibrosis at eighteen months when an infection seized his frail body and swept him into a coma. The toddler spent four days in the hospital and began intensive physical therapy and high doses of antibiotics under the watchful eye of LeRoy Matthews and his protégé, Carl Doershuk, at Cleveland's Babies and Children's Hospital. Under their care, he recovered and thrived—and Dresing and his wife joined the Cleveland chapter of the CF foundation.

The Cleveland chapter, Dresing would learn, received all of its financial support from the Cleveland Health Fund, which also supported several other organizations, including ones for kidney disease and arthritis. The advantage of this arrangement was that the Health Fund took care of the fundraising while the Cleveland chapter used the money they received on the staff and equipment to make their care center the best in the country.

With the Cleveland Health Fund's money, the Cleveland chapter was strong and independent, flush with funds, the wealthiest of all the chapters. And its members disliked the national foundation, particularly now, after Tulcin's request threatened their sovereignty. Shaken by her visit, parents urged Dresing to run as a regional trustee, attend Atlanta meetings, and guard Cleveland from the national marauders. Dresing graciously agreed, and after also campaigning for the position in Ohio, Kentucky, and West Virginia, whose chapters he would also represent, he won the position.

In 1972, there were some forty-seven regional and national trustees. About twenty were regional, like Dresing, elected by a group of chapters. The rest were national, and most came from the medical or scientific advisory boards. Four times a year, both sets of trustees convened in Atlanta with the local staff for three days of meetings and nights filled with eating and drinking at local restaurants. Ostensibly, the goals were to discuss fundraising, chapter management, and public and government relations, and to hammer out budgets for care centers, medical education, and clinical and basic research. But Dresing could see that this was just as much a venue for the trustees to squabble over how money should be allocated. Most disturbing to him, however, was that no one was talking about funding research or a cure. The visit boosted his confidence in and allegiance to his hometown chapter and the care his son received. But he couldn't ignore that the outlook for other, less fortunate children was grim.

The national foundation was supposed to receive 55 percent of all funds each chapter raised. But Dresing knew that chapter organizers kept local funds local—none complied with the 55 percent rule. Each year for the past five, the chapters had raised a total of $12 million—a pathetic sum, Dresing thought. And less than a million dollars of that had been spent on research.

Tulcin, too, found the lack of focus on new research at these meetings painful to witness. She had now attended more than half a dozen of these trustee meetings, and she was fed up. The foundation was broken and needed a fresh start.

Near the end of Dresing's first meeting, both Dresing and Tulcin gravitated toward the bar in the Atlanta restaurant, away from where the trustees were gathered. Neither was interested in socializing. Yet Dresing, seeing Tulcin sitting several stools away, picked up his drink and, despite their explosive first meeting in Cleveland, sat down next to her. After a couple of drinks, the tension from their previous encounter in Cleveland dissipated. They began sharing their ordeal with this disease. She told him about her concerns with the foundation and the lack of research, and the lessons learned from O'Connor and her father.

Dresing quickly realized he had misjudged her. They shared the same visions and worries—they both knew that without research there would never be a cure. The foundation needed great scientists to discover the root of the disease and federal dollars to fund them; a DC location would be better, providing them with sight lines to both the NIH and Capitol Hill. Tulcin was a remarkable fundraiser, a hard worker, and a class act—Ann's good health hadn't made her complacent or less driven to cure this disease. She was smart and compassionate, and a potential ally. If the two of them ran the foundation, he realized, they could change the game.

Over the next two years, Tulcin and Dresing became close, as did their families. They shared birthdays and anniversaries and even vacationed together sometimes. The whole time, however, they kept their eyes on the goal—figuring out how best to fix the foundation.

In 1974, McCreery asked Tulcin to become his vice president. She accepted. And after McCreery completed his second term in 1976, the national board of trustees elected Doris Tulcin as the foundation's new president. She immediately recruited Dresing to be her VP, making him the head of her executive committee, and simplified the name of the organization to the Cystic Fibrosis Foundation. Then they went about pursuing their first goal: moving the foundation's headquarters to Washington, DC.

To relocate the headquarters, they needed the trustees' consent. So, from 1977 into 1978, Tulcin sent Dresing around the country to lobby every trustee in their own home—to plead with them, parent to parent, to move the foundation. The meetings were tense and unpleasant. Atlanta had provided a guilt-free respite four times a year from the often-constant care of their sick children. Now Tulcin and Dresing wanted to take that away. And what angered the trustees more was that

Tulcin and Dresing also wanted to clean house and reboot the entire foundation—its personnel along with its mission.

But Dresing and Tulcin didn't need to wait for the trustees to vote on the Washington move before they began making inroads in DC and laying the groundwork for the foundation's new direction. The first challenge they faced was figuring out how to navigate the political landscape. With whom should they meet? How exactly did one book time with a senator and ask for money? But surmounting this initial challenge would be the easy part. Once they secured more funding, they also needed to determine the best way to use it. They needed a visionary leader who could synthesize all the science known about the disease, create a research agenda, and push it forward.

In 1974, a scientist named Robert Beall, a stout biochemist with a ruddy, freckled complexion and a shock of reddish-orange hair, was trying to find his calling. Beall had just completed his postdoctoral work at Case Western Reserve University in Cleveland, an experience that made it clear to him that academia was not where he wanted to be. He had no desire to spend his life working in the lab on a small slice of a big problem and never see that science translated into a living person. He was ambitious; he wanted to work on health issues facing society and change not just one or two lives, but hundreds of thousands. He decided to try health policy instead, as it would leverage his scientific acumen and also satisfy that desire to create sweeping changes. And the best place to work in health policy was the National Institutes of Health in Bethesda, Maryland.

He applied for and won a position at the National Institute of General Medical Sciences. There, he was asked to participate in the genetics program, but at the time, genetics didn't look that exciting and the field, in his opinion, was moving quite slowly. Scientists were just learning how to cut and paste DNA together, the first tools of genetic engineering; Beall couldn't imagine how this nascent field was going to evolve and revolutionize medicine. He chose to move on again, this time to the National Institute of Arthritis, Metabolism, and Digestive Diseases (NIAMDD).[3]

This was a new era in health policy, when Congress was earmarking tens of millions for big, mandated health programs with lofty goals. In December 1971, Congress passed the National Cancer Act,[4] which launched the "war on cancer"; in July 1974, the National Diabetes

Mellitus Research and Education Act[5] was pushed through; and in January 1975, the National Arthritis Act passed. The acts mandated that the NIH study these diseases in depth, set up disease-focused centers, and educate the public, in the hope of preventing the disorders as well as helping those afflicted with them live longer.

After completing an executive training program at the NIH in 1975, Beall began working as director for the Metabolic Diseases Program at NIAMDD, which included diabetes along with endocrine, digestive, and kidney disorders, and was responsible for implementing the recommendations of the recent National Commission on Diabetes.[6] Scientists at the institute had predicted a huge, looming epidemic of diabetes (predictions that would prove correct), and hundreds of researchers were devising health policies to stem the rising plague.[7] Beall was intrigued by the challenge. He was young, ambitious, and eager to become a major player. One of his tasks as per the new act was helping to establish National Diabetes Research Centers devoted to studying every facet of diabetes, something the National Cancer Institute had already done for cancer. That meshed well with his personal goal of producing a critical mass of science on the disease that would inspire new medicines and therapies.

One wintry, overcast day in 1976, as Beall was focused on the diabetes centers, his institute director, Donald Whedon, came to see him and asked whether he would like to attend a scientific meeting on cystic fibrosis.

Never heard of it, Beall said briskly.

Whedon shared the basics. It was a rare disease. Hereditary. Only affected about 30,000 people in the United States. The cause was unknown and it killed kids before they reached their mid-teens.

It sounded terrible, Beall thought. But only 30,000 affected? He was focused on diabetes, a high-profile disease that even then affected tens of millions. Without the right approach, one he would help sculpt, he was convinced diabetes would quickly reach epidemic proportions, overwhelming the healthcare system. If he changed the trajectory of diabetes, he could save a lot of lives.

But the director was persuasive, so a few weeks later Beall flew to San Diego to hear about this strange, deadly malady.

The meeting was held at a resort called Vacation Village and run and organized by the local chapter of the Cystic Fibrosis Foundation. Beall was astonished to see that the chapter was just a small grassroots operation. Parents brought potluck-style food and ran the projector while a handful of scientists shared the tragically small amount of information available. Beall was unexpectedly moved by these desperate parents. This community had no resources, no science, and no hope. But Beall saw that as a challenge. Here was a disease where he could make a real difference. And he sensed his work could have a greater impact on this disease than on diabetes. He wanted to help—but his hands were tied. Congressman Flood had helped the foundation secure funding for patient care but not research. And Beall's institute had no money for cystic fibrosis research. Unless Congress approved funds and issued a mandate that the NIH study cystic fibrosis, this disease would remain in the shadows.

Fortunately, such a mandate was just what Tulcin and Dresing were now in DC trying to make happen.

TULCIN AND DRESING HAD HIRED CONSULTANTS TO HELP THEM address the funding shortage for this disease and teach them how to lobby Congress. Early on, these advisors connected them with Representative Joe Early. A young Democrat from Worcester, Massachusetts, Early was a low-key, working-class, cigar-puffing, rough-tough in-the-trenches guy who had a reputation for getting things done and eventually served six terms in the state legislature before entering Congress. He was a staunch supporter of medical research, making him a smart pick, and he quickly became a passionate ally of the foundation. Others soon joined the fight: Lowell Weicker, an influential senator from Connecticut, followed by Representative Tom Harkin from Iowa and Senator Tom Daschle from South Dakota.

By late 1976, with Tulcin and Dresing's hard work having earned the foundation support from these influential congressmen, interest in their cause snowballed. In House Report No. 1219, the appropriations committee concluded cystic fibrosis research was a priority and ordered the NIH to undertake a study, to be completed within the next year, exploring the state of CF research and direct future research and treatment.[8]

Recalling Beall's emotional reaction to the San Diego meeting, NIAMDD Director Whedon assigned him as the institute's representative for the CF study and introduced him to Tulcin and Dresing, who provided guidance to get it underway.

Putting most of his other projects aside, Beall spent the next year learning everything he could about the disease: its physiology, the state of the existing science, the current therapies and interventions, and the activities of the foundation and other private entities.

Connecticut governor Abraham Ribicoff, the friend of Tulcin's father who had spoken at her first big fundraiser twenty years earlier, knew that Tulcin was now president of the foundation and that she wanted to learn more about the NIH research so she could more effectively influence the scientific and medical community. So in 1977,[9] with Dresing continuing his travels around the country to convince trustees to move foundation headquarters out of Atlanta, Ribicoff nominated Tulcin for the advisory council of the National Heart, Lung, and Blood Institute. It was a coveted four-year appointment, always held by a layperson, and provided a deep dive into the workings of the NIH and how to get funding. For Tulcin, what was most important was that it enabled her to keep cystic fibrosis on the radar of the scientific community. That was crucial because in 1977, the NIH's budget for CF research was only $4 million.[10] She wanted that to increase.

The most lasting takeaway from her time on the advisory council was the professional relationship she established with Bob Beall. He impressed her with his smart, innovative, and aggressive ideas. After chatting a couple of times, she asked him if he'd move to Atlanta where, unless Dresing succeeded, the foundation would remain based. Beall refused; he had no desire to relocate. But Tulcin wouldn't give up. If the foundation was in Washington, would he reconsider? Beall was noncommittal, but promised Tulcin he would talk if that day ever came.

As Beall worked on the institute's cystic fibrosis report, Tulcin and Dresing collaborated with a local DC firm to draft critical documents to make their case for increased congressional funding. This included a 113-page manuscript, *Cystic Fibrosis—A Plea for a Future: Roles of the Public and Private Sector*, which provided a primer on CF, the research currently being done in the public and private sectors, the challenges, the current funding, and the opportunities increased funding would provide. Finally, they drafted a "Five Year Strategic Plan—1979–1983."

In 1978, 150 scientists from the NIH and US universities, including Bob Beall, completed a 522-plus-page report: *Cystic Fibrosis: The State of the Art and Directions for Future Research Efforts*. Beall then produced a condensed version to insert into the grant applications mailed to thousands of research institutions around the country. He wanted to pique the interest of every researcher in the US—to convince them to learn about CF and then focus their talents and energy on this disease.

It was good, Tulcin and Dresing thought, to now have some attention from the NIH, and it was great that the grant application instructions contained information about cystic fibrosis. But grants were for young novice researchers and too small to lure older, preeminent scientists with broad visions. Tulcin and Dresing aimed to build a critical mass of CF science, not a smattering of research here and there. They wanted a network of multidisciplinary centers of excellence, each with a couple of dozen researchers investigating different elements of this complex disease, all sharing their data and collaborating. To make this vision a reality and have a shot at curing it would take substantially more funding than Congress would give and more resources than the NIH was willing to devote.

If Dresing and Tulcin were going to create this research network—something recommended in the congressional report—they needed Bob Beall to launch and lead it.

When it came time for the National Board to vote on relocation in 1979, the motion passed, but barely, and it created a terrible rift between the foundation and its seventy chapters. Tulcin and Dresing would heal that wound later. Now their priority was snagging Bob Beall.

Once they told Beall the foundation had a new brick-and-mortar office on Executive Boulevard in Rockville, Maryland, just a few miles or so from the NIH, and a completely new staff, he agreed to become the new director of science and medicine. He began at the foundation on January 1, 1980. Together, the three planned a whole new era in cystic fibrosis research—with dozens of research centers devoted to finding a cure.

CHAPTER 13

The Hitman Cometh

1979–1982

Baseball is about talent, hard work, and strategy. But at
the deepest level, it's about love, integrity, and respect.

—Pat Gillick

By his fifth birthday, Joey O'Donnell was thriving, his relative health
a testament to Kathy's zealous adherence to his physical therapy.
Though still small for his age, the blue-eyed, brown-haired boy was
raring to go to kindergarten, giving Joe and Kathy hope that he might
defy his terrible prognosis. And aside from routine cleanouts, when Joey
went to the hospital for high doses of IV antibiotics to fight the infection in his lungs, the family had had a three-plus-year "break" from the
disease.

Joe and Kathy were adamant that Joey live a normal life. He wasn't
going to be coddled. They were not going to keep him in when it rained
or stop him from playing, fearful he'd tire. They let him live. They sent
him to public school in their small town of Belmont where the family
had purchased a new home, and he rode the bus with his friends. And
everyone else followed their lead—family, friends, teachers, and other
children—treating Joey not as a child with special needs but as just
another little boy.

While it isn't unusual for a child to fall sick soon after starting kindergarten, when Joey got sick, unlike his classmates, he had to go to
the hospital. Kathy was afraid that he would fall behind, so she asked
their family friend Bubs Martin, a first-grade teacher, to go to the hospital and tutor him. But after the first session, Bubs called Kathy. Joey
doesn't need a tutor, she said. "He's so smart. He knows all his letters;
he knows all his sounds. He's just as bright as a button."

The next day Bubs got a call back from Joe. He understood that Joey didn't need her as a tutor, but Joey was fond of her and had been delighted by her visit. Would she mind visiting again? So Bubs became a regular visitor whenever Joey was in the hospital, and she became his friend, dropping in to see him at home and, later, even coming to his Little League games.

Though he was smaller than his peers, Joey had inherited his father's big personality, and other children gravitated to him. He was a charismatic prankster; he loved spinning a yarn and embellishing stories, like his dad. And also like his dad, he loved baseball.

Joe coached the pre–Little League team and Joey played second base. He had great hand-eye coordination; he'd clearly inherited his dad's athleticism and could whack the ball out of the park. His teammates nicknamed him "The Hitman." The first year he played, when he was seven or eight, he had the stamina to run the bases, but during Joey's second season, his lung capacity reined in his ability to run. Still, his spirit was unbridled. Joey continued to play baseball, though he couldn't run the bases, and he also took up soccer. Though he couldn't kick far, he was always out there the whole game.

Joey's "Uncle" Del—Paul Del Rossi, a longtime friend of Joe's— was in the bleachers at the baseball field watching Joey the first time his lungs failed him. Joey was exhausted trying to run the fifty feet to first base, but the game didn't stop. No adults intervened. His friends Woody, Timmy, and Eddie just decided among themselves that Joey would hit and one of them would run for him; the kids took care of their own.

That care and friendship wasn't confined to the baseball diamond. Bubs lived close to Joe and Kathy and she sometimes picked up Joey, his cousin Denny, and his best friend, Woody, and took them to the movies or the science museum. Even at six years old, the boys knew Joey's limitations, and they walked slowly, matching Joey's pace without talking about it.

Joey was up for everything. In the spring of 1983, when Joey was nine, the family went to a local fair, where they ran into a pack of Joey's friends. An obstacle course there was a huge attraction for the other boys; they bought tickets and charged through it. Joey looked at his parents. "I want to do that," he said.

Joe and Kathy looked at each other nervously, then nodded, and Joey took off. It was a long course. There were hurdles to jump over, tunnels to crawl through, under, and around, and hordes of kids of all ages racing through, trying to complete the course in record time.

Joey reached the first obstacle and instead of trying to hop or climb over it, he ran around it. Instead of going over the hurdles, he went under them. Each obstacle he navigated in his own inimitable Joey style. When he finished, he came back panting heavily, completely gassed—with a giant grin.

When Joey started school, Kathy had kept close tabs on him, calling the office at the same time every day to check that he was okay. After the first few weeks of kindergarten, the office staff knew to expect the call—"Everything's fine, Mrs. O'Donnell." At one point Kathy considered getting a pager, so that she wouldn't be tied to the house. But she worried that Joey would get frightened—a beeper might signal to Joey that something might suddenly happen to him, that he was always in danger. So she stayed at home while he was at school, just in case. And indeed, over the next few years, the school called regularly. I think he's sick, the receptionist would say. I think you ought to come in. Or they would call to say he was tired, and should come home and take a nap. Or stay home for a couple of days.

Joe had different fears about Joey and school. Joe had grown up in a tough neighborhood where he'd had to learn how to hold his own among other kids in school and on the playground. But Joey was a tiny fellow and Joe was terrified that his son would be bullied. To counteract potential social stress from school, Joe worked hard to create an oasis at home where Joey and his friends could play in a space that was safe and fun. Joe was always cutting deals with Joey, and early on, he told him, "If you get all A's in school, if you never get a B, if you continue to do your school work—that's your job—we'll get you video games. And you can play all day long if you want." He and Kathy bought three arcade-sized video games and put them in the basement.

The deal worked well for Joey; school work was easy for him and he never had to struggle to earn his game time. And Joe's fears of bullying were unfounded. Joey was a popular kid who had no trouble making friends. He even had a tight gang of four friends that sneaked onto his special school bus and rode home with him every day. His best friend, Woody, plus Timmy, Eddie, and Chris, piled into the house most days at 3:15 PM, and Kathy welcomed them with a buffet of snacks. Afterward they tumbled downstairs to play. The basement also held a jukebox, and a custom train set with an electric train that disappeared into the wall before reentering the room and chugging through a mountain and train stations. There was a pool table (off limits to the boys) and just about every toy imaginable. Outside was a zip line that the boys used to fly over the pool to the other side of the yard, while the O'Donnells' large black-and-brown Akita, Panda, chased them.

Nothing made Joe and Kathy feel happier than sitting in the living room and hearing the boys make a racket playing downstairs. Joey showed great skill in all physical activities, including pinball and video games, and his friends could be heard hollering to him for advice. "How do we get to level seven on *Donkey Kong*?"

Daily therapy sessions continued, but once Joey started school, John Nadeau shifted his visits from the morning to 5:00 PM. When he arrived, Kathy called to the boys that it was time for Joey to come in for his physical therapy, and invited them to come watch TV until he was done. It wasn't long before Joey would ask his friends himself if they wanted to stay while Nadeau "pounded" him. Kathy made it normal for Joey and for his friends, and no one was ever embarrassed. She allowed the kids to see cystic fibrosis, be around Joey, and understand—without getting into any of the medical jargon.

When Nadeau did Joey's chest PT—in the basement, on the sun porch, in Joey's bedroom, or wherever Joey and his friends happened to be that day—someone, if not everyone, was always more than happy to keep Joey company, playing video games or chatting during therapy.

I'll stay, called Timmy, not raising his head from the *Space Invaders* game.

I'm staying, too, yelled Eddie, focused on a pinball game—adding that he had to be home for dinner at six.

Once therapy was done, they would all watch cartoons—*Tom and Jerry* was a favorite—and often head outside to play.

TOGETHER, KATHY AND NADEAU HAD KEPT JOEY'S ILLNESS UNDER control during his early years, through a combination of physical therapy, antibiotics, and digestive enzymes—though whenever he caught some kind of bug, he still always needed expert care. This usually meant a fourteen-day stay at Mass General. Joey's sixth birthday, however, had marked a turning point, after which he began spending more and more time in the hospital.

Mass General became Joey's second home as the lung infections became more frequent, and Kathy was always by his side. He was comfortable there; he knew all the doctors and nurses. And he knew from previous regular cleanouts what to expect whenever he was admitted: heavy doses of antibiotics given intravenously through his fragile veins, physical therapy with an expert who knocked all the gunk out of his chest three times a day, and supercharging doses of nutrients.

No matter how sick he was, when he arrived at the hospital he always asked the nurses, "Who's here?" Joey had his gang there, too—a far-flung group of his sick friends who looked forward to seeing each other. Joey became a popular draw; other kids called in to see if he was "in house" and tried to coordinate their treatments with his, flooding the ward with sick kids and their healthy spirits.

All of his friends in Mass General's children's ward, Burnham 4, knew Joey's routine. For the first three days he'd lie in bed, sleeping, until the antibiotics kicked in. Then he'd perk up, and be out of bed and running around with his buddies. Dressed in their civvies (civilian clothes), Joey and his friends roamed the wards, wheeling their IV stands in one hand while playing pranks and creating elaborate make-believe exploits that extended beyond the children's ward onto other floors.

The nurses worked hard to make Burnham 4 homey and comfortable for the kids. At the end of the ward was the playroom, a large, safe, happy space for the kids to hang out that was always packed to its forty-six-kid capacity; as Joey got older, he was a more frequent visitor there. Inside, kids sat with volunteers and parents immersed in art projects that would later decorate the hallways, or they'd play pool or cards, watch TV, or simply hang out. Joe had donated video games— *Pong, Pac-Man, Donkey Kong*—that the kids could play even with an IV strapped to their arm. If kids were too sick to walk, the nurses rolled their beds and IV poles into the playroom, up to four beds at a time, so that they still felt part of the community. On Wednesday nights, Nadeau and his colleague Jerry Mulcahy sprang for pizza for all the kids, many of whom were patients of theirs.

Joey was rarely alone. Most days, Kathy spent the entire day at the hospital, and Joe arrived in the evening. His grandparents would drop by, and so would his aunt Mary and cousin Denny, and there was a constant stream of other visitors—his friends on the ward and his school buddies, who arrived after school and on weekends. But Joey wasn't content just hanging out with the kids. He was curious about everyone. The secretary, the nurses, the doctors—Joey wanted to know about their lives. Were they married? What did they do when they weren't taking care of him? Did they like going to baseball games? What were their favorite TV shows?

He loved it when Bubs would visit and read to him—and he loved competitive games. One day as Bubs walked into Joey's room, Kathy's father was finishing a game of checkers with Joey. "Don't you ever *let* him beat you," he told Bubs privately, after. "It won't mean anything to him unless he wins on his own." So Bubs would play checkers with Joey, and he would beat her. Seeing the pleasure the game brought Joey, Bubs started bringing other games, small knickknacks, and little puzzles; she couldn't seem to enter a store without hunting around for something Joey would enjoy, or a gift that Joey could give to his favorite nurses.

Bubs and Joey grew ever closer, and often she would just sit beside him as he drew—another one of his passions. They also watched TV together, or just talked. Joey was funny and bright and mischievous, but his comments also revealed he was thoughtful and curious, and had a defined vision for his future. He told Bubs that he thought he would go to Harvard so he could live at home, and that way his mother and father could still do his therapy. He was very sick when he made this proclamation, and she remembers how big his knowing eyes were when they searched hers, to see whether she would agree. On another occasion the two were sitting in the hospital watching *Highway to Heaven*, a TV show that Joey loved to watch and discuss. After it ended, he turned to her and asked, "Do you believe in heaven?"

BY THE TIME JOEY WAS ABOUT EIGHT YEARS OLD, JOE HAD BECOME increasingly involved with fundraising for the Massachusetts chapter of the Cystic Fibrosis Foundation. He had tremendous respect for the recent changes that Tulcin and Dresing had made to the foundation, and he wanted to help them any way possible.

Back in 1980, Congress and President Jimmy Carter had designated a week in September as National "Cystic Fibrosis Week,"[1] the first one coinciding with the foundation's twenty-fifth anniversary. Although the disease continued to receive more visibility and recognition, six years after Doris Tulcin had taken over as president, it was still a frustrating time. The visibility hadn't yet yielded more funding for research, and there was no tangible progress that Joe could see in terms of new treatments. So when someone in the Massachusetts group suggested a public service announcement that would run on TV to bring some desperately needed publicity to the disease, Joe agreed that it was a good idea.

The ten-second commercial was simple. A handsome nine-year-old kid with big brown eyes, dark hair, and a baseball hat, holding a baseball and glove, looked directly into the camera and said, "My name is Anthony. I've got cystic fibrosis, and I'm going to die. And I don't need your sympathy, I need your money." Then the image faded out.

The ad provoked attention—and controversy. It was the first time that most of the general public had heard of cystic fibrosis and many wanted to learn more. But it also sparked a slew of calls from angry parents. Joe fielded many from parents who had wanted to keep the severity of their child's illness a secret. These parents had not explicitly told their family and friends that children with this disease were likely to die in their teens. Many hadn't told their own sick children, either. Now, because the commercial was sometimes aired during cartoons, slotted there to fulfill FCC public service requirements, their children were asking questions.

That included Joey.

The morning after the PSA first aired, Joey came into his parents' bedroom to wake them up. It was early, shortly after 6:00 AM, and he had been watching Saturday morning cartoons. Anthony appeared during a commercial break. Joey scrambled onto the bed and looked his father in the eye. "I just saw our commercial. Am I going to die?"

Joe was completely taken off guard, but he answered honestly. "Yes, you are going to die. Mommy's going to die. Daddy's going to die. Everybody's going to die," he said as Joey stared unblinkingly at him with his big eyes. "How long you live depends on how well you take care of yourself. That's the same for everybody. And so, you need to do your exercise." They never called it therapy—always "exercise."

Joey considered his father's words, and then, seemingly satisfied, nodded and went back to watching TV. But the message resonated deeply. Joey never missed a treatment.

CHAPTER 14

Salt and Water

1970–1981

Chlorine is a deadly poison gas employed on European
battlefields in World War I. Sodium is a corrosive metal
which burns upon contact with water. Together they
make a placid and unpoisonous material, table salt.

—Carl Sagan

Of all the organs in the human body, the one that initially captivated
Dr. Richard Boucher was the kidney. Over the course of his medical education, he'd grown to admire the humble, bean-shaped, rust-colored organ that quietly kept the body's levels of salt and water in perfect balance. After all, these precious commodities were critical for life: too much or too little of either could prove deadly. When exercise, vomiting, diarrhea, or sweating led to dehydration, the kidney held on to salt and water. When too much water or salt was consumed, the kidney got rid of it.

But despite his fascination with the kidney, it was another organ, one that also maintained a delicate salt/water balance, on which Boucher ended up focusing his career: the lungs.

After graduating from Columbia University Medical School in 1970, and completing a two-year residency in internal medicine, the young Boucher joined the Indian Health Service and moved to Eagle Butte, South Dakota—headquarters of the Cheyenne River Sioux Tribe, a community of just over five hundred residents. Boucher thought working near the reservation was a good opportunity to serve a community that was in real need of care. The tribe suffered from a plethora of lung diseases: asthma, tuberculosis, and complications that arose from TB treatment. And the more time he spent treating lung conditions, the less he thought about the kidneys. With his newfound devotion to the lungs came the realization that he needed greater expertise in lung physiology.

After working in Eagle Butte, and another community in New Mexico for another year, Boucher moved to Montreal where, at the Royal Victoria Hospital and the Meakins-Christie Laboratories, a talented group of Canadians, Brits, and Americans fleeing the Vietnam War draft had made major breakthroughs in understanding lung physiology—in particular the role of small airways in various lung diseases. Montreal scientists were developing sensitive new tests that could detect when the minuscule airways were not functioning—an early sign of airway disease.

When Boucher arrived in Montreal, he wanted to study the small outermost airways, called bronchioles, because recent research had revealed that's where diseases like CF and chronic obstructive pulmonary disease (COPD), a progressive inflammatory disease that made it difficult to breathe, began. Boucher wanted to investigate how the airways stayed lubricated and clean, and to master techniques for monitoring the outermost airways to help his patients stay healthy.

The problems in the outer airways arose as a result of the lungs' structure: an inverted tree. The lungs' primary airway splits into two symmetrical branches—one going to the left lung and the other to the right. The branches continue to divide twenty times, getting smaller and smaller, like the tiniest twigs on a tree, as they approach the outer realms. Keeping the large and small airways well lubricated is a thin layer of slippery mucus: the first line of defense against the dust, debris, and bacteria that enter with every breath. Once the mucus traps these foreign bodies in its sticky clutches, the cilia—tufts of tiny hairs that sit on the surface of the epithelial cells lining the airways—sweep the mucus upstream, from small airways to larger ones, like small tributaries flowing into a great river, until the mucus reaches the throat and is either swallowed or coughed up and spat out. And just like in a great river system, the flow of mucus is slow in the tiniest tributaries, but as those tributaries empty into the larger waterways, the speed and volume grow faster and larger.

For this trash clearance system to function properly, the mucus must be sticky enough to capture bacteria and dirt, but thin enough to allow the cilia to sweep back and forth. In a healthy person, mucus is 98 percent water, 1 percent salt, and 1 percent sticky proteins called mucins that are covered in sugars and work as glue traps for snagging bacteria, viruses, and debris. In a CF patient, mucus is drier—between 92 and 94 percent water—and clogs their lungs and nose, as well as ducts in the pancreas and reproductive system. The small airways are most vulnerable to disease when the flow of mucus in these pipes is slow; when there is an infection, it's almost impossible to cough out the bacteria-laden mucus. This is true even in healthy people, but in people with CF, the problem is worse because their more-viscous mucus glues the cilia together, and to the surface of the cell, blocking their sweeping motion and allowing bacteria to pile up in the airways. This trash pileup prompts an influx of immune cells that blast the area with chemicals to destroy the bacteria; over time, as these battles rage, these chemicals cause collateral damage, also destroying healthy lung tissue near the infection.

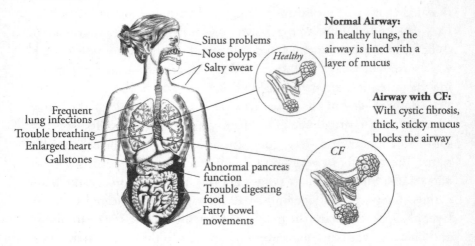

Sinus problems
Nose polyps
Salty sweat
Healthy

Normal Airway:
In healthy lungs, the airway is lined with a layer of mucus

Airway with CF:
With cystic fibrosis, thick, sticky mucus blocks the airway

Frequent lung infections
Trouble breathing
Enlarged heart
Gallstones

CF

Abnormal pancreas function
Trouble digesting food
Fatty bowel movements

Health problems caused by cystic fibrosis.

After mastering the new techniques used to study the small airways, Boucher moved back to the United States in 1977 to the University of North Carolina, Chapel Hill, where he planned to study how allergens cause asthma, a disease that then affected millions of people in the US. His interest was partly driven by his own severe allergy to cats; their dander made him cough, sneeze, and wheeze, but scientists had no idea why. How did the proteins in cat dander (or dust mites or pollen grains) pass through the mucus barrier that protected the airways and infiltrate lung tissue, aggravating the immune system and making it hard to breathe?

Perhaps, thought Boucher, in patients with allergy-induced asthma, the mucus was too dry, like a creek bed in summer, allowing allergens to build up in the airways. He wondered if a little more water would make a difference. Runnier mucus might loosen the irritants, making it easier for the lungs to sweep them out and breathe. But increasing the amount of water in mucus wasn't just a matter of telling the airway cells to release more water. To coax water out of cells and into the mucus, microscopic transporters embedded on their surface had to move electrolytes—including the two components of table salt, sodium and chloride—in and out of the cell. Once salt moves, water follows. It's the reason you can treat a swollen ankle by immersing it in a basin of Epsom salts: the hyper-salty liquid in the basin pulls the water out of the swollen joint, reducing the swelling.

If people with asthma had too little water in the airways, Boucher mused, it could be a sign that the flow of chloride and sodium in and out of these airway cells was disrupted. Fortunately, there was a good way to test that hypothesis. Whenever electrolytes move, they generate electrical activity. Measuring the electrical activity in asthmatic lungs would tell Boucher whether or not sodium and chloride were flowing normally.

In 1979, after two years of research, Boucher's hypothesis proved incorrect: he and his colleagues found no difference in electrical activity in the airways of human asthma patients versus healthy volunteers. The mucus and trash clearance system appeared to be fine.[1] Boucher was disappointed, but later that year, as he was mulling over the asthma experiment, he recalled how pediatricians in Montreal had given his daughter a sweat test during a bout of pneumonia to rule out cystic fibrosis. He knew CF patients had extremely salty sweat—an abnormality that facilitated the diagnostic sweat test. Perhaps cystic fibrosis patients had a problem moving salts in and out of the cells in their lungs. If so, he and his research partner, a young physician and scientist named Michael Knowles, should be able to detect it by measuring the voltage in patients' lungs.

Knowles immediately contacted pediatricians at the local cystic fibrosis center at the North Carolina Memorial Hospital to ask if any parents and siblings of children with CF would be willing to participate in his study. Boucher and Knowles had tested the electrical activity in the lungs of healthy volunteers as part of their study on asthma, but not ones who carried one copy of the CF gene. They had no idea if carriers were different. Better to test them, too, before testing CF patients.

Measuring the electrical activity in the lungs wasn't a trivial procedure. It had to be done in a hospital and involved first sedating a subject and then snaking a long tube about the diameter of a pen into the lungs. Fortunately, there was another location in the body that had similar physiology. The cells in the nasal cavity near the bridge of the nose were carpeted with the same epithelium cells that covered the airways in the lung. And these cells were also covered in mucus. So, at least for this initial round of testing, the nose could serve as a proxy for the lungs.

After Knowles was given the green light, he planned a visit to the center. Into the examination room he wheeled a cart carrying a nasal probe and his strip chart recorder, which scribbled the probe's voltage measurements on a scroll of paper, like a seismograph recording an earthquake. Sitting at a table next to each healthy volunteer, he inserted the first electrode under the surface of the skin near their wrist. It felt like a pinprick. Then he asked the volunteer to look up. Taking the other electrode in his right hand and steadying his arm, he inserted the electrode up the subject's nostril and rested it on the mucousy surface near the bridge of the nose. The strip recorder squiggled an unimpressive line on the paper as the voltage registered between −20 and −30 millivolts (mV): normal.

What they observed in CF carriers was exactly the same magnitude of electrical activity he and Boucher had measured in the asthma patients and healthy volunteers in their previous study. That suggested that the CF carrier parent and sibling volunteers had normal movement of salt. That made sense; these volunteers were healthy and didn't have any issue with salty sweat. Now Knowles and Boucher just needed an adult CF patient to test, too.

While many children with CF visited the clinic, most were too young and squirmy to be administered the test, which required the patient to stay still for some thirty minutes—often longer, depending on how long it took to set up the finicky equipment. And few kids, even those who were used to many medical procedures, would tolerate a researcher poking an electrode up their nose.

Then, one Monday night in 1980, a pediatrician called to say an eighteen-year-old with cystic fibrosis had been admitted to the pediatric ward—a rare occurrence, since few CF patients survived so long. Knowles grabbed his test-gear cart and went to the center. The young man granted permission, and Knowles inserted one probe into his wrist and one up his nose. The electrical activity spiked; the pen zoomed off the narrow strip of paper. The voltage measured −70 mV—more than three times more negative than anything Knowles had seen in his and Boucher's other research subjects.

Intrigued by this bizarre electrical activity, he and Boucher checked the equipment the next day. Then Knowles tracked down eight more CF patients at the Cystic Fibrosis Center of North Carolina Memorial Hospital, where he repeated the measurement. The voltage was abnormal in all of them. The discovery appeared to be real, and Knowles and Boucher were giddy with excitement. They could feel they were on the verge of discovering something, though they weren't quite sure what it was—or what it meant.

What was happening in these cells? Did the high voltage measurement indicate a salt/water imbalance in cystic fibrosis, as Boucher had hypothesized? And what did voltage have to do with the thick mucus in patients' lungs? All the higher voltage was telling them was that the movement of sodium and/or chloride was abnormal in the CF cell, and very different than in healthy people.

They needed more data. They took measurements inside the noses of thirty-two patients with diseases that mimicked one or more CF symptoms. The voltage in all of them was normal—as expected.

Then they measured the voltage in twenty-four more cystic fibrosis patients. Each time the needle shot to the top of the scale and the pen darted off the chart.

In all of the cystic fibrosis patients, the negative voltage was three times that of the healthy volunteers. But these patients were not dying from bad noses. They were dying because of the thick mucus in their lungs, which harbored bacteria that led to pneumonia and other fatal infections. It made sense that these high voltage abnormalities in the nose would also be happening in the lungs, and so could be linked to CF's characteristic thick mucus. If something was blocking the salt from moving into the airways, then water would stay with the salt inside the cell, leaving the airway dry and the mucus thick. But they had to test the voltage in the lungs to be sure the abnormality was consistent.[2]

For their first lung voltage tests, Boucher and Knowles targeted patients without CF who were already undergoing surgery to figure out the baseline voltage for the lungs. After gaining a patient's consent for the test, the pair would station themselves outside the operating room at 7:00 AM, just as anesthesiologists and surgeons were intubating patients and readying them for surgery. They entered the operating room after the patients' anesthesia set in, inserted one electrode just under the patients' skin on the wrist, and stuck another little electrode down their tracheal tubes, measuring the electrical activity at several locations along the airway surface. Just as they had guessed, the voltage was −20 mV, just as it was in the noses of healthy people.

Next: patients with cystic fibrosis. The first they tested was a three-month-old infant who had just been diagnosed. After being sedated, she lay flat on the operating table as Knowles gently inserted the probe through her tiny mouth and three inches down her bronchus, resting the probe on the surface of the airway. Just as it had in the noses of CF patients they'd tested, the pen on the strip recorder shot up, charting a high negative voltage. Knowles and Boucher looked at the chart and then each other, exhilarated. This was important. It offered a clue to the root of this cruel disease. When they took measurements in three more CF patients, they discovered the same thing: the voltage in the patients' lungs, as in their noses, was extremely negative.

These patients' voltage readings were evidence that something was wrong with the salt flow in the airway cells. Scientists knew cystic fibrosis patients had salty sweat; now it seemed like there was a salt problem in their lungs, too. What wasn't obvious was which of the electrolytes wasn't transiting CF patients' cell membranes: sodium or chloride. Figuring that out was important because they needed to know which electrolyte to target if they or anyone else wanted to develop a drug that could fix the problem.

Knowles and Boucher decided to do additional experiments that would measure the movement of sodium and chloride one at a time. First the scientists dabbed a bit of a common blood pressure medication inside the noses of CF patients to stop the flow of sodium into the nasal epithelium cells. They waited a few minutes, then used the voltage probe to monitor the effect of the drug. To their amazement, in what was Boucher and Knowles's second "Eureka!" moment, the sodium-blocker drug rapidly reduced the voltage to normal levels.

So what was happening with the chloride? To figure that out, they rinsed the cell surfaces inside the nose to remove any chloride that was naturally present. In healthy volunteers, this triggered a rush of replacement chloride out of the nasal cells onto the surface, producing a large negative voltage. But when they did the same experiment with CF patients, the voltage remained the same—suggesting that CF patients were missing a pathway or channel for chloride to exit the cell. The highly negative voltage that was characteristic in the noses of CF patients seemed to be due to problems with the movement of sodium *and* chloride.

Now Boucher and Knowles believed they had discovered the cause of the thick mucus: too much sodium moving from the mucus into the cells, and the dramatic drop in the quantity of chloride moving from the cells to the mucus.[3] Both of these problems meant that water flow to the cell surface decreased, dehydrating the mucus until it became dense and sticky—a perfect environment for invading bacteria to colonize and thrive.

The discovery was profound. Boucher and Knowles's team noted that the voltage measurement was a quicker and easier way than Dorothy Andersen's enzyme test to diagnose CF in infants—who, because they didn't sweat, couldn't be diagnosed using the sweat test. But more importantly, they now had a solid hypothesis as to why the mucus in CF patients' lungs was so thick and dry.

But as revelatory as this new information was, the implications for treatment were unclear. Besides, cystic fibrosis didn't just affect the lungs; it damaged multiple organs and affected the sweat glands, too. The work was clearly far from done. Fortunately, there was another scientist, one working on the West Coast, who was committed to figuring out the cause of CF patients' salty sweat.

CHAPTER 15

Salty Boy

1981–1983

Das Kind stirbt bald wieder, dessen Stirne
beim Küssen salzig schmeckt.
The child will soon die whose brow tastes salty when kissed.

—Old European folklore

Like Richard Boucher's team in North Carolina, in 1981 Paul Quinton was also looking for the cause of cystic fibrosis. But his quest was personal. He wasn't only a CF researcher. He was a patient.

Despite the incessant coughing and infections that plagued him throughout his life, Quinton wasn't focused on the lungs. He was interested in sweat, and wanted to discover what made his sweat glands different from other people's. One thing he did know, and had since he was a little kid, was that salt was the key to his disease.

Born in 1944 in Channelview, Texas, a tiny rural town of a few hundred people, with three Baptist churches and one Methodist church on a two-mile spit of road, Quinton arrived while his father was storming the beaches of Normandy. He and his father didn't meet until he was eleven months old.[1] Quinton was always sickly, constantly coughing and sniffing, and visits to his pediatrician brought his mother neither comfort nor cure. "This kid has chronic bronchitis and there is nothing I can do about it," the doctor said gruffly. "He is just going to have to learn to live with it."

In 1946, eighteen-month-old Quinton caught a cold from which his mother claimed he never recovered, and his coughing became incessant. Fortunately, the little infections—the colds, chills, and the occasional fever—didn't impede an otherwise idyllic childhood. Like other kids in East Texas, Quinton enjoyed squirrel and armadillo hunting;

he had a dog and acres of virgin forest around his home to explore. For the children's education and amusement, the family kept at different times a pig and a cow and a goat, horses and chickens, geese and rabbits. When Quinton wasn't wandering or playing with the animals, he spent time tinkering, curious to figure out how everything worked. He wondered how hard a crayfish could pinch, testing its strength with his finger—an experiment that drew blood and screams. Quinton and his siblings were free to play and explore as long as they were home for supper. Being home in time to eat was important, as Quinton was skinny, and his mother, terrified that he would grow up to be the runt of the family, slathered his food with malt, cream, raw eggs, cheese, and butter at every meal to hang fat on his bony frame.

In the '40s and '50s, treatments for Quinton's "chronic bronchitis" were few. Antibiotics weren't widely available until the late 1940s and in the interim his mother's favorite remedy was Vicks VapoRub salve, which left Quinton smelling like a eucalyptus tree. She'd cover the four posts of his bed with a large sheet, creating a tent, and insert a gurgling vaporizer that wafted out the mentholated mist.

Vicks's vapors didn't fix Quinton's ailments, but he learned to live with them and to accept the small differences that set him apart from his brother and sisters. Animals enjoyed licking his salty skin. Salty sweat stiffened his clothes and left a white residue on his shirts and pants that rusted his wire hangers. One summer, while mowing the lawn, he suffered heat exhaustion—like the children in New York City who ended up in di Sant'Agnese's office during the heat wave of 1948. His mother didn't fuss. She gave him liquids and put him to bed, and he recovered. Though otherwise generally healthy, he was plagued with a chronic cough and chest and stomach pains.

When Quinton was about eleven years old, a large van with a mobile X-ray machine began coming to his elementary school to test students for tuberculosis. The disease was feared, and for good reason. In the early 1900s, TB was known as the Captain of Death, killing one in seven people living in the United States and Europe.[2] The disease consumed its victims, destroying their lungs and leaving them emaciated and fatigued. It was spread through coughing and caused debilitating chest pains. Mandatory school screening[3] in the 1950s caught infection early, when it was easier to treat.[4]

When the van arrived, the children clambered on board one by one to get an X-ray. Pictures of TB sufferers' lungs revealed rounded nodules and tubercles, caused by the bacteria, and swaths of tissue damage. Quinton's X-ray sounded alarms: the upper regions of his lungs—the same regions where tuberculosis struck—were covered in spots. However, his TB skin test was negative. Still, the boy was terrified doctors would snatch him from his family and send him to a sanitarium, so every year, on the day the TB machine was scheduled to come to his school, he played sick.

When Quinton was fourteen, his parents moved him along with his two sisters and younger brother to Baytown so that Quinton could attend a better high school. It was an intellectual turning point for the curious teen. His parents had always addressed his endless stream of questions about the universe with religious origin stories. But high school gave him a broader perspective, and in chemistry class his interest in science began to crystallize.

High school marked a shift in his health. Though he had previously been able to dodge the TB machine every year, when it showed up at his new school unannounced, he was forced to have an X-ray. The results were even more startling than the ones from three years earlier, showing large white splotches all over his lungs, and his worried parents took him to a pulmonologist.

The doctor diagnosed him with bronchiectasis—a condition in which the airways become wide and flabby and inflamed. Bacteria had colonized regions of Quinton's lungs, filling them with pus—a foul brew of dead bacteria and dead immune cells. Regions of his airways had puckered and collapsed from immune cells' efforts to fight the bacteria, forming swimming pools where the bacteria stagnated and multiplied, escaping the cleanup crews and amplifying the cycle of devastation. Still, despite the bronchiectasis diagnosis, his incessant coughing fostered suspicions he had TB, and even led the parents of his high school crush to nip the romance in the bud.

During his sophomore year, when Paul was sixteen, his symptoms worsened. Blood vessels in the upper lobe of his right lung burst, causing a burning sensation in his chest and a bloody cough—a sign that the lung was irreparably damaged. To stop the infection from spreading, the only solution known at the time was to remove the infected area—about 15 percent of the lung. After the operation, he remained in the hospital for a week and at home recovering for the next six.

But neither the surgery nor his underlying sickness worried Quinton much until his freshman year at the University of Texas, Austin, when a young woman again caught his attention. Amid fantasies of a relationship and marriage, he began to wonder what it meant to have lost 15 percent of his lung. Would it shorten his life? Was he more prone to illness? He went to the university library to investigate, poring through medical texts to find bronchiectasis. The entry had a footnote: "See in connection with Cystic Fibrosis." He looked up CF—"fatal disease; patients do not survive beyond 6–7 years of age"—and as he began reading the symptoms, the hair rose on the back of his neck. Then a sickening recognition: "Holy shit. This. Is. Me." His symptoms were a perfect match: the lung damage in the upper lobes; problems with digestion; stomach pains from gas. And this was one of the few diseases that caused hypersalty sweat. Yes, he was older than the expected life span for someone with this disease, but that seemed a minor point.

Convinced, he headed home to his pulmonologist and declared he had cystic fibrosis. The doctor eyed him suspiciously. "You are too old to have this disease. People with cystic fibrosis don't live more than ten years at the most. And even if they did, they would have a lot harder time staying alive than you seem to."

Despite doubts, the pulmonologist sent Quinton to Gunyon Harrison, the pediatrician who ran the cystic fibrosis clinic in Houston and had worked in one of the first polio treatment centers in the US.[5] Harrison's nurse came in and did a sweat test. An hour later, she returned and did it again. Another hour later, a third test. Then Harrison came by. "Yeah, looks like you've got it."

"How long am I going to live?" asked Quinton.

"Hell, I don't know," Harrison replied in his thick West Virginia twang. "You're the oldest patient I've ever seen."

Not long after his self-diagnosis and confirmation by Harrison, Quinton traveled as an exchange student to the Pedagogical Institute of the University of Chile in Santiago. After a month, convinced he hadn't long to live, he extended his visit. Together with a few Texas peers he hitchhiked to Pedro de Atacama in northern Chile to Tierra del Fuego—the world's southern tip—on to Uruguay, via Argentina's Pampas, and back to Buenos Aires, where, exhausted and almost broke, he took a train back to Santiago, and eventually made his way home to the US.

After Quinton returned home, Harrison, who had taken a liking to the young man, offered him a summer job and hired him to work in his lab. It wasn't until the summer before graduation that Quinton shared with Harrison that he had been accepted to medical school.

"You are going to medical school?" Harrison yelled, shocked. Quinton nodded. "You're crazy. You don't want to go to medical school." Though only five and a half feet tall, with a slight build, Harrison spoke like a drill sergeant, a consequence of his days at the military academy. If you go to medical school, he said, they'll assign you to the emergency room and you will catch some bug and die. He poked Quinton's chest to make his point. "You want to be a scientist. You go over there and talk to Charles Philpott, at Rice, and see if you can be a scientist."

Harrison had economics as well as the Cold War on his side. Medical school was expensive: $10,000 per year. And in the post-Sputnik era, with the Russians on their way to the moon, Kennedy had cultivated tremendous public support for science and education, with generous stipends for students who attended graduate school. Quinton heeded Harrison's advice, and in 1969 began a PhD program at Rice University.

The fight to unlock the mysteries of cystic fibrosis would need Quinton's drive—and support for research in general. By 1970, there were 115 CF care centers spread across the US, all running LeRoy Matthews's Comprehensive Treatment Plan, and the foundation had made progress securing some funding from Congress to support them, but there was barely any funding for research. And little was known about CF, beyond its being a genetic disease. What caused the lungs to fill up with thick, sticky mucus? What destroyed the pancreas? What caused the sticky, tar-like feces that sometimes tore the intestines of babies and killed them? And why did CF patients have salty sweat?

The link between the malfunctioning lungs and pancreas seemed to be blockages caused by mucus. But what did salty sweat have to do with the sticky mucus in the airways? The question dogged Quinton. There was no shortage of hypotheses in the literature, but so far none of the science had proven replicable.[6]

Sweat glands in human skin, it was known, had special proteins that rescued salt and transported it back into the bloodstream, and Quinton was intrigued by the idea that there might be some rogue factor in his sweat that messed up his salt transport. He decided to figure out what it was, certain that, if he solved the riddle, it would reveal how and why CF destroyed organs, killing its sufferers.

First, Quinton collected a couple of vials of both his own sweat and normal sweat, which he gathered from the son of his department's chairman. Then he tested both sweat samples on frog skin. Because frogs have to sit in pools of water all day to keep from drying out, their skin is filled with transport proteins that capture electrolytes like sodium and chloride that would otherwise be washed away. The high density of these transport proteins made frog skin ideal for studying the transport of salts across membranes, as Quinton wanted to do. Using a device called an Ussing chamber to measure the movement of salts in slices of lung tissue, Quinton tested whether either of the two sweat samples changed the voltage of the frog skin.

When he applied the normal sweat, sodium and chloride ions were carried by transporter proteins into the frog skin cells; the voltage revealed that the salt was moving. But *his* sweat slowed the movement of salt. There did indeed seem to be a mystery factor in his salty sweat. Elated, he shared his discovery with Gunyon Harrison, who called the vice president of medical affairs at the foundation, proclaiming Quinton had made a great discovery. But when Quinton tried to replicate his results, he couldn't. The first experiment was a fluke; the result was wrong. It was a humbling experience.

While the experiment had failed to reveal the root of cystic fibrosis, it succeeded in deepening his interest in salt transport and cystic fibrosis. After completing his PhD, he moved to the University of California, Los Angeles, to begin his postgraduate project: figuring out the basic defect that caused cystic fibrosis.

As Quinton collaborated with scientists at UCLA on a range of various projects unrelated to cystic fibrosis, he also discreetly started work on his passion project by looking at tissue from cystic fibrosis patients that were affected by the disease. By comparing this defective tissue with healthy tissue, he should be able to figure out what was different. But he needed to select a tissue that was easy to collect and study. The lung wasn't an option—harvesting the tissue would require surgery, and the sample was likely to be so ragged and infected and torn up that it would be impossible to compare with normal lungs. The pancreas and the gut presented similar challenges. So that left the sweat glands. Skin harboring the sweat glands was easily accessible, on the outside of the body. And since salty sweat was the key tool for diagnosing cystic fibrosis, it also seemed likely to hold the answers he was seeking.

The trouble with sweat glands is that they're minute, smaller than the head of a pin, and mining them from chunks of skin requires time and patience. The base of a sweat gland looks like a tangle of translucent, hollow spaghetti, from which a single strand only a third to half the diameter of a single human hair[7] rises, carrying droplets of sweat to the skin's surface. Separating the sweat gland from the skin is like extracting a transparent strand of hair from a bowl of clear Jell-O. And to complete his experiment, Quinton needed lots of sweat glands from both healthy people and those with CF.

For samples from healthy individuals, Quinton started loitering outside the operating room at UCLA, scavenging slivers of skin cast off during surgery. But one day a surgeon suggested an easier way. When dermatologists did hair transplants, he told Quinton, they took plugs of skin from below the hairline—a region that never gets bald—and inserted them into the bald pate of the head. The bald skin cores were then tossed in the trash. The surgeon was sure that the dermatologist would be willing to save them instead. He was right, and the scalp cores, a rich source of sweat glands, provided all the healthy tissue Quinton needed.

Every week, beginning in about 1976, he would take a tube into the dermatology offices, collect the scalp plugs, cover them in mineral oil to keep them moist, and refrigerate them in the lab. Once he was ready to remove the sweat glands, he immersed the skin plugs in a bright-red dye. This transformed the invisible sweat ducts into pink spaghetti, easy to see through the eyepiece of a dissecting microscope. With pipettes, forceps, and tweezers he had custom-crafted for the task, Quinton then cut the sweat gland away from the skin—tedious, painstaking work.

To get CF sweat glands, Quinton followed in the footsteps of many other great scientists: he mined his own body for precious tissue to use for his experiments. Using plugs of his own skin, he began excavating sweat glands just as he had from the normal samples and putting them through the same process.

By the time he had developed the tools for dissecting these structures and mastered the art of collecting sweat glands from tissue, he had been at UCLA for eight years—mostly working on projects unrelated to CF as he climbed up the academic ladder from postdoc to research associate and then assistant professor—and had scars all over his arms and legs. He had extracted hundreds of sweat glands. But collecting glands was only half the challenge. He still needed to identify exactly how his sweat glands were different.

The work he had done at UCLA, including using X-rays to analyze microscopic volumes of liquids, had earned him an offer of a tenured position at UC Riverside, which had recently launched a new medical school. He accepted, and in 1981 Quinton moved to his laboratory.

Now, finally, he could focus exclusively on his disease, and on measuring how salt moved back and forth through the slippery microscopic sweat gland. But when he tried to apply the X-ray techniques he knew to measuring the amount of water within a small section of the sweat gland, it proved tricky, impractical, and frustrating. Also, removing sweat from a gland with a pipette narrower than a human hair—a necessary step for tracking the movement of salt in the duct—was nearly impossible. Quinton was faced with the possibility that he had spent years developing a technique that wouldn't actually help him solve the mystery of his salty sweat.

Then a colleague who heard his woes at a conference in Brussels suggested that he measure salt transport using voltage, as he had years earlier in his experiments with sweat and frog skin. The suggestion made sense.

In early 1982, as Quinton was just gearing up to measure voltage inside the sweat duct, a colleague at UC Riverside mentioned a paper published the previous year in the *New England Journal of Medicine* by Boucher, Knowles, and their collaborator, John Gatzy, who had discovered something weird about sodium in the airways of CF patients.* Quinton immediately secured a copy from the library. As he read it, he was overcome by a sickening feeling that he'd been scooped. It seemed that they'd already discovered why people with CF had thick mucus and that sodium was to blame. (Boucher's finding on chloride had been omitted from this publication for the sake of brevity. So even though his team had discovered the chloride defect as well, Quinton wasn't aware of these findings until they were published until 1983.)

Boucher's North Carolina team had found that sodium from the mucus in the lungs was flowing into the epithelial airway cells and the water was following, dehydrating the mucus. But the closer Quinton read the paper, the more he wondered about its conclusion.

Salt is the combination of a positively charged sodium atom and a negatively charged chlorine atom that stick close together like opposite

---

* Boucher and Knowles packaged up both observations, raised sodium transport into the airway cells and the almost nonexistent transport of chloride from airway cells into the mucus, and sent the paper to the *New England Journal of Medicine* (*NEJM*). The *NEJM* liked the paper but said that it was too long. The editors told Boucher and Knowles to cut it in half. Because the sodium abnormality was potentially treatable by the antihypertensive drug, that part of the story was published in the *NEJM* in 1981. The chloride permeability observations were subsequently published in *Science* and the *Journal of Clinical Investigation* two years later.

poles of a magnet. What if the problem wasn't with sodium moving into the airway cell, but with its partner, chloride? What if chloride was stuck and couldn't get out?

If something was stopping chloride from moving out of the airway cells, then its negative charge would attract most of the positively charged sodium, preventing it from leaving as well. The result—water flowing to stay with the salt, inside the cell, leaving thick mucus outside—would be the same whether the problem was sodium moving into the cell or chloride that couldn't move out. But the difference mattered. If a drug developer wanted to treat the disease, it was vital that they targeted the correct part of the salt molecule: sodium or chloride. So Quinton decided to continue his sweat gland study.

The other matter on Quinton's mind was that, where salt was concerned, the airway cells that Boucher's team studied did the opposite of his sweat glands. In airways, the job of the epithelial cells was to secrete enough sodium and chloride for water to follow, and keep mucus wet and flowing. A sweat gland's job was the reverse: to absorb the sodium and chloride, allowing the body to reclaim those precious electrolytes. (What remained of the sweat, now mostly water, was moved up the sweat duct to the surface of the skin where it evaporated, cooling the skin and the body.)

First, Quinton tested a normal sweat gland. Like a physician performing surgery on an insect, he peered through the microscope at the gland's pale pink, translucent duct. Holding the duct with a tweezers in his left hand, he sliced off the gland with a fine blade, leaving a 2 mm stretch of the hollow duct. He then flushed the slice of duct with salty water using a glass pipette, and placed an electrode inside for measuring the voltage. It was like cutting off a short segment of a straw, pushing salty water through it, and then sticking in a metal wire—although this "straw" was narrower than a strand of hair. Once the electrode was properly inserted, he placed the salt water–filled sweat gland duct in a fingernail-sized dish of salty water that held the second electrode and immediately looked up from the microscope at the voltmeter. The needle wavered around −7 mV; a sign that the sodium and chloride ions were flowing freely.

The next day, he arrived at the lab early and repeated the experiment. But this time he used a sweat duct extracted from his own malfunctioning body. When he placed the new gland in the dish of salty water and inserted the two electrodes, the voltage spiked, the needle swinging wildly to –5 mV, then –7, then –10, –20, –60, and –70 before finally settling around –76 mV. It was a dramatic shift, the voltage more than ten times more negative than what he'd seen with the normal sweat gland.

Clearly, Quinton's sweat gland was not rescuing salt from the sweat and bringing it back into the body the way it was supposed to, leaving his sweat super salty. The realization actually triggered Quinton to holler "Eureka!" as he ran down the hallway of his building to share the news with colleagues in other labs.

But, as exciting as it was to witness the difference between his sweat glands and those of other people, what wasn't clear from this first experiment was *why* his sweat gland wasn't functioning properly. There were two possibilities. Because positively charged sodium and negatively charged chloride were partners, the laws of physics dictated that the two couldn't be separated from each other. If sodium moved into the cell, then chloride would dutifully follow, and vice versa. One possibility was

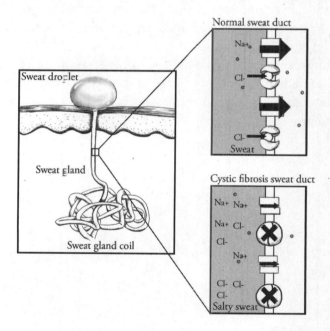

In the normal sweat gland, transporter proteins in the membranes of the cells rescue sodium and chloride ions from sweat and bring them back into the body. In people with CF, those transporter proteins do not do their job and the sweat remains overly salty.

that something was blocking the movement of sodium from the duct into the cells lining the sweat duct. The other possibility was that something was blocking the movement of chloride.

To distinguish between the two explanations, Quinton ran some more experiments. What he discovered was that sodium could move back and forth between his sweat gland duct and the cells—but chloride could not. And because the chloride couldn't move from the sweat gland duct back into the cell, sodium was held captive in the duct, too, raising the concentration of salt in his sweat.

Richard Boucher's team at UNC had revealed disruption in the movement of salt in the lungs and nose; now Quinton had done the same in the skin. While sodium transport was a problem, the chloride transport defect seemed more widespread in the body.[8] Quinton wrote in a paper published in the journal *Nature* in 1983 that this chloride transport defect "may be closely associated with the fundamental disturbance in this disease."[9]

Chloride, everyone now agreed, was the electrolyte causing problems in the nose, lung, sweat glands, and possibly the other organs as well.

Finding the cause of cystic fibrosis had been an obsession of Quinton's since graduate school. He'd had a gut feeling even then that it had to do with salt. And somehow, he'd survived four decades—long enough to prove it.

CHAPTER 16

Birth of an Advocate

1984

Do not wait for leaders; do it alone, person
to person. Be faithful in small things because
it is in them that your strength lies.

—Mother Teresa

By the time Joey was ten years old, Joe and Kathy let him advocate for himself and other cystic fibrosis patients, working with local reporters to tell his own story of the disease. He was a genuine, eloquent ambassador for CF, and happy to talk to television news reporters looking to cover the disease or a minor advance in its treatment. One of the earliest news segments showed Kathy performing physical therapy on Joey, deftly clapping his back and flipping him from side to side to reach all parts of his lungs as he lay head down on an inclined board. It was obvious in the news clip that, while the therapy was tough on both of them, there was a tremendous intimacy and warmth during these sessions; the clip ended with Kathy hugging Joey tightly, Joey coughing up some gunk, and the two of them breaking out in laughter. Another report showed the little boy looking mischievously at his father as they engaged in a fierce game of table hockey—which Joey always won. Then the story transitioned to Joey looking squarely into the camera as he told the reporter his favorite subjects—math and recess—and how his life differed from those of other kids his age. He explained how he needed to cough up the stuff in his lungs, and that was why he had chest therapy three times each day. Sometimes he would get tired riding his bike and would need his mother to pick him up. But there was no sign of self-pity. Joey was a happy, optimistic kid. "I hope they find a cure and as long as they have medicine and therapy, as long as they keep me well, that's all I really care about."

While Joey definitely had good days and weeks, reminders of his illness were never far away. Even as he was lighting up the TV cameras, he was incredibly sick and had developed congestive heart failure. His lungs were so jammed with hard mucus and scar tissue that his heart wasn't properly able to pump used blood into the lungs to get reloaded with oxygen. As his heart tried harder, pressure rose in its right side, and blood backed up into the liver, making it swell. The pain in his abdomen sent him back to Mass General under Allen Lapey's care, but beyond giving him oxygen and medications to remove the fluid, and continued physical therapy, there was little to be done. Patients with right-side heart failure typically improved with this treatment, but Lapey and Joey's parents knew this was only a temporary fix.

Although his hospital stays became more common after this point, he never complained about staying at Mass General or when he was stuck at home unable to play with his friends. He enjoyed spending time with the nurses in the children's ward, one of whom called him "Slick," a reference to how he slicked back his hair, just like the actor Ricky Schroeder did in the '80s sitcom *Silver Spoons*. Joey embraced the name and cultivated a distinctive style—little sports jackets and button-down shirts. But he was so skinny that Kathy would have to alter all his clothes, pinning the jackets and adding elastic to the waistbands of his pants.

Both at the hospital and at school, when he was able to attend, Joey always rose to the high standards of behavior and academic achievement his parents set. This, together with his resilience, sense of humor, and penchant for mischief, was what drew people to him. One day when Bubs was in the ward to visit him, Joey picked up the microphone in the nurses' station and asked over the intercom: "Does anyone know what an over-the-shoulder boulder holder is?" Hysterical cackles erupted from the little boys throughout the ward, and Bubs stifled a giggle. Joey was, of course, talking about bras. It was typical ten-year-old boy humor, and Joey had a deep reservoir. On another occasion, Joey ordered pizza for all the kids, and then asked Bubs if she had any money. When she told him she did, he asked if she would mind paying for the pizzas.

There were, of course, many times when he or his hospital buddies were too sick to play. When Joey's health plummeted and he lacked the energy to move, these friends would come into his room, sit on his bed, and just hold his hand—their presence comforted him, no words necessary. And Joey would do the same for them. It was the space between their words that spoke volumes about how these children cared for each other, and the deep understanding that connected them.

By the fifth grade, Joey was frequently out of school for weeks at a time, falling more and more behind. Each new infection destroyed more lung tissue, stealing his ability to breathe. His arms were a patchwork of yellow, green, and black bruises where IVs had punctured his skin. His oversized jackets covered his arms and baggy clothes disguised his weight loss after hospital visits, but as his school friends grew bigger and stronger, he grew weaker and frailer. Soon he had to be carried up flights of stairs at school.

Kathy worked hard to make sure what time Joey spent at school was as normal as possible. On one occasion Joey's class went on an overnight trip to Grotonwood Camp, thirty miles northwest from their home in Belmont, Massachusetts. Kathy and Bubs's older sister, Connie, rented a motel room nearby so that Kathy could sneak over to the campground early in the morning and complete Joey's therapy before the other children woke.

Although Joey recovered from his infections, the time between them was shrinking. He tired more easily, couldn't ride his bike as far or as long, and was more inclined to sit inside and draw at the kitchen table with Kathy at his side. Joe and Kathy could see that there might be a time when Lapey's therapies and all their vigilance and care wouldn't be enough.

ONE BRIGHT SPOT IN THE O'DONNELLS' LIVES WAS JOE'S BUSINESS, which was booming just in time to aid with Joey's medical expenses. By 1980 his simple mom-and-pop concessions operation was bringing in a couple of million dollars each year selling candy to all the big theaters—Red Stone, General Cinema, Dick Smith. Up until that point he had focused on just retailing the candy, snacks, and hotdogs, but the more he thought about it, and the more local businesspeople he met, the more new and varied opportunities arose.

One thing Joe brought to the table that was rare among other Boston concessioners was an MBA, which earned him a certain credibility among partners. It didn't hurt that it was from Harvard. Joe began distributing the concessions rather than just selling them, delivering them to theaters up and down the Northeast. And he began buying stakes in other ventures. He partnered with or purchased ski resorts; he became the majority owner of the Suffolk Downs racetrack (because, he said, it was a good place to sell hotdogs); he owned or co-owned restaurants, movie theaters, water parks, amusement parks, and lots and lots of real estate. As these endeavors created more net worth, he kept investing it in new enterprises or expanding his share in existing ones. But as much as he worked, Joe always made time to raise money for CF.

A month or so after Joey had been diagnosed in 1974, Joe had received a cold call from a parent named Dick Barnett who had heard that Joe's son was born with cystic fibrosis. Joe had never met Barnett, the executive director of the Massachusetts chapter of what was then still called the National Cystic Fibrosis Research Foundation, but Barnett had learned about Joe, and encouraged him to get involved with the organization. Joe began helping out with fundraising immediately.

At Barnett's urging, Joe teamed up with other local parents to run several fundraisers each year, including an annual walkathon, film premieres, and other events. Eddie Andelman, a sports radio talk show host with whom Joe would become close friends, asked him to help run an unpretentious event called Chow Chow Bambino, in which people came to eat and celebrate. This event and the others the pair co-hosted over the years were the opposite of the stodgy "rubber chicken dinners," with their mediocre food and boring speeches, for which nonprofits were notorious. Joe and Eddie wanted people to have fun so that the participants would return the following year and bring lots of their friends, growing the circle of supporters.

In 1976, Joe joined the board of the Massachusetts chapter of the newly renamed Cystic Fibrosis Foundation. When he first joined, the chapter had a high turnover of executive directors—a new one every year—and no clear mission. So frequent was the turnover that no one ever got the chance to get the hang of the job, and the fundraising initiatives were left up to motivated parents like Joe. But when Tulcin and Dresing took over the foundation, moved it to Washington, and centralized the organization—specifically tasking the individual chapters with fundraising—Joe was confident that things were looking up, and briefly joined the board of the national foundation to support their efforts, including, in 1980, hiring director of science and medicine Bob Beall.

From the early '80s onward, Joe began coordinating more events for the Massachusetts chapter. He would conceive an idea, use his connections to secure participants—restaurateurs and entertainers—and then turn the logistics over to the chapter's staff. And the more events he ran, the more he realized how rare it was to be able to run events that consistently brought in as much money as his did—at least $100,000 per event. He owned properties where he could host the events, and his burgeoning business relationships were spilling over into friendships and an expansive network of connections. With plentiful resources and friends eager to participate, he wondered what other events he could launch to ratchet up the amount of money he was bringing in.

As Joe became more involved with fundraising and his reputation as a businessman grew, word of his work reached Doris Tulcin, who made an appointment to visit the young man. Joe didn't know much about Tulcin except that she was a very successful fundraiser, she had a sick daughter, and her father had helped launch the foundation in the 1950s. Most importantly, he knew that Tulcin had, as president, been responsible for whipping the meandering foundation back into shape, uniting all the chapters, and hiring a VP to lead the research.

Joe was instantly impressed by Tulcin. She was a gladiator: a great CF advocate, the foundation's leading fundraiser, and a tough woman—all rarities among foundation leaders. The two met in a side room of one of Joe's restaurants for dinner. But Doris wasn't there to socialize. She was running a fundraising campaign for the foundation and was there to ask Joe to make a significant commitment. He was already giving time and effort as a national board member. But she needed cash to fund a series of dedicated CF research centers that Beall was going to launch.

In most business dealings, Joe would have waited to hear the ask: How much did Tulcin want from him? But instead he jumped in and promised her $100,000 of his own money. It was an enormous sum for the O'Donnells at the time, but for Joe it was easy to say yes; his fundraising for CF was the most important work he did. And now that Beall was in charge of research, Joe was confident that giving money directly to the national foundation would do more good, faster, than funneling it through the Massachusetts chapter.

In 1982, Paul Del Rossi—Joey's "Uncle Del," Joe's childhood friend and, at the time, president of General Cinema—was living with Joe and Kathy in a spare room after a failed engagement. Neither man had a lot of time to relax, but once in a while, after Joey had gone to sleep, the two would head over to a popular East Boston neighborhood pizza joint called Morelli's. The conversations would swing from management philosophy to Del's ex-fiancée to Joey and ways to raise more money.

One evening, while the two men were sitting at Morelli's, Joe mentioned he wanted to launch a new event—a cocktail hour followed by a limo ride to a movie premiere. Del wanted to help and suggested they hold the reception in his office next to the movie theater. Both Joe and Del were in their thirties and had never organized a fundraiser from scratch before. Joe had conceived events, and pulled together the players, but he had never handled the logistics for any of them. Neither of them had done anything that the Massachusetts chapter hadn't organized. But with Del running movie theaters and Joe the reigning king of concessions, the pair were well positioned to use film premieres as a way to raise money for CF.

They started small. Del called 20th Century Fox Studios, which generously offered *The Verdict*, a courtroom drama set in Boston starring Paul Newman. They sold tickets to the premiere to thirty friends and showed the film in a little screening room next to Del's office at General Cinema headquarters that had just enough seats. Debbie Soprano, Del's secretary, went to the local supermarket and bought some wine, soda, cheese, and crackers. Joe bought the popcorn. The post-premiere reception was held in Del's office.

For the past eight years, Joe had been a creative force orchestrating events from behind the scenes. That's the way he preferred it. But now he had to take a stand in front of the crowd. And he was even more effective there. He was a natural fundraiser: charismatic, warm, never pushy. He made genuine connections with people, and his motives were unquestionable—he watched his son struggle every day. He was careful and honest with his pitches to potential donors and never oversold the impact their money would make.

"We don't know the cause of this disease," he told the audience of thirty in the first fundraiser. "Without knowing that basic fact, we have no hope of ever curing it. We have no chance, zero chance, of curing this thing unless we figure out the cause. Your money might help us do this. Or it might go down the toilet. But we don't have any other way. We have to try."

The amount raised in that first movie premiere was modest, but Joe's appetite had been whetted and he was already planning the following year's event in his head. He knew from talking with Tulcin that the money he was raising was more needed than ever. The Cystic Fibrosis Foundation was on the verge of big changes.

CHAPTER 17

Out of Many, One
1978–1984

Usually, when you're taking over a team, you're restructuring, you're tearing it down, you're building it up again.

—Anne Donovan, legendary basketball player and coach

W hen Bob Beall started at the newly relocated Cystic Fibrosis Foundation on January 1, 1980, he envisioned launching a network of research centers focused exclusively on cystic fibrosis, just as he had done with the NIH for diabetes. Each one would be filled with a multidisciplinary team: chemists, biologists, physiologists, geneticists, electrophysiologists, molecular biologists, scientists with expertise in mucus, others with deep knowledge of the lungs and pancreas and gut and every other body part that CF destroyed or disrupted. At the time, the foundation didn't have the resources for such an endeavor, but Doris Tulcin and Bob Dresing told Beall not to worry about the funding. They wanted him focused on finding the best institution to host their first center and on recruiting scientists to staff it. It was Tulcin and Dresing's job, in the next seven years, to raise the $15 million needed to support the research centers that Beall came on board to lead and direct.

Doris Tulcin stepped down as president in 1982, passing the reins to a reluctant Bob Dresing. Tulcin's husband's bouts of depression had become more severe and were lasting longer, and she felt the foundation was now in the right hands—Dresing's and Beall's—to make lasting changes. But she had no intention of leaving the foundation. She just wanted to go back to doing what she did best: raising money.

Dresing had been working as Tulcin's vice president as a volunteer, not drawing a salary even though he was spending up to three weekdays and even weekends in Washington. It was too much time away from home and his three children, and his business partner was tired of running the company alone. Dresing was stretched thin and stressed. But Tulcin convinced him to accept the position anyway. She was a great salesman and wouldn't take no for an answer. Plus, Dresing wanted her vision to succeed and knew he could transform the foundation into a powerful organization capable of changing and saving lives. He had one condition: full autonomy. He told Tulcin, "I want to be CEO. I won't come to all the board meetings, and I'm not going to satisfy the whims of the volunteers. I either run it or I don't. If I do a good job, fine. If I don't, fire me. I'll report to the executive committee about what I'm doing. And I'll keep you informed. But don't get in my way!"

Dresing kept his businesses well ordered and efficient. When he looked at the foundation, he saw confusion and waste. The seventy chapters were like a dysfunctional armada. Though they all had the same mission, each ship sailed independent of the others, crewed by a jumble of volunteers. Each chapter had its own captain—the executive director—and a staff who leased their own buildings, hired their own staff, and ran their own fundraisers. Worst of all, the directors running the chapters were not contributing the required 55 percent of their funds to the national foundation. The foundation needed an admiral.

Dresing was a polarizing presence, arrogant and fiery. If you didn't agree with him, you were against him. He hated bureaucracy and meetings and listening to the demands of volunteers who had no idea how to raise money, run a foundation, or do anything significant that would change the foundation's course. The vote to elect Dresing CEO and president was contentious. Friends, including his own son's doctor, voted against him. Others told Tulcin that making Dresing president was a terrible mistake. The vote was close, but in the end he was confirmed.

Dresing began his presidency by uniting and centralizing the foundation. He traveled the country, visiting many chapters—during which he stripped each of them of their charters and gave them new guidelines drafted by the revamped national foundation that they had to accept to be rechartered. The meetings were combative and impassioned. Angry parents snatched the microphone out of Dresing's hand, walked out, flipped him off. "You are ruining, destroying, this foundation!" one father screamed at him.

Dresing didn't care, and he didn't back down. There wasn't a cure, and if the foundation continued on its current myopic course, focused only on the immediate health of sick children, there never would be.

"You have your opinion. I have my opinion. And my opinion is we're losing," Dresing told chapter after chapter in his stump speech. The only way they were going to beat this disease was if they consolidated the organization and got the money where it needed to go. Exactly where the money was allocated would be something that the scientific community decided. "Right now, we're in the black hole. And I can't live with that. My conscience won't allow me to."

Dresing didn't negotiate. Or compromise. You were either in or out. "If you don't like it, leave," he told parents. Go and raise money for cystic fibrosis your way. Backyard cake sales and galas raise publicity, not money. The foundation wasn't doing that anymore. "My kid has this disease. The clock is running. I put my head on the pillow every night and all I see is kids dying. And you're telling me that you're going to run your own show. Go run your own damn show. We're running this one."

Parents knew there had been no progress. Their children were dying. As much as they hated Dresing—his manner, his style, his ideas—they knew he was right.

Dresing's only goal was to raise money to fund a cure—and he didn't want to waste a penny that the foundation raised. So he centralized everything under the umbrella of the national foundation. He eliminated redundancies such as publicity and bookkeeping departments; national headquarters would run awareness campaigns and oversee finances from then on. He insisted that the rechartered chapters have new bank accounts that would be swept into the national account every week. There would be no opportunity to scuttle funds for local issues. National negotiated the lease for each chapter and would hire and train executive directors for each one.

Executing this plan of Dresing's was a young man from Louisville, Kentucky, named Richard Mattingly, who had previously run the "Bluegrass" chapter that included Kentucky and West Virginia. Mattingly had come to work for the foundation immediately after graduating from college at twenty-one. Through a college program during his senior year in 1977, Mattingly had worked in the Kentucky governor's office and spent time in Washington, DC, learning, among other things, how politicians raised money, made friends and connections, and bartered favors. But Mattingly didn't want a career in politics, so in 1978 he applied for a position at the Kentucky/West Virginia chapter of the Cystic Fibrosis Foundation through a contact he made in the governor's office. There he was trained to raise money.

Mattingly made friends easily and had a penchant for storytelling, which helped him connect to people—an essential trait for a fundraiser. And he was masterful at organizing small and large fundraisers, both in the cities and in rural areas. In 1978, he successfully pulled off the largest-ever fundraiser for cystic fibrosis in the state of Kentucky, and in 1979, the local board made Mattingly, just twenty-three, the executive director. The promotion caught the attention of Dresing and Tulcin, who gave him an award in 1980 for being one of the most successful executive directors in the country, and began grooming him for a position in the national office they would soon open in Washington, DC. In 1983, Mattingly moved to Washington with his wife, Caroline, to take it.

To enact Dresing's consolidation plan, Mattingly spent more than two hundred nights on the road in 1985 and 1986. He traveled from chapter to chapter, taking away their checkbooks, firing redundant personnel, and adding those who remained to the national payroll. It was a grueling slog. Confrontational. Emotional. Parents were still furious after the foundation's move from Atlanta to DC. Now they were losing both their independence and control of the money they raised in their own community. Who the hell gives you the right to come and tell us how to run our chapter, angry parents yelled at Mattingly during his visits. When tensions became unmanageable, Dresing provided reinforcement. It was an ironic twist; he was now in the same position that Doris Tulcin had been in the first time they met in Cleveland. But mostly, Mattingly was on his own, trying to rally both parents and staff with assurances that they were building a stronger program, providing better care, and funding cutting-edge science.

When Dresing and Mattingly had completed the restructuring, headquarters was in charge of billing, payroll, leases, and expenses. With all the data and finances centralized, management became straightforward. Now, finally, Dresing could look at the revenue coming to national and think about how to fund Beall's research centers.

WHEN BEALL ORIGINALLY JOINED THE FOUNDATION AS THEIR DIRECtor of science and medicine, charged with creating new CF research centers, he focused first on the current care centers—the medical outposts where children received specialized care. He decided that centers would no longer be funded equally, because—as the data from Warren Warwick's patient registry indicated—performance varied widely. So did individual centers' focus. At some, like LeRoy Matthews's center in Cleveland, patient care was exquisite, with life expectancy at the far right of the bell curve—but the research was nonexistent. At others, the treatment was mediocre, but research into the disease was innovative. Beall decided each center would be graded—on research, training, and patient care—and funded accordingly.

Unsurprisingly, parents lashed out, terrified that funding cuts to their local center would jeopardize their children's health. But Beall felt it was the only way to force change at the centers and to motivate the laggards to adopt Matthews's Comprehensive Treatment Plan. This, he expected, would lead to better care for all the patients. The doctors at the top-ranking centers were on a national committee to guide others to reach the same standards, but without financial motivation for changing care practices, this system wasn't working as well as it should. Beall felt deeply that parents should know that life expectancy at each center varied wildly. They should know that doctors at another center might give their child a better, longer life. And they should demand those same standards; their children deserved that.

By 1984, the public's perception of cystic fibrosis was changing. The disease and the foundation were both more visible; annual presidential proclamations continued to declare the third week in September National Cystic Fibrosis week. And the NIH was beginning to provide a little funding for researchers working on CF early in their career.

But Beall wanted more: entire research centers, within academic institutions, each supporting between thirty and fifty researchers who would collectively examine every facet of cystic fibrosis. Together they would amass new science, fueling the discovery of new medicines. It was a research agenda equivalent in scope to the one that Harry Weaver had orchestrated for polio. Such a 360-degree assault on the disease, Beall believed, was the only way to make progress quickly.

Beall needed big money to reel in exceptional, seasoned researchers. He told Dresing and Tulcin he wanted to offer grants between $300,000 and $500,000 to big universities, head-turning sums that would start the rumor mill grinding. But when the three of them approached the foundation's board members, they fiercely guarded the purse. The funds were earmarked for the care centers and the existing research teams based there. Setting aside that much money for a new research initiative could imperil their children's treatments.

Rather than try to raise the extra money through events and dinners, Dresing and Tulcin decided to launch a capital campaign: an intense period of fundraising designed to bring in large donations for a specific goal—in this case, Beall's Research Development Program. No other voluntary health foundation had ever done this. Tulcin would run it. She began by contacting the executive directors of all the chapters and gathering intel on locals who had made large contributions to the foundation. Mary Weiss, the mother of three sons with cystic fibrosis and a passionate advocate and volunteer in Florida, connected Tulcin with a donor who had been generous with the Palm Beach chapter. There, Tulcin secured the first $1 million gift for the campaign.[1] This "easy" first million confirmed Tulcin's hunch that they could secure the $15 million they needed to drive Beall's program.[2]

Running on a parallel track, Beall began approaching institutions with a strong history of research and quizzing them about what sort of program they could build for cystic fibrosis if he gave them a four-hundred-thousand-dollar grant. The size of the grant turned heads, just as Beall had hoped. The department chairs began pitching their facilities and the star-quality faculty that could staff such a program.

After Tulcin returned from Florida with her first million, her and Dresing's next stop was the governor of Alabama, Forrest Hood "Fob" James, whose son had died from cystic fibrosis and would need no schooling on the horrors of the disease. Earlier, in 1981, Beall and Tulcin had reached out to the University of Alabama, Birmingham's president to talk about launching a potential center there, and which accomplished scientist might lead it.

Tulcin and Dresing met in Montgomery at James's statehouse office, where the governor welcomed them. Over the course of an hour the duo shared their vision of the world-class research program they wanted to launch at the university. James listened, then asked how much they needed. Dodging the question, Tulcin cleverly asked how much money he could give them.

Two million, James said without skipping a beat.

Neither Tulcin nor Dresing were quiet people. But James's offer struck them dumb.

Giddy with excitement, they rose from their chairs—quickly, before James changed his mind—and walked straight into the governor's closet. Sheepishly, they backed out and left through the other door. Outside on the steps of the Capitol, they jumped and hugged in excitement. Their first center was going to become a reality at the University of Alabama. Now it was up to Beall to staff it with the talent he envisioned.

Within two years, Beall had also awarded contracts to the University of North Carolina at Chapel Hill, where Boucher's team was based, and the University of California, San Francisco, with the money Tulcin's campaign raised.

Over the next five years, Tulcin, with Dresing and without, crisscrossed the country, meeting one-on-one with high-value donors whom chapter directors identified for her. As she moved around, engaging people in the cause and soliciting large donations, she formed a new network of moneymakers to support the research centers. And as more money came in, the power dynamics began to shift. No longer did Beall need to claw his way into institutions. Instead, if an institution wanted foundation money, Beall could dictate exactly what he needed. The more money Tulcin raised, the more centers Beall built—nine by 1994.[3] With nearly one hundred driven researchers now filling the centers and focusing exclusively on cystic fibrosis, the stage was set for developing better treatments for the lungs, the gut, the pancreas, and everything that the disease disrupted—as well as finding the cause of the disease itself.

PART 2

CHAPTER 18

The Gene Hunters

1980–1984

Bad times have a scientific value. These are
occasions a good learner would not miss.

—Ralph Waldo Emerson

In 1980, biologist Lap-Chee Tsui (pronounced Choy) was sitting at his lab bench at Oak Ridge National Laboratory in Tennessee, flipping through the classified ads of the journal *Science* in search of a new job, when he noticed an opportunity to work on cystic fibrosis at the Department of Genetics at the Hospital for Sick Children in Toronto, Ontario—commonly known around Toronto as "SickKids."[1] He had never heard of the disease, which wasn't surprising, given that it was nearly nonexistent in East Asian populations.

Tsui was born in 1950 in Shanghai, China, just a year and a month after Mao Zedong had proclaimed the country to be the People's Republic of China. His parents left the mainland in 1953, fearing affluent families like theirs would be persecuted under the new communist government, and moved to Hong Kong where Tsui was raised and educated. In 1974, he moved to the US, where he began his PhD degree at the University of Pittsburgh.

There, Tsui studied how hundreds of proteins self-assembled into a twenty-faced shell of a virus called a bacteriophage—one of the most plentiful organisms on the planet.[2] It was a fun project: pure intellectual acrobatics. But after earning his doctorate, he wanted to focus on human health, so he moved to Oak Ridge to work on cancer for his postdoctoral fellowship. To his surprise, however, the work wasn't fulfilling—it was basic science, distant from tangible medical applications.

While he perused the classifieds for other postdoctoral opportunities, he used his time at Oak Ridge to learn the latest techniques in the burgeoning field of molecular biology: cutting and pasting DNA, scanning it for particular genetic sequences, and otherwise working with it, the related molecule RNA, and proteins. So when he saw the ad in the May 9, 1980, issue of *Science* to work on the genetics and biochemistry of cystic fibrosis, he was especially keen to learn about this disease. Tsui yearned for a project where he might witness the rare translation of laboratory science to medicines at the bedside of a sick patient. He decided there were enough clues in the current CF research to track down the gene that caused the disease, and that his skills as a molecular biologist were a good fit.

Convinced the project would be a challenging and worthwhile endeavor, he immediately applied for the job, and was invited to Toronto to interview. About four months later, on November 25, 1980, Tsui received an official offer for a postdoctoral position at SickKids, and by January 1981, he and his wife had moved to Canada. His mission, though not explicitly stated in the offer letter, was to find the cause of the disease.

Geneticist Manuel Buchwald, who had placed the advertisement in *Science* and hired Tsui, was excited to begin collaborating with him, hoping his new colleague would lend a fresh perspective to studying this disease. The son of a refugee from Nazi Germany, Buchwald was born in Lima, Peru. He had studied in the US and then followed his Canadian fiancée to Toronto, where he joined the SickKids biochemistry department in 1970.[3] He had chosen to work on CF because physicians everywhere considered it the most devastating, severe, and hopeless genetic disease, and SickKids in Toronto was the best place to study it.

SickKids had been the place of origin, two decades earlier, for Cystic Fibrosis Canada, a nonprofit dedicated to the care of CF patients and to funding research that would lead to a cure. In a story similar to that of the Cystic Fibrosis Foundation, the groundwork for the organization was formed when Canadians led by Doug and Donna Summerhayes gathered parents of children with CF at the Hospital for Sick Children on May 23, 1959, just four years after Sharples had launched the American counterpart. On July 15, 1960, Cystic Fibrosis Canada was officially launched; and just like the CFF in the US, they launched specialized CF clinics in the 1960s, including the one at SickKids.[4]

Sometime around 1967, Mary Corey,[5] a scientist at SickKids, launched the Canadian Patient Data Registry. Like Warren Warwick's registry, the Canadian registry was packed with clinical statistics on each patient. It was a vital bank of knowledge for anyone who wanted to study the disease or track the progress of patients who were being treated using different strategies. In 1973, as Buchwald began regularly visiting the clinic to collect blood and tissue samples, he encouraged the clinic physicians to document a detailed family history for each child. Who in their nuclear and extended family was affected? What were the symptoms of all the affected relatives? How severe was their disease? The more Buchwald knew about the genetics of the disease, the easier it would be to eventually find the gene that caused it.

Buchwald's collaborator on the CF project was biochemist John "Jack" Riordan. Both scientists were trying to ferret out differences between proteins found in the blood of CF patients and those found in the blood of their healthy parents and siblings.[6] The search was motivated by the hypothesis that there might be a mysterious "CF factor" in the blood that caused all the nasty symptoms (an idea that eventually proved false). When their search failed, Riordan and Buchwald suspected instead that cystic fibrosis was caused by the absence of a protein—by a mutation in the DNA that prevented CF patients' bodies from making it. If, in comparing all the proteins in healthy cells with those in sick ones, they could find a protein that was missing from the patient group but present in people without the disease, they could then use that protein to track down the gene responsible.[7] Such a strategy could work because for every protein there was a gene that provided a blueprint.

Inside every cell in the human body (except sperm and egg cells) are two sets of twenty-three different chromosomes—one set from each parent—that carry the instructions to build a complete human. Chromosomes are made up of DNA, and segments of that DNA encode genes—the instructions to make a protein. In a two-step process, microscopic "machines" in the nucleus of the cell read a gene's DNA and transcribe it into an intermediate molecule called messenger RNA. That "message" then moves out of the cell's control center, called the nucleus, where it is translated by another machine into a string of amino acids— the building blocks of proteins. That linear chain of amino acids,[8] like a beaded necklace, then twists and writhes into a complex three-dimensional shape that is unique to each protein and helps it perform a specific function.

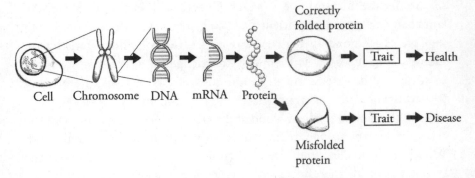

The process through which a cell's DNA translates
to physical outcomes in humans.

Proteins are the workhorses of the cell. Some are hormones that spur growth; others are neurotransmitters that permit neurons to engage in conversation. Muscles are built from proteins, and in red blood cells a protein called hemoglobin carries oxygen throughout the body and escorts carbon dioxide out. What a protein does, and how, depends entirely on its shape. In DNA code that has been altered through mutation, the sequence of amino acids can change, and the shape of the resulting protein can shift. If the shape shifts, the protein can fail to function.

If Buchwald and Riordan could find a protein that CF patients were missing, it would be possible to work backward from the protein to spell out the DNA sequence of the gene and figure out on which chromosome that gene was located. This approach had already worked for several blood disorders, like sickle cell anemia, thalassemia, and hemophilia B. Knowing the code and locations of these genes had enabled researchers to develop prenatal tests to determine if a developing fetus carried a disease-causing mutation in the gene. Buchwald and Riordan hoped this approach would work for CF.

The pair searched exhaustively for differences between healthy cells and those from cystic fibrosis patients that could explain the lung infections, devastated pancreas, and salty sweat. But their efforts yielded nothing, and in the late 1970s, both took sabbaticals to learn new techniques and refresh their perspective. Riordan spent time in California, where he met Paul Quinton, still at UCLA. By then Quinton had mastered the art of dissecting sweat glands from tissue and schooled Riordan in how to mine his own skin for normal sweat glands. Riordan left

UCLA with small scars that bound these friends for life. From Quinton's lab, he moved on to another in California, where he learned to nurture the cells that made up the sweat glands and grow them in dishes in the lab. Riordan returned from sabbatical invigorated and with a new plan: to collect sweat glands from CF patients and figure out how they were different.

Riordan began by going to SickKids' CF clinic, which happened to be a few floors down from the biochemistry department. He asked the doctors to procure skin samples from their patients, as well as the patients' healthy parents and siblings. By this point, the nurses and doctors knew Buchwald and Riordan well, and they spread the word. Patients and families were always happy to donate tissue, hopeful it would yield clues to a cure, and together donated dozens of skin samples. The doctors and nurses also literally put skin in the game, contributing their own healthy tissue in solidarity.

Riordan separated the sweat glands from the patients' skin samples as Quinton had showed him. Then he went one step further: he separated the tiny structures into individual cells, a process similar to disassembling a LEGO structure into single bricks. Next, he grew the cells in a nutritious broth that allowed them to multiply outside of the body. After that, he did the same with sweat glands from the healthy family members. Whatever was different between these cells, he felt, ultimately led to sickness and death. These new experiments were almost identical to the ones Riordan and Buchwald had done prior to their sabbaticals using blood samples. But while, by 1980, no one had discovered anything abnormal in the blood of cystic fibrosis patients, there was clearly something peculiar going on with CF patients' sweat glands—they were releasing too much salt.

To figure out exactly what was different, Riordan decided to examine whether there were differences in what genes were turned on in the sweat glands of CF patients versus healthy people. Although every cell in a human body contains the same set of instructions—the same genes—written in the alphabet of DNA's building blocks (nucleotides Adenine, Cytosine, Guanine, and Thymine, or A, C, G, and T for short), not every gene is active, or turned on, in every cell. What makes a brain cell different from a liver cell, a muscle cell distinct from a skin cell, is which genes are active—turned on and making proteins. Each cell has its own job to do and turns on genes accordingly. The genes active in skin cells, for example, are concerned with making and maintaining that particular tissue. Riordan wanted to know which genes had been turned on in the sweat gland cells—and what proteins they were making.

If a gene was active, then it made the intermediate molecule, messenger RNA. The RNA was then translated into a protein. Using new molecular biology techniques he'd learned during his California sabbatical, Riordan captured the messenger RNA in both the sick and healthy sweat gland cells and converted it back into DNA—the genes from which the RNA was made—producing two collections of genes (one from the CF sweat glands, one from the healthy sweat glands). A collection of genes culled from sick and healthy cells is called a library—in this case, a complementary DNA, or cDNA, library. Creating a library from the sweat gland cells simplified the hunt for the gene. Rather than having to examine all 21,000[9] or so genes in human DNA, the search could be confined to just the ones that were active in the cells being studied.

Riordan was sure that one of the genes required in the CF patients' sweat gland cells was missing or broken. Why else would they release so much salt? Perhaps a mutation in the DNA had altered the spelling of the gene, which altered the messenger RNA, which then altered the protein's shape and disrupted the movement of sodium and chloride in the cell—ultimately causing sweat that was five times saltier than normal. The challenge was finding that mutant gene.

That's what Riordan and Buchwald hired Lap-Chee Tsui to do.

When Tsui arrived in January 1981, he was given a laboratory bench in Buchwald's lab on the eleventh floor of the Elm Street wing of the hospital, where he began to study proteins and RNA harvested from CF patients. His work attracted a lot of attention from curious researchers, who would pop in from neighboring labs to glimpse some of the new molecular biology techniques he was using. Tsui also met regularly with Riordan and his technicians, who worked on the third floor of the Gerrard Wing. But even with Tsui, Buchwald, and Riordan all working on different facets of the same problem, none of them could find a difference between the proteins produced in the sick and healthy. Nor could they identify a gene that was turned off only in the CF patients. None of the biochemical and molecular strategies for finding differences between DNA or proteins were working.

Tsui was not the type to vent his frustrations. A calm, introspective man, he was intent on problem solving and not discouraged by the lack of progress. He was often found sitting in his tiny office behind a back-log of journals, reading, trying to learn the latest tactics and techniques and acquaint himself with new fields of biology. At the time, the field of human genetics was moving fast, but many geneticists around the world were hitting the same roadblock: they couldn't track down the gene that caused the disease they were studying. But while he was perusing a 1980 issue of the *American Journal of Human Genetics*,[10] Tsui came across an article describing a revolutionary approach for creating a map of the whole human genome—of all chromosomes, one through twenty-two, and also the sex chromosomes, X and Y—and a strategy for identifying individual genes responsible for particular diseases.

What was different about this new approach was that it enabled geneticists to zoom in on the exact location of a disease-causing gene—without knowing anything about what the gene looked like or what protein it coded for.

The DNA from all humans, regardless of ethnicity, is 99.9 percent the same. But when you're comparing any two people's DNA—those As, Gs, Cs, and Ts—the sequence will occasionally, every thousand nucleotides or so, vary: one person may have a G where another has an A, or a C might be replaced by a T. One person may carry the sequence GATT**A**TTC, for example, and another person might carry GATT**C**TTC. The genetic locations where a person's DNA sequence can differ are called polymorphisms: *poly* meaning many, and *morph* meaning forms. These minute differences, which are inherited from one generation to the next, are essentially invisible; most do not confer a visible trait like red hair or dark skin or alter the function of a protein, but their locations in the genome can serve as signposts. Every person's genome, with three billion nucleotides, has about three million polymorphisms.

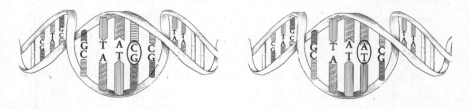

The code for these two pieces of DNA is identical except for one location where one nucleotide, cytosine (C), is replaced with another, adenine (A). This is called a single-nucleotide polymorphism.

The revolutionary paper Tsui read was based on the work of Hong Kong–born geneticist Yuet Wai Kan and his Belgian-born colleague, Andree M. Dozy, who discovered that a specific DNA polymorphism was associated with sickle cell anemia, a blood disease that tends to affect people of African, Mediterranean, Middle Eastern, or Asian origin.[11]

In the now landmark 1978 study,[12] Kan and Dozy suggested that sickle cell mutation was almost always inherited with this polymorphism located close to the gene. You could therefore test whether someone carried the sickle mutation[13] by testing for the polymorphism, which served as a proxy for the actual mutation. It was like saying that a rock star always traveled with their manager, and so even if you didn't know the rock star's location, if you found the manager, you knew the rock star was nearby. It was a surprising finding that laid the foundation for locating mutations that caused other hereditary diseases.

Kan and Dozy also described in their paper how this DNA marker could be used in prenatal diagnosis.[14] By removing a little bit of amniotic fluid during an amniocentesis and testing whether the fetus's DNA was positive for a particular polymorphism, Kan could reliably predict if the child developing in the womb had sickle cell disease.

Two years after Kan and Dozy published their work, a paper in the *American Journal of Human Genetics*[15] described how scientists might create a map of the whole human genome using the same type of genetic polymorphisms[16] that Kan had used.

A map is as critical for gene hunters as it is for treasure-seeking pirates. The genome is enormous: three billion letters. For comparison, that's about 1,000 times more characters than the King James Bible.[17] Imagine two stacks, each of 1,000 bibles arranged end to end, millions of pages—and the impossibly mind-numbing task of looking at each letter, one by one, in search of a single difference, whether typo or deletion, between the two. In 1983, scientists were already able to decode, or sequence, DNA—that is, determine the order of As, Cs, Gs, and Ts in a specific piece of DNA—but doing so was slow and laborious. At that time, it took months to sequence just a few hundred nucleotides, too glacial a pace to let scientists contemplate sequencing the entire genome. (Beginning that endeavor would have to wait another decade.) All twenty-three chromosomes and the genes they carried were uncharted territory.

But now gene sleuths had polymorphisms to serve as "mile markers," helping them navigate that vast, unknown space and figure out if they were getting close to a specific gene—or a disease-causing mutation.

Inspired by Kan and Dozy's article, Tsui recognized that the first step to finding the gene and mutation causing CF was to find polymorphic DNA markers near the gene. This, in turn, would allow him to identify the region of a chromosome harboring the disease gene. Most children with CF, but not *all* their healthy siblings, would inherit the markers along with the mutant gene from their parents. And the closer the marker was to the gene, the more often the two would be found together. Once Tsui found a marker in children with CF more often than in their siblings, the next step was to scour this segment to try to find the gene.

The SickKids' clinic for CF research had banked blood from each member of the twenty families who visited regularly, and if Tsui could obtain previously discovered markers from other scientists, he could scan the patients' DNA for them. By 1983, geneticists who discovered novel markers in their own labs often traded them with others to expand their collection. Discovering them single-handedly was much more time consuming.

As Tsui contemplated the necessary logistics—acquiring markers from others, discovering more of his own, and then using them to test DNA samples from patients and their families—he realized he needed more laboratory technicians and funding. He knew that assistance from the Canadian Cystic Fibrosis Foundation was crucial. The foundation's committee members listened, baffled, as Tsui talked enthusiastically about mapping the gene and these mysterious things called markers. Mapping the genome was a mind-boggling concept; the idea of DNA signposts was utterly incomprehensible. But Tsui's passion and confidence won him support. He seemed convinced he could find the gene, and his request was modest: funding for one research technician. So the Canadian CFF gave him the money.

Next, Tsui reached out to CF clinics all across Canada, from the eastern tip of Newfoundland to the outer realms of British Columbia, requesting blood samples and family histories of other CF patients. As geneticists across the country complied, Tsui's collection of families grew from twenty to fifty.

As other scientists learned of Tsui's efforts, many were skeptical he would succeed, because most of Tsui's fifty families were small, just the parents and a minimum of two affected children, though a few had up to five affected children. This was a challenge because the power of this new mapping technique increased with the size of the families, as Harvard geneticist James Gusella explained to a spellbound audience at the 1983 meeting of the American Society of Human Genetics. Gusella was searching for the gene responsible for Huntington's disease, an inherited condition that progressively cripples the body by destroying brain cells and tends to strike sufferers in their thirties and forties, when they had already begun families and unwittingly passed the gene to their children.

Collaborating with geneticist Nancy Wexler,[18] who had collected hundreds of blood samples from a sprawling Venezuelan family with 3,000-plus members, many of whom had Huntington's, Gusella's team tracked the gene responsible to an address on chromosome 4, near a marker called G8.[19]

The G8 marker was so close to the gene that the two were inherited together as the mutation was passed through generations. Therefore, G8 could be used both to determine if an adult in a family with a history of the disease was a carrier, and as a prenatal test.[20]

The Huntington's breakthrough in 1983, five years after the birth of the first baby conceived through in vitro fertilization (IVF), indicated to parents who had one child with a genetic disease, like cystic fibrosis, that by combining the prenatal test and IVF, they might be able to conceive another child without that disease.

The families in Tsui's study were smaller than the one Gusella and Wexler had used. But Tsui now had DNA from more than fifty families, which he was sure would make up for their smaller size. Tsui was inspired by Gusella's success, and for the first time since his move to Canada, optimistic about finding the gene.

CHAPTER 19

Lucky Number Seven

1984–1985

If Aristotle were living today, he would no
doubt be a molecular biologist.

—William Stockton, *New York Times* science editor

At the dawn of the molecular-biology revolution, academics were running a high-stakes race to unlock the genetic origins of disease—and win the prestige, accolades, and tenure that such discoveries brought. So it wasn't surprising that Lap-Chee Tsui, Jack Riordan, and Manuel Buchwald had competition in their quest to find the gene for cystic fibrosis.

Robert Williamson, a Cleveland-born chemist, had moved with his family at age sixteen to England. There, he studied chemistry at University College London, and then earned his PhD in biochemistry. He kicked off his research career in 1959, just six years after James Watson and Francis Crick's discovery of the structure of DNA.

By 1974, Williamson, like Yuet Wai Kan, Andree Dozy, and many other scientists, was working on thalassemias—a group of inherited blood disorders caused by mutations in the genes for hemoglobin. More than one in ten studies on human genetics before 1985 focused on hemoglobin, the red-pigmented, life-giving protein in red blood cells that carries oxygen from the lungs to the rest of the body.[1] Hemoglobin's popularity was a natural and logical extension of early work by Linus Pauling, later the recipient of the Nobel Prize in Chemistry, in which he discovered a difference between the hemoglobin protein in healthy people and those with sickle cell disease.[2] In 1956, biochemist Vernon Ingram[3] advanced Pauling's discovery by showing that the sequence of amino acids that created the normal protein also differed from the sequence that created sickled protein. Then, in 1960, physicist Max Perutz cracked hemoglobin's atomic structure:[4] a four-leaf-clover shape with two leaves made from alpha globin protein chains and two from beta globin protein chains. By 1970, the genes that code for hemoglobin were the most studied genes on the planet, and for those interested in studying the molecular origins of disease, blood disorders were the natural starting point.

Williamson was intrigued by the most severe form of thalassemia— alpha thalassemia, or hydrops fetalis—which led to death in the womb or shortly after birth.[5] Analyzing blood from afflicted infants showed that the alpha globin protein was missing. The alpha globin messenger RNA was also missing, which suggested that the gene had never been turned on. Williamson wanted to find out what that problem was. Was the gene itself missing? Turned off? Mutated?

The only other person he knew who was working on this question was Kan. Williamson and Kan often discussed their progress by phone, even though they were both racing toward the same goal. When they both discovered that alpha thalassemia was caused by missing DNA that would have coded for alpha globin, they agreed to publish together and share the credit for the discovery.[6,7]

In 1976, Williamson moved to St. Mary's Hospital Medical School in London, where he continued to work on the globin genes. He loved it there; he felt the school was small, quirky, and independent, yet still reverberated with the discoveries of penicillin and the structure of antibodies that had brought the school two Nobel Prizes.[8] The atmosphere encouraged daring work for the intellectual challenge, but as St. Mary's was a hospital, with patients in residence, it also elevated the urgency of converting that bold science into practical application.

In 1980, inspired by the power of mapping to find diseased genes, Williamson began brainstorming which diseases were good candidates to seek out, and cystic fibrosis seemed like a natural choice. Its inheritance pattern seemed to suggest the disease was caused by a single gene, it was the most common genetic disease among Caucasians, and at St. Mary's he regularly met patients with it. And as word spread that Williamson was looking for the CF gene, the beloved leader of the UK's Cystic Fibrosis Research Foundation Trust, Ron Tucker, who knew most of the families and patients in the community, paid an unexpected visit to his lab.

The UK foundation, which simplified its name to the CF Trust in the early 1980s, had a curious origin story. At a cocktail party on the River Thames in 1963, Mrs. Percy Lovel donned a wildly unusual hat that happened to catch the attention of a pediatrician named David Lawson. Dr. Lawson listened to Mrs. Lovel as she poured out her story about her granddaughter, who was sick with CF.[9] He understood. He also had a child with CF. From that day Dr. Lawson began meeting with other parents to forge a research foundation led by a board of pioneering CF physicians and scientists. Several months later, on February 20, 1964, the trust had its inaugural meeting. It was created as a charitable organization whose only function was to raise money for research—something made easier in the UK thanks to universal healthcare, which already provided for basic care of sick children. So although the US foundation was, at the time, raising more money overall, both foundations were spending comparable amounts on research.

Some twenty years later, the trust's director was knocking on Williamson's door. "You don't know me," Tucker said to Williamson, "but people tell me that what you are doing is the way forward for CF." He offered Williamson a million pounds to build a research team to focus on CF.

A stunned Williamson accepted. He wasn't placing all his bets on cystic fibrosis; he was also after the gene for Friedreich's ataxia,[10] a rare genetic disease that damaged the nervous system. But with the hefty backing of the Cystic Fibrosis Trust, the search for the CF gene was now at the top of his research queue and dwarfed all other gene-mapping projects in the lab.

Once the shock of this new funding wore off, Williamson recruited more scientists and students, and began to figure out what he would need to start hunting for the gene. Like Tsui, he needed DNA from families who had children with cystic fibrosis. With the help of one of his postdoctoral researchers, Kay Davies, he decided to collect it from every family in the UK that had three affected siblings with CF.

For Davies and Williamson, the road trip from London, to Taunton in Somerset, to Manchester up north, was a humbling one, in which they were overwhelmed by the generosity and the commitment of the families they met. Though these parents didn't know Williamson, Davies, or the pediatricians who helped them take samples, they welcomed them into their homes and brought their often incredibly ill children to give them blood samples. As the pediatrician took blood and other data and measurements, many parents confided in Williamson and Davies, sharing terrible, painful stories about doctors who told them, "Forget that child," or "You have a three in four chance of having an unaffected child. Try again!" The parents existed in a gray zone between life and death, knowing that none of their sick children would live long. Yet even with this pall over their lives, the parents and children were ready to help, optimistic that any research was a good sign.

Back at St. Mary's, Williamson and his team began screening the DNA they had collected from ten families with the markers they had identified in previous experiments in his own lab. First, they tried markers from chromosome 4, but none seemed linked to cystic fibrosis. Williamson didn't have markers for every chromosome, and for some he only had a handful—the equivalent of having just a couple of road signs on a transcontinental highway. He needed more markers on all of the remaining chromosomes. He began reaching out to colleagues in other laboratories around the world, but available markers were scarce; few had been discovered, and those who had them rarely wanted to share. If he wanted more, he and his team would need to discover them themselves. So members of his lab began hunting for the genetic markers to guide his quest.

In the race to find the CF gene, Williamson's team was off to a strong start. But an ocean away in Toronto, Lap-Chee Tsui was about to make a connection with a biotech company that would accelerate his hunt for the gene and lead him to the first major breakthrough.

IN LATE 1984, TSUI WAS BUSY SCREENING DNA FROM HIS CANADIAN families, using markers collected from random chromosomes. The work was proceeding at a slow, steady, and generally unsatisfying pace when Tsui's boss and collaborator, Manuel Buchwald, received an unexpected call from Dr. Helen Donis-Keller, who led the human genetics department at a well-known Boston biotech company called Collaborative Research, Inc. After completing her PhD and a fellowship at Harvard, she'd accepted a job at biotech company Biogen, and then two years later moved on to Collaborative, which was now spending millions of dollars to map the human genome by identifying hundreds of polymorphisms on each chromosome. If they could find the harmful gene mutations that caused specific diseases, or at least nearby polymorphisms, they could develop diagnostic kits for those diseases—a lucrative prospect. Donis-Keller thought that cystic fibrosis would make a good first case, and she suggested that Collaborative and Tsui work together. She assumed it would be an easy decision for Tsui—her company had more than two hundred markers (compared to the one hundred Tsui had either discovered or borrowed from colleagues) and a cadre of scientists that would help cut his search time in half.

She was right. Tsui had never worked with a private company before, but he had no reason to distrust Donis-Keller and he was eager to get access to new markers. In return, Tsui would share the DNA from the Canadian families, and both teams would scan the DNA. Donis-Keller hoped that either her team or Tsui's would pinpoint a marker inseparable from the disease—one only found in sick kids. Such a marker could be worth hundreds of millions, possibly more than a billion. Cystic fibrosis was the most common genetic disease in Caucasian populations—about one in twenty-nine Caucasians carried a single copy of the mutant CF gene, or in terms of the US population, one in thirty-five Americans—roughly ten million people.[11] Presumably some significant portion of this group, particularly those interested in having children, would be interested in knowing whether they carried a bad gene. But testing individuals to figure out whether they were carriers was only one market. There was also prenatal testing—scanning fetal cells collected during amniocentesis. The Tsui–Collaborative deal was one Donis-Keller hoped to replicate with hundreds of other researchers studying the three thousand genetic diseases known at the time.[12]

Figuring out how closely a marker is linked to a disease requires tracking the inheritance of the polymorphism within families with cystic fibrosis. Tsui's strategy was to identify how often the patients with CF inherited a particular polymorphism from their parents and compare it with how often that same polymorphism was inherited by healthy siblings. So the DNA sample of every patient and family member had to be screened with every marker Tsui had in his collection. The goal was to try to find DNA markers that were inherited more than 50 percent of the time by individuals with CF, but inherited less frequently by healthy siblings in the same family. If many of the children with CF, say 95 percent, inherited a particular marker from their parents, it would mean that the marker was reasonably close to the CF gene—though to confirm the two were linked, Tsui and his peers would have to make additional statistical calculations.

The method worked because of the particularities of genetic inheritance. A marker on the same chromosome as a disease-causing gene is inherited with the mutant gene *unless* a common event called recombination, or crossing over, occurs, in which two paired chromosomes swap genetic information. One opportunity for this is during the production of sperm and eggs that are needed for reproduction.

Because crossovers can occur anywhere along the length of a chromosome, they sometimes separate markers and disease genes that were previously next to each other on the same chromosome, meaning that a patient could have a disease but not the nearby marker associated with the disease in other patients. However, the smaller the distance between a marker and a gene, the lower the chance a crossover will separate them. When a geneticist doesn't find many patients who lack a particular marker—indicating very few crossovers have happened between them as the gene and marker were passed down from generation to generation—it's evidence they are close to the gene.

By 1984, Tsui's lab had grown to four technicians who were testing hundreds of DNA samples from patients and their parents and siblings for the presence of each marker, while Tsui sat in his office—the size of a VW Microbus—and calculated, based on the principles he learned from genetics textbooks, the likelihood that the marker and the CF gene were linked. By the time the first batch of markers arrived from Collaborative in August 1985, Tsui and his team had tested all one hundred of his own markers, and were ready to move on to Collaborative's.

It was good timing. Tsui was leaving town for a conference in Helsinki and his technicians could begin experiments with the new markers. By the time he returned, they had news: a marker in the first batch from Collaborative appeared linked to the CF gene.

The technicians had tested for the marker in just a subset of the families so far, yet the results were already encouraging. Tsui could not believe their seemingly instant luck, so he urged his staff to screen more families as quickly as possible. Once they had the results, Tsui decided to analyze the data formally using the "LOD score," genetics-speak for the "logarithm of the odds," a statistical measurement of the likelihood that a marker and gene are located near each other, versus inherited together simply by chance.[13] The higher the LOD, the greater the likelihood of linkage. Among geneticists, 3.0 was the magic LOD score. It meant there was a 99.9 percent chance that the marker and gene were linked. When Tsui crunched the numbers, he got a LOD score of 2.8. The results were ambiguous.

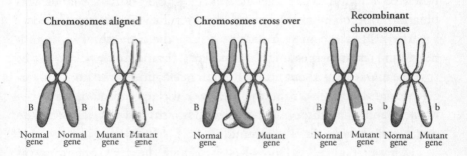

Chromosomes aligned | Chromosomes cross over | Recombinant chromosomes

B   B   b   b   B   b   B   B   b   b
Normal Normal Mutant Mutant   Normal   Mutant   Normal Mutant Normal Mutant
gene   gene   gene   gene   gene   gene   gene   gene   gene   gene

Let's say a mutated gene is usually linked with the marker b, because they are close together on the chromosome. If a crossover happens between the mutant gene and marker b, then these two previously linked locations can be separated and unlinked. In children who receive a crossed-over chromosome, and in all their descendants, the mutated gene will no longer be linked to marker b. When the gene and the marker are far apart, they get separated like this often. When they are extremely close together, they may stay linked for thousands of years, through hundreds of generations.

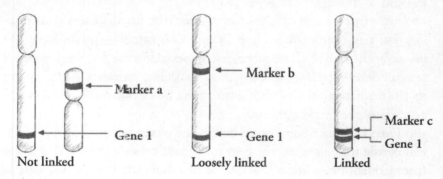

Marker a

Gene 1

Marker b

Gene 1

Marker c

Gene 1

Not linked   Loosely linked   Linked

Left: The gene and marker a are on different chromosomes and are not linked.
Middle: The gene and marker b are on the same chromosome, but far apart.
They can easily be separated by recombination, so they are only loosely linked.
Right: The gene and marker c are right next to each other
on the same chromosome and are tightly linked.

The technicians still hadn't tested all fifty of the families in their collection; perhaps they just needed more to pass that 3.0 threshold. A couple of days later the technicians handed more data to Tsui, who redid the analysis. This time the LOD score hit 4.0. The odds were now ten thousand to one that the linkage wasn't real. The marker was present in too many of the sick children for it to be there by chance.

The LOD score analysis didn't just give the strength of linkage; it helped the researchers estimate the marker's distance to the gene. In this case the marker was about fifteen million nucleotides away from the target—a significant trek. Still, the discovery was an enormous leap. The whole genome had three billion nucleotides. Tsui had just narrowed the search zone by a factor of two hundred.

Ecstatic, Tsui called Donis-Keller to share the news. She, too, was skeptical, and sent one of her team from Boston to verify the claim—which he did almost immediately.

Tsui was excited and asked which chromosome the marker was located on—information Collaborative hadn't supplied when they sent the markers originally. That would mark the starting line for the gene hunt. "It might be on chromosome 7," Donis-Keller said, claiming that she wasn't sure. She told Tsui that Collaborative was working with a team in France to find out; it shouldn't take more than a week. Sit tight, she said.

After three weeks of silence in August, the usually calm and trusting Tsui was feeling not just impatient but suspicious—unfamiliar emotions. With the CF clinic just a few floors below his lab, he knew the price of delaying the work. Children were dying. Irritated by the feeling that his collaborator was toying with him, he decided to figure out the position of the marker himself.

Within a week he discovered the marker was in fact on chromosome 7, which meant that's where the gene for cystic fibrosis was, too. Tsui knew that once the news broke it would be like dropping the starting gate at a horse race; every gene hunter on the planet would rush in. To him, that was a good thing; the more people who knew where to begin the gene hunt, the sooner patients would benefit. Delighted that he had confirmed Donis-Keller's hunch, he called her to report that the marker and therefore the gene were both on chromosome 7. But rather than congratulate him, she was angry he had done the experiments himself rather than waiting for her to provide the answer.

Their disagreements didn't end there. Tsui wanted to prepare their results for publication in *Science* but share the news first at the annual meeting of the American Society of Human Genetics, which was being held in Utah in just over a month. Donis-Keller wanted to keep the location a secret. As she later told a reporter, "We were nervous about making a mistake. We wanted to feel certain it was on 7. It was very early for us; it was our first gene, and we were new to the field."[14]

To Tsui, Donis-Keller's behavior wasn't motivated by uncertainty. A marker that was fifteen million steps away from the gene was too distant to serve as an accurate diagnostic marker for the disease; it couldn't be patented. Tsui suspected Collaborative was guarding their turf until they had secured a marker closer to the gene, to retain their lead in the hunt. The partnership between Tsui and Collaborative soured.

Somehow the news of Tsui's discovery leaked, and by September rumors were swirling at genetics meetings in France and Germany that the CF gene was somewhere on chromosome 7. Tsui wasn't at either meeting but his competitors were. Now they knew exactly where to start hunting for the gene; Tsui had been left behind on his own discovery.

By the time rumors of the CF gene's location reached Bob Williamson's lab, he had already rejected chromosome 4 as a possibility and was considering 18, 19, and 20, as he had markers on those chromosomes at hand. Now, his team started looking at chromosome 7 instead.

Another geneticist, Raymond White at the University of Utah, co-author of the seminal paper on genome mapping a few years earlier that had inspired many a gene hunter, Tsui included, also began scouring chromosome 7 using his collection of markers.

When Tsui presented the data about his linked marker at the Utah meeting a few weeks after the French and German meetings, he felt awkward. His contract with Collaborative prohibited him from saying anything about the location of the marker on chromosome 7. He wasn't secretive by nature, and he hated withholding data important for human health. When he finished speaking to wide applause, the first questions were the ones he was dreading most: "Where was the marker?" "On which chromosome?" Before he could answer, Donis-Keller told the audience they were figuring that out.

Collaborative's attempt to hide the gene's location was just one in a string of early attempts to wring cash from the genome. After discovering the location of the marker, the company filed for patents on it and any other markers they might find closer to the gene.[15] Collaborative chairman and CEO Orrie Friedman infamously stated, "We have fifty-four markers on chromosome 7. We have mapped it in a way no chromosome has ever been mapped—we really own chromosome 7."[16] The arrogant remark stirred anger and distrust in the scientific community. Patenting a swath of a chromosome was like a land grab in the Wild West—claiming territory without knowing exactly what was there in hopes of extracting riches. Collaborative wasn't alone, and no one was shy about the desire to profit. Some fifty other companies were also trampling the same ground, using DNA probes to identify markers close to disease genes of interest.

With the CF marker, however, Collaborative's plan had failed. Once the news of the marker's location leaked, they quickly lost their lead; both Williamson and White submitted papers to *Nature* with new markers on chromosome 7 that were much closer to the gene than Collaborative's—within an estimated 1.5 million nucleotides, rather than Tsui's fifteen million.

Tsui,[17] White,[18] and Williamson[19] all published their results, including their markers' chromosome 7 location, in the same November 28, 1985, issue of the journal *Nature*. Peter Newmark, a *Nature* editor, noted in his editorial that White and Williamson likely benefited from the rumors of the gene's location on chromosome 7—restoring credit to Tsui for discovering the location first.[20] That same week, Tsui's original paper reporting the linkage[21] appeared in *Science* with his Collaborative coauthors, revealing the marker but not its location.

The excitement over the markers lay in their immediate applications in genetic testing. White and his coauthors had written in their *Nature* article that their newly discovered marker might be useful in prenatal diagnosis. Indeed, within a few months, Bob Williamson published a report in the medical journal the *Lancet* describing a prenatal test for cystic fibrosis[22] using the markers that White's team had discovered. The test was not useful for screening the general population, given the high cost of testing DNA at the time and the low probability—roughly 8 in 10,000—that carriers would meet and marry. But parents who had one child with cystic fibrosis had a one in four chance of having another with the disease. And now, with White's markers, they could find out whether future children would suffer from CF while those children were still in the womb.

This insight was not without cost. If the fetus tested positive for the disease, the parents had to wrestle with whether to terminate the pregnancy and try again to conceive a healthy child or, perhaps more daunting, continue the pregnancy knowing that their child would likely die young. Despite the limitations of the test, and the fact that all three teams were still far from the gene itself, the news of its location inspired hope. Finding the gene was the first step to finding a way to repair it and eradicate the disease forever.

CHAPTER 20

Kate

1986

I will love my children no matter the diseases they carry, but I am beginning to think that they shouldn't have to carry mine.

—Lee Cooper, biotech entrepreneur

By 1986, as Lap-Chee Tsui, Raymond White, Robert Williamson, and others around the world were focusing on chromosome 7, pursuing the gene like SWAT teams zeroing in on a suspect, eleven-year-old Joey was dangerously ill, spending more time in the hospital than at home. Joe and Kathy barely saw each other. As usual, they tag-teamed at the hospital, crossing paths only to share a cup of coffee in the hospital cafeteria before Kathy headed home.

Joey's gang—Woody, Timmy, Eddie, Chris, and Andrew—visited him frequently, as they always had. The boys strode out of the elevator and onto the fourth floor swaggering like action heroes, forgetting where they were the moment they saw Joey. Whenever Kathy spied the boys approaching, she welcomed them with bowls of popcorn and dished out other snacks and candy just as she did at home.

During one visit, Joey gingerly got out of bed and, holding his IV with one hand, led the boys on an adventure. Grinning ear to ear, he motioned the gang to the elevators and pressed "B."

Where are we going? asked Eddie, nervously.

The morgue! Joey answered, eyes wide.

The other boys giggled and made spooky ghost noises.

Really? Eddie said, aghast and intrigued at the same time.

When the elevator reached the basement, they peeked out to make sure the coast was clear. It's this way, Joey said. The boys stifled excited, nervous giggles. Holding his IV like a magic staff, Joey tiptoed toward the morgue's entrance. They all gasped as they saw a cart next to the wall with a lumpy form shrouded beneath a blue sheet.

Let's take a look, said Joey, bravely leading the pack.

Just then, the boys saw a burly hospital employee coming toward them. They screamed and took off back toward the elevator. Eddie, the largest of them, picked Joey up and Timmy grabbed the IV. The elevator doors opened and they rushed inside, banging on the buttons and watching the doors close as the man approached. Safely headed back up to the fourth floor, they laughed themselves silly, Joey more quietly than the others; it was impossible to have a real belly laugh with so little air left in his lungs.

When he returned to school between hospital stays, his friends had no idea how sick he was, only that he often needed piggyback rides to go up the stairs to their classroom or to play a game of tag.

Joey was generally in better health during the warm summer months, and because he was always so curious about Dr. Lapey's life outside of the hospital, the doctor invited the whole family to enjoy a weekend at his cabin on Naushon Island, just a few miles off Cape Cod. Naushon was a sliver of land with no paved roads, just hiking trails that snaked through Beech Forest and around small lakes. Lapey taught Joey, who was feeling well and in good spirits, to fish, rowed him around in a little skiff, and explored the forest with him on short hikes. It was an idyllic weekend. But it was what happened at night that made the greatest impression on Lapey.

As Joey lay down to sleep, the relentless coughing began. He coughed all night long, with each cough breaking Lapey's heart. He had never heard anything like it. Though he'd been treating children with cystic fibrosis for almost fifteen years, he never really understood the intensity of the incessant nocturnal cough that possessed their bodies night after night as soon as their heads hit the pillow. What Lapey was experiencing was a typical night in the life of Joey O'Donnell—a typical night for many of his patients and their parents. And it was something that, until now, sitting next to Joey, he had never fully appreciated.

Lapey watched Joey's body shake and convulse with every hack, every few minutes, every hour, all night. For the first time, he truly understood why these children were permanently exhausted: they never got real rest. Neither did their parents, since the house only grew quiet at night if the child received a lung transplant—or died. But for Joe and

Kathy, this night was different. With Lapey just steps from Joey's side, they slept soundly, for two full nights, for the first time in years.

Lapey always asked parents about night coughing, because when it increased it signaled that more aggressive treatment was needed. But now he realized that parents eventually became deaf to the constant hacking and didn't even register it anymore. It became normal. But to Dr. Lapey, Joey's cough was striking—a wake-up call that this beautiful boy was sicker than he appeared. How long, he wondered, would it be before science could offer some answers?

THE O'DONNELLS HAD ALWAYS WANTED A LARGE FAMILY, BUT AFTER Joey was born, their plans changed. Both parents carried the CF mutation, so their chances of having another sick child was one in four. Joe and Kathy agreed that those were terrible odds. They knew they couldn't endure watching child after child suffer from this disease, and refused to consider more children until a genetic test could guarantee a healthy baby.

The couple knew other Irish Catholic couples who'd had multiple children affected by the disease—one family had five children, all with CF—and had watched them die, one after the other. These couples were observant Irish Catholics who were opposed to abortion and believed that God would take care of their children. Joe and Kathy considered themselves every bit as Irish Catholic as those other couples, but having more than one sick child wasn't an ethical choice for them. For the O'Donnells, terminating a pregnancy was a choice they hoped they would never have to make, but it was a relief to know they would have that option.

By 1985, however, Tsui, White, and Williamson had published their results in *Nature*, and Williamson was preparing his *Lancet* article describing a prenatal CF test. The outlook for parents of a child with an inherited genetic disorder who hoped to have healthy children had begun to change. Finally, the gene hunt, which had seemed so abstract with its mysterious markers and linkages and DNA maps, was delivering something tangible that could help families and change the course of medicine. By the middle of the following year there were fifteen genetic tests for inherited diseases, including blood disorders such as sickle cell anemia, hemophilia, and thalassemia; Duchenne's and Becker's muscular dystrophy; Lesch–Nyhan syndrome; adult polycystic kidney disease; Huntington's disease—and cystic fibrosis.[1]

Some tests, like the one for sickle cell anemia, detected the actual mutation that triggered the disease.[2] For cystic fibrosis and Huntington's, for which researchers had not discovered the genes, the tests detected markers linked to the defective gene that could correctly indicate an individual's likelihood of having the disease with 95 to 98 percent accuracy.

Lapey knew well how desperately Joe and Kathy wanted more children and called them the moment he read about the new test. He explained to Kathy that scientists had discovered a DNA marker linked to the mutation that caused cystic fibrosis that was absent in healthy children. Although it wasn't *the* gene, this sliver of DNA was so close to it that they were inherited together and provided an accurate prenatal test.

Kathy, then forty-two, called Joe from the hospital. There was now a test for CF, Kathy told him. It was only available in Connecticut, but they could easily travel to get it done. And how wonderful it would be for Joey to have company.

Soon after she conceived, Lapey arranged for the test.

The test leveraged a new form of prenatal diagnosis, called chorionic villus sampling, which could determine whether a fetus was carrying one or two copies of the CF gene, or any other disease-causing gene. It was a less risky procedure than amniocentesis, which was routinely used to detect other genetic disorders, and could be done earlier in the pregnancy—at between eight and ten weeks. To collect the needed tissue, a needle was inserted through the abdomen and into the chorion— the part of the fetus that forms the placenta, not the baby. The procedure was over in less than ten minutes and the results were returned within a couple of weeks. Both villus sampling and amniocentesis carried some risk, so these procedures were only recommended when there was a chance the child might inherit a genetic disease or when the mother was over the age of thirty-five; Kathy knew she was two for two.

Kathy was nervous and didn't want to leave Joey, who was in the hospital with another nasty lung infection. But, as usual, Lapey reassured her that he'd watch out for Joey. He gave her a hug. "It would be wonderful for Joey to have a little brother or sister."

On August 21, 1986, the day of the test, Joe chartered a private plane to Connecticut. They took a cab to the clinic, where staff whisked Kathy in and performed the procedure. Soon they were back in a cab, on the plane—and back at the hospital with Joey. They didn't mention where they had been. There wasn't any point unless the test was negative.

One month later, Dr. Lapey received a letter from the genetic testing service.

> We analyzed DNA from Joseph and Katherine O'Donnell and their son, Joey. We found Mrs. O'Donnell to be heterozygous and informative for 2 met H polymorphisms and one met D polymorphism. We found Mr. O'Donnell to be heterozygous and informative for the same polymorphisms. Joey is homozygous for all of the polymorphisms tested.

> Translation: Joey had two copies of the mutated gene.

> We analyzed DNA from the O'Donnell fetus obtained by chorionic villus sampling on August 21. The fetus is heterozygous for the 2 met H polymorphisms and the met D polymorphisms, and is therefore an <u>unaffected</u> carrier of CF.

The genetics lingo was confusing, but Lapey understood the word that mattered. The baby was unaffected. The baby would be healthy. Like Joe and Kathy, she would be a carrier of the deadly mutation—capable of passing the gene on—but thanks to genetic testing, would be able to prevent the disease from affecting her children.

With the guarantee of good health, Joe told Joey that he would soon have a sister. Rather than excitement, panic flashed across his face.

"Will you still come and see me?" Joey asked. Joe and Kathy realized that he assumed that the new baby would be sick like him, demanding his parents' full attention. If there was a tiny sick baby, how would they care for him, too?

Joe put his arm around his son. "Your sister is going to be healthy," he said. "She doesn't have this disease."

"When I have to come to the hospital, I'll just get a sitter," Kathy explained, laughing to lighten the mood.

As Kathy's belly swelled, however, the family's stress mounted. Joey was far sicker than they had ever seen him and he spent weeks at a time in the hospital. Back at home, his lungs would remain clear for just a few days before they heard the signs of infection brewing again in his small chest. His breathing grew shallow, he couldn't take more than a few steps without resting, and his strength waned until he couldn't rise from his bed. Kathy was sick with worry. She had no appetite and was barely putting on weight. The baby drained her stamina and Joey was getting weaker, draining her soul.

Despite his worsening health, Joey grew more and more excited that he would soon have a little sister. Kathy overheard him talking to his friends, telling them that his mom had a "bun in the oven," and he began suggesting names. They settled on Kate.

CHAPTER 21

To Screen or Not to Screen Newborns

1985

What are some of the things that we would expect a balanced community health system to do? One, I think it would assure that every child born in this country has the optimal chance for a healthy start in life. I don't believe that any nation could justify not giving children a chance for a healthy life.

—Surgeon General David Satcher

n 1985, as prenatal genetic testing was just gaining a foothold, Philip Farrell, a pediatrician and scientist at the University of Wisconsin, Madison, was grappling with the imminent launch of a groundbreaking study into the effectiveness of a different kind of testing: newborn screening.

He and some of his colleagues had observed that the earlier children were diagnosed with cystic fibrosis, the better they fared. It was intuitively obvious, and he had observed it himself in the infants and toddlers he treated in his clinic. But that was just anecdotal evidence—no one had proven it to be true.

At that time, children were diagnosed with CF at around one year old (boys at around nine months and girls around thirteen months,[1] a gender gap that was reflected in girls' shorter life spans in the United States, Canada, and the United Kingdom). By that age, children were often already severely malnourished and sick with lung disease. Many were deficient in vitamin E, an antioxidant; without it, their red blood cells exploded, causing anemia. The children were salt poor and dehydrated. Some suffered from kwashiorkor—a protein-energy deficiency common to famine-struck regions that distended the stomach and caused fluids to pool in the legs and feet. Some children were so emaciated that their growth crawled and their muscles wasted away as their bodies mined them for nutrients. If treatment for the disease could begin earlier—as soon as a few weeks after birth—could the correct nutritional intervention put the children's growth back on track and keep lung infections at bay, helping to preserve the lungs?

Because cystic fibrosis was the most common life-threatening genetic disease in Caucasians, Farrell wondered: Were the benefits of early diagnosis dramatic enough to recommend that every single child, including non-Caucasians, born in the United States be screened at birth, so the sick ones could be diagnosed before malnutrition and lung infections caused catastrophic damage?

Newborn testing was not a new concept in the United States. Beginning in 1963, all babies born in US hospitals were tested for a metabolic disorder called phenylketonuria (PKU) using a simple test that Robert Guthrie,[2] a Minnesota-schooled bacteriologist, had developed a couple of years earlier. Like CF, PKU was a recessive genetic disease, with one bad gene coming from each parent. Babies that inherited the disease couldn't metabolize phenylalanine, an amino acid found in protein-rich foods like human breast milk, cow's milk, cheese, and meats. With the body unable to process it, the amino acid would rise to toxic levels, damaging[3] the brain and causing seizures and intellectual or developmental disabilities.[4] It was vital to test babies immediately after birth because the symptoms cropped up within just a few weeks of breastfeeding or drinking formula—and the effects were irreversible.

The test was quick and easy.[5] A nurse pricked the baby's heel, collected a few drops of blood on a piece of filter paper, and sent the sample to a lab where they measured phenylalanine. If PKU was diagnosed, the fix was simple: a diet[6] stripped of protein-heavy foods and rich in fruits and vegetables.[7] Research confirmed that as long as these children stuck to this diet, they would grow up healthy, indistinguishable from their peers and siblings without the disease.

Over the next two decades, the number of newborn screening tests rose only glacially, to just under five, as their effectiveness remained controversial.[8] In 1983, when Farrell was just beginning to plan his study, expert physicians, several from the Cystic Fibrosis Foundation itself, wrote a report[9] arguing that the benefits of newborn screening were unproven.

Farrell had been mulling the value of newborn screening for cystic fibrosis since his days training under Paul di Sant'Agnese at the NIH. Like Joey's Dr. Lapey, Farrell had been a lucky recipient of one of the NIH's clinical associate and research trainee positions, working there from 1972 to 1977. Di Sant'Agnese had been impressed by Farrell's graduate studies in vitamin E deficiency and recruited him to the NIH to explore both the impact of vitamin E deficiency in children with CF, and what the treatment should be. Di Sant'Agnese was bothered that physicians were aggressively pushing vitamin E supplements without proof that they were effective. But that was the case with many treatments at the CF centers in the 1960s, which were based on hunches and observations, not fact.

Di Sant'Agnese refused to follow robotically. I don't think you should treat something in a patient until you understand it, di Sant'Agnese once told Farrell quietly as they sat in the lab together. "You should always study a problem before you try to manage it clinically."

Farrell thrived at the NIH, learning from di Sant'Agnese's grandfatherly bedside manner and astute clinical insights. Di Sant'Agnese was often underestimated by colleagues, who were fooled by his modest demeanor, his frailty, and his partially paralyzed right arm, but Farrell knew none of these physical deficits touched di Sant'Agnese's vigorous mind and thorough clinical and analytical skills. Farrell also benefited from the man's national and international network, and frequently met renowned scientists visiting the NIH. That's where he met di Sant'Agnese's formidable friend and colleague Harry Shwachman.

At that point Shwachman had already mainstreamed di Sant'Agnese's sweat test and become the leading CF physician in the country. He was self-assured and quick to jump in with a new treatment if he suspected it made sense—the opposite of di Sant'Agnese, a devotee of evidence-based medicine. Shwachman was in charge of the clinical labs at Boston Children's Hospital and passionate about establishing a cystic fibrosis diagnosis as early as possible, because he wanted to follow LeRoy Matthews's Cleveland model: aggressive treatment with everything. Warren Warwick's registry had confirmed that Matthews's approach worked. And Shwachman, who saw hundreds of children each year at Boston Children's Hospital, had shown[10] that infants diagnosed before three months of age, versus those diagnosed later, had fewer infections and healthier, relatively longer lives. But di Sant'Agnese wasn't a fan of throwing "everything but the kitchen sink" at the disease without testing the interventions first.

Shwachman and di Sant'Agnese frequently attended the same conferences, and though they had very different styles and demeanors, they had a deep respect for each other. At a Washington, DC, meeting at the National Academy of Sciences in 1975, Farrell was sandwiched between the two physicians as they discussed the pros and cons of the sweat test. It was only as good as the physician who was astute enough to order a sweat test, mused di Sant'Agnese. The test might be good technically, but the flaw was that they were relying on family physicians and pediatricians to realize the child might have CF. During a break in the conference, Shwachman looked Farrell in the eye and told him, "You must get into screening newborns for CF."

The comment lodged in Farrell's mind. Then, in 1976, while pondering a job offer to join the neonatology unit and direct the CF clinic at the University of Wisconsin in Madison, he remembered Shwachman's perspective that newborn screening was the only way to get this disease under control—diagnosing patients early enough to provide aggressive treatment. With di Sant'Agnese whispering in one ear about the flawed sweat test and the importance of testing interventions before treating, and Shwachman urging him to explore newborn screening in the other, it was, perhaps, inevitable that Farrell would eventually launch the study he did.

When Farrell accepted the University of Wisconsin's offer, the difference he saw between the sick children in the neonatology unit and the kids he saw in the CF center was stark, revealing the urgent need for screening newborns. In the neonatal unit, babies with meconium ileus—the sticky, tar-like intestinal blockages indicative of cystic fibrosis—were diagnosed immediately, treated aggressively, and remained healthy longer. Adjustments were made to their diet, like adding enzymes at each meal—as Kathy had done with Joey's applesauce—to help these children digest and extract nutrients from their food. And physical therapy began before bacteria infected the lungs. By contrast, in the CF clinic he saw a steady parade of children referred by physicians who, despite being what he considered well trained, had failed to diagnose the disease earlier. By the time these children saw him, they were weak and malnourished, their infected, clogged lungs hungry for air. Just as di Sant'Agnese had predicted, these pediatricians hadn't recognized CF, making the availability of the sweat test irrelevant.

However, a new test, developed in 1979,[11] made it possible to figure out if newborns might have CF without relying on the physician to
order the sweat test. Doctors could instead analyze the blood for a protein called immunoreactive trypsinogen[12] (IRT), which was made in the
pancreas. In infants with CF, the levels were higher than normal. Physicians in both the Australian state of New South Wales and in Colorado
had already embraced it for newborn screening. It seemed to Farrell like
a reliable screening tool. Perhaps that was the method he should use in
his clinical trial to test the effectiveness of early diagnosis?

The trial Farrell wanted to perform was complicated; nothing similar had ever been attempted for any disease. Gearing up for it took
several years. Once he had the necessary approvals, Wisconsin's state
division of health developed a brochure on newborn screening that
pediatricians and obstetricians gave to expectant mothers and new parents to obtain consent for the trial—which, like the PKU test, involved
pricking their baby's heel, collecting blood, and analyzing it. Letters
were sent to every primary care physician in Wisconsin, and the trial
was also publicized in the *Wisconsin Medical Journal.* The plan was for
all newborns to receive the IRT test soon after birth, but the results for
only half the newborns would be revealed. However there were ethical
issues[13] to consider. Did the withholding of neonatal screening from the
control group violate any ethical obligation? Were these children being
deprived of a test that could improve their health, or even save their
life? Ultimately Farrell's team, the University of Wisconsin Institutional
Review Board, the Cystic Fibrosis Foundation, and the National Institutes of Health decided the control group wasn't disadvantaged, because
the value of the screen was unproven.

Farrell calculated that if one in every 2,000 babies had cystic fibrosis, as was generally believed at the time, then it would take four or so
years before the trial identified enough children to begin comparing
whether early diagnosis had a tangible impact. The trial would have
taken less time had Farrell included multiple states, but the complexity
of doing so outweighed the benefits. The trial was restricted to hospitals in Wisconsin and began on April 15, 1985. Between 1985 and 1994,
650,341 babies were born in the state. All of these children received the
IRT test. Half were randomly assigned to the early diagnosis group. Of
these, when a child tested positive, the result was confirmed with a sweat
test when the child was three months old.[14] For the control group, the
results of the screening test were banked in a computer until the child

turned four years old.[15] Cystic fibrosis was only diagnosed in the control group if a child's pediatrician recognized the disease through hallmark symptoms like lung infections or digestive issues, because someone in their family suffered from it, or if, when a child turned four and the banked screening result was unveiled, it was positive.

It took nine years to collect all the data the scientists needed to draw reliable conclusions—longer than expected because the frequency of babies born with CF during that time was closer to one in 4,000—and another three, until 1997, before Farrell published the dramatic results.[16] In the early-testing group, cystic fibrosis was detected in the infants before they reached twelve weeks old; they were then given special diets to keep them well fed and medications and therapy to treat pulmonary disease. In the other group, the children weren't diagnosed until much later—at seventy-two weeks on average, or more than a year after the other children. The advantages of testing early were indisputable. The kids diagnosed at three months were better nourished, taller, and heavier, with larger head circumferences—a measure of normal development. Their lungs were also in better shape, with less tissue damage. In the control group the damage was already visible by the time the children were diagnosed. These children were malnourished, smaller, and weaker—and this caused health disparities that persisted a decade later.

Even though the results were convincing, it took a while to implement change. Fortunately, the CDC and CFF became vociferous advocates and a national plan was implemented. In 2005, eight years after Farrell's study was published, only five states required hospitals to screen newborn children for cystic fibrosis. But by December 1, 2009, all fifty states, and the District of Columbia, had passed legislation that every newborn be screened.

Back in 1985, when Farrell was about to launch his ambitious study, he was paying close attention to the competitive gene hunt, by then an international affair. Although the IRT test was good, no diagnostic test would be more definitive than detecting the mutation in the relevant gene. A DNA test—whether for screening newborns, as the IRT test did, or fetuses developing in the uterus, as the new prenatal marker test did—would be the ultimate test of whether a patient had the CF mutation, leaving no ambiguity in the diagnosis.

CHAPTER 22

Michigan
1985

The man with a new idea is a Crank until the idea succeeds.

—Mark Twain

The groundbreaking publications about the markers on chromosome 7 that flanked the CF gene from the labs of Lap-Chee Tsui in Canada, Raymond White in the US, and Robert Williamson in the UK meant that anyone could now join the search. If chromosome 7 were a transcontinental highway crossing the southern United States from Santa Monica, California, to Jacksonville, Florida, the scientists had narrowed the hunt to a stretch in Texas, somewhere between San Antonio and Houston. The genetic markers, like green highway signs, had defined an interval, and the gene lay somewhere in between.

In late 1985, after the papers' publication, Tsui reached out to his friend and colleague Aravinda Chakravarti at Tsui's alma mater, the University of Pittsburgh, to talk about refining his gene-hunting approach—tracking the gene using clusters of associated markers that had been passed down from generation to generation since the mutation first arose. Chakravarti had originally moved to America to earn his PhD in human genetics from the University of Texas Health Science Center at Houston after a rich mathematical undergraduate education at the legendary Indian Statistical Institute. Then, in 1979, he moved to Seattle for what turned out to be a fleeting and unpleasant postdoctoral fellowship, which he quit before applying for teaching positions around the country. In the end, an accomplished Chinese geneticist, Ching Chun Li (known as C.C.), who had led the department of biostatistics at the University of Pittsburgh for several years, finally offered Chakravarti a job teaching.[1] C.C. took Chakravarti under his wing, training him in the art of teaching and also collaborating and publishing papers with him. Chakravarti eventually began developing mathematical methodologies for his gene- and mutation-hunting colleagues studying hemoglobin disorders. At a genetics meeting in 1985, he met Tsui, who had at that point been working on CF for five years, and the two became close friends.

Tsui was still searching for the gene using polymorphic markers, but because Chakravarti knew much more about population genetics than Tsui did, Tsui wanted to run the idea by him for reassurance that the method would work. By coincidence, Chakravarti had studied a similar problem about markers and the beta globin gene, which codes one of the proteins that comprises hemoglobin.[2]

Chakravarti encouraged Tsui, telling him that, yes, he should be able to locate the CF gene by finding markers that were always associated with the disease. The reason, Chakravarti explained, was that markers carried historical information. Each one arose in an ancestor long ago and got passed from generation to generation. But as time passed, the share of DNA from those distant ancestors kept shrinking: every person got half their DNA from their mother and half from their father; one-quarter of their DNA from each of their four grandparents; an eighth from each of their eight great-grandparents, and so on. People with cystic fibrosis, Chakravarti suspected, were all descendants of a single individual in whom the mutation first arose—a "CF Adam" or "CF Eve." If so, every living CF patient should share that tiny slice of DNA carrying that mutation—along with the surrounding markers that were present when the mutation first arose. Find the markers, and you could find the gene.

The markers Tsui, White, and Williamson had found narrowed the stretch to 1.5 million nucleotides, but these scientists still had a long way to go. The only predictable strategy for finding a gene was to read the DNA in the interval between the markers, decoding the DNA one nucleotide at a time. Researchers had to decrypt in this stepwise fashion, A to G to A and so on, for hundreds of thousands of nucleotides—an approach called chromosome walking—until they came across a genetic sequence that resembled a gene.

Dr. Francis Collins, a new faculty member at the University of Michigan, had a different idea. Rather than walking in a straight line from the easternmost and westernmost markers toward each other, what if you started *jumping* from both ends? It was a strategy he and his mentor, Sherman Weissman, had conceived while Collins was at Yale University. By jumping, they could cover more ground and, theoretically, find the gene faster.

Collins, tall and lanky, with a mop-top Beatles haircut, Buddy Holly glasses, and a chevron mustache, hadn't begun his career in biology. His PhD project at Yale University in New Haven, Connecticut, had focused on quantum mechanics,[3] which, while intellectually stimulating, struck him as glaringly disconnected from issues of human concern. That wasn't the case upstairs in the laboratory of Donald Crothers,[4] whose investigations into the three-dimensional structure and function of DNA[5] were at the forefront of a new field called molecular biology.

In the early morning hours, when many night-owl graduate students filled the labs, one of Crothers's students would regale Collins with spellbinding tales of cutting and pasting DNA fragments from one species into another. The profound applications of recombinant DNA technology had immediate implications for understanding life and treating disease. Suddenly quantum mechanics no longer seemed quite as interesting to Collins.

This surprised him. His interest in biology was, he thought, permanently blunted thanks to a high school biology course that involved little more than rote memorization. But after the other grad student's fantastic stories had roused Collins's curiosity, he cautiously sat in on a biochemistry course—and was immediately amazed by the elegant principle-based understanding of life that was emerging from experimental research.

The alphabet of life was made up of four simple chemicals—A, C, G, T—that spelled out the instructions for building organisms. Precise rules governed how this information was transmitted from one generation to the next. The more Collins learned, the more he felt magnetically drawn toward a new pole. Before long, he had applied to medical school, convinced that even if he was ill-suited to medical research— his true interest—he could at least fall back on clinical skills and have a meaningful career.

He was accepted to the University of North Carolina's medical school, starting classes in 1973 while completing his quantum mechanics PhD thesis on the weekends. His distaste for memorization made him particularly anxious about his first courses—neuroscience and cell biology—but learning about the human body was hypnotizing. Still,

Collins's final metamorphosis into a biologist didn't occur until December. That's when an austere, seemingly unapproachable pediatrician named Dr. Henry Neil Kirkman gave six lectures about human genetics. He was not a charismatic teacher, but his lectures captivated because he did something none of Collins's professors had done: he incorporated his patients into the lecture, revealing an inspiring bedside manner. Collins was moved by this model physician: knowledgeable and professional, but also warm.

Among the patients Dr. Kirkman brought to class was an eloquent adolescent boy with sickle cell disease, who described the excruciating pain and the many life-threatening crises he'd suffered because of the inherited blood malady. There was a little girl with Down syndrome, a genetic disorder caused by an extra copy of chromosome 21; and a third child who suffered from neurofibromatosis, a condition in which tumors sprouted throughout her nervous system. Thanks to Dr. Kirkman, Collins discovered the sweet spot for his research career: medicine rooted in genes. In genetics, he could intertwine his gift for mathematics with his passion for biology, calculating the probabilities of genetic diseases from basic principles. And he could interact with patients—a human connection he cherished.

It wasn't surprising that Collins was enraptured with all the possible ways genetics could be leveraged in future medicinal treatments and therapies. By 1977, as Collins was nearing graduation, genetics and recombinant DNA technology were completing a decade of rapid innovation and evolution. In 1968,[6] scientists discovered a type of protein that could cut DNA at specific sequences. By 1972,[7] Stanley Cohen and Herbert Boyer, scientists at Stanford and the University of California, San Francisco, were using these molecular scissors to cut slices of DNA, each carrying different genetic traits, and then join them with the DNA of bacteria to endow the microbes with novel qualities.[8] In early 1976, Boyer cofounded the first genetic engineering company, Genentech,[9] which soon identified the human gene for insulin, cut and pasted it into a bacterial cell, and began mass-producing human insulin—the first therapeutic medicine made from recombinant DNA. A year later, in 1977, two teams of chemists had each developed new and different chemical procedures for reading DNA.[10]

In short, in less than a decade, scientists had learned to read DNA, identify specific genes, and cut and paste DNA together in unique arrangements to create one-of-a-kind human-made organisms. This capacity to rewrite the genetic blueprint and create newfangled life forms would change the course of biology, human history, and life on earth. By the time Collins graduated, the air was thick with talk of applying the new recombinant DNA technology to human genetic disorders—not just predicting who would inherit diseases, but actually doing something about it. That was something Collins was eager to do.

The problem was that, when he was ready to graduate in 1977, there was not a huge selection of universities with departments for either human or medical genetics, because until the mid to late 1950s, these fields still suffered from the stigma of the eugenics movement and forced sterilization laws.[11] The other perception stunting the field was that genetic diseases were rare, had little impact on population health, and thus were of limited importance to medicine. It had taken several decades of discovery to begin to overcome those psychological road-blocks. And as those ideas started fading, it was becoming clear that application of genetics—whether to engineer bacteria to manufacture insulin or diagnose genetic diseases in the womb—could dramatically improve human health.

Collins knew that pediatrics was the logical path for a physician interested in medical genetics, because there a doctor could focus on children born with disabilities and inherited disorders. But Collins's passion was adult medicine.

Seeking to figure out where he might fit into this amorphous discipline, he began attending the annual meetings of the American Society of Human Genetics, to get a feel for the landscape of this rapidly evolving field. His first was in Vancouver in 1978,[12] where he was captivated by the discoveries of geneticist Yuet Wai Kan, who had just discovered how to determine whether a fetus would inherit sickle cell disease[13]—work that would inspire Tsui's gene hunting strategy a year or two later and lay the foundation for mapping the entire human genome. Collins immediately started looking for medical genetics programs where he could learn laboratory techniques for studying DNA.

Yale University accepted him for its medical genetics training program, but soon after arriving he became painfully aware that he was unprepared—he had never worked in a biology lab. He'd never held a pipette, used an autoclave, or grown bacteria in a dish. To help him choose a project to work on, he dropped by the offices of various faculty, who described their projects. Rather than helping him choose, it just confused him more. Eventually, a faculty member told him to just pick a lab and dive in.

Collins chose the lab of Sherman Weissman, a distinguished biologist and one of the cosigners of a famous 1974 letter to *Science*[14] that urged caution in this new field, in which biologists could merge the DNA of species separated by billions of years of evolution and write the code of life from scratch. Weissman's team had won the race to decode the DNA sequence of SV40, a cancer-causing virus that infected humans and primates, publishing his May 1978 paper just a week before the competing team.[15] After a couple of Collins's exciting, high-risk projects at the lab failed—a combination of inadequate guidance and Collins's lack of experience—he was ready to quit and leave Yale, until Weissman introduced him to a colleague who worked with him on blood disorders, French Canadian hematologist Bernie Forget. A welcoming, generous mentor[16] who had learned the basics of molecular biology in France before coming to Yale, Forget convinced Collins to remain at Yale and work in his lab instead.

There, in March 1982, Collins began his hemoglobin project. At the time, the hemoglobin protein and gene were the most thoroughly studied on the planet, and Forget suggested Collins investigate the cause of a benign condition called hereditary persistence of fetal hemoglobin. In most people, the body switches from producing fetal hemoglobin to adult hemoglobin at birth. In people with this condition, however, the fetal hemoglobin gene never turns off. For healthy individuals, this means very little; they simply produced both forms of hemoglobin. But for people with sickle cell disease, or other blood disorders with abnormal adult hemoglobin, the persistence of fetal hemoglobin was a gift—the two conditions together seemed to cancel each other out.

Forget suspected that the key to finding the root of the disease was studying the DNA of families in which the fetal hemoglobin gene didn't turn off. He arranged for the collection of blood samples from Jamaican and Greek families, and Collins began sequencing the samples' DNA. A team at the University of Wisconsin[17] had already determined the location of the fetal-hemoglobin gene, so Collins was able to go right to the gene and read the code of three patients with high levels of fetal hemoglobin. He noticed they shared a change in a single letter in their DNA—a C was changed to G—unlike more than one hundred individuals without the condition. Most intriguing was that this mutation wasn't in the gene itself. Rather, it was in a region of DNA adjacent to the gene called the promoter—a "switch" that turned the gene on or off. Collins had discovered that this tiny mutation—a substitution of one nucleotide for another—flipped the genetic switch, leaving the fetal hemoglobin on, which explained its continued presence in these individuals' blood.[18] It was an important discovery: one of the first genetic variations discovered in the switch region of a gene that had a profound impact on a person's health.

Now that Collins had some experience, Weissman felt Collins was up for a tougher technical challenge—one that had nothing to do with hemoglobin. The burgeoning field of human genetics was rife with obstacles, but the greatest was figuring out how to travel the genome efficiently. Without this, finding genes was onerous and time consuming—the distances were simply too large.

Until 1985, chromosome walking—either stepping from one chemical nucleotide to the next, or in slightly longer strides of some one- to five-thousand units—was the only way to explore DNA. It was a painfully slow and laborious process. There had to be a way to leapfrog over huge expanses of genetic terrain.

Weissman had been wrangling with the challenge of moving around the genome long before Collins came to Yale, and had previously written a grant application to support the development of a new "jumping" technology that would enable rapid transit on chromosomes. His approach was inspired by the way that, when a portion of a chromosome—or any type of linear arrangement—is deleted, two distant points are brought closer together. For example, let's say you have a chromosome with seven markers, K through Q:

K_____L_____M_____N_____O_____P_____Q

If you clip out the region between L and P, then K and Q are brought closer together:

K_____LP_____Q

Another way to make the jump, instead of clipping out the center, was to transform linear DNA into DNA circles.

In creating the DNA circle, K and Q, which on the chromosome were separated by a string of 100,000 nucleotides, are brought next to each other. Put very simply, by decoding just the new DNA junction where K and Q were connected, scientists could jump from reading the DNA surrounding the K marker 100,000 steps "forward" to the DNA surrounding the Q marker. This technique, Collins and Weissman believed, would allow the scientists to cover a lot of genetic ground much more quickly.

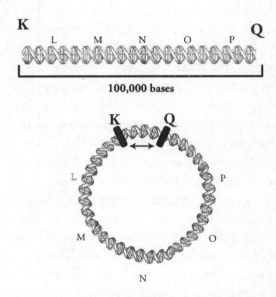

When a 100,000-nucleotide string of DNA is arranged into a circle, it s possible to "jump" over the intervening sequence and travel faster from one end to the other.

The risk, of course, was that they might, in the process, jump over the sought-after gene by skipping the L-to-P segment. But there was a potential safeguard: once you made a jump to a new location, you could use genetic markers, like the ones Tsui had used, to determine approximately how many more steps there were until you reached the gene—or if you had jumped over it entirely, and if so, how far you needed to backtrack.

Collins spent his last year at Yale creating a library of DNA circles that he could use to jump along a chromosome. He and Weissman published a description of the chromosome-jumping technique,[19] pitching it as a potentially powerful tool for gene hunters to travel along a stretch of DNA toward their target gene without reading every single letter of the genetic code. But it was untested. To prove the new technique valuable, Collins still needed to hunt with it for a specific gene.

When Collins left Yale in the summer of 1984, he accepted an offer to join the faculty at the University of Michigan, which in 1956 had been one of the first US universities to establish a human genetics department.[20] He had been recruited by a physician named William Kelley, who himself had been recently recruited, in 1975,[21] to chair the Department of Internal Medicine at the School of Medicine. Kelley was a charismatic and dynamic leader with an eye for young talent and a gift for acquiring funding. He'd made a flashy debut on the genetics scene when he and his colleagues discovered in 1969 that an inherited neurological disorder called Lesch–Nyhan syndrome, which caused intellectual disorders, muscle spasms, and self-mutilation, was triggered by a missing enzyme. Reporting on their research, Kelley and his colleagues explained how the lack of that specific enzyme could cause abnormal compulsive behavior—which suggested that, in the future, this and other behavioral disorders might be treatable, even curable.

The discovery cemented Kelley's belief that the future of medicine lay in molecular genetics and gene therapy. He planned to fill his new institute on the Ann Arbor campus with experts leading these new fields. At the time, though, Michigan's medical school was second rate, and Kelley failed to convince top-tier researchers to accept posts. So he switched his strategy. Instead of hiring the most senior scientists, he decided to hire a crack team of ten promising but untested young researchers to fulfill his vision instead. Kelley and Thomas Gehlerter, whom Kelley had invited to lead his new division of medical genetics, had heard about Collins from colleagues at Yale and were impressed by his work on fetal hemoglobin and chromosome jumping, a technique they anticipated would be valuable in mapping genes and traits, and offered Collins a position. After Collins accepted, Kelley convinced the Howard Hughes Medical Institute (HHMI) to fund his new facility. It was a controversial move on the HHMI's part—they typically only funded senior investigators—but Kelley assured them Collins and his future recruits would be a good bet.

In addition to Collins, Kelley also recruited cardiologist and molecular geneticist Jeffrey Leiden, who would go on to become the CEO of Vertex Pharmaceuticals; Craig Thompson, who later led Memorial Sloan Kettering Cancer Center; Gary Nabel, who would launch the Vaccine Research Center (VRC) at the National Institute of Allergy and Infectious Diseases before becoming chief scientific officer at Sanofi; and James Wilson, who would become a pioneer of gene therapy. In the mid-1980s, there was no inkling of the heights these recruits would later attain, but Kelley had the sense that these researchers were ambitious, hungry to make a name for themselves, and willing to dive into high-risk projects—the kind that could yield the sorts of transformative science that would boost their careers and the reputation of the institute and university. It was an agreeable setup for these young men, all in their late twenties and early thirties. They became friends, and their growing families mingled. They supported one another as they figured out how to submit papers to journals, write grants, recruit graduate students not much younger than themselves, and run their labs.

Soon after his arrival at the University of Michigan, Collins recruited his first graduate student, Mitchell Drumm, whom he met during a fierce faculty-student volleyball tournament. It was a fortuitous meeting. Collins needed graduate students to test that his chromosome jumping method was a faster way to gene-hunt, and Drumm needed to choose a lab and mentor to guide him through his PhD. Drumm signed on to Collins's lab for a three-month trial. The two hit it off and began what would become a five-year collaboration and a lifelong friendship.

Drumm grew up in the small town of New Philadelphia, Ohio, located on the Tuscarawas River. As a child, he loved wandering on a friend's farm or in the woods, and animals were his passion. He initially planned on becoming a veterinarian, but after being assigned in college to the large animal pre-vet program, which focused on cows, pigs, and sheep, Drumm didn't find it quite as gripping as he had hoped, and switched his undergraduate major to genetics.

He was, in a way, primed for that. When he was growing up, his mother worked at a school for children labeled as having intellectual disabilities. The town was small—everyone knew everyone—and many of his mother's students would come by the house daily: children with Down syndrome, fragile X syndrome, cerebral palsy, and other development problems. Some would even spend the night if their parents were struggling. Later, during the summers, Drumm was a lifeguard at the swimming pool at the city park, where he worked with many of the same children, teaching them how to swim. Such close contact with these children piqued his curiosity, making him wonder why children with Down syndrome looked more similar to each other than they did to their own family members.

When he chose to study human genetics at the University of Michigan for graduate school and began working in Collins's lab, he knew he had found his calling. He loved tinkering with equipment, building his own, and learning to manipulate the molecules of life during those early days of the field.

Collins's lab was small, a single eight-hundred-square-foot room the size of a small New York City studio with painted cinder block walls, four laboratory benches, and an office to the side. Through the window there was a view of the forest, which allowed Drumm to watch the wind rustle the leaves and gauge surfing conditions on Silver Lake.

When Drumm joined the lab, his initial project was to work with Collins to create a "jumping library" of the genome. If that worked, he would then test the value of the library by using it to find a disease-causing gene. However, while chromosome jumping was a method that sounded good and was elegant on paper, it was trickier to execute in the lab.

Crudely speaking, creating the library was a matter of taking DNA from normal human cells, chopping it up into stretches approximately 100,000 letters long, and then creating DNA circles. But this was technically tricky. After the DNA had been cut up, it was prone to breaking into smaller fragments that would then rejoin in an order that differed from their original sequence along the chromosome. The DNA was also vulnerable to being "eaten up" by contaminating enzymes from the cells.

Collins insisted on being heavily involved in making the DNA library, which meant instructing and coaching as Drumm executed each step. A poorly constructed library would be missing chunks of chromosomes—and so would doom the entire hunt, for whatever gene they eventually chose to pursue, from the start.

Making the library involved quite a bit of physics—Collins's forte—as well as engineering and designing new devices, Drumm's strength. To do this they used electrical fields to manipulate the DNA pieces without breaking them. Once the long, snaking DNA molecules had been tamed into circles, Collins and Drumm used standard molecular biology techniques to clip out the junctions, which they'd later use to fish out the piece they needed. In the end, they had a collection of more than three million "jumping junctions," representing the entire human genome.

Creating this library took more than a year. And once they were sure it was complete—before Drumm could test chromosome jumping—they needed to select a gene to hunt for, and a point on a particular chromosome to begin the search. Collins had a list of diseases that intrigued him: Huntington's on chromosome 4; neurofibromatosis, which hadn't yet been tied to a specific chromosome but which he'd encountered in patients during his weekly shifts in the institute's genetics clinic. Then there was cystic fibrosis.

Cystic fibrosis resonated with Drumm. A neighbor in his hometown of New Philadelphia had just had a baby who had been diagnosed with the disease. When Drumm learned more about CF, he told Collins that he was particularly interested in working on it. Collins was equally intrigued with pursuing the disease after recalling an incident from his residency at North Carolina Memorial Hospital, where he'd met a twenty-year-old nurse with chronic bronchitis and pneumonias who had just tested positive on her sweat test. Collins recalled how surprised he had been that the woman was diagnosed so late in life and remained as healthy as she had. He and Drumm both agreed that CF was a good target for them.

By the end of 1985, Collins knew from the three publications in *Nature* that Tsui, White, and Williamson had identified markers on chromosome 7 that flanked the gene and narrowed the search zone to just 1.5 million nucleotides—still a huge distance. Now, however, with the neighborhood harboring the gene nicely defined, Collins and Drumm decided that, with the jumping technique, they had a shot at finding the gene first.

Using the closest marker to date, met, which White's Utah lab had discovered, Drumm fished out a DNA junction that would provide them with a starting point for their walking. But it was slow going. Collins, Drumm, and a technician named Jeff Cole all worked together at the laboratory bench, tag-teaming the experiments.

After many failures, and more than eighteen months, their work paid off. Although they hadn't found the gene, they had found a new marker, which they named CF63, that was 100,000 nucleotides closer to the gene than any marker previously published. Collins and Drumm published an article describing that successful jump in *Science*[22] on February 27, 1987, which Collins concluded by writing that chromosome jumping should be useful for identifying genes for all human diseases, and that, as the maps of chromosomes acquired a denser collection of markers, the technique could become even more helpful.[23]

The paper was widely read and appreciated by many in the genehunting community. One person who was particularly intrigued by Collins's work was Bob Beall of the Cystic Fibrosis Foundation, who called soon after reading to invite Collins to an exclusive Guidance,

Action, and Projection (GAP) conference on CF that the foundation held annually, so Beall and Robert Dresing could get to know him better.

Within days of meeting him, Beall and Dresing invited themselves to his Michigan lab, and just a few hours after arriving and talking there, they enthusiastically agreed to fund Collins's research, offering to help in any way possible. They became regular visitors, dropping by every few months, and Beall would ring him up on the telephone and leave messages like a neglected mother asking her son to call, adding, "No pressure!" Collins learned quickly that "the Bobs" could supply resources to keep his experiments moving at a fast clip. Unlike with NIH funding, which took months to approve, Collins only had to explain what he needed and the funds would appear. That was critical because his lab was small, their resources stretched thin, and he didn't have money to work specifically on CF— just general funds for chromosome jumping. So the foundation became a vital source of support.

As Collins and his team worked, expectations were continuing to grow in the larger scientific community that identifying disease-causing genes would offer logical routes to cures, either by replacing the sick gene with a healthy one, or providing the body with a needed protein that the damaged gene couldn't produce.

Diseases like cystic fibrosis, caused by a single gene with a predictable inheritance pattern, provided an opportunity to change the course of human medicine. If scientists could find the gene that caused CF, that breakthrough might provide a model for treating, or even curing, hundreds of more complex conditions for which neither genes nor therapies had been discovered.

CHAPTER 23

Joey's Long Goodbye
1986

Old people die with achievements, memories. Children
die with opportunities, dreams. They carry the hopes
of all of us when they go off. Probably a child's death
is more intolerable for us than for the child.

—Frank Deford

It was November 1986, and Kathy O'Donnell's belly was round. It wasn't huge yet, but it was clear that baby Kate was growing well. When Joe arrived home one cool fall evening, twelve-year-old Joey was sitting at the kitchen counter doing his math. Normalcy was something that Joe and Kathy had always emphasized in Joey's life. He was a normal little boy. He loved to play baseball, video games, and soccer. He hung out with his gang of close buddies. Played in the rain. Had crushes on girls. Despite his health, he studied hard—because that's what he was expected to do. And every year Joe would get his own copy of that year's math book so that he and Joey could work through it together. Joey hated homework. Kathy hated Joe's adherence to it. But Joe kept it up; it only took half an hour each day, and every year Joey would be ahead of the class.

Joey hadn't been to school in a month, and he was on oxygen all the time. Joe stared at his son: the dark circles under his eyes, the shallow breathing, the way he was slumped over his textbook. But Joe had kept insisting on homework—because it was normal.

That night, recognizing Joey's exhaustion, he let his guard down. "Okay, forget that," he told Joey. "You don't need to finish it."

When Joey had started getting dramatically sicker a few years earlier, after suffering from congestive heart failure, seeing children

185

at the hospital with CF that were much sicker than Joey would make Kathy panic, thinking that was going to happen to him. But Dr. Lapey would always try to cheer her up. Kathy knew that Joey had survived this long in part because she and Joe had the means to give Joey all the care he needed. Most other families didn't have that luxury. To keep herself moving forward as his health declined, she told herself that as long as she continued to care for him, as diligently as she always had, he would be OK. But this year it had been harder and harder to believe that.

In the past year, Joey had been admitted to Mass General four times, the most visits since the year he was first diagnosed. Each time he came home, his lungs would be cleaned out and initially he was breathing better. But three or four days later, the infections were back. Nothing was helping. That year, unlike any one before, Kathy saw that Joey had been so sick that it was difficult for him to be happy. Then, at the beginning of November, Joey went in to Mass General for a fifth time, and stayed for a week.

As he had during every hospital stay of Joey's, Lapey gave Joey oxygen, physical therapy, and aerosol treatments to help liquefy the junk in his lungs, as well as the one thing he couldn't easily get at home: IV antibiotics. At the time there was a new antibiotic in clinical trials called ciprofloxacin. It hadn't yet been approved by the FDA, but Joey had received it during his previous few stays for compassionate use, a drug of last resort; the bacteria in his lungs had never met ciprofloxacin before and were vulnerable to it. Each time, Joey's health had stabilized, he'd perked up, and he was able to leave the hospital. The same thing happened again this time.

But Joey had barely been home two weeks when both Joe and Kathy knew it was time to take him back. He was spending most of the day sleeping and his skin was tinged blue, signs that the oxygen he was breathing wasn't getting into his lungs and the carbon dioxide levels in his blood were rising, making him drowsy. When he slept his breathing was quiet; the mucus was staying put. His body was working hard, his chest heaving as he tried to bring air in.

On November 21, 1986, Joe and Kathy drove Joey back to the hospital, and Lapey readmitted him. While the geneticist's prediction of Joey's "slow and painful" death remained as fresh as the day it was spoken, Joey had cheated death multiple times. His miraculous recoveries fostered hope. Maybe, thought Joe, this would never end. Maybe his family was different. Maybe they would thwart the ugly legacy of this disease. He and Kathy were optimistic. But as Joey had gotten sicker, Joe began bargaining with God. "Give Joey an extra year, a week, another day"—until he was simply praying for his little boy not to suffer any more.

Lapey often said that Joey had nine lives. Whenever he administered antibiotics, Joey became reenergized and got better. If he arrived at the hospital sick and struggling to breathe, oxygen would bring him back to life. Many times Joey had been so oxygen starved that he turned blue, but after Lapey's usual treatments, Joey would run around Burnham 4 the next day like nothing had happened—exploding with more life than a whole handful of healthy kids his age.

But this time was different. This time, Joey wasn't getting better. At twelve and a half years old, he weighed barely fifty pounds. He was bluish again, hot with fever, and working hard to breathe. The clobbering doses of antibiotics, even the new one, were no longer able to extinguish the escalating bacterial infections destroying his lung tissue. The mucus in Joey's lungs was so dense and immovable that almost no oxygen could penetrate.

On the night of November 22, Joe and Kathy broke from ritual. Joe felt uneasy and didn't want to leave Joey alone, so he decided to stay at the hospital while Kathy went home for a few hours. And the next day, on November 23, Dr. Lapey told Joe and Kathy that the time had come. Joey had been asleep almost his entire stay. His body was starving for air that no one could deliver, and his chest silently heaved. The only option left was a ventilator, which meant Joey would need to spend the rest of his life in an intensive care unit amid a tangle of tubes and machinery—and even that might buy him only a few more days. The three of them had discussed that option long before today and no one, neither Lapey nor Joe nor Kathy, wanted that for Joey. But there was one thing that Lapey could offer Joey—a comfortable death free from fear.

Lapey was unenviably practiced in orchestrating these last moments. But with Joey, to whom he had been so close for almost a dozen years, it was particularly heart wrenching. Making sure Joey was free of pain and reassuring him he was okay, just sleepy, was especially important for the childlike preteen, who was brave and even now had faith that he would recover and go back to playing with his friends. Just a few months earlier, a reporter had asked Joey what he would tell a friend if they were diagnosed with cystic fibrosis. Joey had answered, "I would tell them to hang in there because they are finding new discoveries, new things, every other day."

Among all the children Dr. Lapey had treated, Joey stood out. He had never seen Joey cry or feel sorry for himself. He never said, "Why me?" He never gave up.[1] He always believed he would get better and that, one day, there would be a cure. It was Lapey's duty as Joey's doctor, and friend, never to shake that faith. So, on that last day, Lapey told him, "I think we're going to give you some medication because you're tired. You really, really need to sleep so that you'll be stronger." With that, he gave Joey morphine to let his exhausted little body rest.

For a healthy child, morphine is relaxing. For Joey, it was not. The morphine calmed his breathing and absolved his body of the drive to expel carbon dioxide, allowing it to rise quietly in the blood and gently guide his body into a coma. Cradled in the safety of Joe's arms, holding his mother's hand, the carbon dioxide naturally lulled him to sleep as they whispered, "We love you, Joey."

"I love you guys, too," Joey answered.

"Are you in any pain?" Joe asked him quietly.

"Uh-uh," Joey murmured, shaking his head. "I'm just really tired. I'll see you guys tomorrow."

At five thirty that afternoon, in his parents' arms, he peacefully passed away.

JOEY'S DEATH GOUGED A HOLE IN THE HEART OF HIS BELMONT COMMUnity. The Mass at St. Joseph's Church on Wednesday, November 26, just before Thanksgiving, drew more than a thousand: doctors and nurses from the hospital, other families whose children had CF, volunteers from the local foundation chapter, friends and family from all over the state, teachers and families from his elementary and middle schools, and a river of teary children.

They packed the church, kneeling in the pews, as Father Rodney Copp delivered the Mass from behind Joey's tiny coffin, which rested in front of the altar.

Neither Joe nor Kathy thought they would lose their boy so soon, and many friends and family thought the O'Donnells would never lose the fight. Their determination and Joey's verve, his charisma and his joy for life, fooled everyone. But in the end, his tiny body just gave out.

When it came to the foundation, his business, and his family, Joe always led the way. On this day, he buckled. He asked Paul Del Rossi, Joey's "Uncle" Del, to give the eulogy.

Del told the congregation of a recent moment Joey had shared with his parents, where Joey had told them that "time on Earth was like the blink of an eye and that heaven would be a special place without therapy and that his parents would meet him there," adding that "they would get there by hyperspace—his favorite trick in one of his many video games." And Del shared a poem, "How Many? How Much?" by Shel Silverstein about living and loving to the fullest, and being loved in return.

"By these measures," said Del, "our friend Joey lived his *many* good days with a magnitude of love for each and every one of us."

THE O'DONNELLS WERE INCONSOLABLE FOR MANY MONTHS. KATHY was so grief-stricken she barely ate, and gained no weight during her final trimester. Joe came home from work early, every day, so that he and Kathy could be together and comfort each other. And every night until Kate was born they went out to dinner alone, just the two of them.

Together, the couple only once visited the Mt. Auburn cemetery where Joey was buried. Witnessing each other's pain and despair wasn't something either of them could bear, and they agreed to visit separately from then on.

They had spent the last twelve years sleeping with one eye open, aware of every breath Joey took as he coughed through every single night of his life. Now, the house was deafeningly silent, and it tore them apart inside. Yet, when they were surrounded by friends and family, *they* consoled others. Kathy invited Joey's friends to their home some weeks after the funeral to talk to them about his death, encouraging the bewildered boys, eyes puffy from crying, to talk and ask questions.

"I didn't even know he was that sick, Mrs. O'Donnell," said Eddie, tears streaming down his face.

"Me neither. I mean, we were just talking and hanging out with him," said Timmy. "He just had that cough like he always did."

"He always came home from the hospital," said Andrew. "I don't understand why it was different this time."

As they left the house, Kathy gave each of them toys—Transformers, Star Wars models, video games—that belonged to Joey, fun things drenched in memories of her son, items that were too painful for her to keep but might provide solace to his friends, things they could enjoy and cherish.

Against all expectations, a few weeks later Joe and Kathy hosted Christmas at their house, as they had done since Joey was two years old. Kathy wanted to keep the tradition. But there was no Christmas tree that year. The decorations were all ones that Joey had made and Kathy couldn't bear to hang them up. Instead, a Christmas tree–shaped arrangement of poinsettias occupied a corner as some twenty family members filled the house, infusing it with the warmth and understanding of the people who had loved Joey the most, and who felt Joey's absence like they did.

JUST SHY OF THREE MONTHS LATER—ON FEBRUARY 25, THREE WEEKS early—Kate was born. Her birth was a milestone moment, not just for Joe and Kathy, but also for Boston's cystic fibrosis community, who all knew the O'Donnells. It was further proof that families with a history of cystic fibrosis now had the option to have healthy children. And Kate was, indeed, healthy, eating and sleeping like Kathy and Joe never knew a baby could. She barely even cried. It was almost as if she could sense Kathy's fragility—as if she knew Kathy couldn't handle it. In the dark shadow of grief, Kathy was amazed that she could feel such joy.

For a few months after Joey's death and then Kate's birth, the O'Donnell home remained a draw for Joey's friends, who would still come over to play with the games the O'Donnells kept after Joey died. Kathy was happy to see them. She knew what they were going through, and that visiting was helpful for them. They had to get through this in some way, too. After Kate was born, Joey's classmates, especially the girls, came to see her, curious whether she resembled their beloved friend.

But after a while, the visits eased off.

Joey's fight was over. But Joe and Kathy's, against the disease that had taken their child, was not.

CHAPTER 24

Mad Pursuit

1987

Medicine is not only a science; it is also an art. It
does not consist of compounding pills and plasters;
it deals with the very processes of life, which must
be understood before they may be guided.

—Paracelsus, a Swiss physician and alchemist

As the O'Donnells struggled to learn to live without Joey, who had
shaped their lives for so long, those searching for the gene they
believed would lead to a cure were feeling a surge of hope. Since publishing the report in *Nature* two years earlier revealing the location of
the CF gene on chromosome 7, Tsui had recruited two new talents to
join his lab and help find the gene itself. His first hire, in 1986, was
Johanna Rommens, a newly minted PhD chemist from New Brunswick
hungry to work on a medical problem. The twenty-six-year-old lacked
experience in human genetics but was eager to learn, and had received
rave reviews from her doctoral advisor.

Tsui's second hire, in 1987, was geneticist Batsheva Kerem, who was
born, raised, and educated in Israel, and had studied genetics at Hebrew
University. Petite and soft spoken, with an intense gaze framed by short-
cropped light brown hair, at just thirty-two years old she already had
publications in two of the world's leading science journals, *Nature*[1] and
Cell[2]—an impressive accomplishment for such a young scientist. Her
husband had just completed a residency in pediatrics, and the couple
was searching the globe for a university and hospital where Kerem could
work in a human genetics lab and her husband could specialize in pediatric pulmonology, treating children with breathing and lung diseases,
including cystic fibrosis.

The couple had written to institutions in Europe and North America, and Tsui was the first to reply. He needed an experienced geneticist to clone the CF gene and Kerem looked like an excellent candidate. After meeting Tsui in Germany at a conference in 1986 to discuss the project, she agreed to move to Toronto. Tsui was also soft spoken and passionate, and he struck Kerem as kind—something she valued as a mother of two children moving to a lab with a highly competitive project.

The SickKids pediatric pulmonology program was also a fortuitous match for her husband, Eitan, who would be just a few floors below treating children and performing medical research in the CF clinic. When Kerem arrived in September 1987, Tsui was stunned to learn she had two young daughters. He had no idea she was a mother and couldn't imagine how she would juggle her lab work with childcare. But he refrained from saying anything, and quickly learned that she was exquisitely efficient. She and Rommens also dovetailed effectively in personality and work habits.

Tsui expected Rommens and Kerem would be key players in helping him find the gene. This was a big project and Tsui needed energetic, talented scientists with the intellectual firepower to generate and interpret all the genetic data. He had several serious competitors: Bob Williamson in London, whose million-pound grant from the CF Trust had bought him a small army of students, postdoctoral researchers, and technicians; Ray White in Utah, who was well funded by the HHMI and also had a large team; and, as of late February 1987, newcomer Francis Collins, who was using a method he helped develop—chromosome jumping—to start looking for the CF gene.

Since his 1985 publication, Tsui had been examining markers from the Canadian families to narrow down the mutation's location. It was slow and grueling work, but the progress was steady.

In April 1987, Tsui was no longer collaborating with Donis-Keller and Collaborative Research, Inc. The agreement between them ended organically after finding the first marker linked to CF, and after they had published two papers together in 1985, there was no further correspondence or collaboration. Free from the company's or anyone else's gag rules, Tsui could now talk freely about his work, and on April 29, 1987, he flew from Toronto to London to give a talk on mapping the cystic fibrosis gene. The audience was the venerable Royal Society—a fellowship of eminent scientists and luminaries.

It was an exciting opportunity and he was keen to share his prog-ress, though he was bleary-eyed and mildly disoriented after the long flight. But shortly before Tsui took the stage to present at Carlton House, home of the Royal Society, Bob Williamson approached him holding out what looked like a preprint of a journal article. Lap-Chee, you ought to see this before other people see it, he told him.

Tsui's entire talk was about the new markers that his team in Toronto had linked to the gene. But a brief glance at Williamson's paper suggested the London team had won the race. The first part of the title read "A Candidate for the Cystic Fibrosis Locus," suggesting that Wil-liamson had nabbed at least part of the rogue gene.[3] Still foggy and jet-lagged, Tsui skimmed Williamson's data—but he didn't see proof Williamson had identified the gene. Yes, the new marker that William-son had found was exceedingly close to the CF gene—that was clear—but it wasn't obvious whether this segment of DNA caused disease. There was no information about the actual DNA, nothing about the potential gene, and no description of the mutation within the gene that caused the disease. To Tsui, it didn't seem as if Williamson had found the mutation.

All Tsui wanted to do was sit down and read the entire paper from start to finish. But he now had to present work that could very well be obsolete. Nevertheless, compartmentalizing his sense of unease, he walked up to the podium and spoke with uncanny equanimity.

Meanwhile, Williamson approached the organizers, requesting a few minutes to show some of his own slides. Talks at the Royal Society were invitation only, so Williamson's request was unorthodox, but the material was appropriate and the organizers acquiesced, allowing him to present three slides. The less-than-five-minute presentation generated a large response and many questions.

Williamson graciously invited Tsui, still reeling from the news, to come to his laboratory where a press conference was happening later that day. Tsui briefly visited, but left before the TV and newspaper report-ers arrived. The next day the papers carried photographs of Williamson and his team holding up a piece of X-ray film showing a sequence of DNA, which appeared as black and white dashes against their bright white lab coats.

The day after that, as Tsui and Williamson sat in a pub located in the basement of St. Mary's Hospital, Tsui pointed to the photo in the paper and teased Williamson, "Is this how you are announcing the

DNA sequence?" Williamson blanched, obviously not yet aware that his sequence was on the front of the newspaper before it had appeared in a journal. Tsui was kidding, but the look on Williamson's face suggested that he feared someone would read the sequence and decode the whole gene before he had a chance to finish. He asked his colleagues to collect all the papers in the bar, a meaningless gesture given there were copies all over London.

That evening, as Tsui traveled to Oxford, he purchased a couple of copies of the newspaper at the train station so that he could get a better look at the photograph that Williamson had whisked away. Tsui didn't think that Williamson had the gene. But many others around the world would, as soon as Williamson's "candidate gene" for cystic fibrosis was published in *Nature*.

As Tsui predicted, Williamson's publication on April 30, 1987, made headlines in America, Australia, Canada, and Britain. The UK's *Guardian* announced, "Scientists have pinpointed the site of the mutant gene that causes cystic fibrosis, the fatal disease which affects 6,000 children and young adults in Britain. A treatment is expected to be generally available within 10 years."[4] In another story, CF Foundation medical director Bob Beall told the Associated Press,[5] "This is the next major step in our ability to identify the gene, and subsequently the basic defect in cystic fibrosis." Robert Dresing, still president of the Cystic Fibrosis Foundation, told a reporter from the *Chicago Tribune*,[6] "This seems to be hot, and there's very high optimism that this segment might well contain the gene . . . If it's true, it could be confirmed in . . . a matter of months." Members of Williamson's team began receiving job offers from other universities and soon left to launch labs of their own.

The preprint of the scientific manuscript that Williamson had given Tsui when the two had met at the Royal Society sent shock waves through the Toronto lab, leaving Tsui's team defeated and despondent. After reading the newspaper accounts of the discovery, many labs around the world halted their CF research. The National Institutes of Health placed a moratorium on funding for all CF gene hunts—including Tsui's work—arguing that Williamson had already succeeded.

When the journal with Williamson's paper finally arrived in the mail a week or so later, Tsui and Rommens scrutinized every detail, including the images, which hadn't been included in the draft. The piece of DNA that Williamson's team described was closer to the gene than anything that Tsui's team—or anyone else—had discovered. The news was particularly demoralizing for Rommens. She had dragged herself into the lab every day, sapped of energy and enthusiasm, ever since Tsui had returned from London; now, after Williamson's publication, her despondency only increased.

When Kerem arrived a few months later, in September 1987, she and Rommens asked Tsui whether it was worth continuing. At any moment, Williamson could publish the pièce de résistance—the gene, the mutation, and proof it caused the disease.

Like a baseball coach giving a pep talk to a losing team, Tsui reassured them that their work wasn't in vain, refusing to give up the hunt. After all, Tsui hadn't seen any proof Williamson had the gene. Yes, the region of the chromosome Williamson had identified was inherited along with the disease. But so much evidence was missing. Williamson hadn't shown that the gene was turned on and active in the regions of the body that the disease affected—lungs, pancreas, sweat glands. He hadn't found a mutation that only the sick carried. And he hadn't explained how the gene he'd found actually caused the disease. These were crucial pieces of evidence. To Tsui, all that the UK team had was an address on the chromosome. And even if Williamson *had* found the gene, there was still plenty of exciting science to be done. They did, after all, have fifty families to study. And so, he told his team with his trademark cheeriness and optimism, "there's nothing to worry about except our own data."

Williamson's paper had scrubbed out much of the competition as researchers moved on to other genes. At the University of Michigan, the newspaper reports rattled Collins, Drumm, and the rest of their small lab. But once Collins read the published research in the journal, he, like Tsui, wasn't convinced Williamson had the right gene, and insisted that his team keep going.

Meanwhile, as spring bled into summer and faded into fall, frustrations and pressure mounted within Williamson's team. From the beginning, the gene hadn't "looked" right. The genetics data pointed to the location, but when Williamson's team examined the DNA, they couldn't find a mutation that appeared in the sick children but not the healthy ones. Because the location seemed correct, Williamson's team had ignored other data that suggested the stretch of DNA that they believed was a gene did not include the mutation that caused CF.

Six months after the April publication, Williamson admitted publicly at the International Congress of Pediatrics in Paris that he had
made a mistake. His team had not found the gene, but rather just a segment of DNA nearby. The gene that caused CF was still undiscovered;
the prize was still out there. Williamson called his friend Don Fredrickson, director of the NIH, and told him that the moratorium on funding
should be reversed. Shortly afterward, the money began to flow to Tsui's
group and others working on CF.

Now that the hunt was back on, the challenge was to make it go faster.
Tsui was using chromosome walking to find the gene, with each step
covering roughly one to five thousand letters. The approach was tedious
and slow, but it was also extremely safe: there was no doubt that the
team would eventually find the gene. How long that took just depended
on how quickly Kerem, Rommens, and the rest of the team could complete their genetic analyses of the fifty families. The chromosome-
jumping technique Collins was using was quicker, leapfrogging over
large sections of genetic terrain, one hundred thousand letters at a time,
but it was difficult and time consuming to determine whether the jump
was toward the gene or away from it. And of course, they risked jumping over the gene and missing it entirely.

The ideal situation was to have walking and jumping going on
simultaneously, with some members of the team jumping, while the
other members marched toward the gene, slowly and steadily, from each
new jump. And neither Collins's nor Tsui's lab had the people power to
take on both tasks alone.

In late 1987, at a meeting of the American Society for Human
Genetics in San Diego, Tsui noticed Collins sitting nearby after presenting on his latest chromosome-jumping study, which he and Mitch
Drumm had done at Michigan. Tsui sat down next to him, and in the
warm sunshine outside the convention center, the two began discussing
what their labs were doing. Tsui suggested they team up: once Collins
took a jump, then Tsui's team could "walk" from the new start point.

For Tsui, the whole Williamson debacle had been stressful and
unpleasant, and had taken a toll on all the members of his lab; he wasn't
anxious for someone to get there first. Though he was confident his
team would get the gene, working with Collins might be a tad faster
than working alone—and with competition from other labs around the
world, not to mention the constant reminder of the patients just a few
floors below his lab, every moment saved was precious. And to Collins,

joining forces seemed smart. Tsui's team had discovered new markers in the past few months that were even closer to the gene than Williamson's, and Tsui's invitation felt easy and comfortable.

The two labs were 450 kilometers apart—a five-hour drive—so the teams first met midway between Toronto and Ann Arbor in London, Ontario, for an intense daylong scientific exchange. Members of both labs packed into the windowless, fluorescent-lit conference room of a run-down Holiday Inn to exchange data and markers and create a game plan for what each team would do. Roughly speaking, Collins's team would do the jumping and Tsui's would do the walking. Kerem would complete all the genetic analyses, testing the DNA from the Canadian families to determine who carried each marker.

In between Holiday Inn rendezvous, the labs kept in touch by phone and fax, sending scrolls of data that could be seen and shared by each person in the lab. Transparency was key; if anyone got the sense that the other group was holding back data, the collaboration would crumble. Drumm was the point person in Collins's lab, and Rommens took point in Tsui's. Though the Cystic Fibrosis Foundation was funding both labs, Beall spoke primarily to Collins. The two shared a similar sense of humor, which was important as the intensity and pace of the research ratcheted up. Beall's phone calls to Collins became more frequent. "I'm sitting by the phone waiting for you to call to tell me you found the gene, but you're not calling, so what's going on," was the basic gist. It was a bit tongue in cheek, but Collins knew he was serious.

Tsui was the mapmaker, his team charting every fragment of DNA that Rommens and Drumm were decoding. Each segment of DNA that was sequenced was a couple of hundred nucleotides long, and Kerem helped Tsui by aligning each one manually with the previous stretch of DNA to ensure continuous coverage of the chromosome. A marker at the tail end of each segment was then used as bait to fish out a new piece of DNA from Collins's jumping library that—they hoped—would be closer to, or the site of, the gene.

Rommens and Drumm then had the maddeningly tricky, time-sucking task of figuring out whether each new jump was toward or away from the gene—or had gone over it entirely. In addition to the jumping library, Drumm had also made a finer resolution DNA collection, called a walking library, that took smaller steps and so allowed the researchers to get a peek at the landscape of the genome they were jumping over. Sometimes, after jumping to a new location with a piece of DNA

DNA piece #1: AGCTGG......TTAACGTATGATA
DNA piece #2:   TATGATACACC......AAAGCCCCTAGT

                              Overlapping   *Bait for*
                                sequence   *jumping library*

A simplified version of how the sequence at the end of one segment
of DNA was used as bait to fish out an adjacent piece of DNA.
Each overlapping region was at least 1,000 letters in length.

that Drumm had sent from Michigan, they would walk in the wrong
direction for weeks, making her wonder whether jumping was saving
any time at all. But once Rommens confirmed a jump was going in the
right direction, the cycle continued. Drumm would send DNA from
the jumping and walking library to Rommens, Rommens would use the
DNA to find new markers, and Drumm would use the new markers to
fish out a new segment of DNA closer to the gene.

Now that the team was very close to the gene, the LOD score—that
statistical measure used to gauge whether a marker was linked to the
gene and its distance from it—was, for rather technical reasons, no longer useful. The only way to figure out if they were any closer to the gene
was to screen the DNA of families with collections of markers, called
haplotypes, to find out who carried which ones; this was Kerem's job.
Did the healthy siblings carry them? Just the sick kids? The sick kids
and their parents who were carriers? If all the children with CF carried
a particular haplotype, it meant that particular sliver of DNA had been
passed down for hundreds of generations unchanged, and likely contained the mutation that caused the disease.

As new sequences were decoded, both labs were examining the
genetic code for potential genes and whether those genes had the right
characteristics to be the one that caused cystic fibrosis. But looking for
genes wasn't simple, because they were written with the same four-letter
alphabet as DNA that doesn't code for anything recognizable. It was
like searching for a coherent sentence among millions of letters just randomly strung together.

The first test for identifying meaningful DNA was to compare it
with the genetic code from other species—a test appropriately called a
zoo blot. Genes that perform important functions in a cell are usually
conserved during evolution. So even if two species parted ways millions

of years ago, they still might carry the same genetic instructions for important proteins. If a swath of DNA was found in multiple species, it was a good sign that it might be a gene, but even then it wasn't proof. The genome, as Tsui's and Collins's teams learned the hard way, is a hoarder—hanging on to DNA even when it is no longer important. The teams were frequently tricked by seductive sequences of DNA that appeared to have been precious enough to be retained for hundreds of millions of years, only to discover they were just genetic detritus accumulated during evolution. The false positives took up so much time and energy that Tsui became increasingly suspicious of each promising lead. But there was little choice besides pressing forward.

Getting a positive zoo blot was just the first step. For the next critical piece of evidence, Tsui had to show the gene was switched on and actively making proteins in the sweat glands, the gut, the lungs, and the pancreas—the places where the disease hit hardest. The team knew from Jack Riordan's studies of cDNA libraries, made from the cells of his own sweat glands some five years earlier, that relatively few genes were turned on in the sweat glands, so the right gene had to be on this short list. This was the point at which many promising candidate genes were dropped from consideration.

If a section of DNA code passed these two tests—it was present in other species, and switched on in the right parts of the body—then it signaled that it was a candidate gene. But Tsui wanted more evidence than that before he was willing to believe they had *the* gene that caused cystic fibrosis.

There was one more key requirement to prove it was responsible for causing CF: the gene had to contain a mutation carried by both the patients and their parents. To determine this, the team used the DNA from the fifty families that Tsui had been collecting since he first came to Toronto seven years prior to perform a haplotype analysis—Kerem's assignment. Kerem looked at the DNA of both sick and healthy children for specific collections of polymorphisms and followed their inheritance from parent to children while looking for markers that differed between patients and their healthy siblings.

This evidence Kerem was seeking was unusual. Typically, proof that a gene caused a disease would be sought by comparing the behavior of a healthy protein to the mutated one and showing, using biochemistry, that the mutated protein couldn't do its job. But in the case of CF, no one knew which protein wasn't working. That's why Tsui was using genetic analyses to track down the location of the gene. Only then could they figure out which protein was broken.

Accumulating all this evidence was complex and time consuming, but there was another reason why this gene hunt was proving so tricky—one that had nothing to do with CF, but rather the current understanding of the human genome. In the late 1980s, geneticists believed the number of genes in the genome to be around 150,000—about six times the final number actually discovered during the Human Genome Project.[8] Based on this assumption, Tsui and Collins had expected the interval they were traveling through to be rich with genes—a genetic rain forest, with one gene following another in series. But it wasn't. The terrain was barren, the absence of genes unsettling. It seemed that in this stretch of the genome there must be only two genes, including the wrong one that had been misidentified by Bob Williamson's group, and the actual CF gene, which they were still seeking.

By late 1988, there was someone working in Tsui's and Collins's labs at all hours of the day. Rommens and Kerem worked particularly closely and cooperatively, like right and left hands. Kerem's children limited which hours she was available, so the two women would tag-team: Rommens, young and single, would stay late to complete experiments and start new ones that Kerem would then continue the next morning. Every Friday afternoon, Tsui, Rommens, and Kerem, along with several of the technicians and students, would cram into Tsui's office to plan the following week's experiments.

As 1988 sped into 1989, both Rommens and Kerem could sense that there was something exciting happening with the DNA and markers they were analyzing. The first sign was that a segment of DNA they had just decoded early in the new year was also present in other species. There was similar genetic code in chickens, cattle, and pigs, suggesting that the DNA they were looking at was important.

But was this DNA segment turned on in the sweat glands? When Rommens compared the part of the gene to Riordan's library of active sweat gland genes, she found a match. Encouraged, she then looked at gene libraries for other organs destroyed or disrupted by cystic fibrosis—the lungs, pancreas, nasal polyps, colon, and liver—to see which genes were active in those tissues.[9] And just as she hoped, the DNA matched a gene present in those libraries as well.

As Tsui, Kerem, and Rommens were testing the gene-worthiness of this sliver of DNA, the rest of the team in Toronto, and Collins and Drumm in Michigan, were working hard to track down the remaining stretch of the gene. It was a complicated process, because genes aren't written in the DNA as continuous strings of letters—they're modular, divided into small segments called exons that are strung along the chromosome. There appeared to be at least twenty-four exons in this gene, interrupted by intervening sequences called introns (incorrectly thought of at the time as junk DNA)—an arrangement that meant the gene was distributed across a total of 250,000 nucleotides. In tissues where a gene is active, machines in the cell transcribe the DNA and then stitch together just the exons. Once this particular gene was edited, it formed a tight message of just 6,129 nucleotides, which translated into a protein[10] that was 1,480 amino acids long.

In Tsui's and Collins's labs combined, ten or so people were working on the project. Many were sequencing DNA. Each piece of the gene had to be sequenced several times to make sure the DNA was correctly decoded—otherwise it would be difficult to distinguish human error from a disease-causing mutation. Other lab members, like Kerem, were working on the haplotype analysis for each family. Tsui and Collins were analyzing all the data from the students, postdocs, and technicians in the labs. Everybody had a role.

In late March of 1989, when Toronto and Ann Arbor were still in the deep freeze of winter, the labs were steaming with activity. In Toronto, visiting scientists and researchers from other labs pitched in to help sequence and decode the gene, eager to participate in such an exciting project. With the majority of the gene in hand, Rommens began sequencing a version of the gene taken from a sweat gland of a cystic fibrosis patient side by side with the same gene taken from an unaffected relative, looking for any mutations in the patient's DNA that might be responsible for the disease.

It was 6 PM on Tuesday, May 9, 1989, when Richard Rozmahel, a talented undergraduate student who'd joined Tsui's lab for a research project two years previous and stayed on as a technician, rushed into Tsui's office with something to show him. It was a computer printout revealing the latest sequences from the side-by-side gene sequencing effort, and he dropped it on Tsui's cluttered desk before Tsui could clear the surface. Rozmahel had been reading the DNA sequence from the healthy subject and the one from the patient when he noticed the patient's DNA seemed to be missing three letters—TTC—that encoded an amino acid called phenylalanine. If the deletion was real and not a sequencing mistake, it meant that the CF patient was missing one amino acid at position 508 of the protein.

Tsui's reaction was more restrained. Even with all the evidence Kerem and Rommens had gathered—the zoo blot, the gene being turned on in all the right organs, and now a mutation that was present in a CF patient but not a healthy subject—neither Tsui, nor Kerem or Rommens, was entirely convinced they had the right gene. And Tsui was even less convinced that the tiny mutation Rozmahel had showed him, just three missing nucleotides, was the cause of the deadly disease.

Missing one amino acid was equivalent to missing a single bead in a series of hundreds. What could one missing amino acid possibly do? Also, what looked like a mutation could just be a polymorphism—a natural variation in the genetic code between individuals. The same mutation would need to be present in the DNA of many more—some 70 percent of patients, based on Kerem's analyses—for Tsui to believe it was real. Additionally, both parents would need to have only a single copy of the mutation, and healthy siblings could have, at most, only one copy. So Kerem began the critical task of testing all the individuals in the fifty families to determine whether they, too, lacked the TTC at position 508.

WHILE KEREM WAS TESTING THEM, ROMMENS AND DRUMM WERE continuing to decode the rest of the gene. As the code emerged, Rommens delivered it straight into the hands of biochemist Jack Riordan. Though he had been the one to hire Tsui eight years earlier to work on finding the CF gene, Riordan had also been working on a family of proteins called ABC transporters, found in organisms as varied as humans, mice, and bacteria. These were proteins embedded at the surface of the cell that were responsible for shuttling chemicals in and out.

Now, to help figure out what type of protein this gene produced, Tsui asked Riordan to look at the gene sequence. Riordan was a biochemist specializing in proteins and their three-dimensional structures; he had more experience than anyone else on the team interpreting how the string of amino acids would bend and fold to form the protein their gene coded for.

Riordan began looking for clues to the type of protein it might be, its shape, where it might live in the cell, and what it might do.

The only hint to date of what kind of protein the mutant gene might code for had come six years earlier from the work of Richard Boucher and Paul Quinton, whose studies suggested that the problem was the movement of chloride in and out of the cell. That suggested that the gene they were seeking might code for a protein that resembled a tube or channel.

Using a newly created database of proteins called SWISS-PROT—a curated collection of protein sequences on CD that was periodically updated with researchers' latest discoveries and sent to labs worldwide—Riordan typed in a partial sequence of amino acids in the new protein and watched with amazement as the As, Gs, Ts, and Cs began to align with the sequences that encoded known proteins: pigment proteins from the eye of a fruit fly, and transporter proteins that moved molecules of sugar and vitamin B12 in and out of the bacterium *E. coli*.

Just as human families share physical features, protein families do, too. Proteins that perform a common function tend to look similar. Riordan was the world expert on ABC transporters, and as he read through the sequence of amino acids that Tsui's and Collins's labs had decoded, he noticed something that no one else on the planet would have detected. This particular arrangement of amino acids formed a couple of serpentine structures that zigzagged in and out of the membrane, like thread passing in and out of a thick fabric—meaning that the protein was a transporter. Riordan was sure of it. The ABC transporters he studied were present in the membranes of bacteria and used energy to pump lethal drugs like antibiotics out of the cell, like a water pump emptying a leaky boat. Riordan guessed that this new protein, like other related transporters, would be shaped like a tube and inserted into the outer membrane of a cell, like an air valve in a bike tire. If this new protein was a transporter, could it be responsible for transporting chloride in and out of the cell? And if this transporter protein was missing an amino acid and so folded incorrectly, such that it couldn't do its job, could that explain the salt imbalance associated with cystic fibrosis?

As Rommens fed Riordan the DNA sequence, he analyzed it and translated it into a growing chain of amino acids that made the protein. His challenge was to figure out how this chain folded, twirled, and twisted into a complex three-dimensional structure to perform its function. It was a puzzle, like figuring out how a single stretch of wire could be shaped into a donut. Certain amino acids liked contact with water—these would be either on the surface of the cell, or inside the cell. Other amino acids were oily and hated water—these lived within the cell membrane. Some amino acids formed bridges linking distant parts of the protein. Other stretches of the proteins formed spirals and sheets, like the lamination of a croissant. Riordan's knowledge of physics and chemistry helped him figure out which atoms were attracted to each other and liked to nuzzle close, and which ones were repelled by each other, forcing segments of the amino acid chain apart. These immutable atomic preferences guided the folding of the protein into the same structure every time it was made in the cell.

As Riordan was investigating the structure of the protein, he kept thinking about how minor the mutation was. The protein was 1,480 amino acids long, and it was only missing an amino acid at position 508. It didn't seem like a catastrophic change that would destroy the structure or function of this enormous protein. But every protein was different. If a change or mutation or deletion happened in the region of the protein that was critical for its particular function, then even a small change could render the protein useless. If it occurred in a part of the protein that wasn't so critical—imagine a flaw on the handle of a can opener, instead of the blade—then even large mutations might be tolerable. Neither Riordan nor anyone else had a clue what effect, if any, the deleted TTC would have on the protein, but he knew it was possible for a mutation at a vulnerable location to muck up the whole design—like a missing step in a complex origami.

While others were thinking about the shape of the protein and its possible role in CF, Kerem's mission was testing all the individuals in her set of fifty families to see which ones carried the TTC deletion. One by one she took the DNA of each parent and child in each of the families and tested their DNA to see which version of the gene they carried: the one with the amino acid at position 508, or the one without it.

When she had completed the tests and tallied the results, she saw that 70 percent of the children and young adults with cystic fibrosis had two copies of the gene that were missing the TTC. Kerem was also able to show that, in patients with two copies of the mutation at position 508, their mother and father both carried one gene with the 508 mutation and one normal copy. The finding was exhilarating, if blunted by an uncomfortable question: Why didn't 100 percent of the patients carry this mutation? Were there other mutations hiding in this gene that had triggered this disease in the remaining 30 percent? There was more work to be done. For now, however, the 70 percent result was impressive. It was proof that, at least as far as the Canadian population was concerned, this was the primary mutation leading to this disease: everyone who carried two copies of the mutation was sick.

CHAPTER 25

The Gene

1989

The genes are the atoms of heredity.

—Seymour Benzer, physicist and behavioral geneticist

In June 1989, Lap-Chee Tsui and Francis Collins were at a Yale University workshop to discuss mapping the human genome, when Batsheva Kerem faxed over the update that 70 percent of the cystic fibrosis patients in their fifty-family data set tested positive for the 508 mutation.[1] When Tsui and Collins walked into the room where Tsui was staying, data was spewing from the portable fax machine the Canadian CF Foundation had given him to stay connected to the lab around the clock. Now, scattered on the floor was proof they had, indeed, found the right gene and a mutation that caused the disease.

Johanna Rommens was so excited by Kerem's data, she didn't sleep that night. She was sure they had the gene. The evidence was now, in her opinion, irrefutable. The protein the gene coded for was made in the organs damaged by the disease, like the lungs and sweat glands. It wasn't made in places like the brain and skin, which appeared untouched by CF. Only sick children carried two copies of the mutation, which they named F508del: *F* stood for the amino acid phenylalanine, *508* was its location, and *del* was short for deletion. Parents only had a single copy of F508del. And the function of the protein also seemed to fit the story: per Jack Riordan's work, the protein resembled a transporter embedded in the surface of the cell to allow some type of molecule to travel in or out.

Collins was convinced, too. That was the night he declared joyously that they had got it. His impulse was to scream and shout—though he held back, in part because Bob Williamson's postdoc Martin Farrall, who was still searching for the gene, was in the adjoining room. Tsui, however, remained stubborn and refused to believe it. He was skeptical, and appropriately so, in the long shadow of Williamson's erroneous announcement. There was still a small chance that this mutation, the missing TTC, was just a marker that was linked tightly to the CF gene—not part of the gene but close to it. Tsui wanted more data, from more families. And he wanted mathematical proof this was the disease-causing mutation.

Tsui wasn't sure how to make such a calculation, however, so he called his friend Aravinda Chakravarti, the geneticist who had initially encouraged Tsui to track the gene using markers.

When Chakravarti answered, Tsui told him to come to Toronto as soon as possible. Chakravarti's voice rose with excitement. "You've found it, haven't you?"

But Tsui refused to tell him over the phone, remaining tight-lipped, and insisting he get on the next flight. Chakravarti guessed that Tsui had made a major breakthrough, and perhaps even found the mutation that caused CF. After the Williamson fiasco, Chakravarti knew that the stakes were too high to delay his trip to Toronto, so he bought a ticket for the next day.

There, Tsui picked him up at the airport baggage claim.

"Now tell me!" Chakravarti demanded as soon as he jumped in the car, skipping over any niceties. Tsui was driving him wild with the suspense.

Tsui shook his head and smiled, and Chakravarti could see that his friend was thoroughly enjoying seeing him squirm impatiently. "No, no, you need to come and see the evidence."

Tsui drove straight to SickKids and the two headed into the lab, where a half-dozen or so people were working. Tsui's lab was tiny. Some of the technicians and students didn't have desks and worked on metal carts—the type found in dim sum restaurants—that could be wheeled around the lab or out in the hallway. They entered Tsui's office, where journals and textbooks and papers and photocopies of references were stacked a foot high on the two small desks—a visual assault for the habitually tidy Chakravarti, whose desk had to be free of all detritus for him to think.

Tsui pulled out a piece of X-ray film with columns of black and clear bands showing the DNA sequence. Then he pointed to the mutation—the missing TTC.

Chakravarti took the film and held it up to the light to examine the pattern of bands. Then Tsui passed him more pieces of film, revealing the mutation in other patients and families. Without saying a word, Chakravarti examined them all, his heart rate increasing as he looked at the data from each family.

Chakravarti understood why Tsui needed him. Tsui's team had been using surrogates—the markers—to find the gene. What proof did Tsui have that this missing TTC at position 508 was actually the mutation that caused the disease? How much longer should he search for other mutations? He needed Chakravarti to make a statistical calculation that would prove this missing TTC was not an innocent bystander—that it was, in fact, the mutation causing CF in these patients.

This wasn't a complex calculation for Chakravarti. But it was the first time anyone had tried to calculate the p-value (the odds of being wrong) so precisely in human genetics. Chakravarti did the math by hand in just a few hours. He figured out the chance of this not being the mutation that caused the disease was 1 in 10^{57}: one in an octadecillion.[2] Those were great odds. It was also compelling that no other candidate mutations had even been found.

He was ready to celebrate and then catch a flight home. But Tsui shook his head and, with his characteristic gentle laugh and nod, told Chakravarti that he should stay in Toronto to help him and Kerem write the genetics paper. It was the most important of the three papers that had to be written, because it was the one that would prove they had discovered the right gene and the disease-causing mutation within. First, however, the two went out for steak and drank lots of wine.

Because of the three extraordinary elements that had been required to find the gene, there were three scientific papers to be written. The first would describe how the Toronto and Michigan teams found the precise location on chromosome 7 through a combination of walking and several jumps across the DNA. The second would detail how, once the team had pinpointed the location, they determined the entire sequence of the gene and translated it into a protein—a transporter protein—that they found was present in the membranes of cells in organs affected by cystic fibrosis. The final paper would hold the genetic analysis of all the patient DNA and the proof that the missing TTC was the mutation that caused the disease.

Tsui wasted no time preparing the manuscripts. The day after Chakravarti arrived, Kerem joined Tsui and Chakravarti so the three could write one of the three papers together. To clear space, Tsui moved several stacks of paper from his desk to the floor, and for the next few days, the three of them sat together, with Tsui writing the first draft by hand on lined yellow pads, leaving holes for the genetic tests on families that Kerem was still running in the lab. Only after they had a solid first draft in place did Chakravarti fly home—where he redid all his calculations just to be sure.

As Tsui, Kerem, Riordan, Rommens, and Collins worked on the remaining two papers to submit to one of the world's most prestigious science journals, most team members in both Toronto and Michigan were frantically running experiments, confirming results, and resequencing parts of the gene to make sure the code was free of errors.

Writing a single paper typically would take months, but the case of cystic fibrosis was unusual. Every part of the story was new, and they needed to publish fast, lest someone else, perhaps Williamson, scoop them. Together, these three papers told an epic story of a revolutionary strategy for gene hunting based on inheritance patterns of genetic markers—research that Tsui had begun alone in 1983. Tsui's approach was a radical departure from the traditional strategy: nabbing the protein first and then using it to find the gene. Then there was Collins's chromosome-jumping technique, which accelerated the final stage of gene hunting.

There was also the unorthodox proof of causality. Typically, biologists would compare the functions of the normal and mutated proteins in vitro in a lab to show how they differed: how one worked, and the other didn't. Here there was no physical proof; that would undoubtedly come later, as other scientists probed the function of the mutant CF protein. Instead, Kerem had demonstrated causal proof by showing that only sick children carried two copies of the mutation—while each of their parents only carried one—as Chakravarti provided the mathematical rationale.

Finally, there was the new protein. Previous work suggested cystic fibrosis was caused by problems shipping chloride in and out of the cell. Now Riordan had a protein that seemed a perfect match—a donut-shaped channel that could transport chloride. Tsui gave the protein the clunky name Cystic Fibrosis Transmembrane Conductance Regulator—CFTR for short—reflecting the team's uncertainty as to whether this protein was the actual channel for chloride to pass in and out, or regulated another protein that did so.

The papers were scheduled for publication on September 8, 1989. But a reporter who caught wind of the discovery called Collins's lab in late August asking whether they had indeed found the gene, and a graduate student, unprepared for the ambush, spilled the beans.

The news broke one week before the peer-reviewed article appeared in the journal, breaking scientific etiquette for unveiling a new discovery and catching both the Toronto and Ann Arbor teams completely off guard. "Cystic fibrosis gene found"[3] and "Cystic fibrosis finding should improve treatment, testing,"[4] reported United Press International (UPI). "Researchers score genetic first—Sick Children's hospital team isolates cause of cystic fibrosis,"[5] the *Toronto Star* trumpeted. "Researchers isolate gene that causes cystic fibrosis,"[6] wrote the *Los Angeles Times*. "Cystic fibrosis: hunting down a killer gene,"[7] was the headline in *Newsweek*. The *New York Times* ran the most underwhelming headline: "Scientists develop new techniques to track down defects in genes."[8]

It was a chaotic unveiling. After months of round-the-clock work in the lab, and with the manuscript submitted and ready for publication, Kerem had left Toronto with her family for a three-week camping trip in the United States. It was while they were driving in Montana that she and her husband heard on the car radio that the cystic fibrosis gene had been discovered.

Kerem was stunned and confused; the discovery was supposed to be a secret until the publication came out on September 8. Did Bob Williamson's group scoop them? It wasn't clear. Desperate to find out, she checked the family into a hotel that night so that they could watch the news. When they turned on the television, Kerem saw Tsui, Rommens, Riordan, and Collins standing together answering questions from reporters. When Kerem and her family returned to Canada, they saw that the Toronto newspapers had mentioned an Israeli scientist was part of the group. But Kerem was absent from all the pictures.

With the discovery now public, the CF Foundation and the Howard Hughes Foundation, which had supported Collins, were each keen to make their financial role in the discovery clear and participate in the press conferences that were to be held the next day. Lawyers from the University of Michigan and SickKids in Toronto scrambled to file patents on the work, arguing over how much of the future royalties would go to each institution. The Hospital for Sick Children held the first press conference in the morning after the television news carried the story and before Kerem was back in town. Then HHMI chartered a plane to shuttle the scientists from Toronto down to Washington, DC, where a second press conference was held that afternoon. At the press conference, Bob Dresing, in his role as CF Foundation president, hailed the gene discovery as the "most significant scientific breakthrough" for the disease in the past forty years. "We can finally look into the eyes of children and young adults with cystic fibrosis and tell them that the door to their future has been opened."[9]

Collins told UPI that he was already working on a way to correct the genetic defect that caused cystic fibrosis using gene therapy. It may be possible, he said, to introduce the healthy gene into the lungs via a spray. But Collins cautioned, "It would not be fair to people with the disease to create a sense of optimism that now a cure is just around the corner."[10] If the disease was a mountain and the cure was the summit, Collins explained on a morning talk show, then the discovery of the gene had only brought them to base camp.

Back in Toronto, SickKids organized a surprise celebration for the hometown heroes, inviting all the researchers and the CF center's families and patients, many of whom had participated in the research, providing their DNA, family history, and other clinical data. When Tsui's team, including all the research technicians, walked into the room, the crowd erupted with cheers, clapping, and crying. Parents and patients descended on them, mothers and fathers embracing Kerem in tears, overwhelmed by what she and the others had done for their children. For Kerem, who had missed both press conferences, this was powerful, emotional, unforgettable.

It was an enormous victory for all of the researchers involved, in Canada and in the US—a scientific tour de force that set the trajectory for the rest of their careers. An avalanche of media attention followed; for the next year, a steady stream of dignitaries and camera crews dropped by Tsui's and Collins's labs. All the main researchers received invitations from institutions all over the world to speak about the gene. One of Tsui's graduate students joked that, in the year following the papers' publication, he saw Tsui on television more than in the lab. They all basked in the glow of progress and the sense that they had struck a significant blow against this disease that, despite plaguing humanity for millennia, had gone unrecognized until just a few short decades before.

When the papers were finally published a couple of weeks later, the discovery earned them the cover of *Science*. The cover photograph was of five-year-old brown-haired, brown-eyed Danny Bessette, sitting cross-legged and smiling. He had cystic fibrosis and carried two copies of the F508del mutation that the scientists had just discovered. But the disease had not yet ravaged his body, and he looked healthy and innocent. The assumption was that now, with the discovery of the gene, Danny would be cured and go on to live a long, healthy life. *Science*'s editor Daniel Koshland underscored this sentiment in his editorial as he placed the discovery of the gene in context: "In this issue of *Science*, there is a story that does not begin at the beginning or end at the end, but has a very happy middle. The beginning is the basic research that made it possible to search for a genetic needle in a haystack of DNA nucleotides. The end is a cure for a fatal disease. The middle is the finding of the cystic fibrosis gene, a milestone of major importance."[11]

THE YEAR AFTER THE DISCOVERY WAS PUBLISHED, THE TORONTO team—primarily Kerem and Rommens—began finding additional mutations in the gene that were responsible for CF in the 30 percent of patients who did not carry the deleted TTC at position 508. They had predicted that there were at least six other mutations in the Canadian population that remained at large, and as they sequenced the gene in these other patients, they were finding them. Now, however, the lab was also receiving samples from patients all over the world. One set of blood samples in particular caught Kerem's attention. Physicians in her home country of Israel had sent blood from CF patients who, Kerem soon discovered, did not carry the common TTC mutation, nor any of the mutations they'd found in the other 30 percent of Canadian families.

In September 1990, a year after the publication of the three papers and six weeks after the birth of her third daughter, Kerem and her family returned to Jerusalem. Now that the F508del mutation was known, dozens of researchers were flocking to study it. Kerem didn't want to compete; she preferred to carve out her own niche. Perhaps, she thought, she would figure out what mutation was causing CF in the Jewish and Arab patients who suffered from this disease.

She began her research by launching the Israel Genetics Center for CF, where she collected blood samples from most of the six hundred or so CF patients in the country. As she tested their DNA, she found that about 30 percent of the CF population in Israel carried the common F508del mutation. But she also discovered, over the next year, that among Israel's Ashkenazi Jewish population with CF, 60 percent had a different genetic alteration: a single-letter change, from G to A, at position 3978 in the CFTR gene. This seemingly minor misspelling had a dramatic impact; it led to the amino acid tryptophan being replaced with a stop signal—a type of error known as a nonsense mutation. It was as if someone had put a period in the middle of a sentence, before it was complete. That meant that this mutation, named W1282X, produced a protein that was too short and didn't function properly,[12] and it was just as deadly as, if not more than, the one she had helped to discover in Canada. And seven years after Kerem's publication, in 1999, Israel introduced population carrier screening for individuals who had no history of the disease in their family but might be carrying W1282X or one of the other common mutations.[13]

As Kerem was probing this newly discovered gene mutation that caused CF, the Cystic Fibrosis Foundation's director of science and medicine, Bob Beall, was busy thinking ahead. Though the scientific community, patients and families, and Beall himself all believed that the discovery of the gene would rapidly lead to more effective treatment, if not a cure, Beall knew that wouldn't happen immediately. In the meantime, he had tens of thousands of sick children and young adults to worry about. To Beall, the key to improving CF patients' chances was reducing the lung infections that were their usual cause of death. One way to do this was to keep the lungs mucus free—and suddenly, it seemed that there might be a new drug that could help.

CHAPTER 26

Runny, Like Water

1989–1994

That is what science is supposed to be about—not an
academic exercise for the ivory tower, or racking up
publications, grants, offers of tenure. It's about using the
tools and technology available to make lives better.

—Mona Hanna-Attisha, pediatrician and
public health advocate

In December 1989, Cystic Fibrosis Foundation leaders Bob Beall and
Bob Dresing were still elated from the discovery of the CF gene, and
had focused on gene therapy as the next step to a cure, when Doris Tulcin received a call from a friend, the medical director of genetic engineering firm Genentech. He encouraged her to come meet with him in
San Francisco and see what they had developed. All he would tell her
was that "it might be important for cystic fibrosis." Tulcin, who at that
point was still devoting her time to fundraising for the Cystic Fibrosis Foundation's research development program, called science director
Bob Beall, and the two of them headed out to see what the company,
then just a couple of converted warehouses in South San Francisco, had
to show them.

Both Tulcin and Beall had reason to be excited. Genentech was a
pioneering biotech firm and the first company to use recombinant DNA
technology to create medicines that were manufactured in microbes.
Now, it seemed, they might have something promising for patients with
CF.

Tulcin and Beall were led to the lab of Steven Shak, an adult pulmonologist with dark hair parted neatly to one side, a thick, triangular
mustache, and round thin-rimmed glasses. Like a magician performing

a trick, he revealed a tube filled with CF's hallmark thick mucus, collected from a patient. Holding it in his left hand, he popped off the top of the tube with his thumb and grabbed a pipette with his right to squirt in a couple of drops of a clear mystery liquid. Recapping the tube, he gave it a quick swirl and then held it up so Tulcin and Beall could see. Within a minute, the nasty, viscous mucus transformed; it was now as runny as water.

IN 1976, ONE OF THE PIONEERS OF GENETIC ENGINEERING, HERBERT Boyer, recognized that by identifying and sequencing human genes that coded for essential proteins like insulin, replicating their code— a process called gene cloning—and then inserting those instructions into fast-growing bacterial or mammalian cells, he could manufacture unlimited quantities of any human protein, which could then be sold as a drug. To this end he cofounded Genentech, an abbreviation for *Gene*tic *Engi*neering *Tech*nology, to manufacture therapeutic human proteins on a massive scale.

With just twelve employees, Genentech's first target was producing a synthetic version of human insulin. In 1978, pharmaceutical giant Eli Lilly needed pancreases from fifty-six million animals each year to meet the US demand for the hormone.[1] Boyer figured that if they could coax bacteria to read the human insulin gene and manufacture enough to meet that demand, they would not only eliminate the need to extract the hormone from animals, but could then apply the same approach for other biological molecules. Just as he hoped, once the human insulin gene was added to a bacterium's chromosome, the microbe began unwittingly cranking out the human protein along with its own bacterial ones. They published their proof of concept in 1979[2] and later teamed up with Eli Lilly to manufacture the insulin itself—the first genetically engineered therapeutic human protein. The FDA approved the synthetic human insulin in 1982 and it went on sale in 1983.

Next, the company began making other human proteins. On October 18, 1985,[3] Genentech received approval for its first in-house drug, Protropin—a treatment for the ten to fifteen thousand children in the US who suffered from a severe growth hormone deficiency. Before Protropin, pediatricians used growth hormone extracted from the pituitary gland of human cadavers; Genentech's revolutionary new method of genetically engineering proteins provided an easier, faster, and safer way to make these medicines.[4]

Shak, a pulmonologist at New York University who had treated all types of lung disease, including cystic fibrosis in young adults, had moved to Genentech a year later, in 1986, after seeing an advertisement for an opening in their molecular biology department. Shak applied to Genentech against the advice of the department heavyweights at NYU, who considered industry a step down from the ivory tower. But Shak was keen to work on solving significant medical challenges and to change the world. He moved his wife and young daughter to San Francisco, close to the Genentech offices, where he began developing a treatment for asthma.

In 1986, many scientists at Genentech were busy working on an injectable clot-dissolving drug called TPA—tissue plasminogen activator—designed to treat heart attack patients by reopening blocked coronary arteries. Shak was walking through the halls of Genentech on his way to his lab in 1987 when his thoughts swerved from sticky clots to the sticky secretions gumming up the lungs of the young adults with CF whom he used to treat at NYU. If Genentech could develop a drug to dissolve coronary plaque, could he find another substance that would dissolve or thin the impenetrable plugs of mucus, allowing patients to clear their lungs and breathe easier?

One of the attractions of Genentech—key to luring its brilliant academicians to industry—was the freedom employees had to spend 10 percent of their time pursuing passion projects and new ideas. Driven by this new question, Shak drove to the library at the University of California, San Francisco, and began thumbing through dusty tomes of research summaries. He discovered that, back in the 1950s and 1960s, LeRoy Matthews—the maverick leader of the Cleveland CF care center who had revolutionized treatment—and a few other CF scientists had explored the biochemical makeup of the goop in CF patients' lungs and discovered the main ingredient that made it so viscous: DNA. During the warfare that erupted in the lungs between immune cells and the invading microbes, the dead cells would explode, spilling their insides— including their DNA. The DNA was long and sticky and thickened all the liquid in the lungs like flour thickens a roux.

At the time, British scientists suggested that an enzyme called DNase, which cut up DNA like scissors cut twine, might help thin the mucus.[5] The DNase, which they sourced from cows, had worked in the test tube, but when used in patients, several had nasty allergic reactions, and the treatment was discontinued and essentially forgotten. But what if, Shak wondered, he could clone the *human* DNase gene, paste it into a cell, and manufacture the enzyme at Genentech? The human version of the protein shouldn't cause an allergic reaction.

He returned to Genentech and immediately checked a couple of databases to see whether the human DNase gene had been cloned yet. Nobel laureates William H. Stein and Stanford Moore and their colleagues had determined in 1973[6] the sequence of amino acids that made up cow DNase but not the human version. So he met with his boss to ask if he could drop everything and focus exclusively on decoding the precise genetic sequence of human DNase, explaining how it might help CF patients. His boss agreed.

Realizing that the human and cow gene were probably similar, Shak used the cow gene as bait to find its human counterpart—a painstaking process that took almost four months, even with Shak spending close to one hundred hours in the lab each week. With help from the gene cloning experts at Genentech, he added the human gene into bacteria, where it manufactured human DNase. Then he spent the next six months working with a biochemist at Genentech named Art Levinson to add the human DNase gene into hamster cells, in order to mass-produce it.

When Shak finally made some DNase, he was anxious to test it and measure how the viscosity of the mucus changed when he added the enzyme. He contacted physicians at the Stanford CF center who collected sputum (mucus and other sticky material generated in the lungs) from several patients, then gave Shak tubes filled with nasty blobs that clung stubbornly to the tubes' sides. Back in the lab, he squirted a few drops of DNase into each test tube of mucus and placed them in an incubator at body temperature. Fifteen minutes later, he picked up the first tube—and watched as the mucus poured easily down the tube's side. If this same thing happened in patients' lungs, it would help them cough up the sickening gunk in their lungs and breathe easier.

Heart pounding, Shak ran, tube in hand, to his boss's office, where he tipped the tube side to side, explaining in a verbal avalanche what he had discovered. Then he ran down the halls, dashing in and out of every room and sharing the news.

His excitement hadn't faded by the time of Tulcin and Beall's visit. After the jaw-dropping demo, Shak gave Tulcin and Beall a tour of Genentech's gleaming state-of-the-art manufacturing facility, including the enormous steel vats where cells were dutifully manufacturing growth hormone. If the foundation could help—by organizing the protocol for the clinical trial, identifying possible patients and sites to conduct the trial, and engaging CF physicians to lead it—Genentech would build a manufacturing facility just to make DNase for these patients.

Beall was sold; a mucus-thinning drug that could potentially reduce the number of lung infections was the kind of potentially transformative medicine that the foundation had been anticipating and hoping for. But as Beall completed the tour, he turned to Shak. "The one thing you've got to know, when you work in CF, is that the kids are the most important thing," he said. "You have to do these trials right the first time." Shak understood. These were children; there was no room for error.

As his next step, Shak assembled a team of two dozen researchers at Genentech to run a small DNase factory and purify the enzyme for use in patients. Then the company applied to the Food and Drug Administration to grant DNase "Investigational New Drug" status; only with a nod from the FDA could anyone begin trials. Approval was swift, thanks to the Orphan Drug Act,[7] which Doris Tulcin had lobbied hard for eight years earlier and which Ronald Reagan had signed into law on January 4, 1983. Although Shak had experience treating individual CF patients, designing a clinical trial was completely different. How would they deliver the DNase drug into the lungs? How would they measure the impact of the treatment? How could they prove that this drug was working and helping the patient?

To tackle each of these issues and figure out a way forward, Shak and a couple of colleagues flew to Seattle to meet with two of the Cystic Fibrosis Foundation's advisors: Bonnie Ramsey, director of the CF center at Seattle Children's Hospital; and Arnold Smith, a microbiologist at the University of Washington. After discussing what medical endpoints were important for patients—greater lung function, fewer lung infections and hospital stays—Ramsey led Shak and his team into the clinic to talk with patients and families about their needs and struggles.

Since medical school, Shak had been inspired by the teachings of physician William Osler, cofounder of Johns Hopkins University, who stressed the importance of focusing on the patient, not the disease. Listen to the patient, Osler had urged, and they will tell you what we need to do. Though Shak had treated patients with CF before as an adult pulmonologist, he had not done so exclusively, and as he sat in the clinic, he was overcome by the desperation of the patients' parents. These children were drowning in their mucus-clogged lungs. And no other drug had ever been approved for treating this disease. If the trial went well, Shak's DNase would be the first.

THE FIRST PHASE I STUDY WAS DONE IN BETHESDA AT THE NATIONAL Institutes of Health, not far from where Paul di Sant'Agnese had trained a generation of CF physicians and researchers. There, a small number of very brave young volunteers had gathered to inhale, at first, just a single breath of DNase. Then, if all went well, they would inhale a little more.

The major goal of this phase I study was to make sure the drug was safe, and Shak was nervous. Having patients inhale the protein had been his idea. He had supervised the drug's manufacture at Genentech, and now these sick young patients were going to take it. He feared that when the drug liquefied the mucus, it would flow in the wrong direction—back into the lungs, making breathing impossible—rather than being coughed up and cleared. If that happened, the treatment could kill.

A few days after the trial began in the summer of 1990, he flew to Bethesda, where the trial coordinator led him into the research center to meet four young adults who had already inhaled several doses of DNase. Shak sat next to one woman and nervously asked: How was she doing? Did she feel any different? Could she tell if anything was changing in her lungs?

After two or three days of treatment, she said, when she took a deep breath it felt like the air went all the way down to her toes.

Once phase I trials proved the drug safe, phase II trials with 181 CF patients began. Ramsey, Seattle Children's CF center director who had helped Shak establish the clinical goals of the trial, was at the helm. The purpose now: explore whether the drug was effective. Did it actually thin the mucus, making it easier to cough up? Genentech was so confident the answer was yes that they began constructing a $37 million manufacturing facility before the FDA had even granted the go-ahead for the final phase.

The phase III trial was the most important one, because it would quantify how effective the drug was and provide the evidence needed for FDA approval. The agency wouldn't approve a drug that improved patient health just a little. They wanted to see dramatic proof that the drug improved the health of many patients, that the volume of air the patients could inhale had jumped drastically, and that patients therefore spent less time in the hospital for cleanouts. The trial launched in

1992, with almost 1,000 participants taking part at CF centers all over the country.

Two of the volunteers were Isabel and Anabel Stenzel, twin sisters studying at Stanford University. The women were an unusual pair; they carried both the F508del mutation, inherited from their German father, and a second, mutated copy of the CF gene with not just one but two ultra-rare mutations from their Japanese mother.[8] (By 1992, scientists at Johns Hopkins University and SickKids were regularly sequencing the genes of patients who did not carry the F508del mutation and had discovered dozens of what would eventually be some two-thousand-plus mutations that could cause this disease.) Their father, a physicist fond of statistics, calculated that the chance of half-German, half-Japanese identical twins being born with CF was approximately one in 1.8 billion. When they were born in Los Angeles in 1972, their life span was pegged at sixteen years. But with five hours of dedicated physical therapy every day, the young women had survived. Despite being severely malnourished and prone to lung infections, they left home at eighteen to study at Stanford, living together in the same dorm room and diligently continuing their therapy early in the morning and after their classes. Both girls were very sick, yet they hoped to fulfill their dreams of graduation. Still, they didn't expect to have a career or a family, because they were unlikely to live that long.

During the girls' freshman year, a neighbor from back home in Los Angeles called to tell them about a miraculous treatment Genentech was testing in clinical trials. Isabel had heard about the in-development drug more than a year earlier at a meeting of Cystic Fibrosis Research, Inc., a Palo Alto–based nonprofit organization separate from the Cystic Fibrosis Foundation, back in the fall of 1991, when Shak had come from Genentech to speak at the organization's monthly gathering of families and patients. He'd showed Isabel and the other thirty or so parents and patients photos of the same demonstration he'd given Tulcin and Beall the year before: test tubes of sticky mucus that liquefied after treatment with DNase. The audience had erupted with questions. Parents of sick children and teenagers wanted to know: When would the drug be available? How could they get it? Was it available for compassionate use during the trial? As the patients hammered him with questions, Isabel remembered, Shak had begun to tear up.

Isabel and her sister knew that without an intervention soon, neither would make it to graduation. So at nineteen years old, they tried to enroll in DNase's phase III clinical trials. The problem was, the trial didn't accept all CF patients. With studies on experimental drugs, physicians are reluctant to include very sick patients because, if the patient became sicker, it would be difficult to determine whether it was due to an adverse reaction to the drug or just the normal course of their disease. To be included in the DNase trial, a patient had to have a lung function—called a forced expiratory volume (FEV1)—of 40 percent. Isabel had an FEV1 of 53 percent and easily qualified. But Anabel, whose condition had been worsening the last two years and who was always the sicker of the two, was stuck at 38 percent.

The girls knew that there were ways to boost FEV1, even if only briefly. So in the car on the way to the trial center, Anabel used emergency inhalers designed for asthma to open her airways. At the center, Isabel pounded on her sister's chest and back desperately, trying to clear her lungs enough that she could draw a deeper breath. While waiting for the nurse, Anabel did jumping jacks to move any blockages.

When the nurse came in, Anabel blew into the spirometer—the device that measured lung function. Thirty-eight percent. She tried again. Still thirty-eight. And again. And again. Both girls started to panic.

As a last, desperate measure, Anabel took another hit of the inhaler, exhaled with every muscle in her body—and hit 40 percent. The girls screamed, high-fived, and hugged. It had taken almost twenty tries, but one exhalation proving lung function of 40 percent was enough to get her into the trial.

Phase III was a randomized clinical trial, and neither sister knew whether they would receive the drug or the placebo. As they left the clinic, they were each given a box of small vials, each holding three milliliters of liquid. During the study, the girls were instructed, they would need to empty a vial a day into their nebulizers, which transformed the liquid into a fine mist, and inhale the contents into their lungs. They were each also given a notebook to chart their use of the medication, the physical changes they experienced, and if they suffered any unpleasant side effects.

The girls returned to Stanford, and the next morning in their dorm room, they did their treatment. The first day, nothing changed. The second day, the same. But the morning of the third day, Anabel sat up in bed and told her sister, "Oh my God, I feel so much clearer. This stuff's just pouring out of me, I'm so liquid-y." The mucus clogging her lungs was much thinner and her breath smelled like the bacteria *Pseudomonas*, which infected the lungs of CF patients, a sign that she was ejecting at least some of the dangerous bacteria. Isabel could almost hear a gurgle in her sister's chest, as if everything was unplugging—but she herself, on the other hand, didn't notice anything in hers. There was no improvement in her symptoms. By the end of the week it was obvious that the drug worked—but while Anabel had the real thing, Isabel had gotten the placebo.

By the time the six-month-long trial was complete, Anabel's lung function had gone from barely 40 percent to 55 percent. Once the necessary data was collected and no life-threatening adverse effects had been reported, everyone in the trial was then offered the drug, and Isabel went from 53 percent to 70 percent.

The phase III clinical trials involved 968 patients—3 percent of all the CF patients in the US—and was completed on December 2, 1992. By March 1993, the results of the trial were public, and Genentech filed a note with the FDA to approve the product. In August, the drug, named Pulmozyme, was approved, and in December, just a few days before Christmas, it was available for sale. Taking a drug from conception to patient typically took at least ten years. Genentech did it with Pulmozyme in less than five—something that never would have been possible without the assistance of the foundation, which had helped recruit patients and provided expert care during the trial.[9]

The approval was great news for Genentech. They began selling the drug to CF patients, and then, recognizing that Pulmozyme might also help people with non-CF-related lung diseases like chronic bronchitis, which was common among smokers, began setting up clinical trials for them as well.

In the meantime, Pulmozyme quickly began changing the lives of CF patients. Now that Isabel and Anabel were no longer using up all their calories coughing and trying to breathe, they began gaining weight—in Isabel's case, thirty pounds—and by the time they graduated from Stanford, both girls had grown four inches. With their physical development no longer stunted, they began to look like women rather than prepubescent girls. The energy boost they got from Pulmozyme was extraordinary. Anabel could finally take exercise classes, go hiking with her geology class, and go camping. Isabel started taiko, a high-energy style of traditional Japanese drumming, and continued to study the art form for the next three years, through graduation.

Pulmozyme didn't stop lung decline, but it slowed it down, and it improved quality of life. It postponed Anabel's need for a lung transplant, and it gave Isabel, and so many other CF patients, new hope: if they could just ride the wave of modern technology, staying alive long enough for the next breakthrough, perhaps they could survive long enough to see a cure.

CHAPTER 27

Venture Philanthropy—
Funding Drug Development

1977–1999

If you're going to live, leave a legacy. Make a
mark on the world that can't be erased.

—Maya Angelou

B onnie Ramsey, the woman who directed the CF center at Seattle
Children's and was the lead investigator for the phase II Pulmozyme
trials, hadn't planned on a career in cystic fibrosis—in fact, after a miserable experience during her Harvard Medical School residency at Boston Children's Hospital in the late 1970s, it was one disease she'd hoped
to avoid forever.

At the time, Children's was still the stomping grounds of CF-
treatment pioneer Harry Shwachman, and had one of the largest centers
for cystic fibrosis in the country. It was 1977, just three years after Joey's
birth, and what young Dr. Ramsey saw there was a ghastly illness. Once
the patients became teenagers, they spent most of their lives in the hospital, in a ward called Division 37. There, all the CF kids were bundled
two to four to a room, where they congregated together, supporting the
sickest of their group. At the time, none of the physicians had realized
that keeping these children together enabled dangerous bacteria to hop
from child to child—though it should have been obvious, as they spent
the whole night coughing. As Ramsey sat in the ward one night, listening to the chorus of coughing, she felt as if she had slipped back in time
to a nineteenth-century tuberculosis ward.

That year, an outbreak of flu had sickened many kids with the disease, and the ward was packed. Life expectancy for CF patients then

was about sixteen years. But the flu accelerated death, and each night for weeks the teenagers watched another friend die. As Ramsey made rounds, she could hear the children whispering, stressed and frightened, about who would be next. Many children just lost the will to live. After most of the teens in Division 37 had died, a new population would then fill the wards and the deaths would start again. The tragic cycle continued for the rest of the flu season.

Most faculty and physicians, feeling helpless, left the residents to deal with the dying patients in Division 37. Dr. Shwachman was different. Though he often ignored the residents, he always had plenty of time for people devoted to the care of these critically ill children—people like Joey's therapist John Nadeau—and for the children themselves. As Ramsey accompanied him on rounds visiting patients in her role as resident, she glimpsed his humanity and devotion to these unlucky children. During one thirty-six-hour shift in the CF ward, she passed time flipping through a patient's files Shwachman had left with her. As she read through letters sandwiched between the medical notes, she realized that this crotchety old man was more than the child's doctor; he was the family's priest. He had a deep relationship with the parents of his patients, and they looked to him as a guide beyond the realm of medicine.

Ramsey saw that firsthand one evening when angry parents accused her of not doing enough for their dying daughter, and demanded she call Dr. Shwachman. Ramsey and the other residents hated calling him, but it was clear that the parents didn't trust her judgment, so she obliged. Some thirty minutes later he walked into the room, looked at the patient, and listened to her lungs. Then he stood back and told her parents softly, "There is no more. We just have to make her comfortable." With those simple words, he hugged the mother and walked out. Even though Ramsey disliked him as a supervisor, she admired how he connected with these families.

Still, her respect for Shwachman's work didn't change how she felt about the disease. *Who would want to be involved in this illness?* she wondered one early morning as she and fellow residents, exhausted and overwhelmed by the parade of death, sat crying on the stairs.

After her two-year pediatrics residency, Ramsey moved back home to Seattle. She'd planned on pursuing a fellowship in the area of childhood blood disorders and cancer—but the requirements specified by the American Board of Pediatrics had changed, mandating a three-year-long residency. So she completed her third year at Seattle Children's. It was then she realized she didn't want to pursue oncology; instead she began a fellowship in general pediatrics the same year her daughter was born.

In 1978, during her fellowship, she began working with a compassionate pediatrician named Edgar K. Marcuse, the chief of the medical clinic. He was fond of Ramsey and recognized her talent, so he invited her to come and work in the outpatient pediatrics department for a year. But just a few days after she had begun working there, Marcuse asked her if she would mind working in the hospital's CF clinic instead. She agreed; she'd had a lot of experience with the disease in Boston, and despite her traumatic experience, she knew plenty about how to care for these children. Marcuse told Ramsey she would work with a supervising physician. But within the first month, the clinical director quit, leaving Ramsey alone in a clinic packed with sick children.

She called the hospital director, Jack Docter, to explain the situation. Docter was also the director of Seattle Children's CF Center, and had trained with Dorothy Andersen at Columbia. But once he became hospital director, he'd stopped seeing his own patients. Ramsey lacked privileges to admit patients to the hospital because she was still in a fellowship program, but she could still care for them, as long as an attending physician cosigned her charts.

Docter thought for a moment, then told her to finish seeing the patients. He would sign the charts each day until she completed her fellowship.

Ramsey, who looked about ten years younger than her thirty-one years, stepped into the waiting room and explained that the doctor had left. Some mothers exchanged panicked glances, then looked at Ramsey. Are you leaving us, too? one asked.

Seeing their stricken expressions, Ramsey shook her head and smiled, motioning for the next patient.

After a few weeks of this—Ramsey seeing the patients, Docter signing the charts—Docter asked Ramsey if she would lead the clinic for the next year, until the newly hired head of pulmonology arrived. Docter would provide some backup and continue to sign all the medical charts. It was an honor to be offered such a responsibility, but also daunting. Ramsey wasn't a pulmonologist, and she would need to teach herself the art and science of CF medicine. So over the next year she would spend hours on the phone calling other care centers around the country, asking them how to handle various symptoms and what the best treatments were. She quickly learned which physicians were kind and would take her calls and patiently explain what to do. Two leading CF physicians, Dr. Carl Doershuk at Case Western and Dr. Mary Ellen Wohl at Boston Children's Hospital, were her favorites. Before long, she was taking care of all Seattle Children's CF patients, in the clinic and in the hospital.

After that first year had passed, Docter invited Ramsey to his office for a chat. He told her that the new head of pulmonary would soon arrive. Ramsey began wondering what it would be like to have a supervising physician after spending the past year running the clinic solo.

Then he told her that she should run the CF program herself.

Ramsey leaned forward. Had she heard him correctly? Was he making her director of the CF center after just one year? Docter explained that the incoming physician would lead the division of pulmonary diseases; there were many other lung disorders aside from CF. "But I'm giving the clinic to *you* because I think *you* are going to make a difference in cystic fibrosis."

Working with cystic fibrosis during the 1980s was no less grim than a few years earlier, when Ramsey was at Boston Children's. The treatments weren't aggressive enough; once a child's lungs were infected, they remained in the hospital for months, hooked up to an IV that dripped antibiotics into their veins. Once physicians and parents realized their child had contracted an infection that would kill them, they became passive, focused solely on minimizing the suffering. All they were doing, Ramsey felt, was keeping these adolescents comfortable until they died.

One year in the mid-1980s, Ramsey had a particularly bad week, losing two patients she adored—a fifteen-year-old girl and a twenty-year-old young man. They were some of her oldest, and she had seen them in the hospital every day for several years. After the second funeral, she decided she couldn't keep treating these kids only to watch them die, one after the other. If she was going to stay in this field, she needed to drive the progress she was so desperate to see.

Her first opportunity came in the early '90s, through the Cystic Fibrosis Foundation. Although Warren Warwick's survey some twenty-five years earlier had proven LeRoy Matthews's treatment strategy was the most effective, the national foundation had lacked the needed clout in the '60s and '70s (before Dresing centralized the foundation and reformed the medical care practices at all the centers) to enforce his protocol. As a result, the landscape of CF medicine remained a hodge-podge of treatments. Maverick physicians like Shwachman, Matthews, and Warwick stubbornly did their own thing, and resisted the idea of standards. And because the performance of each CF center had previously been known only to the foundation, the centers had no incentive to reform medical practices or strive for better outcomes for their patients.

Now, however, after Tulcin, Dresing, and Beall's reforms, the centers depended on national headquarters for their funding—and with this new leverage, national was able to improve CF treatment. They could take a bird's-eye look at the disease and make sweeping recommendations they felt would benefit all the sick children. To do this, the foundation had chosen a consensus group of physicians—including heavyweights like Paul di Sant'Agnese, Shwachman, Matthews, and Doershuk, along with Ramsey, the only female physician in the group—to review the data the foundation had been collecting over the years and use it to improve treatment and dietary recommendations.

The first issue they tackled, at a 1991 conference, was nutrition. Shwachman's team had decided that because CF patients couldn't absorb fat, which made them gassy and their stool foul, they should follow a fat-free diet. And because many physicians followed Shwachman's lead, CF patients in the United States were skinny as rails. When Ramsey began her residency in Division 37 in 1977, the kids were so horrendously malnourished, they looked to her like concentration camp survivors. In Canada, however, physicians had adopted the opposite strategy, insisting patients eat lots of fatty and sugary foods, hoping that more nutrients would be absorbed and nourish their fragile bodies. It was a better approach; the Canadian kids with CF were healthier and survived almost a decade longer.

If there was one thing everybody could do to help these kids, thought Ramsey, it was feed them. After looking at the data, the consensus group of physicians now agreed on their first recommendation: kids with CF had to be bombarded with calories, in any way possible. If, like Joey, they needed a G-tube to help them put on weight, then so be it. They also needed supplements and vitamins to help their bodies survive the abuses of the disease.

As Ramsey left the conference and flew home, she felt a sense of calm she hadn't felt in years. She had helped institute nutritional guidelines that would make all CF patients a little healthier.[1] And through her contributions at the conference, and her position as director of the Cystic Fibrosis Center at Seattle Children's, Ramsey was gaining respect and influence at the foundation.

Then came the opportunity to help Genentech's Steven Shak with the Pulmozyme trials, introducing him to patients, helping him define treatment goals, and finally leading the phase II trials in 1991. The drug improved breathing, and patients taking it had fewer episodes of infections requiring antibiotic treatment and hospitalization. Patients also reported feeling more energetic and a little more optimistic that they would spend less time in the hospital. Ramsey watched, amazed, as her patients coughed and coughed after inhaling the drug—finally able to empty their chests of the toxic muck inside. This was the first and the only effective therapy[2] for cystic fibrosis patients, and Ramsey had played an important role in its clinical trials. But Pulmozyme wasn't the only cystic fibrosis drug she helped bring to market.

Before the Pulmozyme trials began, Ramsey was working with her mentor, Arnold Smith, who specialized in infectious diseases, on developing a new antibiotic treatment for the lungs. Ramsey and Smith had known each other at Boston Children's Hospital, where Smith had been trying to figure out the best doses of antibiotics for CF patients, and had moved to Seattle around the same time. Smith was on the board of the CF Foundation, and during one of his visits to Bethesda in the early 1980s he had heard from an adult patient and fellow board member that, rather than taking his daily antibiotic, tobramycin, intravenously, he had been inhaling it. (Most patients went to the hospital for the IV treatment, but some, like this man, were able to administer them at home with a nurse's help.)

Smith was surprised. No inhaled antibiotic had yet been approved by the FDA. But this man and several other patients he knew were emptying their liquid antibiotics into their nebulizers and inhaling them into their lungs. The man told Smith it helped him breathe and seemed to wipe out the infections faster. And he wasn't alone, Ramsey and Smith would later learn; between 3,000 and 5,000 CF patients were inhaling their IV antibiotics, a dangerous practice because the preservatives in the IV formulation could damage their already-compromised lungs.[3]

The concept of aerosolizing antibiotics wasn't new; LeRoy Matthews and Harry Shwachman had both previously proposed the idea. However, they hadn't been able to get it to work. They'd lacked the

technology to vaporize the liquid antibiotic into the superfine mist necessary to let it plunge deep into the small airways where the bacteria
lurked. Other physicians had tried, too, but results had been inconsistent and the trials small. So efforts to develop an aerosolized antibiotic
had fizzled.

When Smith returned from Bethesda, he told Ramsey about the
patient. He was skeptical whether the inhaled drugs would be effective,
he said, but this patient was adamant about the benefits, so he felt obligated to investigate.

Until that point, the dosage of tobramycin necessary to wipe out
lung infections had only been calculated for intravenous use. Before
any sort of trial could begin, Smith had to figure out how much inhaled
tobramycin was needed to do the same. To determine the necessary
dose, he typically would have exposed bacteria growing on a petri dish
to an escalating range of concentrations.[4] But things were tricky with
CF patients, because the bacteria infecting them weren't exposed, lying
on a surface, like with an infected cut on the skin. They weren't inside
the lung tissue, either—an area that could be targeted with IV antibiotics. These bacteria were in the airways, mixed with viscous phlegm that
provided a layer of protection from the drug. And when Smith collected
sputum from a patient and tried to destroy all the microbes in it with
tobramycin, he found it almost impossible. The sticky phlegm appeared
to bind to the drug and block it from reaching its target. So he kept
ramping up the amount. Eventually, it worked.

When Smith told Ramsey that he'd had to boost the concentration of the antibiotic twenty-five-fold to destroy the microbes, she was
shocked.[5] Tobramycin belonged to a much larger family of antibiotics
called aminoglycosides that were typically given intravenously at the
hospital. The drug had to be carefully monitored because at high doses
it caused deafness and kidney damage. Infusing a patient's body with a
dose that was twenty-five-fold more concentrated was clearly too dangerous. Yet if they could confine the drug to the lungs, perhaps they
could keep the rest of the body safe.

Aerosolizing the antibiotic to blast it deep into the airways was
going to be challenging. Tobramycin was an elephant of a molecule, and
figuring out how to make it fly was tricky. There were refrigerator-sized
machines in hospitals that could be used to transform the liquid into
an aerosol for the early trials. But if these were successful, the patients
would need a portable device for home use.

In late 1986, Smith flew to Bethesda and met with Bob Beall; he needed $1.6 million to complete phase I and phase II clinical trials to ensure that the drug was safe and effective when inhaled. He also had another ask. Ramsey and Smith were physician researchers and not in the business of building and marketing medical devices. So, if these trials went well, they'd need help identifying a company that would collaborate with them on the phase III clinical trials, find a suitable nebulizer for patients' home use, and commercialize and market the drug.

Beall agreed to fund Smith and Ramsey immediately. He also agreed to procure the tobramycin needed to run the trial. After checking in with Bob Dresing, Beall called the drug's manufacturer in Hungary and ordered $1 million worth. A few days later the two Bobs picked up their precious shipment from Newark Airport, several kilograms of white powder, then drove it back to CFF headquarters in Bethesda.

Back in Seattle, Smith and Ramsey coordinated with seven CF centers around the country to run the trials. The first phase I trial, focusing on drug safety, began in March 1989.[6] To avoid bias, neither the patients nor their physicians—including Ramsey and Smith—knew if or when their patients were on the drug. Only a few people at the trial headquarters had the codes that revealed which patients got what.

The seventy-one-patient phase II trial, which Ramsey led alone, tested whether the drug actually killed the microbes, reduced the infections, and either kept patients out of the hospital or made their stays shorter or less frequent.[7] Although Ramsey tried not to scrutinize her own patients in the trial, she could sense the drug was working.

When, after two years, trial analysts finished crunching the data, the results were exciting. The inhaled drug stayed in the airways and didn't trespass into the bloodstream where it could cause side effects. There were no reports of hearing loss or kidney damage. The drug was safe and effective.

Once the phase II results were known, Ramsey and Smith told Bob Beall that they were ready to plan the phase III trial and would need the commercial partner they'd discussed.

Beall knew that finding a partner wasn't going to be easy. This new application for tobramycin didn't look like a moneymaker. Tobramycin itself had been approved to treat patients since 1975,[8] and generic versions of the drug were already available beginning in 1991.[9] There also wasn't a single inhalable antibiotic currently on the market—both a good and a bad thing for prospective pharmaceutical company partners.

The company that took on this challenge and patented a new formulation of tobramycin would, under the Orphan Drug Act, have no competition from other inhalable antibiotics for seven years. But the lack of similar existing products was viewed by many in the industry as a bad omen. Perhaps others had already tried to manufacture an inhalable antibiotic and failed. Perhaps trying to create this product was a waste of money. Plus, even if everything went well and an aerosolized version of tobramycin emerged, there was no guarantee the market was big enough to make the effort worthwhile. This was, after all, an orphan disease, defined as affecting fewer than 200,000 people. Even within the orphan category, CF was a rare disease, with only 30,000 potential customers. The number of patients the drug would help was small, compared to a common condition like chronic obstructive pulmonary disease, or COPD, which affected some 16 million Americans.[10] But Beall wasn't deterred by these perceptions. He was determined to lure industry into the war against CF, because there was a limit to what academic scientists, like Ramsey and Smith, could do.

Genentech had discovered and developed Pulmozyme from the beginning; the foundation had only helped coordinate and run the trials and market the drug. But inhaled tobramycin was very much an in-house project. So while Ramsey and Smith were working on the plans and logistics for phase III, Beall was promoting the phase II data, hoping to hook a company. The foundation had already funded the riskiest portion of the project—the preclinical studies that Smith had done in his lab, and the first two clinical trials that proved that this new delivery route for the antibiotic was safe and effective. All that remained, according to Ramsey and Smith, was formulating the drug and testing it in a small nebulizer that could dispense the drug in a final phase III trial.

Wilbur Ganz, CEO of a new startup named PathoGenesis, heard about Ramsey's trial and was interested in commercializing inhaled tobramycin. He recognized that the collaboration with the foundation would be advantageous for PathoGenesis; once the drug was approved, the foundation could also help with marketing and distributing it to patients—another huge savings.[11] Pharmaceutical companies often spent as much money marketing new drugs to physicians—through sales-force visits, advertising, conferences, and junkets—as they spent on the actual research to develop it. Most patients who would use inhaled tobramycin were seen at the CF clinics, so there was no need to advertise. The seven-year exclusivity period guaranteed by the Orphan Drug Act, plus the other incentives and tax credits that made developing orphan drugs more financially palatable, also made commercializing the drug appealing.

Ganz hired a headhunter to find a scientist who had worked with aerosolized antibiotics and had some CF experience. It was a short list that led them to Bruce Montgomery. Montgomery had become a hot commodity at age thirty-six when he developed and aerosolized pentamidine for preventing pneumocystis pneumonia,[12] the big killer at the beginning of the AIDS epidemic in San Francisco in 1983.[13] While the drug's approval in 1989[14] made Montgomery somewhat of an overnight celebrity, it also made him obsolete at the University of California, San Francisco, where he'd spent the last five and a half years. He was a pulmonologist working in HIV, tasked with treating pneumocystis, but his aerosolized pentamidine was so effective that he didn't have any more patients.

By September 1989, Montgomery had landed at Genentech after talking with his friend Steve Shak, where, at Shak's behest, he worked on DNase, later known as Pulmozyme. Montgomery's specialty was translational medicine: taking basic scientific discoveries and figuring out how to translate them into a medicine that could be tested in clinical trials. He'd solved how to administer the DNase protein with a nebulizer, creating Pulmozyme. Then, working together with Bonnie Ramsey and another investigator at the NIH, he designed the phase I clinical trial, before going on to work on an asthma drug, Xolair, and coinvent a psoriasis drug called Raptiva.

Ganz hired Montgomery from Genentech on December 1, 1993, to get inhaled tobramycin ready for phase III trials. But before PathoGenesis could begin work on the drug, a few more issues needed to be resolved. First, Montgomery had to broker an agreement between Ramsey, who was representing Seattle Children's, and Beall, who was representing the foundation. The foundation believed that they held the rights to all the data from the phase I and II tobramycin clinical trials and the early preclinical work that Arnold Smith had done in the lab, because they had paid for it, to the tune of $2.6 million. The lawyers for Seattle Children's argued that the data was theirs because Ramsey and Smith were their faculty and they did the work. Neither group would compromise, which meant that PathoGenesis couldn't license the work and move forward with the phase III trial.

The two groups had been in a stalemate for a year when Montgomery, after three weeks at the company, suggested to Beall and Ramsey that PathoGenesis pay a 5 percent royalty on sales if the drug went to market—half to Children's, half to the foundation. His solution appealed to both sides. The royalties would be fair payback for all the time, money, and effort both Seattle Children's and the foundation had invested. If the drug were approved, and sales were good, both parties would receive a stream of income they could invest in other promising research. Requesting royalties in exchange for an investment was standard practice in the corporate world. But this was the first time a health nonprofit had ever raised money through philanthropy and invested it in a for-profit company in return for future royalties.

Another major issue the parties needed to deal with was price. Montgomery estimated that just the raw quantities of tobramycin that Ramsey used in her phase II trials would cost patients more than $60,000 per year—*without* the standard markup a pharmaceutical company would impose on the product. This was during an era when almost no drug was priced at more than $30,000 a year.

Montgomery assured Ganz that they could make the treatment less expensive, in part by finding a nebulizer that delivered the drug into the lungs more efficiently; the ones Ramsey had used in the early trials only delivered some 5 percent of the medicine. One of the newer, more efficient nebulizers on the market would cut the quantity of the drug needed and, also, its cost. Montgomery also promised to find a manufacturer that would make the drug more cheaply.

The final issue was the drug that Ramsey had used for the first two trials had to be made daily for the patients. Montgomery needed to create a stable formulation that could survive on a pharmacy shelf until sold.

Solving all of these problems took about eighteen months. But by 1995, just over a year after the FDA approved Pulmozyme, PathoGenesis was ready to move forward with phase III.

With the foundation's infrastructure and Ramsey and Smith's trial experience, Montgomery knew the clinical trial would be efficient—the two scientists could lead the program after helping to recruit patients. They only needed 520; because tobramycin had been on the market for close to twenty years and its effects were well studied, the phase III trial didn't need to be as large as the one done for Pulmozyme.

The CFF's involvement was a huge advantage. On average, in drug trials, half the designated trial sites fail to recruit enough patients.[15] But because the foundation had spent decades nurturing its network of CF care centers, the centers' doctors were quickly made aware of any clinical trials for cystic fibrosis. And the patients and families were willing to try any drug that might improve their condition. Patients as young as twelve understood that unless they participated in medical research, it was impossible to create new medicines. Many had volunteered for the Pulmozyme trials; others were, in 1993, just beginning to enroll for gene therapy trials. Now there was a potentially more effective antibiotic to try.

Phase III studies began in August 1995, with patients volunteering for the three-month trial at sixty-nine CF centers across the United States, and it wasn't long before Ramsey, Smith, and the team at PathoGenesis recognized that this medicine provided a huge bump in patients' quality of life.[16] In some cases lung function expanded by up to 11 percent—air volume equivalent to a couple of eight-ounce cans of soda.[17] There were 25 percent fewer admissions to the hospital in the group receiving treatments, compared to the group that got the placebo, and for those who did end up in the hospital, their stays averaged about five days, compared to the placebo group members' eight. When participants' sputum was analyzed, it contained significantly less of the nasty *Pseudomonas aeruginosa* bacteria—proof that the drug was successfully destroying these bugs as Smith's earlier experiments had predicted.

PathoGenesis submitted the phase III data for Tobramycin Inhalation Solution, which they named TOBI, to the FDA on July 10, 1997,[18] and the drug was approved in December 1997. It was a huge milestone for the foundation and for Ramsey personally. Quality of life improved dramatically for her patients, who could now take their antibiotics at home with their families rather than spending months of their life in the hospital tied to an IV. Ramsey got an even better reward for all her efforts when she bumped into the mother of one of her patients while bringing a medical student to the CF clinic. I just want to thank you, Dr. Ramsey, the mother said. "I want you to know that you and the TOBI you helped develop have changed my child's and our family's lives. My daughter is now at home and able to be a normal child."

CHAPTER 28

The Gene *Is* the Medicine
1989–1991

The advance of genetic engineering makes it quite conceivable
that we will begin to design our own evolutionary progress.

—Isaac Asimov

In 1989, as both Pulmozyme and TOBI were in early stages of development, the CF Foundation was increasingly investing its money in gene
therapy, now that the CF gene had been found.

Gene therapy was a wholly new approach to treating disease. The
idea was to give sick individuals with mutated genes a normal version of
the gene. In the case of CF patients, the mutated CFTR gene couldn't
produce a functioning protein. But if researchers could deliver a healthy
version of the CFTR gene to specific cells in the patient's body, the
cells could use these new instructions to build working CFTR proteins.
If the gene was delivered to the right cells and remained stable, these
instructions could remain in the body for years. The process, in theory,
was analogous to the old adage about giving a man a fish, which fed
him for a day, versus teaching him to fish, which would feed him for
a lifetime. Giving a patient a protein, like insulin, was a quick fix; giving the patient a gene provided the body the instructions to make the
protein itself.

However, the challenge with gene therapy was that unlike giving
patients a drug—a capsule filled with a powder, liquid that could be
swallowed, or an insulin injection—delivering genes was tricky. Genes
couldn't be encapsulated in a pill and swallowed, because the enzymes
and acids in the stomach would destroy the DNA. They had to be
placed into the cells in which they would normally be active—like a

hand-delivered package. Fortunately, nature had developed such a package: the virus.

In the late 1980s, the most promising gene therapy strategy was to use specially designed new viruses as a delivery vehicle to get healthy DNA into targeted cells. Viruses are experts at hijacking cells, injecting their own genetic material, and usurping the cell's machinery to replicate and produce a new generation of viruses. So it wasn't a stretch for scientists to tweak the viruses—removing the disease-causing DNA and replacing it with a therapeutic gene—and use them to deliver the cargo to the "sick" cells. In 1980, a graduate student at Stanford University had successfully used a virus to introduce a healthy version of a mutated gene into cells from a patient with another genetic disorder, "fixing" the defective cells.[1] And while repairing a diseased cell in the lab was a long way from curing a human being, it was proof of the concept. The same approach might work for CFTR genes in patients. It was an elegant, simple-sounding solution. The body would accept the new gene. The disease would be cured.

CF Foundation president Bob Dresing and medical director Bob Beall were both believers, confident that this new breed of researchers they called the "gene therapy jocks" would be able to "solve" the disease. Dresing had by that point completely reorganized the foundation to drive medical research toward a cure, and when the Toronto and Michigan teams published their discovery in the September 1989 issue of *Science*, Beall immediately began funding gene therapy experiments.

One of Beall's grant recipients was James Wilson, an ambitious young scientist who occupied the lab next door to Francis Collins at the University of Michigan. Wilson was rapidly making a name for himself in gene therapy, which in 1989 was less than a decade old.

Wilson's interest in medicine crystallized as a graduate student in 1977, working in the lab of William Kelley at the University of Michigan (the same William Kelley who would recruit Francis Collins, as well as Wilson, a decade later). Kelley suggested to Wilson that he look for the mutation on the X chromosome that caused Lesch–Nyhan syndrome, a rare, inherited behavioral disorder that was caused by a shortage of an enzyme called hypoxanthine-guanine phosphoribosyl-transferase (HGPRT) in the brain. Many suffering from Lesch–Nyhan couldn't walk, suffered from muscle spasms, were impulsive, and mutilated themselves.

Kelley gave Wilson the funds to jet around the country and the world to meet physicians and collect blood samples from patients with the disease. In a few cases, Wilson brought patients back with him for additional studies. That included a fourteen-year-old named Edwin, whom Wilson brought three times from the institution where he lived in North Carolina, to the Michigan campus, where he'd stay for a month working with several researchers.

As Wilson got to know Edwin, they became friends, and he learned that many of the hallmark Lesch–Nyhan behaviors—spitting, hitting, and abuse—were stress triggered. Because these patients had difficulty speaking, they were often labeled "mentally retarded"— which Wilson discovered wasn't true. Edwin was sassy and aware. On campus, Edwin participated in clinical research and gave blood. Then Wilson would take him to football games. The lab team doted on Edwin and treated him like a king, which made returning to North Carolina traumatic for him. Wilson would take him back, sedated with Valium for the plane journey, and his mother would meet them at the airport. Every time, Edwin would wail and cry in distress. It wrenched Wilson's gut.

The third and final time that Wilson brought Edwin back to North Carolina, he had good news. He had identified the mutation in Edwin's DNA that caused his terrible disease. For Wilson, it was a scientific and intellectual high; he could now tell the world that the reason this young man behaved this way was because he had a mutation in his HGPRT gene that disrupted the structure of the protein and destroyed its function. When the plane touched down, Wilson struggled to help Edwin out of his seat. He was particularly agitated and squirming in his restraint as he made his way down the stairs.

As they reached the ground, Wilson looked eagerly toward Edwin's mother, who was waiting tensely, her brow furrowed, to take her son. Wilson had been anticipating this moment for weeks and was excited to share the discovery with her.

But once he finished explaining the research, Edwin's mother paused and, without a lot of emotion, asked him, "Well, how will this help Edwin?"

The question hit Wilson hard. He had no idea, and he had to tell her so.

But Wilson *did* know how it could help; it just hadn't been done before.

The O'Donnell family in Falmouth, MA, in 1984.
Joey is ten years old.

Kathy and Joey O'Donnell, 1983. Joey
is nine.

While Joey was young, Joe coached and played in the
Hosmer Chiefs, a semiprofessional amateur baseball
team. Joey was the bat boy. This is 1980 and Joey is
six years old.

The O'Donnell family in 2019. From left: Mike Buckley and two-year old Blair, Kate O'Donnell, her sister Casey and five-month old JD, Joe and Kathy O'Donnell.

Sixteenth-century Dutch professor of anatomy Pieter Pauw is believed to be the first to describe cystic fibrosis after performing a public autopsy of an eleven-year-old girl.

"On January 16th 1595…I conducted an autopsy on an 11-year old girl said to be bewitched… Inside the pericardium, the heart was floating in a poisonous liquid, sea green in colour. Death had been caused by the pancreas, which was oddly swollen…with a hard and woody texture. When it was removed the interior was found to be brightly coloured, a kind of hard white viscous mass. The little girl was very thin, worn out by hectic fever." (Quote from R. Busch, "On the History of Cystic Fibrosis," *Acta Univ. Carol* 36 (1990): 13–15.)

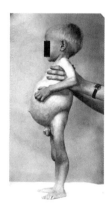

Left: Dorothy Andersen, MD, PhD, the first to realize that some patients diagnosed with celiac disease actually suffered from an entirely different disorder, which she called "cystic fibrosis of the pancreas" in her seminal 1938 paper.

Right: A child that Andersen diagnosed with cystic fibrosis of the pancreas in the early 1940s. At thirty-four months old, he looks half his age and is severely malnourished—with skeletal limbs and a hallmark cystic fibrosis potbelly—and so weak that he can't stand without help.

Paul di Sant'Agnese, the first to notice that people with cystic fibrosis had abnormally salty sweat. Dorothy Andersen first hired di Sant'Agnese to diagnose cases of the disease.

Harry Shwachman, the formidable Boston Children's Hospital doctor and fierce advocate for children with cystic fibrosis. He was chief of the Division of Clinical Nutrition, which eventually became the largest cystic fibrosis center in the world.

Paul Quinton, pioneer of cystic fibrosis science, who diagnosed himself with CF in college and figured out why his sweat is so salty.

Doris Tulcin, the woman who turned the ailing Cystic Fibrosis Foundation around and transformed it into a fundraising machine, with her ally and successor as CFF president Robert Dresing (left) and master fundraiser Joe O'Donnell (right).

Bob Beall, visionary scientific leader who, during his tenure as the foundation's director of science and medicine, launched specialized labs around the country exclusively focused on cystic fibrosis, and as president and CEO of the foundation, launched its partnership with Aurora Biosciences Corporation.

Mike Boyle, CF physician who figured out a strategy for testing whether Vertex drugs would work for patients with rare mutations, with patient Sarah Foose. He took over as CEO and president of the foundation in January 2020.

Preston Campbell (right) with chair of the Cystic Fibrosis Foundation's National Board of Trustees, Cam McLoud (left). Campbell was hired to oversee the foundation's medical programs, including the 113 care centers; helped create the Therapeutics Development Network; and, in 2015, took over as president and CEO after Bob Beall.

Bonnie Ramsey spearheaded the effort to launch the Therapeutics Development Network, which was key for all the clinical trials.

Lap-Chee Tsui led the CF gene-hunting team in Toronto that included Batsheva Kerem, Jack Riordan, Johanna Rommens, and Pittsburgh collaborator Aravinda Chakravarti.

Robert Dresing, far left, and Bob Beall, far right, present three of the gene hunters with the foundation's Paul di Sant'Agnese Award.

From left to right: Dresing, Lap-Chee Tsui, Paul di Sant'Agnese, Francis Collins, Jack Riordan, and Beall.

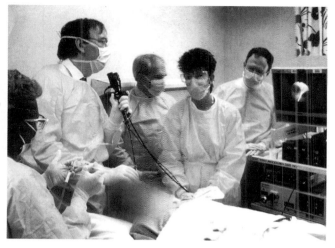

The first gene therapy trial for cystic fibrosis at the National Institutes of Health. Gerry Mc-Elvaney inserts a bronchoscope down the throat of a cystic fibrosis patient and delivers a dose of therapeutic virus, as gene therapy pioneer Ron Crystal watches on the monitor.

Above: Aurora's first proprietary high-throughput screening facility, a fully automated system of conveyer belts and robots that rapidly retrieves chemicals, housed in the racks in the background, for testing on cells.

Left: Joshua Boger, founder and CEO of cystic fibrosis drug pioneer Vertex Pharmaceuticals, at the company's original Cambridge location.

One of the keys to Aurora's high-throughput screening was its iPhone-sized test plates, each of which contained 3,456 miniaturized wells in which cells could mingle with a particular drug. When the desired chemical reaction occurred, a fluorescent reporter dye would emit a different-color glow, indicating success.

Vertex Pharmaceuticals scientists and Cystic Fibrosis Foundation drug development leaders at the San Diego campus on September 19, 2007.

Sitting from left to right: Amanda Kuchta, Fred Van Goor, Melissa Ashlock, Paul Negulescu, Viji Arumugam, Catherine Quan, Jinglan Zhou, Sabine Hadida, —. Standing from left to right: Chris Penland, Bill Burton, Kim Straley, —, Dennis Hurley, Bob Beall, Diana Wetmore, Bonnie Ramsey, Tom Knapp, Ashvani Singh, Preston Campbell, Frank Accurso, Peter Grootenhuis, —, Jeff Wine, John Joubran, Ray Frizzell, Brian Bear, Denny Liggitt, Peter Mueller, —, Erica Salmon, Jason McCartney, Heather Clark.

Edwin needed a healthy version of the HGPRT gene. While working on figuring out Edwin's mutation, he'd read about Richard Mulligan, a graduate student at Stanford University, who in 1980 had "fixed" defective skin cells from a patient with Lesch–Nyhan disease by using a virus to deliver a healthy copy of the mutated gene.[2] Mulligan had taken a healthy version of the HGPRT gene from a bacterium, inserted it into a virus, and then used the virus to infect the malfunctioning cells from a Lesch–Nyhan patient, delivering the healthy gene. The bacterial gene was so similar to the human version that it repaired the defective cells in a test tube.

Others had coaxed cells to accept and read foreign genes before. But Mulligan's work captured Wilson's attention because it was one of the first demonstrations that involved a real human disease. It was a discovery that would change the course of medicine.

After the episode with Edwin's mother, and having met so many other patients with Lesch–Nyhan and other disorders, Wilson knew he wanted to *treat* patients with genetic diseases, not just find the mutant gene. Gene therapy was the logical approach. To Wilson, it sounded simple: just pick your gene and put it in the cell. But some cells were easier to access than others. Treating Lesch–Nyhan would involve inserting the healthy HGPRT gene into brain cells, and Wilson knew that with current technologies that was far too difficult. He needed an easier disease.

After completing his degree program at the University of Michigan, Wilson went to Massachusetts General Hospital for his residency.[3] There he met a little girl who suffered from a rare disease called familial hypercholesterolemia (FH), which raised her "bad" cholesterol—LDL—to deadly levels. By age six, she'd had multiple heart attacks and was now waiting for heart and liver transplants. Her case weighed on him even after he moved to MIT in 1986[4] to do a research fellowship with Richard Mulligan, whose work at Stanford had first inspired Wilson's passion for gene therapy. Perhaps gene therapy could help this girl. Patients with FH were sick because their liver lacked a receptor to mop up LDL. If he could transform the liver cells with a gene that encoded the missing LDL receptor, using a virus as the delivery vehicle, he should be able to treat the disease.

After preliminary experiments with rats,[5] Wilson showed he could use the human LDL receptor genes to fix liver cells from rabbits suffering from FH.[6] When "cured" cells were later transplanted into rabbits, the rabbits' cholesterol levels plummeted—initially. Then they rebounded.[7] But just a year later, in 1991, Wilson figured out how to transform the cells in a way that would keep cholesterol low for the long term.[8]

"Moving with a swiftness that far exceeds expectations of just a few months ago, researchers are preparing to use a new type of gene therapy against a lethal hereditary illness," Natelie Angier wrote of Wilson's work in the *New York Times*,[9] just before his rabbit experiments were published in the journal *Science*. "This is the sort of true gene therapy we've all been striving for," said Dr. Dusty A. Miller of the Fred Hutchinson Cancer Research Center in Seattle. "You do the treatment once, and it persists for the lifetime of the patient. I'm very excited by this."

The progress was remarkable—from the discovery of DNA's structure in 1953 to new ways of reading and manipulating the code of life in the 1970s, and now, in the early '90s, the potential to rewrite that genetic code to cure disease.

WHILE PURSUING HIS WORK ON FAMILIAL HYPERCHOLESTEROLEMIA, Wilson had also worked on a handful of other diseases since moving back to Michigan in 1988. His involvement in CF, however, was serendipitous. In the summer of 1989, Wilson was sitting in his office when Francis Collins dropped by to ask a favor. The keynote speaker that Collins had invited to the North American Cystic Fibrosis Conference had dropped out. He wanted Wilson to fill in and talk about gene therapy for CF.

Wilson knew that Lap-Chee Tsui and Collins were close to the gene—but had no idea they had the mutation and were in the final sprint. Still, he worried that such a talk was premature.

Just do it theoretically, Collins coaxed. How would you do it if you had the gene? He added that the audience would be a mix of scientists, patients, patient advocates, nurses, and respiratory therapists, so the talk had to be fairly general anyway.

Wilson agreed. Then, in the last week of August 1989, six weeks before the talk, he learned with the rest of the world that Tsui and Collins had discovered the mutation that caused CF. He was stunned that, with Collins working just next door, the news hadn't leaked to him earlier.

He rewrote his talk, because now, with the gene in hand, gene therapy was suddenly a very exciting possibility. He flew into Tampa,

Florida, knowing little about the cystic fibrosis community or what to expect at the meeting.[10] When he arrived a few minutes late and walked straight to the podium, music was pulsing, and the atmosphere was charged with the emotional energy of a rock concert. It was almost religious. In front of him were 1,500 to 2,000 people energized by the discovery of the gene. As he spoke, he could feel the excitement and anxious anticipation of the crowd as he explained how gene therapy worked and how this might—if they could get the gene into the lung's airway cells—provide a cure for their children, grandchildren, brothers, and sisters.

The DNA sequence of the functional CFTR gene that Tsui's team had painstakingly sequenced was the recipe for making a healthy CFTR protein; anyone with the recipe, Wilson explained, could synthetically manufacture the DNA and try to insert it into cells. And the most popular approach at that time was to use a virus.

Viruses, Wilson elaborated, straddle the line between living and dead. They don't think, or eat, or sleep. The only thing they do is infect, multiply, destroy. And that makes them the ultimate delivery device. They carry a set of genetic instructions—written in either DNA or RNA—that spell out the instructions to make more viruses. But they can't make more viruses alone. To reproduce, they have to latch onto a cell and inject their viral genetic code, which hijacks the cell's own machinery to make hundreds of new viruses. Once the cell fills up with them, it explodes, releasing a swarm of new viruses that then infect new cells and start the process all over again.

But, Wilson said, biologists had figured out how to hack their genetic instructions and de-weaponize them—alter a virus's DNA so that it could infect a cell and deliver new DNA, but then do nothing more. These neutered viruses, unable to multiply or cause illness, became tools for therapy rather than disease. Biologists would insert therapeutic DNA—like the functional version of the CFTR gene—into the viruses and then add them to a test tube or flask of cells they wanted to transform. Just as they were designed to do, the viruses would then grab onto cells and inject their DNA, like microscopic six-legged delivery drones.

As Wilson spoke to the crowd, he knew gene therapy wasn't going to be as simple as it sounded. But nothing he said would have convinced them of that. As far as the crowd was concerned, now that they had the mutated gene, repairing it with gene therapy was the inevitable next step.

Wilson returned from the conference feeling inspired, and he told Collins that he wanted to work with him to make CF gene therapy a reality.

There were, however, two challenges that all gene therapy faced. The first was how to get enough of the healthy gene into the right cells, so it could actually make a difference. The architecture of the lung was complex, making the task particularly difficult with CF. The trachea splits into two bronchi leading to each lung; then these two airways split another twenty times, generating a labyrinth of air passages with a surface area of several meters. How could they get the healthy gene to penetrate these thousands of pipes? How many of these airways would need to be corrected for the patient to feel any benefit?

The second challenge was finding the right delivery method. There were thousands of types of viruses, and researchers had to find the best one for the job. An obvious candidate was a cold virus, which had a natural talent for entering the lungs, infecting the cells, and making

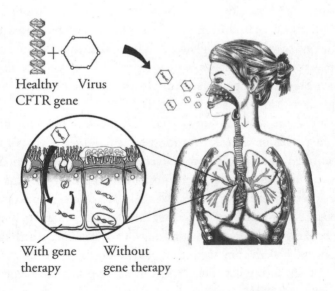

In gene therapy, a healthy copy of the CFTR gene is inserted into a virus. Then, the nasal or lung tissue is exposed to the virus carrying the therapeutic cargo. The virus enters the cells and delivers the healthy gene, which makes a healthy CFTR protein.

the host sick. But delivering the virus to cells in a living creature versus a few cells in a test tube was a monumental challenge. The lung, over millions of years of evolution, had developed effective defenses against inhaled, disease-causing bacteria and viruses. Would their innocuous viruses, like a video game avatar, be able to dodge the immune system, reach the destination, and deliver the gene without getting wiped out or captured? Furthermore, in humans, these airway cells turned over somewhere between every thirty to sixty days. Even if the gene was delivered to the correct airway cells, those cells wouldn't live forever. Gene therapy wouldn't be permanent unless it transformed the progenitor stem cells—which give rise to all of the others. Any gene therapy that didn't transform those stem cells would require regular application—and thanks to the body's natural defenses, even if a virus succeeds once, once the immune system has issued molecular "wanted" posters for its capture, succeeding a second time becomes much harder.

Either hurdle was daunting; the combination, mind-boggling.

Despite these challenges, Wilson and many others believed that if a healthy copy of the damaged cystic fibrosis gene were added to a CF patient's malfunctioning airway cells, the lungs could be saved. Most patients ultimately died from lung infection, so fixing these cells could potentially save lives by restoring the balance of salt and water, lubricating the airways so that the river of mucus could flow again and wash out the deadly bacteria. That seemed logical and appealing, and many scientists, Wilson included, decided to try.

WHILE WILSON WAS PLANNING HIS CF GENE THERAPY EXPERIMENTS, other scientists were doing the same. And, Bob Beall, always seeming to be one step ahead, had the foundation's support to fund them all. It wasn't surprising that Beall was so optimistic—by 1990, just months after the CFTR gene discovery was published, other scientists working on gene therapy for different diseases were already forging ahead with clinical trials in humans, making the path forward for any inherited disease sound straightforward.

On September 14, 1990,[11] W. French Anderson and R. Michael Blaese, researchers at the NIH, began what would be the first-ever human gene therapy trials, for children born without a functioning immune system—a syndrome that left them so defenseless that common, normally harmless bacteria and viruses could trigger life-threatening infections. These children lacked a critical protein—adenosine deaminase, or ADA—essential for fighting bacteria and viruses. The disorder was a variation of "bubble boy" disease—severe combined immunodeficiency—named for the sterile plastic containment in which Texas boy Joseph Vetter lived to guard against germs. For many of these children, a weekly shot of ADA protein kept the disease in check, but they were still vulnerable.

Anderson and Blaese were sure that delivering a normal ADA gene would provide a permanent cure. For the human trial, the NIH team removed immune cells—a type of white blood cell called a T-cell—from the blood of two young girls with the disease. They multiplied the cells in the lab, infected them with a virus carrying the normal ADA gene, allowed the cells to grow and multiply further, and then, nine to twelve days later, returned them to the patients.[12] After a thirty-minute infusion,[13] the first patient ever to receive engineered genes, a four-year-old, was wheeled into a recovery room where she watched cartoons and sipped juice while her newly transformed cells began secreting healthy ADA into her bloodstream. The second patient also fared well. Neither had suffered dangerous side effects, and both, while not fully cured, were better off, with more of the critical enzyme in their bodies.[14] In early 1991, Anderson and Blaese tried gene therapy for two cancer patients with melanoma, and then in April a third ADA patient was given a treatment.[15] With these experiments, public support was growing for this therapy, even though the long-term success wouldn't be known for many years.

Michael Welsh, a physician from the University of Iowa College of Medicine who had been studying the CF protein for at least half a decade, began his gene therapy studies by proving the CFTR gene that Tsui and Collins's lab discovered was the gene responsible for the defect in the cells of CF patients—a critical piece of evidence that was missing from the original three papers. He first took mutant cells from patients' airways—cells in which the chloride channel was dysfunctional, trapping chloride inside the cell—and kept them alive in cell cultures in the lab. Then, he "rescued" or fixed these broken cells[16] by using a virus to deliver the healthy CFTR genes. The cells accepted the DNA and used it to make a normal CFTR protein, which proceeded to transport chloride—"curing" these cells.

Welsh's result not only proved that the F508del mutation caused cystic fibrosis, but also implied, as Welsh and his coauthors pointed out, that if you delivered a healthy CFTR gene to the cells in a living person, just as he had in the test tube, the gene therapy might work as a treatment. Although Welsh's team did their experiments first,[17] Wilson, who had been doing similar experiments with malfunctioning cells from a CF patient's pancreas, rather than airway cells, published his results a week before Welsh's team, on September 21, 1990. Both of these results were published just a little more than a year after Tsui and Collins had published the discovery of the gene.

Wilson's report showed essentially the same result as Welsh's work: viruses delivered the healthy gene, and the pancreatic cells began making the chloride channel and enabling chloride to move.[18]

The publication of these inspiring results on this auspicious anniversary reinforced the idea that gene therapy was both possible and imminent. Emblazoned on the front page of the *New York Times* was the headline "Team Cures Cells in Cystic Fibrosis by Gene Insertion."[19] "One of the things that we're awed by is the speed with which these things are happening and coming together," Bob Dresing told the paper. "We're in the business of trying to save as many CF kids as possible, and we just took a giant step forward in that direction."

"We are talking about years, not decades any longer," Beall told *Science* magazine,[20] echoing Dresing's optimism. And Wilson cautiously admitted to the *Times* that, "At this point, it's impossible for me not to be optimistic about cystic fibrosis."

Still, using gene therapy to directly transform airway cells in the lungs of a living, breathing person was trickier than what Anderson and Blaese had done by engineering white blood cells externally and then returning them to the patient. It wasn't possible to take lung cells from a patient's airways, transform them with gene therapy in the lab, and then put them back into the patient's body. Cystic fibrosis would be an entirely new challenge.

At next year's North American Cystic Fibrosis Conference (an annual conference sponsored by the foundation), the mood of attendees reflected the belief that a cure was near. Wilson attempted to instill a sense of reality as he shared his progress with an audience of parents and scientists in a sardine-packed conference room: "Everyone has been taking out the liver, or the bone marrow, and putting the gene in. But this ex vivo [out of body] approach isn't going to work for CF. We are going to have to put the genes into cells while they are in the body of a patient."

But with Welsh and Wilson's breakthroughs coming on the heels of the Anderson–Blaese trials, expectations were stratospheric that the cure was coming. The rate of progress so far was stunning: the CF gene discovered in 1989, and now, just a year later, the disease cured in a test tube. The seemingly quick succession of results amplified the perception that curing disease was easy, once the correct gene was found.

Eliminating cystic fibrosis seemed just a matter of time. The next step was testing gene therapy in a living animal, to make sure the therapy was safe. After that: humans.

CHAPTER 29

Transforming the Lungs

1992–1993

It is difficult to develop revolutionary medical therapies
in perfect safety and gene therapy is no different.[1]

—Ron Crystal

I n the early 1990s, gene therapy was still new, and its progress felt exhilarating. The public was captivated by the promise of these better medicines. A national survey conducted by the March of Dimes showed that, in 1992, 89 percent[2] of Americans approved of using gene therapy to treat genetic diseases. This new frontier presented limitless opportunities for smart, ambitious researchers; there were thousands of known genetic diseases, and creating a technology that could cure any one of them using a healthy gene could revolutionize medicine.

James Wilson and Michael Welsh's teams had laid the groundwork for CF gene therapy by proving that adding a healthy gene to cells carrying the mutant CF gene could make them function normally. But they weren't the only ones interested in developing gene therapy treatments for CF patients. Richard Boucher and Mike Knowles from the University of North Carolina—who had in the early '80s figured out that CF patients had a chloride transport problem in their noses and lungs—were also working on gene therapies. And another scientist, Ronald Crystal, a physicist turned physician turned pulmonologist, had begun experiments at the National Institutes of Health in Bethesda, Maryland, just a few miles from the headquarters of the Cystic Fibrosis Foundation, similar to the ones Welsh and Wilson were doing.

Back in the early 1980s, as others were embarking on the hunt for the CF gene, Crystal was focusing on a common inherited lung disease called alpha-1 antitrypsin deficiency that led to emphysema by age thirty and early death. Patients lacking enough of this enzyme were prone to wheezing, shortness of breath, and increased lung infections and liver problems. Over the next decade, Crystal's team learned to purify alpha-1 antitrypsin from pooled human donor plasma and inject it back into people who needed it. Crystal performed the first study in 1981, and by 1987,[3] thousands of people worldwide were receiving weekly infusions of the protein to combat their disease. But it was treatment, not a cure. A permanent solution required replacing the damaged alpha-1 antitrypsin gene—a onetime fix that would provide a patient's body with a permanent set of instructions to manufacture the protein themselves.

To meet the challenge of introducing a new healthy gene into the lung cells, Crystal chose adenovirus, named for the human adenoid[4] glands where the virus was first discovered. These medium-sized viruses were masters of infecting the human respiratory tract, causing mild colds in healthy people and a deadlier spiral in those with weak immune systems. Like all viruses, adenovirus had evolved to latch onto a specific type of cell and inject its own genetic material.

To tame these viruses and transform them into benign delivery vehicles, the viruses first had to be "genetically castrated." A Paris lab had recently figured out how to clip out the harmful viral genes responsible for replication and cell destruction and replace them with therapeutic genes. So Crystal recruited a young physician named Melissa Rosenfeld (later Ashlock) who had just joined his lab to lead the development of human gene therapies for lung diseases, and sent her to France to learn how to make therapeutic viruses.

Rosenfeld came to Crystal's lab in 1989 after a tragic experience during her internal medicine residency at Dartmouth's Hitchcock Hospital in Hanover, in which a young woman with advanced CF died shortly after giving birth. The situation seemed cosmically unfair and she wanted to help develop treatments that would save lives. Cystic fibrosis was among the several conditions that physician scientists in Crystal's lab were working on, and Rosenfeld was eager to do something to change the course of the disease and maybe even cure it.

Rosenfeld was also intrigued by gene therapy research, and her timing was perfect. When she joined Crystal's team, they were working on gene therapy for another fatal genetic disease, alpha-1 antitrypsin deficiency, and Crystal had just chosen the adenovirus to deliver the alpha-1 antitrypsin gene. He told Rosenfeld that her mission was to get their gene therapy working in rats—to prove that the adenovirus could safely and effectively transfer genes into the lungs—so that they could try the same approach in humans.

Once Rosenfeld had slipped the alpha-1 antitrypsin gene into the adenovirus, her challenge was getting the virus into the rats' lungs. She did this by making a hole in the rats' tracheas though which she inserted a thin catheter, which she then used to deliver liquid containing the virus directly into the lungs. The procedure took about thirty minutes. Over the next two weeks, Rosenfeld took samples of the rats' lung cells and was thrilled to discover that the virus had made it into the airway cells, delivering the gene; and that, for at least one week after the procedure, the cells were reading the gene, making alpha-1 antitrypsin protein, and secreting it into the airways. It was the first time anyone had used this type of virus to transfer a gene *directly* into a living, breathing animal. It was a powerful demonstration of gene therapy's potential to cure a human disease.

Crystal had been working on alpha-1 antitrypsin deficiency for close to a decade and had helped create the weekly protein infusions then used to treat patients with the disease. And because these patients already had a safe treatment, there was no reason to put them at risk with experimental gene therapy. But there was no equivalent treatment for CF. And, as the race to find the CF gene heated up and Crystal tracked the rumors that Tsui and Collins were close to finding it, he realized that once that gene was available, he could adapt adenovirus gene therapy to CF.

Once the gene sequence was published in September 1989, Bob Beall, who knew about Crystal's gene therapy work, connected him to Francis Collins's Michigan team and encouraged Collins to share the gene sequence encoding CFTR. At the time, Collins was collaborating with Wilson on CF gene therapy at the University of Michigan, which made Crystal a competitor. But one of Beall's conditions for funding was sharing data freely, so the Michigan team faxed over the sequence of the CFTR gene to Crystal.

When Crystal received the sequence, Rosenfeld had been there more than a year. Her new mission was to essentially repeat what she had done in rats with the alpha-1 antitrypsin gene, but with the CFTR gene instead.

Rosenfeld and her colleagues replaced the alpha-1 antitrypsin gene in the adenovirus with the human CFTR gene and a switch. Every gene in the body has an on/off switch, called a promoter, that controls when a gene is turned on to make proteins. The switch the team inserted kept the CF gene permanently on, so that the lung cells would continuously manufacture as much of the protein as possible, to help chloride move in and out of the cells and restore the salt water balance to the airways. Then she gave the rats the therapeutic adenovirus with the CFTR gene. Six weeks after Rosenfeld had delivered the virus, an analysis of the rats' airway cells showed that the cells were still producing the human CFTR protein and in the correct location in the airway—proof that the cells received and read the instructions.[6] What had worked for alpha-1 anti-trypsin worked for the CFTR protein, too. The stage was now set for Crystal's team to move toward a human trial.

Crystal had some familiarity with how the FDA worked; he'd had to work with them to get the protocol for the weekly alpha-1 antitrypsin protein treatments approved. But when he told the committee he wanted to perform gene therapy using an adenovirus to ferry the cystic fibrosis gene into human airway cells, the ten panelists reviewing his proposal stared at him blankly.

The Food and Drug Administration was, at the time, unprepared for the new era of gene therapy. The agency was responsible for regulating and approving all new drugs and the clinical trials to test them, but had no protocols in place for approving a medicine that used live viruses to deliver engineered genes into living humans—an approach that sounded more likely to cause disease than cure it. W. French Anderson and R. Michael Blaese had done the first gene therapy trial a few months earlier, in September 1991, to restore the immune system in two young girls. But their trials had been simpler and less potentially risky than the one Crystal and Rosenfeld were planning: they had first removed the patient's cells from the body and then altered them in the lab with the virus, before returning the fixed cells to the patient. No one had ever put a genetically engineered virus carrying a human gene directly into people before, so there was no approved protocol or safety procedures in place for doing so.

The panel didn't give Crystal the go-ahead immediately; instead, they laid out a series of experiments his team would need to do first in rats and nonhuman primates to gauge the safety of the approach. They also gave Crystal instructions on how the virus needed to be manufactured for the clinical studies. Rosenfeld had mastered engineering the virus for animal studies. But for people, the FDA mandated that the viruses be produced by companies with accredited manufacturing practices, where they would be prepared in sterile facilities to minimize the chance of contamination with other naturally occurring bacteria and viruses. Promiscuous gene swapping is common among bacteria and viruses. And as wildly overblown as it might seem today, one of the many concerns the regulators had was that the therapeutic virus—designed to be incapable of replication—would encounter a related virus during manufacturing or in the human airway, acquire some of its genes, and begin to reproduce, creating an Andromeda Strain virus that would escape the lab and infect the population of Bethesda and beyond—the ultimate sci-fi nightmare scenario.

Fear of an outbreak mandated a special containment facility at the NIH Clinical Center. "Negative pressure rooms" were engineered to suck in the air when the door was opened, to prevent the escape of any potentially dangerous airborne viruses. Patient quarters were designed to be as homey as possible, with a television, exercise bike, and a few other comforts. If the trial was approved, the first two patients treated with these therapeutic viruses would be required to remain in the facility for six weeks after treatment as Crystal's team tested all their body fluids to ensure they were free of the virus itself.

Before any gene transfer could be done with humans, the trials had to be registered and cleared by the Recombinant DNA Advisory Committee (RAC), created by NIH director Donald Fredrickson. This committee was tasked with providing oversight for any research involving recombinant DNA molecules—like these therapies, which were part virus, part human.

When Crystal registered for his presentation in front of the RAC, he was confident his team was leading the effort for human cystic fibrosis gene therapy trials. It was only then that he saw other names on the list that he recognized: Iowa's Michael Welsh, in collaboration with Genzyme, and Michigan's Jim Wilson were also presenting gene therapy proposals.

As Crystal's team had been advancing toward human trials, so had Welsh's and Wilson's. Each had been carefully probing the risks of their procedure, including the consequences of a patient receiving too much or too little of the gene therapy.

In 1992, Boucher's team at North Carolina figured out, using airway cells grown in a dish, that, at least with the virus they were using (which differed from the one Rosenfeld was), they only had to insert the gene into between 6 and 10 percent of cells lining the airways to boost the chloride transport to the same levels found in healthy people.[7] That was good news; it meant the virus only had to successfully deliver cargo to one in ten cells for the procedure to work.[8] But if so little of the CFTR protein was needed, as Boucher's results suggested, then, wondered Wilson, was too much of the protein dangerous? Was too much chloride transport a bad thing? Could it actually harm the lungs—disrupt breathing, or liquefy the mucus too quickly, drowning the patient?

To find out, Wilson's team engineered mice in which the human CFTR gene was switched on permanently in the lungs. Though the mice's lung cells produced much more of the CFTR protein than necessary, the mice showed no ill effects—there were no lung abnormalities, and the animals were fit. They could reproduce, producing offspring without birth defects, and lived a normal mouse life span.[9]

On the other side of the Atlantic, Bob Williamson, who had lost the race to find the gene, was also experimenting with gene therapy, though his approach didn't use viruses. Williamson and scientists at the Universities of Oxford and Cambridge were using fat droplets, called liposomes, to deliver genes to the lungs. Rather than packaging the CF gene in a virus to carry it into animal airway cells, these British teams encapsulated the gene in microscopic spheres of fat, then aerosolized the mixture in a nebulizer.[10] Mice whose native CF gene had been disabled then inhaled the greasy mist while sitting in an exposure container. Because liposomes are made of similar fats as the membranes of cells, these spherical gene-bearing droplets melted seamlessly into the airway cells. Liposomes appeared safer than viruses because they were nontoxic and didn't alert the immune system. The catch was that they lacked the precision targeting of viruses. Still, when these DNA liposomes were tested in mice, the imported gene was taken up by the airway cells in the trachea and fixed the chloride problem. Here was another promising, and possibly even safer, avenue for gene therapy.

With all the teams testing their own therapies, Wilson in Michigan, Welsh in Iowa, and Crystal at the NIH all had completed the necessary studies to begin human trials. Whether they admitted it or not, every team had fantasies of ultimately curing not just cystic fibrosis, but also hundreds of other genetic diseases. This, they anticipated, was just the beginning.

CHAPTER 30

First-in-Human Trials

1993–1995

The great tragedy of science—the slaying of
a beautiful hypothesis by an ugly fact.

—Thomas Huxley

On December 3, 1992, more that 350 people crammed into Wilson Hall, a large conference room in NIH's Building 1—a stately structure with towering columns and the seal of the US Public Health Service emblazoned on its triangular pediment—to hear about the next wave of gene therapy trials as the genetics researchers presented their plans to the Recombinant DNA Advisory Committee. Since W. French Anderson and R. Michael Blaese's successful immune disorder trial, this new type of medicine suddenly looked like a way of treating and potentially permanently curing a huge collection of genetic diseases, and everyone was interested. It was an unusual gathering, mixing researchers, physicians, pharmaceutical and biotech industry scientists, newspaper and TV reporters, and curious or concerned members of the general public, all there to hear who would be conducting trials, for what diseases, and using what experiments. Several venture capitalists were there, too, eager to determine whether gene therapy might be the next supersized investment opportunity.

That day, there were five presentations focusing on new gene therapy trials. Three were for CF: Crystal's, Welsh's, and one of Wilson's. In addition to his CF human gene therapy trial, Wilson was also sharing his plans for a trial focused on familial hypercholesterolemia, which had originally spurred his interest in gene therapy. The final presentation was for a trial linked to breast cancer. Crystal presented, then Wilson, then Welsh.

Crystal's and Wilson's strategies were similar—they were using similar viruses, and both were planning trials that would start by testing the therapy in the nose, then quickly move to the lungs. In Crystal's plan, if that first stage went well, he would test the virus in the lungs forty-eight hours later. Welsh, in contrast, planned to move more slowly. He wanted to test the gene in the nose only; it had the similar epithelial tissue as the lung, but was more accessible and a safer choice if there was a bad reaction to the virus. A little more nasal mucus and inflammation might feel like a bad cold, but the same symptoms in the lungs could hamper breathing. It was also easier to test whether the therapy was working there, by measuring voltage. If these tests proceeded smoothly, Welsh would then return to the RAC with a proposal for testing the virus in the lungs. He had also chosen a slightly different cold-causing adenovirus than Crystal.

The RAC approved all three CF-related trials, but the RAC, as part of the NIH, wasn't a regulatory institution. The final nod had to come from the FDA, and they weren't typically quite as swift. Each team first had to collect the FDA-required data on manufacturing the therapeutic virus, and assemble all the results from their previous animal studies for inclusion in an Investigational New Drug application. It was only at that point that the agency could rule on whether the procedure was safe and give the green light.

Fulfilling all the FDA's requirements and providing the data took close to four months, after which Crystal, eager to claim first place in the race for a clinical trial, was climbing the walls. His first test patient was lined up and the negative-pressure rooms were ready for use. All his experimental ducks were in a row. He just needed the final go-ahead.

In the early 1990s, a couple of FDA reviewers had their own research labs on the NIH campus. Crystal knew the location of his reviewer's lab and, unable to restrain himself, walked over from his own NIH lab to find the man working at his lab bench. The reviewer gave Crystal the official nod, and Crystal dashed over to the iconic clinical center known as Building 10, the largest research hospital in the world,[1] to begin final preparation for the trial. Building 10 had been home to a long list of medical firsts: the first combination chemotherapy and immunotherapy for cancer, the first treatment for HIV/AIDS, the first clinical trial for lithium in bipolar disorders. It was *the* place to test gene therapy for CF.

The next morning, on April 16, 1993, a twenty-three-year-old man lay on a bed in a negative-pressure containment room in the NIH's

Clinical Center. He was healthy and conscious, eager to be the first to try this new procedure—first in the nose, and two days later in the lungs—though he had been warned not to expect much. The man carried two copies of the F508del mutation; both of his CFTR genes were disabled. Crystal and his team were dressed in full surgical gear—dandelion-yellow gowns and face masks—to protect them from the bio-engineered virus.

One of Crystal's pulmonologists, Gerald McElvaney, a tall, intense, but soft-spoken doctor from Ireland, loomed over the patient with the first dose of the virus, prepped by Melissa Rosenfeld, in his hand. With Crystal by his side, McElvaney moved the tube with the virus into the patient's left nostril and up toward the bridge of his nose, and spritzed the mixture onto the surface of the wet, mucousy epithelium. If the treatment worked—if the virus delivered the gene and the cells began making the CFTR protein—then Crystal and McElvaney would be able to detect a change in electrical activity as chloride ions began to move out of the cell. It was the same measurement that Knowles and Boucher had done a decade earlier in North Carolina.

The procedure was quick, over in less than thirty minutes. Then there was nothing more to do except let the patient relax until the next day, when they could take a measurement. In the meantime, to be sure that none of the virus was contaminating the outside world, Crystal's team monitored the man's spit, stool, urine, nose, and pharynx, hyper-vigilant for signs the virus had spread beyond the target tissue.

When McElvaney and Crystal arrived the next day, having counted down the minutes, they could barely contain their excitement as they again donned protective hazard gear and stepped into the patient's room. The patient was comfortable and alert, watching TV. McElvaney sat next to the patient and asked him to tip his head back. Inserting a probe into the man's nose, where he had delivered the virus the previous day, he observed a change in electrical activity, signifying that chloride was successfully moving in and out of cells. At that moment it seemed a sure sign that the virus had delivered the gene and the cell was producing a healthy chloride-transporter protein[2]—evidence they had cured that small patch of cells in the nose. The whole team was filled with a wave of optimism. McElvaney used a fine brush to collect some of the cells and gave them to Rosenfeld to test, so she could confirm in the lab that the CFTR gene had in fact entered the cells, and that the gene was turned on and churning out protein. And indeed, Rosenfeld saw that the gene was producing the healthy CF protein, unlike cells collected before the therapy.

When he woke up the third morning, McElvaney was excited and worried. For him the work was intensely personal, raising his hopes of curing the condition that had sickened and killed so many children in his native Ireland. But he was worried about overselling this therapy to their test population of patients. The patients were volunteers, and he had been honest with them, explaining that even though there was a chance this first trial and dose of the virus could work, it didn't mean there was going to be a cure straightaway. As excited as he was to test the therapy, he couldn't shake his sense that he and Crystal and Rosenfeld wouldn't be able to help this young patient as much as they hoped.

Crystal also had his concerns. For the last four years, most of his lab personnel had been preparing to test this new therapy in people: engineering the virus with the gene and testing it in cells in the lab, then in cotton rats and rhesus monkeys.[3] Those experiments had gone smoothly. The monkeys hadn't shown any obvious clinical ill effects. However, there had been signs of inflammation when the monkeys were sacrificed and their organs examined under the microscope. White blood cells had swarmed to the site where the virus had been administered. And the higher the dose of virus given, the more white blood cells there had been. The animals had tolerated the treatment well overall, but he still felt nervous. Ultimately, he wanted to cure the condition. But his first concern was not harming their young patient, and Crystal kept turning over a checklist of procedures in his mind to ensure everything went as planned.

When Crystal and McElvaney visited the next day, again in the cheerful yellow gowns, the patient was in good spirits and relatively good health; he didn't look like someone sick with serious lung disease. They explained, again, what they were going to do—spray the virus onto the epithelial cells, this time in the lungs—and then McElvaney asked whether the patient would prefer the bronchoscope tube going down his nose or mouth. He shrugged and chose his nose. The anesthesiologist sprayed the patient's throat and vocal cords with lidocaine to numb them and then gave the man a relaxant to ease anxiety.

Once the man was numbed, McElvaney inserted the bronchoscope—which had three tubes or channels, like three straws bound together—up through the patient's nostril, past the bridge of the nose, and down the back of the throat. One channel had a tiny light and camera on the tip that hooked up to a monitor next to the patient's bed, to help McElvaney steer the bronchoscope to the right region of the airway. A second channel was for delivering the virus. The third could be used as a suction device or to collect tissue samples from the airways.

McElvaney watched the screen as he threaded the scope through the narrow opening into the lungs, careful not to hit the supersensitive vocal cords drawn tight like pale pink curtains on either side, lest they spasm. He continued lowering the probe along the undulating red surface of the airways, which were slick with a thin layer of pus—a sign that even though this patient appeared healthy, his airways were inflamed.

When the tube reached the bifurcation point where the trachea split into the bronchi, McElvaney steered the bronchoscope into the right bronchus toward the right lung and then halted. After snapping a few pictures to note the location, so he could find it the next day, he suctioned the pus off a small area of the bronchus. He then dripped 20 milliliters—about 5 teaspoons—of clear liquid containing the virus onto the exposed airway surface before removing the bronchoscope, all the while chatting with the patient in his soft, lilting brogue. Outside the room, a whole crowd of people, including Crystal's team, were waiting, and when Crystal and McElvaney, another physician, and a nurse exited, the crowd applauded. They had just performed the first gene therapy of its kind.

The following day, McElvaney repeated the procedure, guiding the tube to the same location—but this time, after suctioning off the mucus, he inserted a tiny brush through the bronchoscope to sweep the treated area, taking a sample of nearby airway cells. McElvaney passed the samples to Rosenfeld, who took them back to the lab, eager to see whether these cells, like the ones from the patient's nose, had received the new healthy gene and were making the CFTR protein. As RAC regulations specified, the team continued to sample the patient's urine, blood, and stool for viruses. They found none. They also checked cells in the pharynx and the untreated nostril for the virus and found nothing—a sign that these engineered viruses were staying put in the lungs.

The trial continued. The second patient, also a young man, received the same dose of virus as the first, in the nose and then the lungs. But when McElvaney tested the electrical activity in the nose of the second patient, there was no change. And when he then retested the first patient's nose, the electrical activity he had noted two days previous wasn't the same. Rosenfeld's lab work, however, revealed encouraging signs: in airway cells collected from the second patient, she could detect the healthy CFTR gene, the messenger RNA bearing instructions to produce a healthy protein, and the CFTR protein itself. The gene therapy was working in the nose of patient 1 and the lungs of patient 2. The healthy gene was being read and used to make a protein—even if, so far, the results were mixed.

Crystal tried to temper his excitement. This was just a proof of concept, he reminded himself: they had only administered the virus to a tiny, quarter-sized site in the lung, and the minute quantity of CFTR protein now being produced in the volunteer's lungs wouldn't trigger a tangible change in his health. Also, the results they received as they continued to monitor the patients indicated that the therapy wasn't long lasting: after ten days, there was no longer any trace of the gene in either patient's lungs. That was a stark contrast to the results in rats, in which the therapeutic gene had continued to produce CFTR for more than six weeks.[4] But even with all the caveats and inconsistent findings, Crystal's gene therapy for cystic fibrosis, the first attempted in a living person, had—at least briefly, in a tiny fraction of the lung—worked: in a small patch of cells, they had cured the disease.

Soon after treating the lungs of a third patient, however, Crystal's euphoria dissipated. The third patient had received the highest dose of the virus yet, and just like in the previous cases, the delivery had gone smoothly. But twelve hours later, the young woman—just twenty-four years old—began feeling ill. She pressed her temples, which throbbed from a headache, and felt exhausted, though she had been sitting in bed all day. When her temperature and heart rate spiked and her blood pressure tumbled, the nurses called McElvaney and Crystal to return to Building 10. They rushed back to examine the patient, who was suffering from low oxygen; her lung capacity had shrunk. When they took an X-ray, they saw that a pneumonia-like condition had erupted in the lower and middle lobes[5] of her right lung—close to where they had delivered the virus. CF patients were highly susceptible to pneumonia; it could have been a coincidence. Or it could have been the virus.

With Crystal and McElvaney's consent, the physicians at the clinical center responded swiftly to the woman's symptoms with broad-acting antibiotics, antipyretics to curb the fever, oxygen, and fluids, while the two men tried to make sense of their patient's sudden sickness. She had received more virus than the previous patients; had the impotent virus mutated and begun multiplying in her lungs? When the woman's blood was analyzed, they found high levels of interleukin-6—a chemical the immune system produces when exposed to dangerous viruses.

To figure out whether all CF patients might have the same immune reaction, or whether this patient was an exception, the next day Crystal reached out to local CF patients who were not involved in the trials. He asked them to come to Building 10, where McElvaney could take samples of their airway cells for Rosenfeld and the team to test. In the lab, they exposed the airway cells to an equivalent amount of the therapeutic virus they had given the third patient, then measured the cells' interleukin-6 levels. The airway cells from all the CF patients released this retaliatory protein when exposed to virus—evidence that the gene therapy had triggered this woman's symptoms.

Crystal immediately stopped the trial, reported the harrowing episode to the RAC and the FDA, and called his competitors Welsh and Wilson, warning them about the adverse effects. In the meantime, he continued to monitor the patient. Her blood pressure recovered after three days, her fever after six, and her breathing after fourteen. The fluid in her lungs dissipated after twenty-five, and her lung function normalized a month later. The woman's adverse reaction to the virus was frightening, a jarring reminder of the cost of developing new medicines.

While the incident was a setback, Crystal was anything but discouraged. He had recently left the NIH and moved to Weill Cornell Medicine in Manhattan at the beginning of April so he could continue his academic and clinical work while having the freedom to dive into the private sector. Genentech, the San Francisco developer of Pulmozyme, wanted to fund a new gene therapy company targeting CF, and they asked Crystal to launch and run it—something he couldn't do as a government employee at the NIH. Although his NIH trial was aborted, the experiment had proved gene therapy was possible, and Crystal began planning a much bigger study with his new collaborators at Genentech.

While Crystal had been the first to test CF gene therapy in humans, Welsh was close behind, and actually published his results first,[6] in October 1993, almost a full year before Crystal published his.[7] Welsh had planned his trial experiments with his collaborator Alan Smith at another biotech company, Genzyme; Smith had done much of the foundational work figuring out which virus to use and which type of on/off switch to include with the CFTR gene to regulate its activity. Their nose-only strategy had been more conservative than Crystal's, only requiring an ear-nose-throat specialist to insert a little spoon-shaped plastic device Welsh had built up the patient's nose as they sat upright, their head tilted back a little. Once the device was in the right location, the physician dripped the viral soup on the target tissue for about fifteen minutes, for each of the first three patients. The next day, when Welsh measured the voltage in the patients' noses, the results revealed that the therapy had fixed the chloride problem. The news traveled far and quickly. "Big Victory in War on Cystic Fibrosis; Gene Therapy Corrects Key Defect," wrote the Associated Press in October 1993.[8]

But as Welsh and Smith expanded their study and tested more people, the results were dramatically different. In the next batch of patients, there was absolutely no change in the voltage. It was crushing. Sitting alone in his office, Welsh tried to think what they could have done differently the second time. Then he figured it out. The first time, the plastic device must have scratched delicate nasal tissue, creating an open wound that allowed the virus to enter the cells. The second time they ran the experiment, they had just dripped the viral solution onto the surface. The nasal mucus had prevented the virus from reaching the epithelial cells.

Like Crystal's study, Welsh's had shown that gene therapy was feasible, but not easily reproducible. While Welsh was disappointed, he recognized the study had been important, and had revealed the major problem of delivery. The procedure just wasn't efficient enough yet to get enough virus into the tissue to make a difference. The body had evolved defenses against pathogens over millions of years, and getting a therapeutic virus past the mucus barrier wasn't going to be easy. Welsh continued to try various approaches, but after that initial failure, ended up pushing gene therapy to a back burner.

Then there was Wilson. To prepare for his human gene therapy trials, he chose to test the therapy in fourteen live baboons, divided into four groups to test increasing concentrations of the virus. Using a bronchoscope, the therapeutic virus was delivered into a small, well-defined section of baboons' lungs. Some of the animals were then sacrificed on day four after the procedure, and the rest on day twenty-one. This allowed Wilson's team to chart a timeline of when their lung cells began making the CFTR protein and for how long.

What Wilson discovered was troubling. At low concentrations, the virus seemed to do no harm—but also not much good. At higher concentrations, the virus made it into the airway cells and produced the CFTR protein—but the airways became damaged and inflamed.[9] It was clear that the knitted mosaic of cells that formed the surface of a live animal's airway behaved very differently from the tube full of cells in the lab that was supposed to represent them. Wilson also noted that, within three weeks, the human CF gene was barely active in the baboon lung, producing little to no CF protein. It seemed that the DNA carrying the therapeutic gene had degraded inside the lung cells, and was beginning to disappear.[10] For the gene therapy to work, the virus would have to be given multiple times—and it wasn't clear whether this would be possible without the immune system going berserk and causing a deadly reaction.

Wilson shared his primate data at the Williamsburg Conference in 1992, an annual invitation-only event (formerly called the GAP conference) organized by the Cystic Fibrosis Foundation for scientists engaged in cutting-edge CF research. Wilson showed slides revealing the inflammation in the baboons' lungs when given high doses of virus. He told his audience that gene therapy may not go as they had planned. It may not be as easy as they had anticipated to use viruses to get the CFTR gene into the cells of a living person. But even this group of astute researchers didn't want to believe it. The gene therapy train was moving fast and nobody wanted to jump off.

Wilson knew that his data was solid. These were his animals; what he had seen was real. And his concerns grew as he began the first phase of human clinical trials. In a phase I safety study, Wilson tested his therapy—the same one he'd tested on the baboons—on eleven volunteers with CF, hoping the inflammation he'd seen in the baboons was species specific. The results were discouraging. Just like in Crystal's trial, high doses made the patients sick with flu-like symptoms. Also, four days after the virus had been spritzed into the lungs, less than 1 percent of the bronchial cells carried the gene. Forty-three days later, the healthy CFTR gene was undetectable.

Beginning in 1995, the lackluster results from the first human CF gene therapy trials in the lungs had planted a poisonous seed of doubt that such therapy may not be as simple as scientists had anticipated, and people started to lose faith. For his part, Wilson regretted that he hadn't expressed his doubts more forcefully and managed expectations better, so that when the therapy failed, it would be less of a letdown. But the pressure to succeed was enormous, and everyone—the scientists, the patients, the foundation, the universities, and also the public—wanted gene therapy to work.

None of the main players were quite ready to give up, believing that getting gene therapy to work was a matter of discovering or engineering the right type of virus. And one group in particular ended up devoting more resources to this approach than all previous groups combined. In September 1994, the NIH's Recombinant DNA Advisory Committee approved the plans of a Seattle-based company, Targeted Genetics, and researchers at Johns Hopkins University to launch a gene therapy trial using adeno-associated virus (AAV)—a new type of gene therapy vehicle.

A harmless virus, AAV causes no illness, and almost all humans have been exposed to it by adulthood. Targeted Genetics had conducted studies in rabbits and the CFTR gene had remained active in the animals for six months.[11] The company was in the midst of finishing studies with primates and expected to launch a trial with twelve CF patients from Baltimore in 1995. Once the trial was approved, physicians administered AAV to the nasal epithelium and the right lower lobe of the lung. Analysis of some samples provided a glimmer of hope, revealing very low levels of the healthy CFTR protein and improved electrical activity in the nose—and no flu-like symptoms.[12]

The results were interesting enough to spur a larger phase II study[13] with forty-two patients, approximately half of whom got the AAV-CFTR and half the placebo. Those who got the virus inhaled three doses using a nebulizer three times—on days one, thirty, and sixty. There were still no severe reactions to the virus compared to the placebo, so Targeted Genetics launched the largest CF gene therapy trial ever attempted, with 102 patients. The goal: find out if there was a clinical benefit.[14] This phase IIb followed the same dosage and timing regimen. The patients dealt well with the virus; there was no harm caused. But the gene therapy didn't help: it didn't improve lung function or help patients breathe better compared to the placebo, and it didn't reduce the number of infections in the treatment group.

Wilson and Crystal continued to work on gene therapies for other diseases, reengineering the viruses to dampen their aggravating presence on the immune system. Other researchers joined in. Some created new viruses that didn't trigger quite as strong an immune reaction. Others used the fatty globules[15] Williamson used to successfully smuggle in the genes. Williamson was first to run a placebo-controlled study of the liposome-CFTR in a group of patients in 1995. There were no ill effects, and biopsies of the nasal tissue where the liposomes had been sprayed were free of immune cells, suggesting that this first dose didn't trigger alarms. Electrical activity in the nose increased, peaking on day three, but then dropped to the patient's pretreatment level by day seven. They had restored perhaps 20 percent of CFTR function, but only for a brief few days; one month after the treatment, none of the patients could breathe better.

By the early to mid-2000s, most researchers and companies would abandon efforts to use gene therapy to cure cystic fibrosis. The field was too immature and the tools and technology too crude. And for CF in particular, gene therapy no longer looked like the best way forward.

CHAPTER 31

The Beall Curve

1996–1997

The reasonable man adapts himself to the world:
the unreasonable one persists in trying to adapt
the world to himself. Therefore, all progress
depends on the unreasonable man.

—George Bernard Shaw

In 1996, seven years after finding the cystic fibrosis gene and three years after the first human gene therapy trials, scientists were no closer to developing a cure. The insurmountable hurdles of gene therapy were snuffing out scientific interest in it as a panacea for all inherited disorders, and public opinion followed. Cystic Fibrosis Foundation CEO Bob Dresing had grown frustrated with the lack of progress, and, stressed by his son's declining health and increasingly tied up with other commercial ventures, he left the foundation in 1994. Bob Beall, who took over as CEO, was also becoming increasingly pessimistic about gene therapy. Data from the previous three years of human trials painted a bleak picture. It was clear gene therapy would not lead to a cure—at least not in the near term. And, although he couldn't have predicted it, the field of gene therapy was about to collapse in 1999 with the unexpected death of a young patient in one of Jim Wilson's trials that was unrelated to CF.[1]

Eighteen-year-old Jesse Gelsinger was an Arizona teen who suffered from a deadly inherited disease called ornithine transcarbamylase deficiency (OTC), which blocked the body's ability to process and eliminate ammonia, the cleaning-product chemical that arises as a toxic by-product of protein digestion. If ammonia levels rise too high, they permanently damage the brain. Gelsinger's body was so sensitive to protein that even a peanut butter sandwich could spike his ammonia to toxic levels.

As soon as he'd turned eighteen, Gelsinger had volunteered for OTC gene therapy trials being run by Jim Wilson, now at the Institute for Human Gene Therapy at the University of Pennsylvania.[2] Wilson's strategy was to inject a cold virus—similar to the one he had used in the cystic fibrosis gene therapy trials—to treat OTC. As with CF, Wilson first completed preliminary studies in mice and monkeys. He transformed the liver cells of mice[3] and then fed them and untreated mice high-protein foods. Those with the gene survived for several months, whereas the untreated mice with the same diet died immediately.

The mouse trials were compelling, but the work in monkeys was disturbing: high doses of virus triggered fever, a blood clotting disorder, liver failure, and finally death. However, the dose planned for humans was significantly lower, and with some changes, the Recombinant DNA Advisory Committee approved the trial. Wilson's immediate goal was to figure out the maximum amount of virus he could give a patient without triggering the dangerous side effects. He planned six mini-trials, with three patients in each group; the doses would increase in each subsequent trial. It was a slow process.

The first trial began on April 7, 1997.[4] Gelsinger was in group six, receiving the highest dose. The first patient who received it, a young woman, had done fine. So had the other fifteen patients who received lower doses. There had been fleeting fevers and flu-ish symptoms, but nothing frightening.

Gelsinger received the therapy on the morning of September 13, 1999, and by that evening, he looked flushed. His temperature hit 104.5°F. The next day his eyes were pale yellow. The jaundice was a sign that his liver was damaged, or that he was developing a clotting disorder, in which red blood cells exploded, leaking their protein-rich guts into the blood—particularly dangerous for Gelsinger, whose body couldn't process protein.

By midday September 14, Gelsinger was in a coma, his blood ammonia levels ten times the norm. By Wednesday evening, his lungs had stiffened—so doctors hooked him up to a device that mimicked a lung. On Thursday, his kidneys failed. Unable to produce urine, he swelled with extra fluid. By Friday morning, Jesse Gelsinger's brain activity flat-lined. At 2:30 PM that afternoon, the life support was turned off, and Gelsinger died.

Ronald Crystal, then in New York, was still dabbling in cystic fibrosis gene therapy when he told the *New York Times* that the death of patient Jesse Gelsinger was tragic but "would not in his view have a significant impact on the field."[5]

He was wrong. Gelsinger's death all but froze the gene therapy field in the US.

The failed trial came close to killing Wilson's career, too. Devastated that he had made errors in judgment that led to Gelsinger's death, he was barred from leading or working on FDA-approved trials for the next five years, and he stepped down as the leader of the Institute for Human Gene Therapy. He retained his title of professor and retreated into his lab to focus on basic research—discovering many new types of viruses that others would eventually adopt for future gene therapy trials.[6]

As the sparkle of gene therapy dulled, other scientific breakthroughs stole the limelight. In 1996, Dolly the sheep made her wooly world debut and focused the public's imagination on cloning. And the health crisis dominating the news was the AIDS epidemic, which had raged for a decade but had only recently reached broad public awareness. By 1996, 581,429 AIDS cases had been reported in the US,[7] a number almost twenty times larger than the entire population of cystic fibrosis patients. As the world moved on, the CF community was left frustrated and angry. The mutation was known, and the gene had been found, but children continued to die.

Beall had been the foundation's medical director for fourteen years when he took the reins from Bob Dresing as CEO in 1994. He was weary waiting for a cure. Help wasn't coming from the government, which was notoriously tightfisted when it came to funding rare diseases. There were too many other diseases, affecting tens of millions of Americans, that took priority: cancer, diabetes, heart disease, and more. The benefits to society of curing millions with these ills overshadowed CF's tens of thousands, and the economics fell in favor of focusing on these "big diseases," too. Developing a drug was both time consuming and expensive, taking up to fourteen years and costing about $2.6 billion on average.[8] The slim odds of recouping the research and development costs of a CF drug when the market was only 30,000 in the US and maybe 70,000 worldwide was an obvious black mark against the venture. Beall recognized that few pharmaceutical companies would take the plunge and develop a drug for CF unless—as they were seeing with

TOBI, then wrapping up its phase III trial—the early, most failure-prone stages of drug development were funded by someone else.

The solution, Beall realized, needed to come from within the CF community itself—from people who had a real stake in the outcome of the disease. The foundation's network of clinics had facilitated the successful testing and approval of Pulmozyme. And by 1996 it looked as if the foundation's funding, with PathoGenesis pitching in during phase III, would yield a second promising medication in TOBI. Still, both of these medicines treated symptoms of the disease, rather than the root. And that's what Beall wanted to do. He wanted to treat the root of the disease and cure it.

In 1996 the foundation was in good shape, thanks to the work of Doris Tulcin, Bob Dresing, Beall, and now-COO Rich Mattingly over the previous two decades. Money feuds between the chapters and national headquarters had ceased once the foundation was centralized, restructured, and moved to Washington; all the money raised was sent directly to national to fund critical breakthroughs. The discovery of the gene—though only partly funded by the foundation—had legitimized Dresing and Tulcin's reboot of the organization and validated Beall's strategies for accelerating the science. That amplified the community's trust in them. Now, as Mattingly worked with the charters on their fundraising, he could tell parents and donors not what they hoped to do, but what they *had done*.

Under Beall's medical direction, the foundation had spent the 1980s and early 1990s doing things that no other health nonprofit was doing. They'd established relationships with biotech and pharmaceutical companies and catalyzed the creation of drugs to treat CF's worst symptoms: oral pancreatic enzymes that helped patients digest food and extract nutrients; Pulmozyme to break down asphyxiating mucus; and TOBI to more effectively fight lung infections.

The lesson Bob Beall learned from the foundation's involvement in Pulmozyme and TOBI in particular was that they had the power to fund and direct the development of medicines. Genentech had conceived Pulmozyme, but it was the foundation that had coordinated all the clinical trials and handled the drug's marketing and distribution. TOBI was homegrown; except for the last phase III trial, it was almost entirely funded by their little nonprofit.

Perhaps, Beall mulled, he could use a variation of this strategy to lure a company into creating a cure for cystic fibrosis—from scratch.

The board of the CF foundation—a mix of parents, scientists, and physicians—was as frustrated as Beall. They were anxious for him to take more risks with the foundation's money—to invest in the most cutting-edge research that could yield the greatest reward. They opened the purse and ordered him to gamble. And in 1997, Beall figured out where to place his bet.

WHILE THE SCIENTIFIC COMMUNITY HAD BEEN SEDUCED BY GENE therapy, a revolution was brewing in the pharmaceutical industry. In the old way of making medicines, a drug company would license collections of chemicals—also, like collections of genes, called libraries—made by the scientists at industrial giants like Dow Chemical. Some chemicals in each library might have the potential to heal. The job of the pharmaceutical company was to hunt through these libraries, with their millions of compounds, to find a molecule that would treat the malady they were trying to cure.

But this began to change in the 1990s, with the evolution of a new field called combinatorial chemistry.[9] This technique enabled all chemists, in labs big and small, to create an almost infinite variety of their own small molecules by mixing chemical building blocks. Combinatorial chemistry also enabled chemists to create infinite varieties of chemical compounds simultaneously, rather than having to develop one at a time, which yielded hundreds of thousands, if not millions, of novel molecules. Rather than having to license others' libraries, drug company chemists could now create vast collections of their own chemicals to test for any medicinal properties before matching them to a disease.

Another force shaping the landscape of drug discovery was the Human Genome Project, which had begun in 1990 at the NIH. In 1993, Francis Collins had left the University of Michigan, where he had worked on the CF gene and then collaborated briefly with Wilson on early gene therapy studies, and moved his lab to the NIH, where he was invited to become director of the National Center for Human Genome Research (later the National Human Genome Research Institute). By 1994, he had stopped working on CF, becoming an observer rather than an experimentalist.

As director, Collins replaced James Watson, codiscoverer of DNA, who had passionately advocated for one goal: determining the entire sequence of chemical building blocks—all three billion As, Ts, Gs, and Cs—that made up the human genome. Now the ambitious endeavor, named the Human Genome Project, would be Collins's to oversee. Many peers regarded Collins[10] as a congenial collaborator, an accomplished gene hunter, and a compassionate man, capable of shepherding the many unruly egos in the field of human genetics and leading the international sequencing effort. By 1996, analysis of the genome sequence was yielding hundreds of newly discovered genes, many responsible for causing disease and suffering.

With millions of novel chemicals pouring out of the pharmaceutical industry and hundreds of biological targets emerging from the Human Genome Project, the only thing missing was a lightning-fast way to test the curative effects of these molecules. By the mid-1990s, companies were beginning to solve this problem, too. A small biotech startup in San Diego and a handful of other companies around the country were leading the field with sophisticated robots that could measure the effects of an individual compound on a specific biochemical reaction in a miniature test tube or a cell on an industrial scale—not one at a time, like in traditional labs, but by the thousands. This newfangled technique was called high-throughput screening.

It was the perfect storm—there were now millions of drugs, thousands of targets, and automation to figure out more swiftly what each chemical did to each cell.

Beall first learned of high-throughput screening in 1997, while flying to Seattle, from a journal article on the top seven companies leading this new industry. The article described how robots tested potential drug candidates in vast arrays of microscopic petri dishes. In a single day, these robots could screen some 20,000 chemicals; an academic lab, by contrast, could process only a handful in that period. Perhaps, Beall thought, this could provide a new avenue to a cure for cystic fibrosis. Rather than trying to insert a healthy gene into cells to make a healthy protein, maybe it would be easier to find a drug that could fix the mutant chloride channel that already existed in every cell of the patients' bodies.

There was some early evidence that CF's mutant protein could be, if not repaired, then at least encouraged to open and close. Michael Welsh at the University of Iowa, while engaged in gene therapy in the lab, had discovered that when cells with the common F508del mutation were cooled below body temperature, the deformed CFTR proteins could reach the surface of the cell and function, allowing some movement of chloride.[11] Somehow the temperature change altered the structure of the chloride channel, restoring its activity. Regular body temperature, however, was 98.7 degrees Fahrenheit, and obviously it wasn't possible to permanently cool patients to 80.6 degrees Fahrenheit without killing them. But this finding suggested that this mutant protein could be redeemed. Maybe a small molecule with a similar shape to the missing amino acid could tweak the CFTR enough to make it functional.

Yet how would anyone know if a potential drug was having the desired effect and helping the chloride flow in or out of the cell? What exactly would the fancy new robots measure? For this whole system to work, they'd need a sensor that would signal if a drug had activated the chloride channel.

As Beall sat in his office in Bethesda after returning from Seattle, massaging each element of the problem, he received a visit from Alan Verkman, a nephrologist at the University of California, San Francisco, who had been leading the school's Cystic Fibrosis Research Development Program and was seeking more funding for a new project. Verkman was a tall, slim man with a high forehead and a mass of curly brown hair styled in an Einsteinian fashion. He explained to Beall that, while studying chloride in the kidney, he had developed a chemical sensor that would "shout out" when chloride was moving in and out of a cell—a fluorescent molecule that glowed brightly when chloride levels rose.

Beall was interested. He stood up and began pacing as Verkman spoke; his wheels were turning. Perhaps this was the missing link. If Verkman's sensor lit up as the level of chloride increased, maybe it could detect when a potential drug-like molecule fixed the broken CFTR protein and restored chloride transport. Verkman reminded Beall about the new companies using high-throughput screening to scan and evaluate thousands of drugs. It might be possible to find a drug that could manipulate and correct the activity of the mutant chloride channel. Beall was thinking the same. Verkman asked Beall for funding to start hunting for promising molecules and to develop his fluorescent sensor, which Beall granted. Before walking out the door, Verkman also advised Beall to look for companies that could do this on a much larger scale than his university lab.

After talking with Verkman, Beall called his colleague Ray Frizzell, one of the foundation's scientific advisors and an early pioneer of CF research, and told him about his new plan for developing a therapy for CF. Beall asked Frizzell if he had any connections with companies that were using robots to screen chemicals for potential drugs. Frizzell knew of one. His former college roommate, a brilliant inventive guy named Roger Tsien, had just started a company in San Diego: Aurora Biosciences Corporation. "You should talk to Roger," Frizzell told Beall. The name rang a bell. That company had been one of the seven mentioned in the article he read. Hedging his bets, he decided to contact all of them.

No one had ever tried to fix a broken protein. If a company could develop a once-a-day pill that would rehabilitate patients' malfunctioning CFTR protein, then this might be the simplest and safest treatment—and the next best thing to a cure. Beall thought it was worth a shot.

PART 3

CHAPTER 32

The Joey Fund
1989–1999

At the end of the day it's not about what you have or even
what you've accomplished. It's about what you've done with
those accomplishments. It's about who you've lifted up, who
you've made better. It's about what you've given back.

—Denzel Washington

Casey O'Donnell was born June 21, 1989, two years and five months
after her older sister, Kate, and just a couple of months before Lap-
Chee Tsui and Francis Collins's Toronto-Michigan team announced
their discovery of the gene for cystic fibrosis. As they had with Kate, Joe
and Kathy used prenatal genetic testing to make sure their second baby
would not suffer from cystic fibrosis. Casey was born healthy, but, like
her sister, was a carrier. Every one of her cells harbored a single copy of
the mutant CF gene—a trait that could one day be passed to any children she chose to have.

Even as Joe and Kathy began a new phase of life with their new
healthy daughters, some things remained the same. Kate and Casey
attended the same school Joey had, and Joe coached every one of their
sports teams—soccer, baseball, basketball. Joe continued to build his
concessions business while expanding into other ventures, like hotels,
resorts, and restaurants. And the family remained in Belmont, where
they raised their daughters in the same home where Joey had lived.

The girls grew up always knowing they'd had a big brother. His photos decorated the house, home movies revealed his various adventures
and exploits, and Joe and Kathy always shared stories and anecdotes
about him, never shying away from their curious daughters' questions.
Who were Joey's favorite characters? Ernie and Bert from *Sesame Street*.

Action figures? Star Wars, of course. Toy? His bike. They wanted to know everything about him, and recollections were always joyous.

Other things were different. Unlike Joey, the girls didn't need to be within arm's length of a hospital at all times, so the O'Donnells were able to travel and vacation more, spending time at their home on the Cape, traveling to ski resorts with friends, or going to Disneyland. Once a year the family traveled abroad, Joe and Kathy keen to show their daughters more of the world than Joey had been able to see.

Also new to their lives was Joey's Park, which the Belmont community built with money raised from a local walkathon almost three years after Joey died. Neighborhood kids, including Joey's friends, helped design the park, and an army of parents gathered daily between September 20 and 24, 1989, to build it in a modern-day Amish barn raising. The result was magical. The park was a fantasy land right next to Joey's elementary school, Winn Brook. It had dozens of hiding places, a zip line, several turrets, and platforms connected by ramps and slides and chain-linked bridges that rocked and wiggled as children ran over them, screaming and laughing. Kate and Casey were at the park the day it opened, just two and a half years and four months old, respectively, and the place became their cherished play area as they grew. They were proud of it, emphasizing to other toddlers and preschoolers that it was *Joey's* Park, their big brother's park—at least until Kathy gently explained to them otherwise, stressing that Joey's park welcomed all children.

Joe and Kathy had always told their daughters that Joey was in heaven, but exactly what that meant wasn't clear to either of the girls until the day Kate, in first grade, learned about death and finally made the connection that Joey had passed away. The weight of that realization made her erupt in tears.

While Joe and Kathy were keen to establish a connection between their daughters and Joey, there was one connection that no one expected them to keep: their tie to the cystic fibrosis community. After having a child die from the disease, many emotionally exhausted families chose to disengage, anxious to push the excruciating experience into the dark corners of memory and never think of the disease again. The O'Donnells never considered that, even after they had healthy children to focus on. This disease was part of them.

When Joey died, Joe never dreamed he would stay on and work on this cause. He thought he would just curse God and get on with life.

But though he wanted to give up and walk away, he couldn't. Partly that was because he didn't want other families to lose a child to this disease. But the other reason was gratitude. He and Kathy had benefited from the science that the CF Foundation had supported for thirty years. A significant chunk of funding that enabled Tsui to create the early gene maps had come from the foundation. And the markers on those maps had made it possible for parents like Joe and Kathy, who knew they were carriers, to have children not sickened by this disease.

Another motive for staying connected was that throwing himself into something else, among a community of people who understood his and Kathy's pain and what they were going through, was the only way he knew to combat his grief. After Joey's death in 1986 led Joe to skip the film-premiere fundraiser he'd held for the previous four years, Joe decided in 1987 to dive back into raising money for CF. But, still grieving the loss of his son, he wasn't completely clear about what he wanted to do.

Close friends of the O'Donnells suggested creating a foundation in Joey's memory and holding the annual film premiere under the umbrella of this new fund, which could also host a variety of other events during the year to raise money for cystic fibrosis. Joe and Kathy embraced the idea. Fundraising for the newly forged Joey Fund would be a deeply personal commitment, a way to channel sorrow and heartbreak—and Joe had a clear vision for what to do with the money he raised.

After the gene was discovered in 1989, most people, including Joe, were convinced the road to a cure was gene therapy. He agreed with the Cystic Fibrosis Foundation that the only way to develop such a cure—whether it was gene therapy or something else—was to fund science, so that was what he would use the Joey Fund to do. But he didn't want his fund to be an offshoot of the foundation. The Joey Fund was a family affair. He didn't want an administration or board or headquarters or anything that would distract from the only goal he thought worthwhile: raising money for a cure.

To keep expenses minimal, most of the Joey Fund's event planning was done in house. There was no overhead, because there was no office and no staff—just Joe and Kathy. They covered the costs of advertising and invitations. When they ran events, employees of the Massachusetts chapter of the foundation would step in and organize, or Joe would seek out friends like Debbie Soprano, Paul Del Rossi's secretary, who had helped pull together that first film premiere at General Cinema.

Although the Joey Fund was distinct from the Massachusetts chapter, the money raised all went to the same cause. So if Joe needed help with an event, the chapter had staff and volunteers ready to pitch in.

Despite his focus on funding CF science, Joe never forgot about the families in his home state of Massachusetts who were struggling with this disease. Joe and Kathy knew the strain cystic fibrosis caused, both financially and emotionally. Divorce was common. The O'Donnells had been fortunate: they were well off, able to pay their medical bills and give Joey the best care money could buy. Joe's salary allowed Kathy to stay home with Joey and care for him, and funded Joey's therapist, John Nadeau. And Kathy had understood that the best way to help Joe manage his grief and stress was to give him the freedom to work and grow the business. He had always been a workaholic, but Kathy recognized that his work gave him a sense of control that he had lacked completely when it came to Joey's health. So they were able to support each other and make their marriage work through the bitterest times. While they couldn't offer marriage counseling to other struggling couples, they could offer financial assistance, reserving 5 percent of the Joey Fund money for cash-strapped families.

Figuring out which families to help and how wasn't immediately obvious to Joe or Kathy. But the answer emerged organically in the summer of 1990. Moved by Joey's death, Debbie Soprano had joined the Massachusetts chapter of the Cystic Fibrosis Foundation and was asked to visit a mother in the working-class suburb of Waltham who wanted to volunteer for the foundation. The woman and her sick child lived in a sweltering one-bedroom apartment in which curtains hung from the ceiling to divide the space into a kitchen, living room, and dining room. She had a fan in her living room, but even with the fan, the cheese on the platter she had placed in front of Debbie was melting. Here was a working mother struggling to keep a roof over her family's head, trying to keep her daughter healthy—and still making time to volunteer.

When Debbie returned to her office, she called Joe and told him about the mother and the stifling apartment where even she could barely breathe. She couldn't imagine how the sick child was faring. Can we get her an air conditioner? Debbie asked Joe. Is that the sort of thing you want to do with the Joey Fund money?

Joe thought for a moment and agreed that this seemed like a good use of the money set aside for Massachusetts families. Unsure of exactly how to proceed, Joe went to an appliance store, bought the air conditioner, and had it delivered to the woman's apartment.

Word spread by mouth, in those days before email, that support was available from the Joey Fund for CF families. From then on Joe and Kathy received a steady stream of letters—pleas for help. Joe brought them home and, one night a week, the couple would sit at the kitchen counter, pour out a pile of letters, and read through the requests one by one, figuring out which they could accommodate. An air conditioner? Yes. A gas bill? Yes. The mortgage? No. The rent? No. One month's rent? Yes. A mother who couldn't afford a night in a hotel while her kid was getting treatment at Mass General? Of course. A father who lost his job and got another, but the family had a two-month health insurance gap, and the cost to cover it was $2,200? Yes. With each approval, the money went straight out the door. No ceremony. No pictures. No nothing.

By the early 1990s, Joe was emerging as a successful and respected businessman after an emotionally devastating five years that included not just Joey's death but his father's two years later. Business had also been rough. Early in 1986, Joe had purchased Allied Advertising, the nation's third-largest print advertiser.[1] Around the same time, he'd befriended some local real estate developers with whom he invested about $10 million in eight properties. The arrangement was such that every partner was responsible for the debt of all the others if something went wrong. And that's exactly what began happening in 1987, when the tax laws changed and the real estate bubble burst. Many of Joe's partners declared bankruptcy. Others fled the country and never returned. By 1989, Joe was saddled with $141 million in debt.

But as with many terrible things in Joe's life, the disaster turned into an opportunity. Though he felt humiliated, he met with a group of bankers and hammered out a deal to pay $28 million—roughly three times what he originally owed—over the next five years. For Joe, it wasn't a loss but an investment in credibility; everyone he paid back spread word that Joe O'Donnell was an honest and model businessman.

As Joe's business grew, many of his contacts and partnerships evolved, becoming more personal. Once those bonds humanized, everyone who knew Joe and Kathy and learned about Joey wanted to do something for the couple and their cause—they wanted to become part of the solution and cure this terrible disease. The year after Joey died, the Joey Fund organized a Boston Celtics fashion show that lured many of the state's top CEOs, as well as big-name sportswriters from around the country, politicians, and reporters from all the local papers. Everyone who attended wanted a photograph with Joe. Friends and acquaintances brought their children to meet him. Joe shook hands with each of them, placing his second hand over theirs in a warm clasp, looked them in the eye, and asked about their future plans. He remembered the names of children whom he had met before, warming the hearts of parents as he referenced their previous encounters. It was a scene reminiscent of Marlon Brando's turn in *The Godfather*—albeit a kinder one, sans the organized crime element and penchant for murder—with each visitor requesting a brief but intimate audience with the Don.

The respect and admiration was well deserved; Joe had a reputation for treating people well. He was happy to provide guidance and advice to college-bound seniors. He had a broad influence and could help relatives get internships or job interviews, or connect you to anyone, including senators, representatives, and even past and future US presidents. He knew everyone, and he was keen to share his wealth and connections. If he could help, he would. It didn't matter who you were.

Joe was also known for his generosity outside the cystic fibrosis arena. It was important for him to show appreciation to anyone or any institution that had helped him achieve his current status. To that end he regularly raised money for Malden Catholic High School, Phillips Exeter Academy, and Harvard College. Each of the three institutions had given him a full scholarship to attend and he wanted other kids to receive that same gift.

By the late 1990s, the Joey Fund was hosting a dozen or so fundraisers every year. There was the Hotdog Safari, in which people ate as many hotdogs as they desired for a $20 entrance fee. The event began at a restaurant called Busters on Route 1, until it got too big and moved to a nightclub called The Palace, which Joe owned. After five years they moved it again to Wonderland, a dog racing track, and then finally to Suffolk Downs, a racetrack that Joe co-owned that could accommodate up to 100,000 people. Some years Joe would invite a grand marshal to host the event; those included his friend Mitt Romney, who was governor at the time; NFL Hall of Fame coach Bill Parcells; and Boston's mayor Tom Menino. Fundraising events also included an annual billiards night and the Kings Cup bowling competition, which raised money by selling the lanes for an evening to private equity firms for their junior employees to enjoy a night out, competing and bonding with their counterparts at other firms.

The pièce de résistance was always the Joey Fund Film Premiere, which had grown from its original attendance of just thirty to hundreds. Every year it drew a larger crowd. The key to its success, as with the success of all Joey Fund events, was giving donors an experience that deviated from the traditional staid fundraiser format, which Joe defined as a stodgy dinner, with a rubbery piece of chicken and overcooked vegetables, followed by a speaker that lulled people to sleep. Joe wanted everyone to have fun; if they enjoyed themselves, they would return year after year as regular donors.

So, instead of a sit-down meal, the movie was preceded by a casual reception where donors could meander and sample foods supplied by local restaurants, including his own. After a couple of hours of good food and drinks, Joe made some brief remarks about CF—never more than five minutes—as he had done for the past fifteen years. But unlike when Joey was alive, the decade to come would be peppered with moments of optimism, which he highlighted in his talks. In 1989, he mentioned the gene; in 1990, gene therapy; in 1994, he told the crowd about Pulmozyme, which had just been approved; in 1997, he shared the good news about TOBI; and throughout the decade, he would talk about the promise of gene therapy. None of these successes, he told his audience, would have been possible without their support.

When he completed his thank-yous, it was a signal the movie was about to begin, and limos would pull up and whisk everyone away to watch a film that hadn't yet premiered in Boston. While the film played, Joe, Kathy, and a tight group of friends who helped organize the event would sit and eat. Then Joe would return to the cinema as the film was ending and stay until the theater was empty, personally thanking everyone who attended.

When Kate was six, and Casey just three and a bit, they began joining their parents at the annual movie premiere that Joe had launched more than a decade prior. By that point the event had grown. After people had been eating and drinking for two hours, a local entertainer would take the stage. Then the O'Donnell girls would come up to draw names out of a hat and present the winners with a bouquet of roses—a reference to the way a child once mistook the disease name *cystic fibrosis* for *sixty-five roses*.

It wasn't long before Kate and Casey were also joining the fundraising effort. They participated in the Great Strides Walkathons that were the bedrock of the CF Foundation's volunteer fundraising events. And by the time each turned twelve years old, they were no longer allowed to receive gifts at their birthday parties. Instead, their friends were invited to make a donation to the Joey Fund, if they desired.

Neither of the girls were thrilled with the idea at first. After all, opening presents is an integral element of birthday party fun. But Joe and Kathy were emphatic that the girls develop a generous spirit early in life and include public service as a part of their lives. The couple set a tangible example by hosting potluck dinners for Kate's and Casey's classes in their Belmont home and participating in fundraisers and benefits for a range of institutions and causes. And the girls noticed how Kathy, in her quiet way, would regularly take baskets of treats or other food to homes in the neighborhood—often without saying anything or even attaching her name—whenever she heard a family had a sick child. Kathy knew how illness strained a family and wanted to ease that burden for others. She and Joe were adamant about paying it forward, and expected their daughters to be thankful for what they had. After all, the girls were born healthy and into privilege, and their parents were acutely aware that they needed to learn how to leverage that position for doing good.

By Kate's sixteenth birthday party, the disappointment of giving up her presents had worn off, and she was eager to see how much money she could raise. She decided to celebrate at Kings Bowl America, and the raucous gathering of teenagers brought in about $2,000—a tidy sum that set the bar for future events and stoked her competitive instincts.

The money that was raised through the Joey Fund, minus the 5 percent Joe kept to help Massachusetts families with CF, went directly to the Cystic Fibrosis Foundation, where it supported a growing repertoire of research that president Bob Beall was funding. After the failure of gene therapy, the foundation's board was pushing Beall to invest in radical, borderline-delusional science that might yield potentially life-saving drugs for their children. But Beall already had his sights set on paying a company or several to invent a drug from scratch. He knew that it wasn't going to be easy to convince a pharmaceutical company to take on the project, so the foundation would need to fund the entire initiative. That was going to be expensive, to the tune of tens of millions of dollars, at least. It was going to take ten, maybe even one hundred, times what the chapters were currently bringing in. Yet, though neither Joe nor Beall knew it at the time, Joe's fundraising and business savvy would be critical for accelerating the foundation's ambitions.

CHAPTER 33

A Network for Developing Therapeutic Drugs

1997–1998

If you build it, they will come.

—Field of Dreams

With the successful launch of two drugs, Cystic Fibrosis Foundation president Bob Beall now had a new agenda and strategy: he would work directly with biotech and pharmaceutical companies to design medicines. Anticipating that this new venture would eventually yield the foundation a drug to test, he wanted to address a looming problem that haunted all clinical trials: recruiting patients.[1] Pharmaceutical companies frequently struggled to fill their clinical trials, even for common diseases, and the problem was much worse for rare diseases like CF. Beall needed a system to rapidly identify the right patients for a trial and funnel them to a nearby center where they could participate. And he knew Bonnie Ramsey was the mastermind for the job.

Beall and Ramsey had been friends and colleagues for some time when, in 1997, Beall came to Ramsey with a proposal. He wanted her to lead a new project for the foundation, one she had been pushing him to launch for a decade: a permanent infrastructure for future CF drug trials, through which physicians could recruit patients quickly and perform those trials more efficiently.

In the past, each time Ramsey completed a clinical trial, the entire network of nurses and scientists and physicians disbanded. When another began, they had to reassemble the team, hire and retrain nurses and trial coordinators, and establish new drug testing protocols. Wouldn't it be more efficient, Ramsey had asked Beall after the

Pulmozyme trial, if the foundation had a permanent network of sites dedicated to clinical trials? One staffed with nurses and lab technicians familiar with all the nuances of cystic fibrosis, experienced and comfortable with the complicated breathing measurements and chloride tests—professionals with a deep knowledge and appreciation of the disease and its quirks?

The model Ramsey envisioned for this network was the NIH-funded Children's Oncology Group,[2] which began in 1955 as a cooperative of cancer clinical-trial groups working to find treatments and cures for infants, kids, teens, and young adults. The COG's strategy had three prongs: clinical trials, long-term follow-up with patients, and basic research into the biology of cancer. Over the next fifty years, the network of medical centers participating in COG studies and trials rose to 230 in six countries and involved more than 5,000 pediatric cancer specialists. Within the same period, the survival rate for patients diagnosed with childhood cancer skyrocketed from just 10 percent—a death sentence—to 80 percent, transforming the field.[3] The key to this turnaround was the multidisciplinary patchwork of physicians, nurses, biologists, pharmacologists, psychologists, and other specialists who treated patients and studied them through each new trial and beyond. Ramsey wanted to do the same for cystic fibrosis.

Before Pulmozyme and TOBI, cystic fibrosis treatments had a long history of being tested in small, weak clinical trials at individual clinics.[4] In about 80 percent of these studies, it was hard to tell whether the drug or therapy was successful, because the studies either recruited too few patients or were poorly designed. If the foundation was going to spend millions on basic research and developing drugs, as Beall planned to do, it made sense to create a permanent network of centers, distributed throughout the United States; this was a rare disease and its patients were scattered.

Ramsey thought her words had gone unheard—until Beall's visit. In typical Bob Beall style, he began, "Bonnie, I've decided I'm doing this." The foundation would offer grants to lure industry into designing drugs. But first, he told her, he needed the clinical trials network she'd been imagining for years. It was going to be called the Therapeutics Development Network. Would she develop and lead it?

Ramsey was dumbfounded; Beall hadn't been ignoring her all this time. Rather, he had been incubating her ideas until the time was ripe. Now he wanted her to take a sabbatical from leading the CF center

at Seattle Children's—the foundation would fund her—to launch the
Therapeutics Development Network (TDN), an alliance that would be
able to run back-to-back clinical trials in case multiple drugs needed
simultaneous testing. In 1997, the idea that there would be multiple
drugs for this rare disease sounded outlandish, but Beall had big plans.

It was an overwhelming proposition. Ramsey cared deeply for the
patients at her clinic, and she was uncomfortable leaving them with
another physician. But by building a network of clinical trial sites, she
had the chance to help *all* children with cystic fibrosis. After confirming
that her research nurse, Judy Williams, with whom Ramsey had done
all her clinical studies, would come and help her with this enormous
undertaking, Ramsey accepted Beall's offer.

Ramsey had assumed she would spend nine months conceiving this
network, then possibly another nine getting it up and running. Beall's
time frame was different: he wanted the network *up and running* within
nine months. To focus, Ramsey and Williams moved out of the hospital and into an office at PathoGenesis, the company that had worked
with the foundation to develop TOBI. Then they began brainstorming.
What did they need? Whom should they hire? Which of the foundation's 113 care centers would make the best permanent clinical trial sites?

The first priority was obvious: patient safety. Ramsey began by
establishing a committee of physicians and scientists who would vet
clinical trial proposals submitted by drug companies and ensure that
the protocols were safe. This independent regulatory unit, called the
data safety monitoring board, was run by Arizona-based pediatrician
Wayne Morgan. The unit would monitor every patient in every study,
staying alert for side effects and adverse events that occurred during
clinical trials. And they would investigate any deaths—which weren't
uncommon, as these were sick patients—to figure out whether they'd
been caused by the drug.

With the help of Dick Kronmal,[5] a biostatistician at the University
of Washington who had run some of the biggest cardiovascular epidemiology and treatment trials in the United States, Ramsey assembled
a statistics unit to create a data management system that would crunch
the clinical trial data.

Finally, Ramsey turned to establishing the clinical trial centers
themselves. At Beall's nudging, Ramsey announced the launch of the
new network to all 113 existing Cystic Fibrosis care centers, and solicited
applications to be a part of it. To join the network and receive special

funding from the foundation, a center had to demonstrate expertise in all the standard measurements—lung function, electrical activity in the nasal cavity, sweat testing, and inflammation—as well as other specialties, like interpreting chest X-rays and CT scans. If accepted as a trial center, the care center would need a permanent staff with clinical research nurses, recruiters to identify and contact patients, and specialists to run the trials.

Once the announcement was out, 25 of the 113 centers applied, and in 1998 the foundation selected seven inaugural sites as the backbone of the network. Seattle, where Ramsey was based, would be the mother ship—the central coordinating center.

IN 1998, BEALL HIRED THE FOUNDATION'S FIRST PHYSICIAN, PRESTON Campbell, who had led the CF center at Vanderbilt. The foundation had always had physicians guiding their actions, but never one in a senior management role. Beall had hired Campbell to oversee the foundation's medical programs, including the 113 care centers, and also to collaborate with Ramsey by helping integrate the best seven sites into her Therapeutics Development Network.

Campbell was a contrast to Beall in both appearance and temperament; he was six foot seven inches, with dark hair, large blue eyes, and a pensive disposition. He came across as reserved at first, but when he got talking about what mattered to him, he was animated, warm, and quick-thinking. He grew up in the small town of Johnson City, Tennessee, and went to college at Georgia Tech before attending the University of Virginia's medical school.

One summer during medical school, in 1978, he worked as a counselor at Camp Holiday Trails, a place for children with chronic medical conditions like diabetes, cystic fibrosis, and a host of other very rare diseases. Campbell was in Bunk 8, along with several boys with CF. There he saw firsthand the troubles these children face with their digestion and from the constant coughing that disrupted their lives. To help with their medical care, he learned how to do their daily chest physiotherapy. That summer Campbell also got to know a physician named Dr. Robert Selden, who led the CF center in Charlottesville, Virginia. Campbell was moved by Selden's passion for improving the lives of these children and approaching the disease with scholarly rigor and evidence-based medicine. Campbell felt tremendous compassion for these young people and declared, when he returned to medical school that fall, that he wanted to specialize in cystic fibrosis. His colleagues laughed, telling him there were too few patients for him to make a living.

After medical school Campbell did his pediatrics residency at Vanderbilt, then his chief residency, where he picked up managerial and leadership skills. After getting married, a passion for tropical medicine and a drive to practice in the developing world led him to the Liverpool School of Tropical Medicine and then to Haiti for two years, where he set up a pediatrics program during the country's 1986 coup d'état. Every day in Haiti taught Campbell to think out of the box; he couldn't do procedures the same way he did in medical school, where he had all the facilities and lab tests and technology. Here he had to fix broken bones without X-rays, deliver babies in the dark, and figure out how to save lives without blood tests and cultures.

When he returned to the US, he accepted a job at Vanderbilt in general pediatrics. Most of the cases he saw were pulmonary, largely because Vanderbilt was in the middle of searching for a pediatric pulmonologist. And when the director of Vanderbilt's CF center retired shortly after Campbell's arrival, it left a glut of patients with severe lung problems in his care. It wasn't long before Campbell was asked to direct the CF center, even though he wasn't a pulmonologist by training. He was convinced it was a bad move for his career; once the medical school hired a pulmonologist, all CF patients would likely be transferred to that person's care. But taking the job seemed the ethically right thing to do, so he accepted.

When the new pulmonologist was hired, however, Campbell was asked to stay on and run the CF center—a route into the world of CF strangely similar to Ramsey's. In his first year in charge, the number of children in the CF center doubled to 60; the next, it reached 120; and the third year, it doubled again, hitting 240. The center's rapid growth caught the attention of the CF foundation and Bob Beall, who asked Campbell to share his expertise by joining various committees.

Beall also began making trips to Nashville, alone and with board members, to convince Campbell to accept a job as the foundation's executive vice president of medical affairs. Campbell was both torn and thrilled. He loved his patients and Vanderbilt. But finally, in mid-1998, he accepted the job and moved to Maryland because, like his new colleagues, he didn't just want to treat this disease—he wanted to cure it.

One of his first duties for the CFF was visiting the Children's Oncology Group's data coordination center in California with Ramsey. The

COG had a slightly different mission from the foundation's Therapeutics Development Network; they were working with pediatric cancer drugs that had already been approved and figuring out the best combinations and doses for particular patient groups. By contrast, the TDN would be running clinical trials for completely new, untested drugs. Despite this difference, there was plenty that Campbell and Ramsey could learn. In particular they were hoping to pick up some ideas about how the COG managed data and issues of patient safety that they could then apply to the TDN.

When the TDN began accepting applications to use the new network, Ramsey expected most would come from academics at foundation-supported labs at universities. But after working with Genentech and PathoGenesis, Beall's ambitions had soared. He was determined to lure industry—and he had orchestrated the perfect introduction to the network to pique their interest. He had invited executives from pharmaceutical companies to review the seven sites in the trial network, taking the opportunity to show off the sophisticated, efficient facilities. The takeaway from Beall's tour was simple: design a drug for us, and we will help you recruit the right patients and test it in the perfect setting.

Beall's plan worked. To Ramsey's surprise, within one week of the Therapeutics Development Network opening its doors, industry came knocking. The first trial request came from Seattle's Targeted Genetics, which had presented plans for a CF gene therapy trial at the NIH in 1994 and was now ready to proceed, even though enthusiasm for the gene therapy field was quickly waning.

As more trial requests started coming in, and with the existing centers working well, the foundation expanded the Therapeutics Development Network so that more physicians and scientists could participate in the trials. By 2004 there were eighteen sites; that number would grow to ninety-one. Originally, each center had just one principal investigator in charge of running trials. But as the number of trials rose, so did the number of investigators—to as many as four at the busiest locations, including at least one for trials in young adults (life expectancy for CF was, by 1997, around thirty years old), and one for pediatric trials.

In addition to this slickly efficient new network, Beall had another card up his sleeve to guarantee industry would want to use the foundation's trial network to develop therapies. Of the 30,000 or so patients with CF in the United States, 85 percent were seen in one of the foundation-supported care centers. If industry wanted to create a drug for this disease, it behooved them to work with the foundation. This also meant companies had to play by the foundation's rules. If the foundation found a flaw in the design of a clinical trial, the company had to fix it or risk their drug never being tested.

Beall was trying as hard as he could to make working on cystic fibrosis attractive to pharmaceutical companies. Now, with the Therapeutics Development Network, the companies didn't need to expend resources recruiting patients to their trials; they had a population ready and waiting. The network's trial physicians, who knew the patients well, could draw on the foundation's registry to recommend those who would be a good fit. And companies would not need to train physicians to measure the impact of a potential drug on the patients; the physicians and nurses at the trial centers were already experts. Once a company had a drug worthy of testing, they could trust the trials would be efficient, and the data both high quality and trustworthy.

As Ramsey, Williams, and Campbell were coordinating the first gene therapy trials for Targeted Genetics, Beall turned to a more immediate issue. He wanted the foundation to invest in more companies and fund early research, ideally in exchange for a similar arrangement to the one used with TOBI—one where a royalty agreement was attached to the foundation's funding in case the company actually developed a drug. But funding those preliminary studies wasn't cheap. The foundation had poured $2.6 million into developing TOBI before handing it off to PathoGenesis. How could such a little foundation continuously raise such large amounts of cash? In the mid-1990s, the foundation was spending about $19 million annually on medical and scientific research, but that wasn't enough to spur pharmaceutical innovations. Beall realized he needed a separate stream of revenue.

Now that PathoGenesis had started selling inhaled TOBI, and the company was enjoying its first profitable quarter ever, the CF Foundation began receiving $1 million in quarterly royalties as the drug sold. But even this wasn't of much use. To make effective investments in other companies, Beall needed tens of millions *immediately*. So he put out word that the foundation would sell all future royalties for TOBI to an outside party for one lump sum. Several buyers came forward. But in June 1998, after closed-door negotiations, the foundation sold them back to PathoGenesis for the healthy sum of $19 million—enough to launch pilot projects at multiple companies. Now Beall could fund other new drug ventures. He was betting that a small proportion of these drug deals would pan out and deliver royalties, so that he could continue the cycle of investment and royalties as long as it took to find a cure.

With that $19 million in hand, Beall was ready to start investing. After his conversation with Alan Verkman, Beall wanted to hire a company that specialized in high-throughput screening, which could test tens of thousands of molecules to find one that would fix the defective CFTR protein. But Beall didn't want the foundation to partner with just any company; he wanted the high-throughput sequencing leaders. But when he called the seven companies listed in the journal article where he had first learned about this new technology, only two of those responded. The one that stood out from the beginning was Aurora Biosciences Corporation, the company founded by Roger Tsien and recommended by CFF advisor Ray Frizzell.

Aurora's response came in the form of a phone call from a young scientist named Paul Negulescu. Aurora's founder, Tsien, was intrigued by Beall's proposal and had routed the inquiry to Negulescu. They both wanted to learn more.

CHAPTER 34

Aurora

1998–1999

> The most successful scientist thinks like a poet—wide-ranging, sometimes fantastical—and works like a bookkeeper. It is the latter role that the world sees.
>
> —Edward O. Wilson

When Paul Negulescu first entered college at UC Berkeley, he wasn't expecting to study science. In fact he was a history major, who, during his junior year, took a physiology class and was so unexpectedly inspired that he decided to get a bachelor's degree in both physiology and history. What really spiked his curiosity was the unit on how the kidney worked. It was an organ that did a lot of filtering of salts—sodium, chloride, and potassium—and kept them balanced throughout the body. Negulescu could visualize the proteins that did this as little machines moving these salts in and out of cells.

Captivated by these proteins, he wanted to study them more. So as a graduate student in the late 1980s, Negulescu focused on channels and transporters—proteins that shuttled charged atoms in and out of cells. In his last year of graduate school, Tsui and Collins discovered CFTR, and Negulescu decided to study the role of the CFTR protein in the stomach—one of the areas of the body where the gene was turned on. He applied to the Cystic Fibrosis Foundation for a research grant but, ironically, his application was rejected. The work and research he did to write the grant, however, gave him a broad understanding of CFTR and how it worked, a background that would later prove very useful.

During graduate school, Negulescu spent many hours each day working on his research project in the lab of Roger Tsien (pronounced

Chen), then a Berkeley professor of pharmacology, chemistry, and bio-chemistry. Tsien had recently developed colorful chemical compounds that could detect calcium, chloride, and sodium ions as they moved in and out of cells. He was a wizard with color and had a talent for designing eye-popping dyes that painted otherwise invisible atomic-scale chemical changes inside the cell with fluorescent strokes.[1]

Tsien's focus on color and chemistry had begun in elementary school in New Jersey, when he first discovered the lavishly illustrated *Golden Book of Chemistry Experiments*. He had derived intense satis-faction from performing the experiments, which provided a window into the vivid and violent nature of chemistry, and particularly enjoyed the flamboyant colors of the raw inorganic chemicals he used in the experiments—the copper compounds' array of blues; the chromium compounds' spanning the spectrum from red, orange, and yellow to parakeet green;[2] the manganese compounds' delicate rose pinks and deep purples. But even the experiments that lacked color compensated with explosions and other entertaining phenomena, like the time he dropped a small zinc gear into a beaker of concentrated hydrochloric acid, causing a geyser to erupt out of the flask as the zinc stole the chlo-ride from the acid and released a spout of hydrogen gas.

For Tsien, though, the most impressive experiments were the ones that fed both his curiosity and his eyes, like one in which he poured a dark purple solution of potassium permanganate through a leaf of filter paper into another beaker, transforming the solution, seemingly magi-cally, to a sparkling emerald green. The visual was simple and striking, planting a deep passion for color and light that would guide the next fifty years of his life. Though just eight years old, he carefully jotted the results of his experiments in a small notebook that would eventually end up in the Nobel museum in Stockholm.[3]

Tsien's childhood preoccupation with pretty colors reemerged in graduate school when he began developing dyes to spy on the inner workings of cells. In the body, calcium is critical for transmitting signals between nerves. It makes muscles contract, and triggers cells to split and multiply. "Chameleon" dyes[4] that flipped from cyan to yellow when cal-cium entered a cell allowed Tsien to watch, through the color change, as cancer cells divided in real time. High calcium levels are also a hallmark of certain brain diseases, so scientists worldwide started using Tsien's dyes to chronicle various activities inside other cells. One application of a fluorescent dye Tsien created was color-coding tumors, which enabled surgeons removing them to ensure no cancer cells were left behind.[5] Another dye he created tracked the movements of zinc in the cell.

While many of Tsien's colleagues expected his calcium dyes to earn him a Nobel, what actually won him the prize was his work on the green fluorescent protein (GFP)—a natural protein first discovered in the jellyfish *Aequorea victoria* that emits a gentle green glow when showered in blue or UV light. Tsien had little respect for the natural protein, regarding it as "dim, fickle, and spectrally impure."[6] So he altered the genetic sequence of the jellyfish gene to create a brighter green. He then figured out how to tweak the protein further to create a prismatic spectrum of dyes—cotton candy pinks, sunflower yellows, Caribbean azures—that could tag and track molecules of interest as they moved through the body and in and out of cells. Scientists in thousands of labs around the world adopted his GFP and its rainbow cousins as "reporter dyes" to learn when and where particular genes were active in cells in the lab and in living organisms.

Negulescu got to know Tsien during graduate school when Negulescu began using his dyes to track calcium, chloride, and sodium in his thesis work. Tsien was a patient and humble man with a talent for nurturing young scientists and helping them discover their passions. Though Negulescu was not one of his students, Tsien was generous with him, sharing microscopes, dyes, imaging systems—all of which he had invented and built himself. But because Tsien's lab was small, just three rooms—one for chemistry, one for computers, and one for the imaging system—the handpicked postdoctoral researchers who either helped design the chemical sensors or tested them on cells were all vying to use the various microscopes and instruments during the day. The only time the equipment was available to someone who didn't officially work there was between 2 AM and 6 AM. So Negulescu began his experiments at 2 AM, when Tsien also went into the lab to program the computers to analyze the images of cells and tissues impregnated with his dyes.

The two tiny rooms where he and Tsien worked were adjacent. Whenever Negulescu hit a snag and his microscope or computer glitched, he would poke his head into Tsien's equally claustrophobic room for help. Tsien would come in, utter an "Oh, darn," type in some new code, and get it going again. Through these brief, unplanned interactions, the two got to know each other.

After graduate school, Negulescu moved to the University of California, Irvine, where he continued to work on how ions moved in and out of cells and switched genes on and off. He was on the brink of accepting a job offer from the University of Connecticut five years later when, in 1996, Tsien called him to ask whether he would like to be part of his new startup. Tsien planned to combine all the tools he had invented over the years—dyes, microscopes, software, and robots—into a company that would speed up the discovery of new drugs for pharmaceutical companies. Negulescu, he thought, would be a good recruit— he was already familiar with not only the physiology of the cell and ions' movement in and out of it, but also Tsien's dyes.

The idea to start a company was not actually Tsien's. The seed for the company came from Kevin Kinsella, an MIT-schooled venture capitalist who had been nosing around the technology disclosures at the University of California, San Diego, and had contacted a UCSD professor named M. Geoffrey Rosenfeld for a meeting. Rosenfeld, in turn, called his friends and colleagues Charles Zucker and Tsien (who had moved from Berkeley to UCSD in 1989), to see whether they were interested in meeting Kinsella and pitching ideas for startups, too. When the four got together, Kinsella was most captivated by Tsien's ideas.

Kinsella was drawn to Tsien's Technicolor reporter dyes and had a hunch that the technology would fill an unmet need in the drug discovery landscape of the 1990s. Before meeting Tsien, he had launched dozens of early-stage biotech companies,[7] including some that were using combinatorial chemistry to make and sell mass quantities of new chemicals to pharmaceutical corporations for possible drug development. He had also launched a business to identify new genes that, when turned off or on, could potentially cure disease. But he had no way to connect these technologies. Kinsella was convinced that Tsien's magical dyes were the missing link: a way to detect whether a compound had the ability to change gene activity, calm down an overactive protein, or fix some other defect in a misbehaving cell.

Tsien's vision for his new startup involved exposing diseased cells to a chemical, visualizing the effect with a dye, and then measuring the level of response with a microscope and image processors. He also wanted the entire operation automated, with robots doing the tedious work of adding thousands of chemicals to cells in special dishes and then monitoring each one for the desired chemistry. What excited

Kinsella about Tsien's idea was that this new company would be doing something that no other company at the time could do: testing drugs *inside* cells and at a scale that had never been attempted. By miniaturizing the volumes of these chemical reactions and automating everything, Tsien would be able to do not just high-throughput screening but *ultra*-high-throughput screening that would enable the company to test tens of thousands of molecules per hour.[8] This technology was revolutionary, and Kinsella knew it.

Before the 1990s, drug development was slow. Even in the big pharmaceutical companies, drugs were tested one at a time, one disease at a time, in a test tube. If a candidate molecule performed well in vitro, catalyzing a particular chemical reaction, only then was it tested on living cells. Scientists waited until late in the process to expose living cells to new molecules because cells were finicky and short lived—hard to work with and expensive to keep around. It was also notoriously difficult to measure what a drug was doing inside a cell. Was it performing just the desired chemistry, or was it interfering with other inner workings of the cell? And it was unimaginable that such painstaking work could be done on an industrial scale.

Tsien's new company would be able to test the drugs inside real living cells in the first stage, and if a compound couldn't function inside a cell, or was toxic, it could be discarded immediately. Being able to screen thousands of chemicals quickly was worth the hassle and cost of maintaining vast supplies of cells. Tsien could simultaneously test a single molecule on cells taken from patients with hundreds of different diseases to see if it had immediate treatment potential. Or he could test thousands of different molecules on cells representing a single disease. Or both at once. The possibilities were as infinite as the molecules. Kinsella could see the potential of Tsien's technology and where it fit in the drug development ecosystem, and the two struck a deal. In 1995, Tsien, Kinsella, and a couple of colleagues from the University of California, San Diego, officially founded Aurora Biosciences Corporation in the biotech oasis of La Jolla, California. In 1996, Kinsella was able to raise $18.6 million for the fledgling company after four rounds of venture funding.

When scientific superstars founded startups in this era, the individual would actually have little to do with running the company day to day. That was exactly what Tsien wanted. He was a scientist to his core and had no desire to micromanage Aurora, preferring to continue as a university professor and work with the company in a scientific advisor role, offering guidance and ideas. The first CEO that Kinsella brought in to manage the company was Tim Rink, a quick and critical thinker, a sharp scientist, and a man who would instill in the scientists the value of a good sense of humor and an upbeat vibe. And because Tsien wanted Negulescu to lead all the biology at the company, Rink offered Negulescu a job as director of discovery biology.

When Tsien first described to Negulescu what this new company would do, Negulescu knew that it was the chance of a lifetime. He'd be moving Tsien's inventions into the field of drug discovery to uncover new medicines. Negulescu was fond of Tsien and well acquainted with all the machines, molecules, and software Tsien had built. He knew they were both unique and robust, and it wasn't a stretch to imagine ramping up the drug-testing technology's speed to an industrial level. In academic labs, Negulescu was used to doing a few experiments per week. At Tsien's company, that number could soar to a million or more. Or, depending on the client, Tsien explained, they might design and build a custom screening robot that a pharmaceutical company could use to do the work themselves. It was an irresistible prospect for an ambitious scientist, and Negulescu accepted the job.

As it turned out, Kinsella had perfectly anticipated the demand for Aurora's technology.[9] Within its first year, the company had filled with engineers, software developers, and other IT folks who were building robots and software to screen millions of chemicals for Aurora's clients. Merck, Pfizer, Warner-Lambert, Bristol-Myers, and other pharmaceutical giants were commissioning one-of-a-kind assays and the robotic systems to perform them. In 1997, Aurora's initial public offering raised another $40 million.

Aurora's technology was, as Kinsella had foreseen, game changing. Three elements enabled the company to not only screen hundreds of thousands of chemicals or more each day, but do so faster and more accurately than their rivals. First, they had miniaturized the chemical reactions and the vessels that held them. Rather than working with test tubes that involved volumes of 10 milliliters or so—just under 2 teaspoons—as a scientist might do in a university lab, Aurora's scientists were running the experiments in one or two microliters—one-fiftieth of a raindrop. These microscale experiments were done in iPhone-sized plastic plates containing 3,456 holes, or wells, each of which cradled a few engineered cells and a drug candidate. Shrinking the volume reduced waste and cost, and Aurora had achieved a degree of miniaturization that surpassed all others in the industry.

Another feature facilitating their warp-speed research was a chemical retrieval system known to company insiders as "the vault"—a series of monolithic gray rectangular silos, each housing a million or so unique molecules stored in other 3,456-well plastic plates stacked atop one another. A database listed the name and structure of every molecule in each plate, allowing a scientist to easily retrieve and test one or thousands as needed.

The third element was, of course, Tsien's proprietary fluorescent dyes, which were matched to each experiment and would change color when the desired reaction occurred in one of the wells. With their automated camera system using Tsien's custom software, Aurora's scientists could instantly identify specific wells among thousands that were the right color and showing the anticipated behavior.

When the three technologies were combined, the result was a futuristic version of Henry Ford's assembly line. Conveyor belts carried the plates filled with cells to a robot, which would retrieve a plate of chemicals from the vault, suck up microscopic quantities of the plate's 3,456 chemicals simultaneously, and then squirt them into the wells of cells on the belt. Robotic arms then loaded the plates full of cells and chemicals into an incubator, to give each chemical time to work its magic. After hours in the cozy incubator, the plates continued on the conveyor belt, to where another robot added dye. The last stop was a sensitive fluorescence detector and camera that would note the location of any glowing cells.

The company became successful quickly. But the financial success didn't inspire Tsien, and he wasn't excited by the work Aurora was doing. Most of the projects for the big pharma companies were confidential. And Tsien found the process of drug discovery boring. When exciting science emerged, profit, not publishing, came first.

In 1998, Bob Beall from the Cystic Fibrosis Foundation called, wanting to hire Aurora to find a molecule that would resurrect a malfunctioning CFTR protein. Tsien was immediately captivated, not because he had an overwhelming interest in cystic fibrosis or a driving desire to cure this disease, but because what Beall was proposing had never been done before. His curiosity piqued, Tsien encouraged Negulescu to pursue it. If Negulescu's team succeeded, they would be inventing a completely new category of drug.

At the time there were three main drug categories: chemical agonists, chemical antagonists, and biologics. Agonists bind to a receptor in or on a cell, like a key fitting a lock, and trigger a specific biological effect. For example, our cells have receptors "designed" to fit naturally occurring opioids, like the endorphins produced during exercise, which dull pain and cause a natural high. Drugs like morphine, heroin, oxycodone, methadone, hydrocodone, and opium are all opioid receptor agonists, which work by hijacking the opioid receptors, binding tightly to them lock-and-key style, preventing pain and triggering a powerful—if addictive—euphoria.

Antagonists, by contrast, are drugs that block an agonist from binding to its receptor. Naloxone, a rescue medication for heroin overdose, prevents heroin from reaching its target opioid receptor.

The third type of drug, biologics, includes a wide variety of products from vaccines to blood and blood components to cell and gene therapy. It also includes proteins like insulin and growth hormone (like those produced by Genentech) that are substitutes for naturally occurring substances in the body.

The drug that Beall was envisioning for the mutated CFTR protein didn't fit any of these categories. What Beall was proposing was what Negulescu's team would later refer to as a "corrector"—a drug that would help the misfolded protein with the F508del mutation fold properly, so that the protein would travel to the right location in the cell and open and close like it was supposed to. No one had ever found a drug that could rescue or refold a damaged protein. No one had even attempted it. The implications of such a breakthrough were enormous.

The project would be one of the most difficult Negulescu and his team at Aurora had ever attempted. But regardless of whether it worked, the lessons learned would be publishable. Researchers had discovered that many diseases were caused by misfolded proteins: Alzheimer's, Parkinson's, Huntington's, Creutzfeldt–Jakob, Gaucher's, and many others.[10] If Aurora could find a molecule to refold this protein, maybe it would serve as a road map for treating other devastating conditions. And if they failed, that would be instructive, too.

It made sense for Tsien to connect Beall directly with Negulescu, who would lead the cystic fibrosis project. After all, he knew more about CF than anyone in the company, thanks to his graduate school work on transporters and channels. He knew a lot about the CFTR protein, including how it regulated the movement of chloride. He also had a broad knowledge of many types of ion channels—the donut-shaped pores in the membranes of cells that provided tunnels for potassium, sodium, chloride, and other ions to enter and exit the cell.

Beall didn't mince words when he first spoke to Negulescu: "We're interested in drug discovery for CFTR. Can Aurora find a chemical that will make a malfunctioning chloride channel in cystic fibrosis cells behave normally?"

Negulescu assured him that he knew the CFTR protein well, and he told Beall about a dye that Tsien had invented that was sensitive to changes in voltage. If Aurora found a chemical that repaired CFTR and restored chloride flow, the voltage would shift and the color of the dye would change—revealing the cell had been fixed. Negulescu told him that Aurora would build an automated system for screening tens of thousands of molecules to find one that might work.

Negulescu—a soft-spoken, even-tempered Californian—soon learned, as many had before him, that Beall was all business: driven, methodical, organized, impatient. He never engaged in small talk, though he would catalog one personal fact from everyone he met and refer back to it whenever they spoke. He was mission oriented and laser-like, ending every conversation and meeting with, "What's the next step?"

How long will this take? Beall fired at Negulescu. How long before you know if you are on the right path? Whether this is really an option?

Negulescu outlined a schedule with broad brushstrokes, but Beall wanted a detailed written proposal. What was Aurora's first step? How would they develop a test for CFTR activity? How would they scale up the prototype to scan thousands of molecules for their ability to modulate CFTR?

Negulescu estimated that it would take a team of three people about a year to figure out first, whether it was possible to screen mutant CF cells with the fluorescent dyes to identify possible drugs, and then, to scale up the procedure to test hundreds of thousands of molecules. Beall made a quick call to Negulescu to hammer out a few details. Then, with the board's consent, he approved the project. It would cost the foundation just under $1 million.

Sitting in the cubicle next to Negulescu at Aurora was Tom Knapp, who ran the high-throughput screening. As he sat combing through data, he couldn't help overhearing Negulescu's calls from Beall. Aurora had collaborations with all of the big drug companies for discovering drugs to treat conditions from pain to psychological disorders to diabetes—big health issues that affected tens of millions. But Knapp thought this project sounded exciting and different. It would provide an opportunity for Aurora to lead its first drug discovery project, if they found a worthy molecule. And, of course, a breakthrough for cystic fibrosis could change the course of the disease and save thousands of lives.

RECOGNIZING THAT NEITHER OF THEM KNEW MUCH ABOUT DRUG development, Beall and newly hired medical director Preston Campbell decided that they needed to recruit someone to lead the foundation's side of the drug discovery effort and communicate with Negulescu at Aurora. Beall knew exactly who that should be: Melissa Ashlock (formerly Rosenfeld), who had pioneered CF gene therapy in Ron Crystal's team. After Crystal's CF gene therapy failed, Francis Collins, who was by 1993 leading the National Human Genome Research Institute, had offered Ashlock a position at the institute to work on the human genome. There she'd run her own laboratory, continuing to study the CFTR gene while also developing viruses for other teams at the institute.

Ashlock had a particularly valuable set of skills that made her the perfect choice to lead the foundation's new quest. She learned a lot from Crystal about the logistics of developing new therapies and how to work with the FDA. And she was an accomplished scientist who knew how to assess CFTR gene activity in cells taken from patients—something she had done for the gene therapy studies. Once briefed on Negulescu's 1998 pilot project, Beall and Campbell knew she could help connect Negulescu's team with the scientists who pioneered the CFTR research and familiarize them with techniques to study the protein.

Over the next year, with the help of Ashlock and Christopher Penland, who led basic science research for the foundation, Negulescu's team assembled the pieces. The first task was finding a cell line for screening drugs. By 1999, researchers worldwide had identified hundreds of mutations that could corrupt the CFTR protein. Which mutations should Aurora work on first? Each mutation broke or disabled the CFTR channel differently; were some easier to fix than others? The obvious choice was the F508del mutation discovered by Tsui and Collins; nearly 90 percent of CF patients carried either one or two copies, so fixing it would help the majority of sufferers.

The protein created by the common mutation had two problems. The first was that it was so poorly folded that the cell treated the mutated protein as trash and recycled most of it. The small amount of CFTR that wasn't trashed was like an elevator stuck on the ground floor—it couldn't rise to where it was needed, embedded in the cell's top membrane to regulate chloride transit. This type of problem is referred to as a trafficking defect. The other problem was that, on the rare occasion when the mutated protein evaded the cellular trash collectors and did make it to the surface, it couldn't open and close as it was supposed to. It was as if the cell had a stuck door; the chloride had absolutely no way to get out. Patients with two copies of this mutation had almost no protein with functioning chloride channels, which, in most cases, led to a severe form of cystic fibrosis.

To launch the project at Aurora, Negulescu needed cells. Mike Welsh, the University of Iowa gene therapy researcher, gifted the company mouse cells that his team had engineered to carry the human CFTR gene with the F508del mutation. These could be mass-grown in

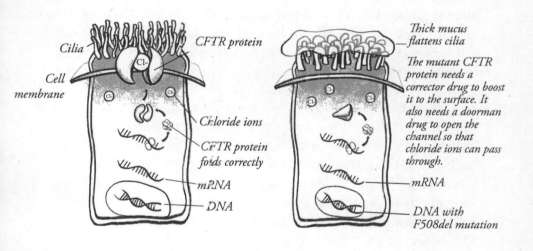

Left: A healthy airway cell with a normal CFTR gene. CFTR protein folds normally, reaches the surface of the cell, and becomes a channel, or "door," that opens and closes for chloride ions to move out of the cell.
Right: A sick airway cell from a patient with the common F508del mutation. The CFTR protein folds so poorly it cannot reach the surface of the cell. This disrupts the flow of water and makes the mucus on the cell's surface thick and dry.

the lab and used for testing drugs. The next step was installing a fluorescent sensor in the cells. The sensor they chose was one of Tsien's creations. When there was little or no chloride transport, the cells glowed orange. *But,* if the channel opened and the chloride was able to move out of the cell, the cells glowed blue instead. As a new drug-candidate molecule was introduced to a cell, a robotic camera system called a voltage ion probe reader, or VIPR (pronounced *viper*), would scan for blue cells—a sign that the added molecule was working.

There was a catch, however. Fixing the deformed chloride channel wasn't as simple as finding a single molecule that would make CF cells glow blue. Aurora needed to discover two different drugs to fix the problem. First, they needed a corrector molecule that could fix the folding of the protein and get it to the surface. Then, they needed a molecule that behaved like a doorman, opening and closing the channel. But Beall had been clear about the need for speed—children were dying—and this serial approach would be too slow. Perhaps, if the cells were set up cleverly, they could find a way to search for both drugs simultaneously.

The challenge was daunting, but Negulescu's team figured out a solution just shy of a year later and called Beall: Aurora now had a working prototype—a cell test—to show the foundation. And when Beall and Campbell arrived at Aurora for the first time in early spring 1999, Negulescu took them straight to see it.

To hunt simultaneously for the corrector drug and the doorman drug, Negulescu explained to them, he and his team had created two separate tests. The first was designed to find a corrector. Once Welsh's F508del-mutation cells were added to the grids of wells in Aurora's rectangular plates, a compound was added to the cells. The mixture was allowed to sit for sixteen hours—hopefully time enough for a potential drug to refold the protein. Then the dye was added. Finally, a compound called genistein,[11] a cancer drug derived from soybeans, was added to make the CFTR protein open. Genistein wasn't itself a good option for a drug both because the molecule was weak, and also because it interfered with other parts of the cell in ways that could trigger dangerous side effects. But it could be used in the laboratory to open the chloride channel in tests where they were looking for correctors. It also provided powerful proof that there *were* chemicals that could force the mutant protein to work. If a candidate compound worked, refolding the protein well enough to allow it to reach the surface, the sensor would emit blue light.

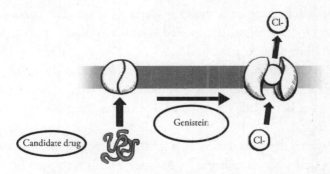

To discover a corrector drug, compounds were added to cells with
the F508del mutation, which were then incubated. If a corrector
helped the protein fold correctly and move to the surface of the cell,
then later, when genistein was added, the channel would open.

The second test, to find a doorman drug like genistein but without dangerous side effects, took advantage of an intriguing characteristic of the F508del protein. Negulescu and his team knew from Welsh
and his team's previous work that F508del was temperature sensitive:
at 80.6 degrees Fahrenheit (17 degrees below normal body temperature), the protein folded better and could reach the cell surface.[12] It still
didn't open and close as well as a normal protein, but at 80.6 degrees
the F508del mutant protein only had one problem rather than two. So
for this test, the cells and potential drug were mixed for sixteen hours at
80.6 degrees, then a dye was added. With CFTR already at the surface,
if the candidate drug was able to open the protein, the sensor would,
again, emit blue light.

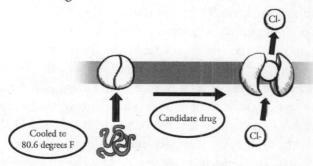

To discover a doorman drug, cells with the F508del mutation were
incubated at 80.6 degrees Fahrenheit, allowing CFTR to fold and rise to
the surface. Then Knapp's team would test various chemicals to see if
any opened the CFTR protein and let chloride move out of the cell.

Aurora's screening prototypes worked well: the dye changed color when the chloride channel opened. And Negulescu had proved to the foundation that Aurora could detect when a "doorman" molecule opened the CFTR channel on airway cells. He'd also showed that they had a reliable test to find corrector molecules. Together these demonstrations gave Beall and Campbell hope that a drug for cystic fibrosis could be found.

But before the foundation invested more money in Aurora, Beall and Campbell wanted to bring in Ashlock and their scientific advisors to vet the next, more expensive, stage: developing a large automated system for testing potential drug candidates.

IN 1999, AURORA'S THREE-STORY BUILDING WAS NESTLED ON A HILLSIDE with a boardroom that faced northeast, overlooking the scrubby desert canyon and Peñasquitos Creek. La Jolla's famous Torrey pines surrounded the parking lot and the air was crisp with fresh Pacific breezes. It was an uplifting location to design lifesaving drugs.

In the company's boardroom, Beall sat at the head of the long oak table with Campbell, Ashlock, and a handful of the foundation's advisors, scientists who had all made breakthroughs in CF research. Among them was Ray Frizzell, Tsien's former college roommate. At the other end of the table were Aurora scientists Paul Negulescu and a chemist who had helped develop the voltage-sensitive dyes,[13] along with a group of cell biologists and engineers.

Beall wasn't there to just hand over the money and wait for the result. As a scientist and now as a businessman, he was always hands-on and detail oriented. Beall's goal at this meeting was to talk through the science, answer questions, discuss the hurdles, and outline the plan.

The group sat at the conference table and brainstormed on all the elements of the project and what they would need to get started. Beall's team explained the types of cells Aurora would eventually need to use for the assays, and their complexities—how the epithelial cells that lined the lungs' airways were knitted together in sheets and that they had a top and bottom. They discussed measuring the electrical properties of individual cells and of living tissue like the airway's epithelial membrane.

Beall's team was there to advise the Aurora scientists and share expertise, but most of the foundation's scientists doubted that the strategy would work. The F508del mutation damaged the CFTR protein so severely that no one could imagine how a drug could fix it; it couldn't even travel to the top side of the cell, let alone open and close as it was supposed to allow chloride out. There was also an element of jealousy that amplified their skepticism; the foundation was going to give millions to Aurora, a for-profit fledgling company with almost no experience with CFTR, rather than academic scientists like themselves. But despite the CFF scientists' cynicism, most of them had never been exposed to science outside the university setting, and they were fascinated and inspired by the atmosphere of innovation at Aurora.

The company had a youthful, scrappy, startup vibe and a radiant confidence that they could build anything and handle any challenge. In the basement was an electronics and machine shop where dozens of engineers were designing and building custom robots to perform various high-throughput screens for other clients. Some of the conveyor belts and roller systems were originally from the food industry—rollers built to move cookies down assembly lines had been adapted to move plates of cells about the size of cookies. The machine shop was noisy; every device had its own metronomic creak or squeak or hum and roll that combined to create a wall of sound.

In another room, Beall's team watched Aurora's high-throughput screening in progress: each series of robots performing its task, one moving plates of cells, another adding microscopic volumes of chemicals, and yet another adding dye. The plate then moved down a conveyor belt to another station, where a microscope-rigged camera snapped a picture for analysis with Tsien's custom software. Once the analysis was completed, Negulescu explained, the picture was then sent to screening leader Knapp and a technician, who would examine the picture for color changes that revealed a positive drug effect.

There was a palpable pride in these homemade systems, and Beall's team could see that the engineers enjoyed their work; there was a lot of talking and laughing and the atmosphere was light. But Beall and the rest of the team also recognized that Aurora was an odd place to design drugs. Most employees were engineers. No one at the company had experience making medicines or curing diseases. There were no medicinal chemists—the people who actually built the drugs. But Negulescu had assured the foundation that Aurora would hire all the chemists they needed once they had developed the working prototype.

If Beall could have had his way, Aurora would have been screening drugs the next day, but there were technical hurdles to overcome.

The price tag for this next stage of the project was $3.2 million. After several conversations with Negulescu, and with the support and encouragement of the foundation's board of trustees and scientific advisory group, Beall told Negulescu to move forward with the project.

Beall and his team had known from the beginning that, even if the project eventually succeeded, the drug that resulted wouldn't be the kind of financial success that a drug for a more common disease would be. So, to keep Aurora's interest, the foundation had chosen to foot the lion's share of the bill, just as they had for the early work on TOBI. It was a good deal for Aurora: whatever happened, they wouldn't lose a dime. And they had a lot to gain if they succeeded, inventing an entirely new type of drug. It was also a savvy move for the foundation: by funding all the work, Beall's team remained in control of the research and its pace. Beall wasn't interested in intellectually intriguing forays that might take the researchers away from the task at hand. For Beall, Aurora had only one mission: find a drug.

Even with funding from the foundation that made hunting the drug free for Aurora, and the tax breaks and other benefits from the Orphan Drug Act, Beall was still worried about keeping Aurora committed and focused. To further sweeten the deal, he included bonuses to be paid out when the company reached certain milestones. But to fund them, he'd need more money.

CHAPTER 35

The Gates Open

1987–2000

Believe in god; fund science.

— Paul Flessner

Paul and Sue Flessner were Midwesterners from farming communities in central Illinois. Paul grew up in Roberts, a farming town of five hundred people; Sue grew up in Germantown Hills, outside Peoria. Their beginnings were humble. They had small-town upbringings, large happy families, and, as adults, a strong community that gave them strength and support.

Their first child, Andrew, was born in 1987, and their next one, Jonathan, in 1989. Andy grew well but Jonathan didn't, becoming ill just a month after birth. Their pediatrician dismissed the Flessners' concerns, but a vigilant nurse advised the couple to take their son from the local hospital to Children's Memorial Hospital (now the Ann & Robert H. Lurie Children's Hospital of Chicago). For several days, doctors were puzzled by Jonathan's symptoms—until May 1, 1989, when they did a sweat test. A doctor at Children's gave Sue the bad news: Jonathan had cystic fibrosis.

Sue immediately called Paul, and the couple met with the doctor. Just as the O'Donnells' doctor had some fifteen years earlier, the physician described the spiraling trajectory of the disease, making it clear that their son would die young. Patient life expectancy, she said, was just eighteen years.* And the current state of Jonathan's lungs was bad: they were pus-filled and infected, and the middle lobe of his right lung had collapsed. Without considering their fragile emotional state, she

---

* Patient life expectancy at the time was actually twenty-nine years.

told Sue and Paul that families that had a child with CF had a divorce rate of 80 percent. Almost as an afterthought, she mentioned that CF was a genetic disease; Andrew should also be tested.

Paul and Sue, recalling Andy's messy diapers and other symptoms the doctor described, knew at once that he would test positive, too. The next day, on May 2, he did. From that moment, their lives changed, as the couple refocused their energy on fighting to get their children the care they needed. Through networking, they discovered that the Cleveland CF center—renowned for their care of CF patients, and the stomping ground of LeRoy Matthews, who had revolutionized CF treatment—was the place to go. When they called the center, the Flessners spoke with Carl Doershuk, Matthews's protégé, who recommended a young doctor, Chris Green, who had just started practicing in Madison, Wisconsin, much nearer the Flessners' home.

A few days later they met with Dr. Green, who examined Jon and began treating him with oral and intravenous antibiotics. Eight months later, when Jon was almost a year old, Green sedated him and gently inserted a bronchoscope deep into his lungs, into the epicenter of the infection, and blasted antibiotics directly on the festering tissue. Not wanting to leave anything to chance, however, the couple wanted a second opinion. Bob Dresing—the president and CEO of CFF at the time—recommended Tom Boat, an experienced physician in Chapel Hill, North Carolina, who was also treating Dresing's son, to examine Jonathan's lungs and determine whether anything further could or should be done. When the Flessners visited Boat, he X-rayed Jonathan's lungs and discovered that Green's antibiotic treatments had worked, wiping out the infections and reinflating the collapsed lobe.

For the next five years, Sue and Paul made the two-hour trek every three months from their hometown in Illinois to Madison so that both boys could see Green. As the boys grew, the couple would combine the visits with family outings, enjoying lunch at a good restaurant and maybe visiting a landmark, museum, or park along the way to keep the mood upbeat. As the O'Donnells did with Joey, the Flessners were fastidious with their boys' chest physical therapy, antibiotics, and nutrition. But Jon's disease was always more severe than his brother's, and even with the attentive care, he was hospitalized four times before age five, two weeks per stay.

Green was reassuring throughout. Jon had been born the same year that Tsui's and Collins's teams discovered the CF gene, and

Green felt certain that in ten years there would be a onetime treatment that would cure Jon and Andy. Your kids are young; they're in great health, Green told the Flessners. Just keep at it, and it's going to be fine. They believed him and soldiered on. Every day, Sue would get up, do the boys' physical therapy and nebulizer treatments, and take care of the kids.

The couple was lucky. They had a tight-knit, supportive family and a great doctor. But early on, Paul had a nagging feeling, after noting how some parents gave up on their kids when they were diagnosed, that they were not doing enough. Both he and Sue wanted to be sure they were doing the absolute best for their children. They had joined the local Cystic Fibrosis Foundation chapter and participated in fundraisers, but Paul wasn't satisfied. He had questions. Where did treatments stand? Were any new medicines on the horizon? Was gene therapy panning out? Was a cure imminent? In 1991, when the kids were four and two, he flew to Bethesda to meet with then-CEO Bob Dresing to get answers.

Dresing updated Paul, describing the medicines that were in the pipeline—Pulmozyme and TOBI—and his plans for raising money and hiring more physicians who could guide the medical side of the foundation. Paul trusted him. Dresing's son was sick, too—the stakes were personal for him—and he seemed to know what he was doing. As he left, he gave Dresing a gift he had made in his wood shop. It was a plaque that said: *Thank you for caring CF*. "When we cure it," said Paul, "I'll come back and change 'caring' to 'curing.'"

A few years later, in 1994, when Paul decided to switch jobs, he knew he could only consider offers in cities with a good CF center. In the end, he was deciding between Denver and Seattle. Paul was a programmer, and an offer at Microsoft in Seattle seemed like a good fit. When the family arrived in Seattle, both boys were in good health. Andy's lungs were functioning at near 100 percent and Jonathan's health was stable. They immediately connected with the local foundation and, just as they had in Illinois, participated in the fundraisers. Bonnie Ramsey, who had guided the development of Pulmozyme and had begun working on TOBI's clinical trials, became the boys' physician.

Five years later, not much had changed. Most US scientists were feeling pessimistic about gene therapy's prospects, even before Jesse Gelsinger's death, although local company Targeted Genetics was showing promising results with adeno-associated virus. The Flessner family had invested a lot of emotion and faith in Targeted and its gene therapy treatment for CF. Paul knew the CEO, H. Stewart Parker, and had taken his boys to visit the company in order to humanize this rare disease, motivate the staff, and show them what was at stake. But although work at Targeted was moving ahead, with other gene therapies falling by the wayside, Flessner was getting worried. He called and met with Ramsey to find out what else he could do.

Since moving to Seattle, Flessner's career had taken off.[1] He had entered Microsoft to help build their PC server business, and during the past five years he'd led a team of 2,500 programmers, focusing more on producing high-quality database software and less on the time required to bring the product to market. The team's focus on quality, along with Microsoft's other server businesses, led to rapid growth in server sales and lured programmers from competing companies. Paul had become a respected voice within the company and a powerful industry leader, and by 1999 was one of the top ten executives at Microsoft. But backgrounding all this success was his sons' declining health.

Not convinced that the foundation was pushing hard enough for a cure, not content to wait for change, and wondering if the foundation needed a completely new strategy, he asked Ramsey if she thought it might help if he quit his job, moved to Maryland, and ran the foundation himself. Ramsey recognized his passion and angst and assured him that wasn't necessary. Bob Beall was now the CEO; she would connect them. She also tried to offer some reassurance, emphasizing the progress Pulmozyme and TOBI represented, and the years they'd added to patients' life expectancy.

But Paul wasn't impressed. Neither of these medicines came close to what he and other parents were expecting after being promised a cure from gene therapy. Neither therapy targeted the root cause of disease. And although he and Sue worked hard to keep their children healthy and their lungs clear, Jon and Andy inevitably needed IV antibiotics every eighteen months or so. Were there any new antibiotics in the pipeline? he asked Ramsey. He knew a drug-resistant bacterium could swing the boys' health from good to critical within days.

A few months after meeting with Ramsey, in late spring of 1999, Paul was standing outside watching the rain while talking to Beall and Ramsey on a conference call. He had been brainstorming how he could drive progress. He asked Beall, "What if I could get you a significant amount of money? Would you have a specific project? Not a general fund. A specific project that we could go after and really try to get things moving?"

Beall explained that he was close to signing a $3.2 million deal with Aurora to screen chemical libraries for a molecule that could fix the broken CFTR protein. After a short pause, Beall then said he had a few ideas about how Paul might be able to help and would get back to him. But in the meantime, Paul spoke to a colleague of his, Steve Ballmer, who in 1999 was Microsoft's president. He then shared the details of that conversation with Ramsey:

From: Paul (MICROSOFT)
Subject: RE: Follow-up from our dinner
Date: July 26,1999 at 10:00 AM
To: Bonnie Ramsey

I spoke to Steveb [Steve Ballmer] a couple of weeks ago. He said he was going to help again this year. He also volunteered to speak to Billg [Bill Gates] on my behalf.

He seemed interested in this. I emphasized that we need the money NOW rather than later. He understands my sense of urgency.

If the CFF were to receive a gift now, this year, of say $50m, what kind of an impact would this have in fighting this disease? What can steve/bill do now to have a maximum impact on this disease in the fastest possible time?

Paul[2]

Paul stressed to Ramsey that Beall had to be honest with the top Microsoft brass—that Ballmer and Gates sat through thousands of pitches from teams within and outside of Microsoft, and that the two were like bloodhounds. With their fine-tuned noses sniffing out faults, contradictions, and weakness, they regularly tore people apart for flawed proposals, shoddy work, or poorly conceived sales pitches for software or services. Better to be honest and realistic about how the money would be used and its possible impact on the disease; otherwise Gates in particular would see right through him. If the foundation could use the money now, then Paul would help them immediately. If they didn't have a strong contender for research funding, he advised, it was better to wait. Gates and Ballmer wouldn't throw money after anything that wasn't thoroughly vetted.

Beall's pitch would already be unusual, compared to the ones the committee usually heard. The William H. Gates Foundation (which would later change its name to the Bill & Melinda Gates Foundation) had been founded five years earlier with the goal of bringing healthcare, education, and agricultural development to the poorest people on the planet.[3] The foundation had a particular commitment to combating HIV/AIDS, tuberculosis, and malaria. Genetic diseases were not their priority.

Beall didn't waste time. He immediately proposed a meeting in Seattle, and William H. Gates Sr.—Bill's dad and the Gates Foundation's cochair—suggested Beall, Ramsey, and Paul Flessner meet with him at his home in northeast Seattle overlooking picturesque Lake Washington on August 31, 1999. That day, the four sat around the dining table, talking about CF and Aurora while the scent of pot roast wafted through the room.

Gates Sr. listened carefully as Beall told him about the CF foundation, the disease, and their plan to find a cure through a new technology called high-throughput screening. Initially, the former lawyer looked slightly hesitant, because the project was beyond the Gates Foundation charter. But after the nearly two-hour-long meeting, Gates Sr. asked Beall to send a detailed ten-page proposal to him outlining the project at Aurora.

The proposal, delivered within a week, explained plainly that the foundation needed to move into the drug discovery business, because a pharmaceutical company was unlikely to embark on such a venture alone. Beall was requesting a $47 million grant from the Gates Foundation to fund collaboration with Aurora Biosciences in San Diego, which would use automated high-throughput robots to screen chemical libraries hundreds of thousands of molecules strong for promising drugs.

Beall didn't have to wait long for a response. On September 14, 1999, he got a call from Gates Sr. telling him that the Gates Foundation was going to contribute $20 million—an absolutely thrilling amount for which Beall was incredibly grateful. When he hung up, he immediately called Flessner to tell him the good news. Paul then emailed Bill Gates:

From: Paul Flessner
Sent: Tuesday, September 14, 1999 8:47 AM
To: Bill Gates
Subject: CF

Bob Beall from the national CF office just called. He said he had just spoken with your dad. He gave me the news that you and Melinda will be donating $20m to drug discovery for CF.

i really don't know what to say....

We have a lot of hard work in front of us to cure this disease. A donation like this will really make a difference.

These resources now provide us the power to get the combinatorial-chemistry and high-throughput screening off the ground. Bob said he would be traveling to Aurora next week to get things going. We are all very excited.

I will stay on top of this. I will make sure this money is put to good use. I want you and your family to be a proud partner in ending this disease forever.

With all of my heart I and my family (and thousands of other families with CF) just want to say...

thank you[4]

The decision to fund the foundation underscored the close connection that Bill Gates and Paul shared. Gates knew of Andy and Jonathan, was a parent, and recognized what was at stake now, not just for Flessner's kids but all kids suffering from the disease. "I am also excited about this," wrote Gates in an email. "It seems provident that I can do my best to help. Your personal involvement and passion for this really touch me as a father. Great science, philanthropy and human caring can all come together here. There are some incredible challenges to solve yet but its [sic] great to know things will be moving full speed ahead."[5]

The $20 million check arrived at the foundation offices later that week in an unceremonious brown envelope delivered by the United States Postal Service.

The gift was scheduled to be announced at the Cystic Fibrosis Foundation Breath of Life Gala and Auction at the Seattle Sheraton on October 23, 1999, but a local Seattle radio station leaked the news: "Gateses Give $20 Million to Cystic Fibrosis Group,"[6] the *New York Times* reported on October 24. Though a few of the gala attendees had already read the news, hearing Beall read the announcement the foundation had prepared about the Gates Foundation left the room pin-drop silent as all eating and talking ceased. The donation was the largest to date in the foundation's forty-four-year history. Beall repeated the number: "Twenty million dollars." This time the entire audience erupted with cheers and shouts and exploded in a standing ovation. The gala itself raised another $2.7 million, due in large part to the generosity of Steve Ballmer—Paul's friend from Microsoft—and his wife, Connie.[7]

From then on, both Beall and Flessner sent regular progress reports to Bill Gates. Just before Christmas 1999, Paul shared that Beall was crafting a new contract with Aurora, which he expected would be signed in the new year. Aurora was now ready for the next big step, the one Gates's donation would help pay for. Negulescu had delivered a proposal for developing drugs that would target the root of cystic fibrosis: the protein that caused all the sickness. However, Beall was reworking it to match his expectations. Aurora had proposed that 20 percent of the money be delivered based on the scientists reaching certain milestones and the rest given up front. Beall disagreed. He was only interested in paying for progress. He amended the document to say 75 percent of the payments would be based on the scientists reaching specific goals, a focus on achievement that Beall felt would drive the scientists to work harder and faster. He expected this new deal to be signed in mid-February 2000.

Early in 2000, Beall, foundation drug development leader Melissa Ashlock and attorney Ken Schaner flew out to the company to hammer out a final agreement with its "chief knowledge officer," John Mendlein, a lawyer with long blond surfer-style locks. Mendlein also had a PhD in physiology and biophysics from the University of California, Los Angeles, was a coauthor on many papers, and had a stack of more than seventy-five patents to his name. Mendlein reflected Aurora's youthful and innovative vibe, and Schaner found him relaxed and funny, though his methods were unconventional.

Mendlein began by projecting the agreement onto a screen in a conference room at Aurora, amending the language in real time as he typed on the computer. As Schaner and Ashlock worked with Mendlein, Beall became increasingly impatient, pacing around the room. By 2 AM he was cranky and exhausted, and insisted the rest could be finalized the next day. At that point he left and drove back to the hotel and went to bed. The others remained, and by the next morning the financial terms of the deal were done—though it would take several more months for Negulescu, Ashlock, and Penland to hammer out the scientific plan. After all, Aurora wanted the money, and the foundation was anxious to give it to them.

When Beall wrote to Gates Sr. on May 30, 2000, he outlined the final agreement between Aurora and the CFF: if Aurora met all their mutual goals, then the foundation would pay up to $46.9 million over the next eight years. This amount would include the Gates's $20 million, the $19 million from TOBI royalties, and some money from the CFF general fund, plus additional funds Beall was sure that they could raise. The foundation was also prepared to shell out another $48 million to help with the cost of moving up to three candidate molecules into clinical trials, although at the time Beall had no idea where they would come up with those funds. The total obligation would be about $95 million, and Beall expected the work to be done within six to eight years—half the time it took a traditional drug company to make a drug from start to finish. At roughly $32 million per drug, he estimated, it would also cost only a tenth of what a traditional pharma company would have needed to bring a candidate drug to the same stage.

Thanks to Mendlein and Schaner, the agreement also included a royalty clause, as TOBI's had. If one of the molecules that Negulescu's team discovered successfully moved through clinical trials, was approved, and became a drug, the foundation would receive up to 12 percent royalties on sales.

A few months after the deal was inked, Paul Flessner flew down to San Diego to join Beall, Campbell, Ashlock, and the rest of the CFF team on their quarterly visit in September 2000. He wanted to meet the Aurora scientists and follow the money. He had never been in a commercial pharmaceutical lab, and when Negulescu took him to the room where the scientists were developing the screening tests, to him, the machines looked pretty crude; the rudimentary setup was clearly a version 1.0. Still, Paul was impressed by Negulescu. He listened to Negulescu's tone, analyzed his body language, and dissected his word choice, gauging how confident he was that this project would work. In Paul's experience, people who grasped a concept well were unfazed by questions. Negulescu was patient, modulating his explanations to match Paul's level of understanding and answering every question in great depth—Flessner was a software expert, not a biologist. Many of the questions focused on how Aurora would deal with the deluge of data from the chemical screens, and the type of software they were using to assess the cells.

Paul could hear that Negulescu loved the science and really believed that this drug hunt would be successful. And as rough as the machines looked, he saw that Negulescu was immensely proud of them. As someone who had helped refocus the culture at Microsoft to become more conscious of quality and spur innovation, Paul also appreciated the culture and vibe that reverberated at Aurora. People were happy to see Negulescu and share their data with him when he walked into a room. He'd witnessed plenty of startups where acrimony abounded between the leader and staff, poisoning the company and dooming the project. Paul sensed that Aurora's team had an openness and eagerness to collaborate.

As Paul was leaving the building with Beall, he turned to Negulescu. "I'm confident that you guys are going to be able to do the scientific part," he said. "My main worry is that you're going to get bought and the project's going to get dropped by whomever buys you." Negulescu just nodded; whether the company was bought or sold was an issue above his pay grade.

Still, as he departed, Paul felt good, excited, but careful not to get too excited. He knew that whatever was going to happen was still ten years away, at best. He and Sue knew from their past experience that science moved very, very slowly. Paul tried to buffer against disappointment by telling himself to keep things in perspective.

Beall's progress report to Gates Sr. in February 2001 echoed Paul's optimism. Aurora was now screening approximately 10,000 chemical compounds per day. Over the next few months, Negulescu's team hoped to identify "hits"—compounds that fixed the chloride channel and made it work. That would be a major milestone, he explained.

The game plan was set, but Beall had no idea what the chances were of finding a drug. He had no experience with actual drug discovery. And there was no precedent for what they were doing with Aurora—looking for molecules that would fix a broken part of a cell. In the meantime, all he could do was wait.

CHAPTER 36

A Tale of Four Families—
CF in the New Millennium

1995–2005

To be deeply loved by someone gives you strength,
but to love someone deeply gives you courage.

—Esther Huertas

Many elements of medicine had changed dramatically in the quarter century since Joey's birth in 1974. In 1999, on the cusp of the millennium, genetic counseling was now standard practice for ethnic groups at high risk for certain genetic diseases. As part of that protocol, Kim and Rob Cheevers of North Andover, Massachusetts, who were expecting their first child, shared their family's medical history during a checkup with Kim's obstetrician. Were there any medical conditions that seemed to run in her family? the doctor asked.

Kim shook her head. As far as she remembered, her parents, aunts, uncles, and cousins had all been pretty healthy. Rob agreed. He didn't recall much sickness on his side, either—until Kim reminded him that his father had a sister who died young during the 1940s or 1950s. It was some type of lung issue, Rob recalled. "No one ever really knew what she died of."

Both Rob and Kim were of Anglo-Irish ancestry, so the doctor encouraged them to undergo a genetic screen for disease-causing mutations common in those populations. The most common, the doctor explained, was cystic fibrosis. Kim knew the disease. She was a nurse at Massachusetts General Hospital and worked in the pediatric intensive care unit; over the years she had cared for many children with CF.

The couple met with a genetic counselor, provided blood for analysis, and then forgot about the test—until they received the results. One carried a single copy of G551D, a rare CF-causing mutation, and the other carried F508del. They were shocked. Any child they conceived would have a one-in-four chance of having the disease.

Sixteen weeks into the pregnancy, Kim had amniocentesis, and the test revealed that both Rob and Kim had passed their mutant gene to their daughter, who would have the disease.

For them, as practicing Catholics, terminating the pregnancy wasn't an option. And Kim was optimistic. Treatments for cystic fibrosis had evolved and the disease was no longer a childhood death sentence. There were new drugs—Pulmozyme and TOBI—and Kim knew from experience that sometimes the disease could be managed. At least half of children born with CF now lived to age thirty. And lung transplants,[1] a complex, high-risk procedure, were becoming more widely available as a lifesaving therapy of last resort. Furthermore, the disease was fickle and varied wildly in severity. Kim knew families where one child had died at age five while the sibling, who carried the same mutation, remained completely healthy. Kim was petrified, but she knew there was a chance that her daughter would be okay.

After a rocky beginning in the neonatal intensive care unit where the newborn suffered from meconium ileus—a bowel obstruction linked to CF—baby Laura thrived, cultivating hope in both her parents. Except for extended recovery from an unrelated ruptured appendix at age two, their daughter was in good health and there were new drugs on the horizon. So the couple started to think about a second child. They wanted a sibling for Laura, and with one-in-four odds, they felt luck was on their side. Yet the amniocentesis proved them wrong: Cate tested positive. At least these two sisters would be able to support each other through taxing times. To care for their daughters, Kim and Rob chose Joey's doctor, Allen Lapey. And the couple soon became dedicated fundraisers for the Massachusetts CFF chapter, believing that the more money the foundation could raise, the greater the chance of developing a drug for their daughters.

FOR ROBERT COUGHLIN, A FUTURE THAT INVOLVED CYSTIC FIBROSIS and public service was inevitable. Coughlin, the youngest of six sons, was a native of Dedham, Massachusetts, a town on Boston's southwest border. In 1990, he was elected to serve on Dedham's school committee when he was still a college student and not even old enough to drink beer. He acknowledged that people might have thought they were voting for his dad, a local politician, with the same name.

Five years later he was elected the youngest member of Dedham's Board of Selectmen, a specifically New England term for a city council. Soon afterward, a high school hockey buddy whose child had CF told him about a fundraiser called Great Strides, the Cystic Fibrosis Foundation's largest national fundraiser, and asked him to be the honorary chairperson. Bob's father had supported the Jimmy Fund—the cancer charity of the Red Sox—because he had worked at Boston's famed Fenway Park. He'd advised his son to pick his own charity and stick to it. So, although he knew nothing about the disease, Bob chose Great Strides as his charity, and accepted the role as chairperson.

The timing was fortuitous. Within ten days of committing to Great Strides, Bob learned that his own niece had been diagnosed with cystic fibrosis. He doubled down on his commitment to the charity, serving as annual cochair from 1995 through 1999—by which time he was married with two young children.

Selectman was a part-time, unpaid position, so Bob had been earning his living at an environmental services company called Clean Harbors. After ten years, however, he wanted to leave the private sector to advance his political career by running for the Massachusetts House of Representatives.

At the time, Bob's wife, Christine, was pregnant with their third child when a new prenatal test revealed that she was a CF carrier. Their surprise and panic doubled when Bob tested positive as a carrier, too; after all, he now knew a lot about this rare disease.

On Christmas Eve, 2001, Christine underwent an amniocentesis at Children's Hospital. The fetus was approaching twenty-one weeks, and the window in which they could terminate the pregnancy was closing fast. On New Year's Eve, they learned that their son had cystic fibrosis.

The doctors and nurses at Children's comforted the couple, pointing to bulletin boards wallpapered with photos of children with CF. Bob was unmoved by the nurses' sugarcoated anecdotes. He knew kids with the disease. He saw how it ravaged their bodies from the inside out. And while he projected outward calm, his mind was in a tailspin. He was furious, angry with God, asking what they had done to deserve this.

With the news about their son tightly coiled around their hearts, the Coughlins agreed with the doctor's recommendation to meet with a social worker before leaving the hospital, even though the idea just made Bob angrier. Why on earth did he need to meet with a social worker? Was she a shrink? He was a politician. He knew people. He could meet the CEO of Children's Hospital or the head of the CF center if he needed to know something. Why did he need to meet this woman?

Lynne Helfand, their new social worker, asked them, Do you have anything big going on in your life right now? Bob looked at her as if she were nuts, but Helfand wasn't put off by his anger and continued to probe. Bob explained he was about to run for the Massachusetts state legislature. But now, because his son was sick, he wasn't going to do that.

Helfand shook her head. Don't change your plans, she told him—if you do, you will resent this baby. You don't know how sick or healthy your child will be. No one can predict that. If you run and win, you could help craft legislation that would help all sick people, not just your son. You could help the hospitals do a better job. It's important work, she added.

Bob got up and motioned for Christine to stand. He'd had enough of this crazy lady.

Sitting tight-lipped in the car on the ride back to Dedham, Bob felt the panic he'd dammed at the hospital returning—until Christine said quietly, You can't get out of this race now. Everything's happened for a reason. You should do it.

The couple drove the rest of the way home in silence. Even though she was pregnant and awaiting the birth of a sick baby, she was giving him permission to run for office.

As Bob Coughlin began talking to people, word traveled fast that a volunteer was now having a child with CF. Bob knew about the O'Donnells' history and was aware of the Joey Fund, so he called Joe, seeking advice. Joe met him in the cafeteria at Children's Hospital. Joe could see himself in Bob—a rising star in Massachusetts, a man ripe with ambition and plans who had been sucker-punched by his genetic lot. Joe took him under his wing.

"Bob," said Joe, "we're going to go raise money and buy a cure for your kid. My Joey died. Your kid isn't going to."

Baby Bobby Coughlin was born during the race in May 2002. Bob ran against incumbent Maryanne Lewis and won—because, he believed, he was on a mission. His job was to make Massachusetts a better place for pharmaceutical companies, for biotech, and for everyone who worked in the hospitals so that they could find a cure for his son, and everyone else who was sick.

He wasted no time. Bobby was just three months old when State Representative Coughlin introduced his first bill, which would permit children with CF to take their pancreatic enzymes in public schools without going to the nurse—a small but significant step to achieving normalcy in these children's lives. When the bill became law, it reinforced Bob's conviction that this was what he was meant to do. He wasn't a doctor or a scientist. But he could contribute to the fight against this disease in other ways.

JENNIFER FERGUSON AND HER HUSBAND, MATT, WERE LIVING IN SAN Diego when she went into labor at twenty-eight weeks, almost three months early, in March 2001. She had attended prenatal yoga, taken her vitamins, abstained from alcohol—everything to make sure she and her child would be healthy. Now this. Her baby, Ashton, was in distress, and she needed an emergency C-section. When she saw him, fleetingly, during a break in her morphine-laced slumber, Jennifer thought that he looked like her: pregnant, with a big, round, distended belly.

When she woke the next morning, nauseated from the anesthesia, doctors told her Ashton had been moved to the adjoining children's hospital, where surgeons had removed two and a half inches of torn intestine from his tiny, three-pound body in a five-hour-long operation. During week three of Ashton's recovery, the doctors told Jennifer that he had CF. She and Matt had been offered genetic screens during her pregnancy but had declined, because neither of them had a family history of the diseases being screened, and because of widespread concerns at the time that insurance companies would deny coverage to people with diseases written into their DNA.[2]

What Jennifer read when she went home and searched the web was depressing: a life expectancy of around thirty, at best. And with Ashton's gastrointestinal complications, he was unlikely to live even that long.

Jennifer wasn't one to sit idle, and within hours she called the CF Foundation to figure out how to navigate her new life. She gave up her job as an events coordinator for Planned Parenthood to focus on caring for Ashton and fundraising for the foundation. In September 2001, Ashton had only been home for about six months when a foundation representative called to ask if Jennifer would speak to scientists at a local company who were working on CF.

The request took her by surprise. She wasn't a public speaker. Besides, what could this little mom-and-pop biotech company do for this disease? The gene had been known for more than a decade, all the attempts at gene therapy so far had failed, and the chance of a breakthrough was remote at best. Still, most people who raised money for medical charities never had the opportunity to meet scientists who worked on the disease, or to see what happened behind the scenes. And perhaps she could make a difference. Maybe making a connection to a child with this disease would drive the scientists harder.

In late September 2001, Jennifer pulled into the serene eucalyptus-lined parking lot she'd been given directions to, lifted Ashton out of the back seat, and walked into the building. A chill ran down her spine, and the hairs on her arms spiked. She had no idea what was going on in this building, but she sensed the place was special. Six-month-old Ashton, however, seemed to disagree. He needed a diaper change, then food—forcing Jennifer to nurse him in the lobby—and then a nap.

When Ashton finally settled into sleep, the company's CF project leader, Paul Negulescu, welcomed Jennifer, and after a quick whirl around the facility and some of the labs, he led her into the lunch room, where the entire staff was waiting. Jennifer gingerly transferred Ashton to the arms of friends from the foundation who had just arrived and stepped up to the microphone. She was filled with excitement after the tour and determined to give the pep talk of her life.

She began by telling the audience how she had to feed Ashton every hour or so because he wasn't gaining enough weight. Every time he ate, she had to give him pills. She described what it was like to give physical therapy to a five-pound baby she could hold in her palm—pounding and thumping his tiny chest—hearing him cough and forcing him to swallow liquid medicines that he clearly didn't want. Ashton, she told them, and thousands like him needed their help—they needed to innovate, to work hard and fast, and to use the foundation's money to create something transformative. Her words caught in her throat, moving many researchers, including Negulescu, to tears. It was clear to her that these people were genuinely interested in what she had to say—that this wasn't just a job for them. She could see her audience was genuinely spellbound, and many lingered afterward to ask questions. She had successfully completed her mission.

Ashton spent six of his first eighteen months in the hospital, his parents never sure if he would survive. In late 2002 the Fergusons moved from San Diego to Jennifer's hometown of Wichita, Kansas, to be near family. Jennifer wanted more children. But not children with CF. So eleven weeks into her second pregnancy, Jennifer's obstetrician performed a chorionic villus sampling, the same prenatal test that Kathy O'Donnell had. It showed this second child was free of CF, but still not in good health. She had Turner syndrome,[3] a genetic disorder occurring in one out of every 5,000 births in which one of the two X chromosomes, essential for complete female development, was missing or abnormal. Her perinatologist warned her there were many ways this pregnancy could go wrong—including a late-term miscarriage that could endanger Jennifer's life. Jennifer and Matt chose to terminate the pregnancy. They needed to focus on Ashton and CF; they couldn't handle a completely new disorder.

To exercise control over the next pregnancy, Jennifer called a company in Chicago that specialized in pre-implantation genetic diagnosis—a combination of in vitro fertilization and genetic testing. Once a lab-fertilized egg had divided three times, yielding an eight-cell embryo, a technician plucked off a single cell and tested it for various genetic diseases. Only once they knew an embryo was healthy did they try to implant it in Jennifer's womb. Over the next year and a half, Jennifer and Matt endured four rounds of in vitro, amounting to $45,000. All failed. Some embryos died naturally before they could be implanted. Others had CF or something else wrong with them. After four rounds of in vitro, only two healthy embryos remained. When these were implanted, neither survived. Jennifer couldn't get pregnant. So, for a little while, the couple just gave up.

Once Ashton began preschool, the Fergusons decided to try again. Jennifer got pregnant naturally and received the results of the genetic tests at thirteen weeks. She was carrying a little girl who had CF. After weeping inconsolably at the news, Jennifer drove to her mother's house and thought. She wanted this little girl. And CF was a disease she now knew well.

Lola was born in Kansas on December 20, 2005, without major problems. An enema cleaned out her gut and Jennifer began giving her enzymes so she could digest her food, mixing the enzyme powder with applesauce and dipping her finger in it, then into Lola's mouth, before nursing her. The baby girl was a novelty. No other child had been born at Via Christi Hospital St. Joseph, Wichita, in which CF had been diagnosed before birth. Nurses and doctors came to Jennifer's bedside to learn how to care for the baby.

"How do you give the enzymes? What kind of breathing treatments do you do?"

Jennifer showed them. Nurses gathered to watch. Doctors peppered her with questions. They had no experience dealing with CF. The attention bolstered her confidence. She knew how to take care of the child.

"Don't worry," she told the nurses. "I've got this."

SEAN YOUNG WAS DIAGNOSED WHEN HE WAS ALMOST TWO YEARS OLD. His birth on May 7, 1998, at Dayton Children's Hospital in Ohio was untraumatic, but he quickly displayed the hallmark signs of a child with CF: small, constantly hungry, telltale messy diapers, and distended belly. He was Katrina and Robert Young's first child, and since the rest of their family was in California, and they had no nieces or nephews for comparison, they just assumed he was small—fifth percentile on the growth chart—but otherwise healthy. There were few clues until their second child was born fifteen months later. Their daughter's growth and eating habits were very different—she had two diapers for each of his ten—flagging that there was a problem with Sean.

Katrina suspected an allergy. Her pediatrician suspected that Robert—at the time in his first year of residency at Wright State University—was just projecting his newfound medical knowledge on his child, and then refused to look at Sean's messy diapers, just telling Katrina to stop giving him juice.

Exasperated, Katrina took Sean to the doctor's partner, who immediately recognized that the boy had CF. A sweat test on May 3, 2001, a few days shy of Sean's second birthday, confirmed it. By that point Katrina and Rob had already matched Sean's symptoms with a textbook description of CF. Though more than a quarter century had passed since Joey's botched diagnosis, Robert and Katrina's experience had been disturbingly similar.

On May 4, the couple took Sean to the CF clinic of the Children's Medical Center in Dayton, where they repeated the sweat test and took a full medical history, though there was no history of the disease in either Robert's or Katrina's family. Then, bleary-eyed from a sleepless night, they met with Dr. Michael Steffan, the pediatric pulmonologist and director of the CF clinic—a big, sweet man with a gentle and soothing bedside manner. Dr. Steffan spent hours with the couple talking to them about cystic fibrosis and how the life expectancy for children like Sean was around thirty years.

Just because your son has cystic fibrosis doesn't mean that all his hopes and dreams can't come true, he said kindly. Thirty years ago, babies were born and no one expected them to live much past elementary school. So many parents didn't put a lot of effort into their education or social skills. They never asked their sick children what they wanted or hoped for in life. They were forgotten because nobody thought they had a future. But medicine was catching up, he noted. There was amazing research in progress. Treatments were getting better and children who were never expected to live beyond their teens were now reaching their twenties and thirties.

As he reassured them, Katrina and Robert listened, a disorienting mix of anxiety, sleep deprivation, and hope churning in their guts. The best thing you could do for your son, Dr. Steffan stressed, is give him a future. Love him the same way you did yesterday—before you knew about this disease. The medicine's not here yet. But I promise you by the time he gets older, the medicine will be better and he'll have a true shot at a future.

CHAPTER 37

The Takeover

2001

> Innovation in an existing company is not just the sum of great technology, key acquisitions, or smart people. Corporate innovation needs a culture that matches and supports it.
>
> —Steve Blank

By early 2001, Aurora, its technology further streamlined, was screening 50,000 compounds a day for molecules that could fix CFTR. It was a brute-force approach: battering the CFTR protein with hundreds of thousands of unique molecules to find a chemical that would open the door and let chloride flow. The corrector screen would look for chemicals that would install the door, and the doorman screen would look for chemicals that would open it.

The CF screen was trickier than the ones Aurora had developed for other clients, so at first Tom Knapp, who led the high-throughput facility, had to run them in lower-capacity plates with just ninety-six reactions. To compensate, he set up three daily shifts to keep screening around the clock. Some of the chemicals they tested were made in house by the growing troupe of chemists on the floor above. But he also ordered hundreds of thousands of novel chemicals from online chemical catalogs, a process as easy as ordering books on Amazon.

It was an exhausting schedule, but Knapp and his colleagues were inspired by CF project leader Paul Negulescu's vision that the technology they were inventing could lead to breakthroughs in CF. Here was a terrible, fatal disease, and the scientists at Aurora had the chance to make a difference; they felt invested in the goal. And unlike the other big contracts that paid the bills, Knapp felt that the CF project was particularly exciting because the scientists at Aurora were the ones driving progress, and so he felt a special sense of ownership and commitment.

By this time, too, Aurora's reputation for developing innovative screening technology had spread through the industry, and several large companies were looking hungrily at the startup as fodder for a takeover—just as Paul Flessner had predicted. And the most interested party was Vertex Pharmaceuticals Inc., a company founded by former Merck chemist Joshua Boger.

BOGER WAS BORN IN 1951 IN CONCORD, NORTH CAROLINA, A SMALL town of about 18,000 people and the site of a large textile mill. His father was a textile chemist who nurtured his son's passion for science. On one occasion, he showed Boger how to make hydrogen gas from tinfoil (it was still actually tin back then) and Drano. After learning the skill, Boger used it to fill up balloons, to which he attached tiny gondolas for carrying mouse "astronauts," so he could explore the impact of "space travel" on mice. The balloons rarely crested a couple of hundred feet as they floated over the forests surrounding his home, but Boger diligently tracked each gondola until its balloons leaked their gas, bringing his intrepid explorers slowly back to earth.

Boger never knew what his father would bring home but loved the anticipation. When Boger was thirteen years old, his father gave him a large bag of brown powder called agar and a collection of twenty glass petri dishes, then explained how to transform the powder into a nutritious brown gel and grow microbes. Boger then began a foray into microbiology. He took samples of bacteria from the throats of his school friends, grew them in the petri dishes, and then tested a collection of mouthwashes he had seen advertised on TV to figure out which was the best bacteria buster.

At age fifteen, Boger spent his summer working double shifts inspecting corduroy at the textile mill; a year later he'd saved enough to buy a car. He took a wad of twenty-dollar bills directly into a dealership where he bought a red Datsun 510—an homage to Grace Slick, the lead singer of Jefferson Airplane, who, legend had it, had walked barefoot into a Mercedes dealer in San Francisco to buy a car. He drove the car to college interviews. Duke and Cal Tech offered him generous scholarships, but he chose Wesleyan University in Connecticut, though it offered no breaks on the tuition.

At Wesleyan, Boger connected with one of the most influential chemists in America, Professor Max Tishler.[1] Tishler had led Merck's entire research and development program[2] from 1957 to 1969, when he left the company soon after a promotion to senior vice president for science and technology, just eighteen months short of retirement. Tishler's work during his legendary thirty-two-year career at Merck had transformed medicine and human health around the world. He figured out how to mass-produce vitamins essential for growth and development that could be added to foods or sold as supplements. Tishler led a team that produced antibiotics like actinomycin D, streptomycin, and penicillin via fermentation. He industrialized the mass production of cortisone, a hormone that eased inflamed joints in arthritic patients. He invented the first antiparasitic agent for chickens—sulfaquinoxaline—which helped the birds fight off intestinal parasites, and enabled poultry farmers to raise the animals in closer quarters. But Tishler's new corporate position as senior vice president had divorced him from his lifeblood: laboratory research, problem solving, and the army of 1,800 scientists, and their families, that he'd come to know so well. So he left to teach chemistry at Wesleyan instead, to be back among scientists and students like Boger.

Boger took basic and advanced organic chemistry courses with Tishler, in which the fiery, energetic professor inspired his students with stories of elusive molecules, vitamins, and drugs, and how chemists had figured out how to economically synthesize vast quantities of them to bring new, life-changing medicines to the people. Manufacturing vitamins and drugs was a practical problem—building molecules that changed the lives of billions. And when it came time for the final exam, on which Tishler encouraged his students to use their notes and collaborate, he gave them five questions and one real-world extra credit problem with an eye-popping prize: he offered $25,000 to the student or team who formulated a series of chemical reactions that would yield an intermediate molecule in the manufacture of vitamin C.

Neither Boger nor any of his classmates could solve it, but Tishler hadn't expected them to. The extra credit was more of a life lesson, conveying that solving these unwieldy challenges was not the domain of the individual scientist toiling away in misery and solitude, but rather a vibrant, collaborative process that could be all consuming, desirable, and fun. After Tishler's course, Boger recognized, he now had the skills to solve problems that were important for the world. It was a powerful epiphany.

After Wesleyan, Boger pursued a PhD in chemistry at Harvard. He was a star student, finishing his studies ahead of schedule, working and coauthoring papers with notables who went on to win the Nobel Prize. When he graduated in 1979, he followed Tishler's example, shocking the faculty by heading straight for industry. Tishler opened doors for him at Merck's medicinal chemistry department in northern New Jersey, helping the twenty-seven-year-old get a toehold. Drug development had become more appealing to Boger as new tools and technologies enabled scientists to describe the precise choreography of small molecules and proteins inside cells that lead to disease, health, or behavioral changes.

Upon arrival at Merck, rather than joining an existing team, Boger launched his own project: developing a new renin inhibitor. Renin was an enzyme produced in the kidneys that could lead to high blood pressure. Boger wanted to develop a new medication that blocked its activity and would also lower blood pressure. An integral part of his process was to first build a physical model of renin that showed its atomic structure. Using this model, he could then figure out the best location to target with a drug.

Within four years, Boger's project had become the largest in the company, with a hundred chemists working on the problem, and Boger published the research in one of the world's top scientific journals.[3] The structure-based strategy he'd used to develop his family of renin inhibitors had yielded compounds a thousand times more potent than any other renin inhibitor made at Merck—and although more research was necessary, these molecules were potent enough to be medicinally useful. The publication was proof he didn't need to be in academia to produce meaningful work *and* pioneer pathbreaking science.

Though Boger was productive, he was increasingly irritated by the bureaucracy of working for a behemoth like Merck. In 1986, the only computers at Merck were mainframes barricaded behind glass walls or their smaller cousins that only collected data from instruments to which they were tethered. Boger believed that computers could be key tools for drug design, so each day he began backpacking one of the original Macintoshes, which he bought in the first ninety days of its release, to and from Merck (the company wouldn't let him leave a computer in the lab overnight). Later, he built his own network, connecting Merck's computers by rigging the phone connections in the basement.

Rumors of Boger's unusual use of computers in chemistry and drug discovery spread throughout Merck, and though he wasn't a senior executive, his boss invited him to a corporate retreat for executives in late 1986 to share his ideas. Not exactly sure of what to say, in a fifteen-minute presentation to Merck's sixty or so vice presidents he pitched his vision for a new 150-person multidisciplinary chemistry department, one filled not just with classical chemists but also biophysicists, computational and computer specialists, and molecular biologists all working together to design drugs. To his surprise, the idea was instantly embraced by his boss, who yelled from the back of the room, once Boger had finished his presentation, "Let's do it!" The next day the company approved the idea and Boger embarked on a recruiting tour. He attended dozens of conferences around the country, boldly sharing his vision as he hunted for the right types of scientists to staff this new department, for which Merck had committed to build a new facility. As he spoke around the country, he acquired a broad audience.

One person who heard him talk in late 1987 was a venture capitalist named Kevin Kinsella—the same Kevin Kinsella who would help Tsien launch Aurora almost a decade later. Kinsella contacted Boger with a tantalizing proposal: he knew how Boger could achieve his goal of creating powerful medicines faster than he could at Merck. When Boger called him back, Kinsella suggested that Boger start his own company.

It was a surprising recommendation. After all, Merck was building Boger a facility where he could do exactly what he wanted, in his own way, with a staff of 150-plus scientists. But as terrific as that was, Boger still frequently faced middle-management resistance to his ideas. For instance, when he ordered a refrigerator-sized graphics computer that required a 220-volt line, he was told the line would take six to nine months to install. Dozens of similar petty aggravations slowed him down. Those wouldn't go away, Kinsella told him.

The other compelling argument, which Boger hadn't considered, was money. Kinsella assured him that the kind of money that Merck was spending on his vision could be had outside big pharma. Not only that, but Kinsella could get it for him. With it, Boger could build drugs faster, and his own way.

The choice wasn't easy, and Boger sought advice from his college professor Max Tishler, who told him to stay at Merck; and his thesis advisor at Harvard, who encouraged him to go solo and start his own company. By the fall of 1988, he remained torn. But over the

Christmas vacation a tragedy catalyzed his decision. On December 21, 1988, Pan Am Flight 103 exploded over Lockerbie, Scotland, killing all 259 aboard[4]—including Boger's friend and closest colleague at Merck, Irving Sigal, an internationally known expert on molecular biology.[5] Sigal was a scientist the same age as Boger and on the same rocket-like trajectory within the company.

When Boger recovered from the shock, he recognized that life was short. His good friend's sudden death filled him with a sense of his own mortality. He couldn't put things off. Who knew what life would bring? He picked up the phone and called Kinsella. He was ready to start his own company. He left Merck, and on January 4, 1989, founded Vertex Pharmaceuticals to create drugs that would treat inflammatory and viral diseases—two disease categories that affected hundreds of millions worldwide. The startup was based in a modest building at 40 Allston Street in Cambridge, Massachusetts, a car repair shop that had been upgraded to office space. As Boger gutted and renovated the facility over the next year, he grew the company from two employees to twenty-five, luring chemists away from big pharma—mostly young men who, like him, were ambitious and bored of the bureaucracy, glacially paced innovation, and power structure in these massive companies.

The fledgling company described itself as "a leader in the use of structure-based drug design."[6] That was industry-speak for designing drugs based on first understanding the structure of a target protein and then fashioning a drug to fit. With a technique called X-ray diffraction crystallography, researchers had learned how to use the two-dimensional images produced as X-ray photographs to construct a 3D model of a particular molecule, just as Francis Crick and James Watson once used Rosalind Franklin's X-ray images to determine the structure of DNA. Once a molecule's atomic structure was revealed, chemists could then construct a drug to interact with the region of the molecule they were targeting. Describing Boger's strategy as "structure-based" was simplistic, however. His approach was rational drug design; beyond just studying molecular structure, he was interested in any data-rich technique that could streamline the drug development process.

Using software and algorithms, Vertex scientists could bring X-ray pictures to life, creating three-dimensional computer simulations that allowed encounters between the protein and prospective drugs. It was speed-dating for molecules. Was a candidate drug a potential match on the screen? Did the structure of the drug fit, lock and key style, into the target protein's active site—necessary for the protein to perform its function? Boger's vision was that such simulations would make screening huge numbers of molecules unnecessary. Chemists could design drugs based on the X-ray photos.

But while this structure-based design yielded some successful drugs, the approach didn't always work. Proteins, even those that can be crystallized and photographed, aren't static entities; they are extraordinarily complex structures that interact with and respond to their environment, always twisting and writhing like a toddler who can't sit still. Sometimes the models and simulations X-ray diffraction crystallography allowed for were just too simple, unable to anticipate how a protein would respond to a drug in the dynamic world of the cell. But the rational approach did at least help narrow the number and types of candidate molecules.

One of the drugs Vertex designed using structure-based drug design was Agenerase, used for fighting HIV, the virus that causes AIDS. By 1998, Vertex had completed phase III trials and teamed up with Glaxo Wellcome to formulate and market the drug, which was approved by the FDA on April 15, 1997—just a decade after Boger had founded the company. Unfortunately, Glaxo's formulation was unwieldy; twice a day, patients had to take eight soft-gel caps, each so large they were almost too big to swallow. Patients didn't like it, and there were other competing drugs available; Agenerase didn't snag the number-one spot. But Boger wasn't discouraged—the company had other molecules in the works. By 2001, in addition to new HIV drugs, Vertex was working on medications for cancer, hepatitis C, autoimmune diseases, inflammation, and neurological disorders. It had six molecules in phase II clinical trials, and partnerships with nine large pharmaceutical corporations.[8]

Boger was certain that designing a well-fitting drug from X-ray snapshots could be a smarter approach to building drugs than bombarding a target with thousands of molecules to find one that produced the desired effect. Regardless of their approach to identifying possible drugs, however, Vertex still lacked an efficient system to test them. Boger had heard rumors about a big pharma company that had subcontracted their drug screening efforts to another, smaller company called Aurora, which had screened their 350,000 compounds in two months; the same work would have taken the client company more than two years. Aurora, he heard, was always innovating, trying to make every part of the drug screening process more efficient.

Boger knew from experience at Merck that the most important tool for drug development was a reliable way to test a newly created molecule and track its exact behavior in the cell. Without that, a promising molecule could be overlooked. Tim Rink, Aurora's inaugural CEO, knew this as well, which is why he'd recruited electrical and mechanical engineers to create the instruments that could actually interrogate these cells and collect multiple measurements simultaneously. This meant Aurora had a powerful army of engineers—something no other pharmaceutical company had. Tsien's innovative instruments could harvest more data from every cell by using more than one reporter dye, each of which emitted a different color reflecting a different property of the cell. While one dye reported on whether the candidate drug transformed its target (like the orange-blue change if a molecule opened the CFTR channel), another might emit a different color if, say, the cell's acidity level changed. Technicians collected this data even if its significance was unknown, because it might someday be valuable.

Boger recognized that Aurora's approach wasn't just about doing things faster or cheaper or replacing humans with robots. It was about gathering more data, learning more about the chemistry in the cell, so that they could produce a better drug. A cell was like a city with many different activities happening at once. One reporter could only tell one story, but a team of them could monitor what went on in utility departments, with trash collection and disposal, city government, and so forth, to paint a fuller picture. Boger was envisioning systems on the order of thirty reporters, each broadcasting their data on a different wavelength or channel. The more reporters, the more knowledge, and the better the science.

Using Aurora's technology, Vertex could also ask much broader questions. Rather than having to ask if a single, known chemical changed the behavior of a cell, he could ask what *types* of chemicals might change a cell's behavior, and then throw 10,000 different chemicals of those types at a cell to find the ones that fit best. Boger's company had money and Aurora was developing the type of information-rich technology that Boger wanted.

After investigating the company from afar for more than a year, Boger first met Paul Negulescu and Aurora's second CEO, Stuart Collinson, at a conference in 1999, not long after Aurora had launched their pilot project with the Cystic Fibrosis Foundation. From that meeting it was clear to Boger that the scientists at this small but mighty company wanted to shift from screening drugs for big pharma to the business of designing them. After speaking with Negulescu and Collinson, Boger visited Aurora a couple of times. Then Vicki Sato, the Vertex president of research, did her due diligence by exploring all the collaborations Aurora had in its portfolio. No one at Aurora was suspicious about the intent of these meetings; after all, the company frequently received clients from big pharma companies.

Boger felt that Aurora's talents were underappreciated by most of their clients. It should become part of Vertex, he thought. Acquiring the company would benefit Vertex and please the Aurora scientists by enabling them to work on inventing important new drugs.

When Vertex acquired Aurora for $592 million[9] on April 29, 2001, the little San Diego startup wasn't so little anymore. It had research agreements with fifteen major life science companies, a staff of three hundred, and more than $60 million in revenue for fiscal year 2000.[10] And with the acquisition finalized, all of Aurora's partnerships were in jeopardy. The leadership at Vertex would be scrutinizing each one to decide which were in its interest.[11]

When they came to the little project funded by the CF Foundation, Boger didn't pay much attention. He was vaguely familiar with CF, but he didn't know anyone with it. And his priority was diseases that affected hundreds of millions, not rare ones with a few thousand patients. What he did know was that the CFTR protein belonged to a family of proteins called ion channels, which were also important in diabetes, cancer, heart disease, and pain. So perhaps Vertex could apply knowledge gleaned from the CFTR experiments to other diseases. And as the CF project wasn't a financial drain owing to the CF Foundation funding all the research, Boger never had to include it when talking to investors. Plus, if it did succeed, it would be a financial win, since the foundation had promised an additional $48 million if Negulescu's team found promising molecules—though Boger knew Aurora's goal of finding a drug that let CF's damaged protein work normally was more science fiction than science.

Sitting in Boger's office and helping him make the decision about the foundation was Vertex's president of research, Vicki Sato. The deal with the CFF made sense when Aurora was just a screening company, she noted. But she wondered whether the deal was worthwhile for Vertex, a drug company. Vertex already had several molecules in phase II clinical trials. Would continuing to partner with CFF just be a distraction?[12]

Boger believed the decision to work with the foundation had to be evidence based and dispassionate. If funding was roughly equivalent to what the company might receive in a corporate partnership research deal, he argued, that would at least neutralize the issue.

But Sato was concerned about the lost opportunities if they made a commitment to CF. It would occupy researchers who could perhaps work on other drugs instead—drugs with bigger market potential. She was also worried that the deal between Aurora and the foundation only covered the initial research stages, the discovery and development of the molecules—the riskiest part of the whole drug-discovery process, but also the least costly. If Vertex did discover molecules worth testing in people, they would need to launch a real clinical development initiative. And if that happened, Vertex would want the foundation to invest in that stage as well, to share the continued risk.

Boger knew that Sato was right. If Negulescu's team actually managed to develop molecules that were fit for preclinical and clinical trials, who would pay for those? And what about the toxicology tests? What about the cost of organizing clinical trials? What about the effort required to make the molecule into a tablet that could sit on the shelf in a pharmacy for months or years? Early on, the foundation had run a survey asking patients what form of drug they preferred. The response was overwhelmingly for some type of pill. Patients didn't want another medication that they had to sit and inhale.

Each one of these steps could cost millions, if not tens of millions. Would the foundation contribute to those? Boger wondered. He also wasn't entirely comfortable collaborating with a nonprofit. He was worried about them meddling in the science or trying to direct it—and this was before he had met Bob Beall.

But it was clear that the scientists at Aurora had a heartfelt connection to the foundation. Many of them, including Negulescu and Knapp, had met patients and their families and were participating in local San Diego fundraisers for CF. Morale would take a hit if the connection to the foundation was severed.

In the end, Boger decided to allow the CF team in San Diego to continue working on the project with the foundation—*if* they could renegotiate the contract between them after finding a promising molecule. Boger didn't think they would ever get that far, but Vertex wasn't footing the bill—and, like Tsien, he was a self-confessed "sucker for exciting science with a mission."

LATE IN 2001, MORE THAN SIX MONTHS AFTER VERTEX HAD TAKEN over Aurora, the executives had closely examined all of Aurora's portfolio and were now reviewing the last remaining element: the agreement between Aurora and the CF Foundation that had been signed back in 2000, shortly after the Gates money came through. In addition to Vertex wanting the CFF to fund a portion of the animal testing and clinical trials, Vertex's top brass—Boger, Sato, and general counsel Ken Boger, Josh's brother—wanted to revise the guidelines for sharing project data with the CF scientific community.

The original agreement allowed the foundation to share the details of the assays that Aurora developed with the entire academic community; if Aurora discovered any promising drug candidates, then the foundation could also share the structure of those molecules as well. Beall had always insisted that when the foundation funded a researcher, the resulting data had to be shared among the CF scientific community. It made sense that if more minds had access to more data, breakthroughs would emerge faster. It was a good arrangement for university researchers. And it also made sense that Aurora had signed this deal, because, at the time, they were a contractor, providing discrete services for a fee. They were not a pharmaceutical company.

But Vertex was. And sharing the assays they developed, or the molecules they discovered, would undermine the company's for-profit model. Vertex's goal was to make and earn revenue from important medicines. The R&D required along the way was proprietary; otherwise, another competitor could use the data to make their own drug.

In early 2002, foundation drug development leader Melissa Ashlock and Eric Olson—whom Negulescu had hired as the new CF project leader so he could focus on integrating Vertex's projects into the San Diego site—encouraged Bob Beall and director of medical affairs Preston Campbell to come to Vertex and meet with Boger and Sato. Beall assumed that the executives needed to be convinced that the CF project was worth their time and that the foundation was a worthy and strategic partner. So, armed with a too-long slide show, Beall began the meeting by outlining the foundation resources that would facilitate Vertex's work: the crown jewel, Bonnie Ramsey's Therapeutics Development Network, where experienced physicians would recruit and monitor patients for clinical trials; and the registry, which provided a trove of data, including patients' medical histories and which mutation they carried. While Vertex could not have direct access to the registry—that was private, he said—the doctors could use it to match patients with the right trial.

None of this was news to Boger. He knew about the foundation's assets. He recognized that the foundation had invested millions in each of these pieces over the last several decades, and that once Vertex had a drug to test, the foundation would help fast-track trials. But that was a moot point unless the foundation was willing to renegotiate the contract.

As Beall wrapped up the presentation and brown-bag lunches were delivered to the conference room, Sato motioned to Beall to come and chat in her office. When Beall emerged about an hour later, he was facing a harsh realization. As much as Boger and Sato were good individuals who wanted to develop lifesaving medicines, Vertex was still a business. The research had to be kept under wraps, Sato had explained to him, and the foundation had to continue to invest in Vertex for the long haul—far longer than Beall had envisioned when he'd first signed with Aurora. At the time, he'd thought the foundation would fund early research and that would be it. Now, he knew that, in order to maintain Vertex's interest, the foundation would have to keep the money flowing at least throughout phase I and II clinical trials, and possibly further. Sato had no idea if all the hard work would yield a drug worthy of testing in patients. But in case it did, she wanted Beall to be prepared to start raising money—a lot of money—to keep a drug moving to the finish line.

CHAPTER 38

Getting the Band Together

2001–2003

> One will weave the canvas; another will fell a tree by
> the light of his ax. Yet another will forge nails, and
> there will be others who observe the stars to learn how
> to navigate. And yet all will be as one. Building a boat
> isn't about weaving canvas, forging nails, or reading
> the sky. It's about giving a shared taste for the sea.
>
> —Antoine de Saint-Exupéry,[1] from *Citadelle*

In early 2001, Paul Negulescu, who oversaw all the biological research for Aurora and was aware of the upcoming Vertex transaction, was occupied with building a team to make the CF drug a reality, now that the leader of the high-throughput team, Tom Knapp, had the screens working well and was analyzing tens of thousands of chemicals every day. Negulescu needed medicinal chemists, who could build a powerful therapeutic drug from the crude, weak scaffold that Knapp might discover. He needed biologists who knew about ion channels, who could study CFTR and prove that a candidate drug (if they actually developed one) was opening it as expected. He needed someone to lead a unit to mass-produce cells, not just for the CF project, but to study other diseases as well. And the list would go on, depending on how the CF project progressed. Negulescu began by hiring two new scientists to join the CF project: Eric Olson, to take Negulescu's own place as CF program leader, so Negulescu could focus on his new responsibilities integrating Aurora's research and workforce into its new parent company; and Fred Van Goor, an expert on protein channels.

Olson was a slim, witty, gentle-natured Minnesotan biologist familiar with CF. His mother was a pediatric nurse, and at dinner she would tell the family about cases she saw during the day—including, occasionally, a child who was diagnosed with CF.

Olson had joined the pharmaceutical industry more than a decade before, in 1985, at the Upjohn Company (later acquired by Pfizer) in Kalamazoo, Michigan, where he worked on drugs to fight new superbugs—bacterial strains resistant to antibiotics. In 1991, he continued this mission at Warner Lambert in Ann Arbor, Michigan, the same town where, unbeknownst to him, Collins and Wilson were working on curing CF with gene therapy. At Lambert, he worked on *Pseudomonas aeruginosa*, an opportunistic bacterium notorious for colonizing the lungs of people with weak immune systems and forming biofilms—slimy 3D matrixes where the microbes lived, shielded from antibacterial drugs. The microbe was the primary cause of death for CF patients;[2] once the biofilms formed, *Pseudomonas* was almost impossible to evict. In the lab next door to his was a virologist friend who was working on HIV treatments while personally battling CF. Her incessant coughing was an ever-present reminder of the urgency of his mission.

In the late 1990s, Warner Lambert, through Olson, began collaborating with microbiologist E. Peter Greenberg, a *Pseudomonas* expert at the University of Iowa and father of a child with CF. Greenberg and his colleagues had discovered that all bacteria, far from being primitive, single-celled loners, communicated and cooperated in "societies" through chemical signaling scientists had dubbed "quorum sensing."[3] Based on this finding, in 1998 Greenberg had founded a small company in Iowa City called Quorum Sciences, where scientists hunted for compounds that could scramble this chemical communication to disrupt and destroy microbe colonies.

While Olson and Greenberg were working together to discover drugs to block bacteria communication, in 1997 Warner Lambert launched a separate collaboration with Aurora. Soon afterward, Olson met Negulescu, who was in town to discuss how the Aurora technology might help Warner Lambert discover new antibiotics. They took an instant liking to one another and talked eagerly about the direction of Negulescu's research. Olson was captivated by Negulescu's descriptions of how Aurora's robots auditioned hundreds of thousands of molecules each day to see which one could alter the target molecule in the desired way. Imagine, Olson mused, if he could use those robots to find powerful new antimicrobial drugs.

Olson's fantasy of using Aurora's technology to find new antibiotics came true in October 2000, when Aurora acquired Quorum[4] and joined the war against *Pseudomonas* and its biofilm citadel. Also delighted was Bob Beall, because the Cystic Fibrosis Foundation had been funding Greenberg's research at Quorum. Now, the company the foundation had promised $46.9 million to just six months earlier would also be working on another drug that—though unrelated to the CFTR protein—would also benefit people with CF.

Olson enjoyed working with Greenberg and was committed to hunting *Pseudomonas*-killing antibiotics. But in 2000, Pfizer acquired Warner Lambert, forging one of the world's most gargantuan pharmaceutical companies, and Pfizer terminated the collaboration.[5] Olson was disappointed by that decision; plus, after sitting through meetings with Pfizer representatives, he wasn't interested in working for them. He had spent more than fifteen years in large pharmaceutical companies and he wanted a change. He wanted to join a small drug company where he would have a pivotal, hands-on role in creating new medicines, where he could nurture a molecule through development, to market, all the way to patients' bedsides. Discovering a drug was a rare event, and he'd yet to work on a successful one. As word got out that Olson was looking for a new gig, Vertex research president Vicki Sato reached out with an offer to lead work on antifungal drugs. But Olson declined; his passion was bacteria. Then Greenberg suggested Olson check out Aurora.

In early 2001, when Michigan was still blanketed with snow, Olson was chatting with Negulescu in bright California sunshine at a picnic table in Aurora's parking lot. There was a lot to admire: the one-of-a-kind robots running hundreds of thousands of experiments every day; the smorgasbord of "reporter" dyes that enabled new ways of spying on the innermost workings of cells. He could barely contain his excitement, thinking of all the awesome tools he'd have there for seeking out new antibiotics.

Negulescu offered Olson a position in early April, just three weeks before the Vertex takeover was announced. Olson knew that Aurora had signed the $46.9 million deal with the Cystic Fibrosis Foundation—a solid source of funding. But like others before him he was concerned that the company, with its unique and tantalizing technology, was prime fodder for a takeover.[6] Though the uncertainty ate at him, the job was too exciting to turn down.

When Olson arrived on campus, Aurora was already Vertex. Although he knew that the employees were primarily engineers, experts at building robots and high-throughput screening, and biologists, he was surprised that there were so few chemists. But that made sense; before the takeover, the company had only collaborated with big pharmaceutical companies, with the exception of the foundation, and hadn't needed any in-house chemists for drug discovery. Their role was limited to screening collections of chemicals and turning over the promising molecules to their client. But now they were part of Vertex—and in the drug development business.

Olson, who had the most experience in drug discovery, and Mark Murcko—one of the first Merck chemists that Boger hired and Vertex's chief technology officer, who had launched many of Vertex's drug discovery efforts in the areas of antibiotics, cancer, pain, and more—were charged with bringing the San Diego team into the world of drug discovery. To transition the West Coast team into the pharmaceutical business, Murcko and Olson began with a daylong symposium for all the Aurora staff that gave them an overview of the drug discovery process. Once a promising molecule was discovered during high-throughput screening, medicinal chemists would take this molecule and produce other, ideally more potent versions of it that biologists would test in cells. Next, toxicologists would test it in living animals for safety and to make sure the drug ended up in the right part of the body for the right amount of time. Then another set of chemists would transform the molecule into the appropriate formulation—a spray, cream, liquid, or pill. Only then would the drug be tested in human trials, in three phases. If the drug was effective, the FDA would approve it. The entire process usually took between ten and fifteen years.

Olson and Murcko then described to the employees how Aurora would change according to Vertex's priorities. For a start, they would be hiring more biologists who could interpret all the screening data Knapp's team was generating. They'd also bring in pharmacokinetic scientists, who tracked the behavior of a drug in a living organism, and medicinal chemists, who built drugs atom by atom, to train and brief the new recruits joining the San Diego facility. And Murcko would visit every other week to learn how each group was progressing. He was an important champion for the San Diego site, since few of the Cambridge biologists and chemists were fans of high-throughput screening, which they considered the antithesis of Josh Boger's rational drug design.

After the merger, as Knapp continued testing chemicals in search of one that would allow chloride to flow in and out of the cell, Olson began drafting a "spec sheet" for the anticipated CF drug—a list of preliminary characteristics a molecule had to satisfy to warrant further development. Just enabling chloride to move out of the airway cells wasn't enough. The candidate couldn't interfere with a protein named hERG—an ion channel that regulated heartbeat. Such an interaction could be deadly. It also couldn't interfere with proteins called cytochromes because these were critical for transferring energy within the cell and also involved with metabolizing other drugs. Tampering with these proteins could trigger dangerous drug-drug interactions. And the list continued.

As in all drug discovery projects, the list of criteria was initially short, six items long. Beginning with the bar too high might rule out promising molecules. Additional specifications could be added once a compound showed real potential.

While Olson worked on the spec sheet and other elements of the chemistry, Negulescu's other new hire, Fred Van Goor, was figuring how to measure the activity of a single CFTR protein—a strategy vital for proving whether a drug was truly manipulating the mutant protein as they hoped.

VAN GOOR, A CANADIAN-BORN ELECTROPHYSIOLOGIST WITH A MOP OF floppy brown hair, had also accepted a position at Aurora just weeks before Vertex acquired it. Van Goor had an intimate understanding of the cell and the biology of ion channels—those donut-shaped proteins, like CFTR, that provided a gateway in and out of the cell. He had completed his PhD at the University of Alberta, where he explored the intricacies of sodium channels in the pituitary gland cells of goldfish. This fascination with ion channels continued when he moved to the National Institutes of Health for postdoctoral research, where he studied how these channels regulated the flow of calcium in the brains of rats. It was good preparation for the project at Vertex; ion channels in both fish and rats were similar to those in humans.

After completing his prolific postdoctoral stint at the NIH, during which he produced some forty publications, Van Goor applied for jobs in late 2000, considering for the first time positions in industry. When Negulescu interviewed Van Goor, he asked if he knew about CFTR. "Sure," he replied. Negulescu knew Van Goor was an electrophysiologist, studying how electrical currents move across cell membranes, and asked if he also knew how to make recordings of a single protein channel on a single cell. "Sure," said Van Goor again— and left the meeting thinking, *I need to learn about CFTR and figure out how to do this single-channel stuff.* After interviews with three different companies, Van Goor chose Aurora, where he'd be working on the CF project. It was a job that was critical to the entire CF effort.

Although the screens that Knapp was developing could detect when the electrical activity of a cell changed by scanning for a color change, someone still needed to prove that it was actually the fixed CFTR protein on the surface of the cell that was opening and closing like it was supposed to. One way to do this was to measure the electrical activity on a single microscopic "patch" of the cell surface containing a single CFTR channel—a technique known as patch clamping. Van Goor would need to master the delicate procedure of measuring the electrical activity of a single CFTR protein as chloride flowed out. This was the gold standard of proof that a chloride channel had been repaired. Only once this proof was in could Vertex contemplate moving to the next stage of the drug development process.

Negulescu, Olson, and now Van Goor all continued recruiting more team members as Knapp kept the high-throughput screening on track. In addition to robots and chemicals, Knapp needed an enormous supply of fresh cells to expose to new chemicals in the expanding number of screens. Vertex San Diego needed someone to establish a "cell core"—a large-scale facility for producing genetically engineered cells for the CF project, and also other types of cells for the many other diseases that Vertex was working on. The man Negulescu hired for the job was Tim Neuberger.

Neuberger was average height, wiry, and spoke at the speed of a bullet train, as if he were always anxious to complete his thought. He grew up on a small farm nine miles west of Parkston, South Dakota. Not ready for college immediately after high school in 1975, Neuberger instead joined the Air Force, and left for basic training at Lackland Air Force Base, Texas. He spent the next four years in Colorado, the Philippines, Spain, and finally Tacoma, Washington.

Military service had whetted his appetite for travel and new sights; afterward, he wasn't interested in returning to South Dakota. So in 1979, when an Air Force buddy encouraged him to move to Albuquerque, where he was based, and attend college at the University of New Mexico, Neuberger didn't hesitate. There he met his wife, and after Neuberger completed his bachelor's degree, the couple moved to Burlington, Vermont, where he earned his PhD in neuroscience in 1989, and then Richmond, Virginia, for postdoctoral work. While in Richmond, he got a call from his friend in Albuquerque: his daughter had just been diagnosed with CF. It was depressing news for Neuberger. His expertise was in neuroscience; he wouldn't be able to work on treatments for her or contribute to a cure.

After completing his research training, Neuberger made the jump to industry, working for a time at a biopharmaceutical company north of New York City developing therapies for people with neurological disorders. There Neuberger learned to nurture vast quantities of cells in the lab—a job that was as much art as science, requiring him to figure out the exact conditions that made the cells happy.

Now at Vertex, his task was to cultivate a variety of genetically engineered cells, including several types for the CF project. It was challenging and deeply personal work; he finally had the chance to join an effort that might help his friend's daughter.

So far, Aurora had been using mouse cells genetically engineered to carry mutations that caused CF. But eventually, they hoped, Knapp would find some promising molecules to test in human cells. To prepare for that breakthrough, Neuberger had to learn to grow human lung cells from real CF patients. Typically, when researchers discover a mutation in a gene that causes disease, they study the disease by engineering an animal—usually a mouse—to carry the mutated gene. In most cases, the mutant human gene produces the same mutant protein in the mouse, and because human and mouse DNA and physiology are so similar, that mutant protein will trigger the disease. But "CF mice" don't follow this trend. Mice carrying two copies of the F508del mutation don't develop cystic fibrosis. They suffer some digestive problems, but their lungs remain perfectly normal and healthy—ruling them out for the early preclinical testing needed to test efficacy before a drug could be given to people in clinical trials.

Without an animal to bridge the gap between genetically engineered cells and humans, Vertex needed to grow actual human cells to test the drug. If the drug worked in these human cells, there was a good chance they would work in people, too. But it was no guarantee; just as success in animal trials did not guarantee success in human ones, a drug that worked in an isolated human cell might not work in a living, breathing human.

Neuberger's partner in creating this cell core was Angela Kemnitzer. She had started at Aurora as an intern, just a few months shy of receiving her undergraduate degrees in biology and economics, shortly before Neuberger's hire. She had no previous experience growing cells, but under Neuberger's tutelage, she worked in the cell core, learning how to brew huge volumes of nutritious broth to feed the cells, grow them in large flasks, and nurture them so that they would be ready to receive and respond to whatever molecules or tests the chemists concocted. Together, they kept the biologists and chemists supplied with more than twenty different types of cells for Vertex's many projects. And if the CF team succeeded in identifying some promising molecules, it would be up to Kemnitzer and Neuberger to then learn to harvest and cultivate human cells from the lungs of patients who had died or been lucky enough to receive a transplant.

WITH THE NEW CELL CORE IN PLACE, THERE WAS STILL A VITAL GROUP of players missing: the chemists.

The handful of molecules Knapp was trying to identify, ones that would help mutant CFTR fold correctly and others that would open the CFTR channel and make the cells glow blue—"hits"—would initially be too weak and imprecise to deliver a therapeutic benefit to sick patients. The point of screening thousands of molecules was to find hits that provided clues about what type of molecule could do the job—a starting point from which the chemists could begin. CFTR was a complicated and uncooperative protein. There were no atomic-resolution pictures of it, so finding a drug to unlock the channel had to be done purely by trial and error—as Knapp was doing. It was like testing hundreds of thousands of keys just to find one that fit the lock—but it was unlikely that it would be a perfect fit. For that, they needed medicinal chemists to modify it, tweaking and augmenting a rudimentary molecule atom by atom until it was so powerful that patients only needed a tiny quantity to open the CFTR channel.

Many at Vertex, newcomers and the original Aurorans alike, were extremely skeptical of the whole venture; developing a molecule that could "fix" a damaged protein sounded like a fool's errand. While Negulescu was one of the few who believed that it was possible to find corrector and doorman molecules, he couldn't gauge the odds. So, in 2002, looking for reassurance, he stuck his head into Olson's office and asked his opinion. What were the chances they would find a molecule that could fix the broken CFTR? In industry only about 1 percent of promising molecules discovered in the first round of screening for a particular disease actually worked. But Olson thought that their odds were much better because, although they still hadn't found any hits worth developing, they knew that proteins with the F508del mutation could be made to work by simply lowering the temperature several degrees. That suggested there were conditions under which the mutant protein could fold correctly and function. And there was already an existing doorman drug—the toxic cancer drug, genistein—that could open CFTR with the G551D mutation; it just wasn't safe. So it wasn't so improbable that they could find another, less dangerous molecule that could do the same thing. Olson guessed they had about a 20 to 30 percent chance of succeeding and finding doormen and correctors in the screens that Knapp was running.

Negulescu still looked dissatisfied, so Olson stood up, grabbed a pen, and moved to the whiteboard. In the F508del mutation, the protein is missing phenylalanine, he said, starting to sketch out the amino acid.

Amino acid phenylalanine, which appears at
position 508 in a healthy CFTR protein.

He tapped the molecule he'd just drawn. When the amino acid phenylalanine at position 508 of the CFTR protein is missing, amino acid 507 gets directly stitched to amino acid 509, and the protein misfolds as a result.

Negulescu nodded.

So, said Olson, it wasn't a big stretch to think that they could find another molecule, with a similar shape to phenylalanine, that could, as the mutant protein was in the midst of folding, wedge itself between amino acids 507 and 509 in the damaged CFTR, forcing the protein to fold more correctly and therefore increasing the likelihood it would move to the normal location in the cell.

Hopefully Knapp would soon identify a hit strong enough to pass to the medicinal chemists. But in 2002, there were still no such chemists at Vertex San Diego; they had never been needed. Now Vertex did need them, and not just for the CF project, but for its other interests as well. So Negulescu began recruiting.

In September 2002, he hired Peter Grootenhuis, a Dutch chemist who in 1998 had moved from the Netherlands with his family to work for a San Diego company called CombiChem that, like Vertex, was using computers to design drugs and robots to synthesize thousands of molecules simultaneously. But by 2002, CombiChem had been bought and sold several times, and Grootenhuis wanted a more stable workplace. He was friendly with colleagues at Vertex Cambridge and also acquainted with Aurora. Grootenhuis was a believer in rational drug design, for which Vertex was world renowned; after meeting Negulescu, he was eager to join the team.

Negulescu hired Grootenhuis to design and analyze small molecules that might be potent pain blockers—a priority for Vertex. But later that year, Negulescu asked him to work on CF instead. Grootenhuis was a doubter; like so many others in the industry, he thought that trying to intercept a misfolded protein and fix it was crazy. Any other project at the company had a better chance of success. Working on CF seemed like a quick way to kill his career.

But the CF project needed a medicinal chemist, and he was the one with the most experience. Previous rounds of screening had not been very successful, so Grootenhuis's first mission was to examine the chemical libraries that Knapp had tested and recommend which ones he should try next. Vertex had many libraries, each with thousands of molecules, but lots of these chemicals had very similar shapes and properties. Testing more collections similar to the ones that had already failed would be a waste of time. Instead, he needed to test molecules vastly different from these in shape, size, volume, and charge. There was an infinite universe of chemicals out there, and the broader the selection of molecules they could test, the faster they could find one that felt promising.

The next round of screening, in 2002, with Grootenhuis's chosen molecules, proved more inspiring. As Knapp plowed through the data, he saw more of the molecules they were testing made the cells glow, and more of these glowed brighter. It was a sign the new chemicals Grootenhuis selected had features that, in the corrector screens, moved the protein where it needed to go, and in the doorman screens, were opening the CFTR channel to permit chloride transit—all clues that they were on the right path.

For future rounds, Knapp and Grootenhuis chose libraries in which all of the chemicals had characteristics similar to those ones that had triggered electrical activity in previous screens. And as he saw more cells test positive, Grootenhuis became slightly more optimistic that they might eventually find molecules that could serve as scaffolds, or starting points, for both drugs—the corrector and the doorman.

THE CLOSER KNAPP AND GROOTENHUIS GOT, THE MORE URGENT THE need became for not just one but a team of medicinal chemists. Grootenhuis had a colleague at CombiChem whom he thought would enjoy the challenge—Sabine Hadida, a Spanish chemist born and raised in Barcelona.

As an undergraduate, Hadida had studied pharmacy at the University of Barcelona and fell in love with organic chemistry—the molecules of life. The art of building molecules fascinated her in general. But what she found most enthralling was the prospect of building molecules that would travel through the body and perform a specific biological function, which led her to medicinal chemistry. Medicinal chemists interrogate a molecule to learn how each atom contributes to the drug's overall function. Does an atom make the drug better or worse? Would a different atom be more effective? Hadida was intrigued by the idea of creating therapies for specific diseases, but didn't know how to begin. Did one study medicine or chemistry or biology? During her fourth year at the university, a professor urged Hadida to earn her PhD in chemistry—the first step if she really wanted to design drugs.

Until graduate school, Hadida's knowledge of chemistry had been largely theoretical: memorizing chemical reactions from books. Working beside her mentor in the lab, she learned how to do experimental chemistry—how to join atoms together and build molecules. The lab focused on replicating natural products, complex molecules built by Mother Nature that were found in plants or bacteria, so that these substances could be easily manufactured in labs. To Hadida it was like being an architect seeing a very complex building, then figuring out how to replicate the design faster and more cheaply.

The catch was that building a molecule wasn't a single-step process. You can't stir a few ingredients together in a flask and then, poof, the flask is filled with the compound. It requires a string of chemical reactions—transforming one compound into another by adding atoms a few at a time—each of which inches the molecule a little further toward the final structure. Plus, after every reaction, the desired compound has to be separated from the rest of the mixture and purified before the next reaction can be performed.

Some natural products were easy to make in the lab. Others seemed impossible to synthesize—only Mother Nature had figured out how to link atoms together in certain ways. But once a chemist cracked the puzzle of how to build a certain type of chemical bond, the potential applications, from industrial to chemical to medical, were endless. It was a tough, almost Zen-like exercise, requiring patience and discipline.

Her degree in hand, Hadida moved to Pittsburgh for postdoctoral studies and then finally to San Diego in 1997 to work at CombiChem.[7] There, her work shifted from esoteric academic pursuits to actual drug development. Clients provided a target—perhaps an overactive protein—that had been photographed using X-ray diffraction, producing an image on the computer. Sitting in a pitch-black room with special goggles, Hadida would examine the target in 3D, probing the atomic nooks and crannies of the molecule the way an astrobiologist might examine a meteorite. The structure was artificially colored to make the regions of the protein easier to distinguish, and manipulating it was like playing with a strange, colorful deep-sea life-form. She could spin the target—usually a protein with thousands of atoms—and see how it moved. Every molecule had its own bonds that could twist and wave and bend individually, giving the protein tremendous flexibility.

Hadida's job was first to find the protein's Achilles' heel—a spot that, if harpooned with a drug, would stop it from doing its job—then to design and create the actual drug. To begin, she would search through chemical catalogs to find a molecule with the right shape and electrical properties to target the vulnerable region of the protein. If she couldn't find one in the catalogs, she'd design the molecule herself. Then she would take the design and figure out the series of chemical reactions necessary to build it. Afterward, she would go into the lab, make it, and test it. If that didn't work, she would find or design something new and try the test again. It was fun; Hadida loved the work and the challenge of beginning every project from scratch. Far from unnerving her, it fueled her excitement and passion. But every molecule she made was eventually handed off to the client. She never had the satisfaction of watching one of her molecules be tested outside of the lab and transformed into a real drug.

When Hadida interviewed at Vertex San Diego in the fall of 2002, she knew nothing about cystic fibrosis, but that didn't matter. What they needed from her was her ability to create a drug. The task would be tricky, however, because with no atomic-resolution photo of CFTR, she wouldn't have the visual cues that she was able to use at CombiChem. But that didn't concern her. If the drug she built worked, it would be an opportunity for her to be part of the entire process of drug development, from atom to molecule, from cell to clinical trial.

When she arrived at Vertex, there were a total of three full-time medicinal chemists, including her and Grootenhuis. Though more chemists would probably be required if a promising molecule was discovered, the team leaders were now all in place to keep the foundation's project moving quickly.

By the summer of 2003, San Diego still didn't have a promising hit—that scaffold that Hadida and her team would build into a drug. Some chemicals that made the cells glimmer didn't work on a second try. Some were toxic. Other molecules didn't have the shelf life to be a useful drug; they broke apart too quickly. But still the team persisted, with almost $47 million for the discovery phases in hand.

Once every three months, a meeting of the Joint Research Committee, which included members from both the foundation and Vertex, was held. During these meetings, everyone convened in the boardroom. Bob Beall sat in his regular seat at the head of the table with CFF folks flanking him; Negulescu and his team faced him at the other end. A week before each meeting, Olson would send the foundation the data they would present in person the following week. Such transparency gave everyone the chance to review the data independently. Ashlock managed this committee and the in-person events from the foundation side, reviewing the progress reports from the Vertex program leaders—Paul Negulescu, Eric Olson, and Peter Grootenhuis. Then she would share the data with the CFF advisors. The chemical structures were reviewed behind closed doors with Vertex chemists by two senior medicinal chemists, CFF advisors from the pharmaceutical industry, whom Ashlock had hired for this purpose. Although they were also on the Joint Research

Committee, they kept the structures confidential from the foundation members, so that none of Vertex's proprietary research would be leaked.

The Vertex team felt the urgency of this mission and were emotionally invested. The longer they worked on the project, the more families they met. The foundation encouraged parents and children with CF, like Jennifer Ferguson and her son Ashton, to visit Vertex San Diego, meet and talk with the scientists, and learn what they were doing. And as high strung and impatient as Beall was, always in a suit and tie, he tried to create an atmosphere that was supportive and comfortable, where people could speak about setbacks as candidly and easily as progress. All Beall wanted was the truth. Had they discovered any promising molecules? What were the problems? Could the foundation and their science advisors help? Where was the project lagging? What was going well? That's why Beall insisted on these quarterly in-person meetings rather than a long-distance call or teleconference. Though he was constitutionally allergic to small talk, he valued personal relationships—and he wanted to hear everyone's ideas and opinions.

Until now the team had been focusing more heavily on the hunt for a corrector to fix proteins with the F508del mutation. That made sense, because close to 90 percent of all CF patients carried it. Approximately 45 percent of patients had two copies of this mutation; another 41 percent carried one copy of F508del and one copy of a second, rarer mutation.

But even if they managed to find a corrector, it wouldn't work alone, as Van Goor had reminded everyone at a meeting in late 2002. Patients with the F508del mutation needed two drugs: a corrector to move the damaged protein to the surface of their cells *and* a doorman drug to open it. So Van Goor suggested devoting equal resources to hunt for corrector and doorman candidates simultaneously, which might help them reach their goal faster. Plus, there was another, though smaller, group of CF patients with the G551D mutation who only needed a doorman. Once they had created a doorman drug, they could immediately test it on this group of patients to see whether the concept of fixing a broken protein made any sense at all. And if it did work, this tiny group of patients would have a medicine immediately.

As the San Diego team was following this course in the summer of 2003, there were changes at the company as a whole. Eric Olson's role expanded. He took the helm as leader of the CF program in both Cambridge and San Diego—his first chance to lead a development team—while also helping to drive the creation of a new antibiotic developed at Vertex Cambridge that was now pushing into animal trials. Before moving to Cambridge, he passed the baton for San Diego's CF work to Peter Grootenhuis, but continued to lead weekly progress meetings with Grootenhuis, Van Goor, and Hadida from afar. Olson had never led development and was paying close attention to the other programs at Vertex Cambridge to see how it was done. It was good training for when San Diego found a molecule worthy of transforming into a drug for CF—something he was confident would happen soon.

Once settled on the East Coast, Olson quickly realized that few people, with the exception of Vertex CEO Josh Boger and Mark Murcko, who was still remotely managing Vertex San Diego, took the CF research seriously. It was the butt of many a joke: it was a money pit; the California team had no idea what they were doing; the program was deadweight that would never yield a viable drug. But rather than lament the lack of support, Olson saw an opportunity.

He decided to wage a personal PR campaign to drum up enthusiasm for the project. Once the San Diego team landed on the right building block for this drug, he would need solid backing to push development forward. He began by holding a monthly lunch with a side of CF education—a lecture on the disease, its symptoms, and the strategy to find a drug. At first, only a handful of people showed up, mostly new employees who hadn't yet been indoctrinated against the project. But he refused to feel defeated. He knew the West Coast team would succeed. The only thing that did worry him was how the foundation would raise the money for the next steps.

CHAPTER 39

Pay to Play

2002–2004

Donors don't give to institutions. They invest in
ideas and people in whom they believe.

—G.T. Smith

Vertex had drawn a line in the sand: once chemist Sabine Hadida's
team had built a drug candidate, they weren't going to fund the
clinical development alone. If the Cystic Fibrosis Foundation wanted
them to continue, it needed to raise a lot of cash, and fast.

This was not the kind of money that could be raised with walkathons and galas. A rough calculation revealed that the foundation
would need well over $100 million to fund their share of a door-
man and a corrector, plus a hypothetical third drug to treat another
mutation.

The foundation's CEO, Bob Beall; his executive VP of medical
affairs, Preston Campbell; drug development leader Melissa Ashlock;
and head of basic science research, Chris Penland; were focused on the
science, which was now progressing quickly. The job of how to raise the
funds fell on Richard Mattingly, the same man who had worked with
former CEO Bob Dresing to implement the changes when Dresing and
Doris Tulcin had restructured the foundation. He was now in charge of
fundraising for all of the foundation's seventy chapters. As experienced
as Mattingly was, raising more than $100 million was beyond his skill
set. And he had absolutely no idea whether that amount was even realistic. To raise such a large sum, he'd need donations of $100,000 and up,
and he didn't know people with that type of money.

But Joe O'Donnell, who was still raising money for the foundation through the Joey Fund, and also for Harvard University, did. And

Mattingly knew that he was the person who should lead the foundation's new fundraising campaign.

By the fall of 2002, Mattingly was calling Joe frequently, and had flown to Boston several times to try and convince him to lead the capital campaign. Joe refused, becoming increasingly irritated as Mattingly tried to muscle him into a role for which he had no time. He knew he was a great candidate for the job. He had the connections. He was good at raising money. And he had credibility—his kid had died seventeen years earlier and Joe still raised more for CF every year. But Joe didn't want to do it. For the past two decades, he'd been busy growing his businesses and raising Kate and Casey with Kathy. He'd expanded Drive-In Concessions of Mass., Inc. into the Boston Culinary Group, which catered to stadiums across the US. He had some 12,000 employees and stakes in ski resorts, hotels, movie theaters, and restaurants.[1]

He was the majority owner of Suffolk Downs, a thoroughbred racetrack in East Boston, and his Allied Advertising Agency was the third-largest print advertiser in the US. He was a busy man.

Instead, he tried to boost Mattingly's confidence that he could do this without Joe's help. Joe told him to go to Harvard and meet with a friend of his, the dean of the business school. But the visit didn't amount to much. The dean just told Mattingly what he already knew: that the key to raising money through a capital campaign was who you know. And Mattingly didn't know the whales of philanthropy, the big donors who could make a difference, the way Joe did.

So Mattingly tried a new tack. The next time he went to visit Joe, he brought a mother, Amy Barry, whose son, Jamie, had CF.

Born in 1995, Amy and her husband, Peter's, second child was an unhappy baby, crying night and day. The couple took Jamie to his pediatrician, who dismissed Amy's concerns, and then a gastroenterologist, who wanted to give the child a sweat test—just to rule out the worst, though extremely unlikely, diagnosis of CF. Amy went home and searched the web to learn something about the disease. She then picked up Jamie and cautiously licked his forehead. It was very salty. She called her doctor and insisted the sweat test be done the next day.

The couple was home when the call came, and after hearing what they already suspected, they sat, terrified, in their New Jersey home. After several hours Amy called her sister Kate in Rye, New York, who drove down immediately. Kate and her husband, Bob Niehaus, were take-charge types. They took Jamie's healthy sibling back to their home for a few days as the Barrys got their sea legs, telling Amy and Peter they would come up with a plan.

Within days, Kate connected Amy with John and Amy Weinberg, whose daughter had CF, and then with Doris Tulcin. Then, without telling Amy, the Niehauses called Bob Beall, expressing their desire to make a significant donation. Within a week, Beall flew to meet them. He talked to them about the disease and exactly where the science stood—the discovery of the gene, and the gene therapy trials that were in progress in 1995 but unlikely, in his opinion, to yield a therapy. He described the new types of antibiotics, including TOBI, that would be more effective at targeting the microbes in the lungs.

As grim as the outlook was, Kate and Bob could see and hear Beall's determination and drive. He was on a mission and was going to find a way to help these sick children—a group that now included their nephew. The couple began donating to the foundation, and Beall kept in touch, frequently asking about Amy, Peter, and Jamie. After Beall tried to lure Bob Niehaus onto the foundation's board, Kate suggested that he consider Amy instead. Kate knew her sister had been in a dark place in the months following the diagnosis. Focusing her energy on the board would be a lifeline—she'd be close to the science and everything that the foundation was doing. Beall suggested to the board that she might be interested in serving, and when Amy confirmed this, the board nominated her officially and she was voted in as a trustee.

Years later, when Mattingly asked her to speak to Joe with him, she was eager to help. She had never crossed paths with Joe or heard of him during her years on the foundation's board, but after a little research, she quickly understood why he mattered. She was comfortable being Mattingly's prop if she could convince Joe to take the campaign's helm.

Mattingly had become increasingly pessimistic that he could rope Joe into leading the campaign. He hoped that Amy, as a desperate mom, might be able to appeal to Joe in a way that Mattingly couldn't. But if not, the visit wouldn't be wasted; he planned to ask Joe for more details on how he'd recommend raising the necessary funds, and then try and take on the task himself.

In October 2002, Amy and Mattingly flew separately into Boston's Logan Airport, where they caught a cab together. When Mattingly gave the cabbie the address for Joe's company, he turned around. "You must be going to see Mr. O'Donnell. Do you know him?"

Upon their arrival, Joe's secretary led Amy and Mattingly into a conference room. When Joe walked in a few minutes later, Amy thought he looked irritated, even angry; he barely glanced at either of them before sitting across from them at the conference table.

Joe *was* irritated. He couldn't remember: Was this the third time Mattingly had come to Boston? Or the fourth? Tapping the table with the end of his pen, he said, gruffly, "Okay, what are we doing here?"

Mattingly was not flustered by his grumpiness. He'd known Joe for years and loved him like family. He coolly launched into why he was there, forgetting to introduce Amy. After some fifteen minutes of talking, Joe interrupted him and looked at her. "And who are you?" A little rattled by Joe's abrupt attitude, she introduced herself and explained that her son Jamie had CF.

After a few more minutes of Mattingly talking, Joe let out a sigh as if already fatigued by the conversation and suggested they go to lunch—even though it was only 11 AM. As they walked to a nearby restaurant, Amy saw people on the sidewalk staring—Joe really was a local celebrity.

At lunch, Mattingly continued the conversation, asking Joe to break down the fundraising process and how to run a capital campaign. If Joe refused to run it, Mattingly said, he should at least teach Mattingly. After another big sigh, Joe grabbed a paper napkin, pulled a pen out, and began to sketch a pyramid. In the old days, he explained, the fundraising model was 80–20: 80 percent of the money raised was from the top 20 percent of the donors. Recently, it was more like 95–5: 95 percent of the donations tended to be raised by the wealthiest 5 percent of the people, while the remaining 5 percent came from the efforts of thousands of other volunteers and donors. Joe called those other volunteers and donors "the Standing Army"; these were the dedicated people who had been raising all the foundation's money for the past fifty years. And while they would continue to raise money during the capital campaign, the money that Mattingly had to raise for Vertex was separate from all that.

Joe continued to talk, explaining the stages of the campaign, all the time scribbling on more napkins to make his point. Capital campaigns began with what was known as a quiet period, the planning stage, when the leaders figured out the length of the campaign—five years, three years, whatever. It was also the time when, after some preliminary fundraising, you figured out how much you would reasonably be able to raise. So you started by planning, but you also needed to privately start fundraising, Joe noted, focusing on the napkin rather than Mattingly. Ideally, you wanted to raise between 20 and 30 percent of the goal during this quiet period, before you publicly announced your campaign.

Both Amy and Mattingly were surprised by this, so Joe explained. One reason to start fundraising during the quiet period was to test whether the end goal was realistic. You didn't want to announce a goal of, say, $100 million if the funding raised during the quiet period suggested you'd never break $70 million. The other reason was psychological. People like giving to successful campaigns and causes. So it looked better to announce a campaign when significant progress had already been made toward the end goal. "You don't want to start a campaign at the bottom of the thermometer," Joe told them.

When he was done, he suggested heading back to the office, and crumpled up the napkins—but before he could toss them, Mattingly asked if he could have them. It was the first time that morning Joe cracked a grin.

Before they left, Mattingly headed to the restroom. Joe, relieved to get a break from Mattingly, turned to Amy, took her hand, and said, "Tell me about your kid. Tell me what he's like." Talking parent to parent, Amy saw Joe transform, shedding the gruff, business-like exterior and focusing on her story. Joe teared up as she told him about Jamie; moved by his genuine interest in her child, so did Amy.

When they all returned to the office, Joe asked Amy if they could detour to his office to view some pictures of Joey before rejoining Mattingly in the conference room. He was thick with emotion as he led her around his office, showing her the pictures and sharing anecdotes of Joey's antics. Joe was a completely different person from the one she had met a couple of hours earlier, and by the time they returned to the conference room, they were both weeping.

Then, without any prompting from Mattingly, Amy stepped in: "Joe, you are the only guy who can do this and we need you. And I need you, and my son needs you, and my husband needs you."

Amy and Joe now had a connection, and her words were convincing. How could Joe possibly spend his time doing other things while Amy and thousands of other families were suffering and struggling with sick children, a life he and Kathy knew too well? But as much as Amy had moved him, Joe still didn't want to say yes. So Mattingly decided to move forward with another plan.

DURING MATTINGLY'S ROUGHLY TWENTY-FIVE YEARS AT THE FOUNDA-
tion, he had met many consistently generous donors. Among those, six
smart business types stood out—men he felt could guide him on how
to raise this gargantuan sum. So Mattingly decided to organize a dinner
with them in New York, to allow everyone to socialize, and then a meeting the next day to brainstorm how best to get this money. They were all
highly motivated to help—they all had children or other relatives with
cystic fibrosis—and some were already seasoned fundraisers themselves.

The group included Joe; Amy Barry's brother in-law, Bob Niehaus, founder of private equity firm Greenhill Capital Founders, previous managing director in merchant banking at Morgan Stanley &
Co., and a fundraiser for his alma mater, Princeton;[2] J. Taylor Crandall, a CFO for the Bass family and a highly regarded venture investor; Thomas Hughes, an executive at Deutsche Bank; and Gary Sabin,
chairman and CEO of Excel Trust, Inc., a real estate investment trust.
The last was John Weinberg, whom Amy's sister had contacted shortly
after Jamie's diagnosis, and who until retiring had spent his career at
Goldman Sachs, which his father had run from 1976 to 1990.

Mattingly still hoped to convince Joe to take the lead in the foundation's new fundraising effort. But if he could not, this meeting might at
least provide some other candidates.

The meeting was tough to arrange. Most of the players had all-
consuming day jobs, in some cases managing hundreds of millions
of dollars. But after months of wrangling schedules, they all met for
dinner in May 2003 at a restaurant in Rockefeller Center: Mattingly,
Campbell, and Beall from the foundation; Bob Niehaus and his
brother-in-law Peter Barry, Amy's husband; Sabin; Hughes; Weinberg;
Crandall; and Joe. Then, the next day, they got down to business.

The group convened in a windowless conference room at Deutsche
Bank in Midtown Manhattan. There, the nine men sat elbow-to-elbow
facing a flip chart. Beall kicked off the meeting by quickly covering the
science and describing what Vertex had done and was planning next.
Then Campbell, in his laid-back Tennessee twang, told them, in simple
terms, why they had to come up with the money: "The science is there.
How horrible it would be not to see this through."

Money was what lay between sick kids and a cure.

Mattingly explained how much money they needed to give Vertex
for what he hoped would be three or so drugs, and then turned the conversation over to the six men.

There were two investment strategies to debate. The first, which Beall had conceived, was to launch a venture capital company to finance Vertex and lure investors. If Vertex developed a drug, the investors would see a return. If not, as with any other high-stakes gamble, they wouldn't.

The second option, which Joe and Mattingly favored, was to procure large donations from philanthropists, with no strings attached, and then invest the money in Vertex. The foundation had no plans to just give the money away, of course. A royalty agreement guaranteed that if Vertex was successful, a percentage of the profits in the low double digits would go to the foundation and could be reinvested in future drug development, clinical trials, or both.

The first option came with the potential for a hefty return that was likely to attract investors. But it was also fraught with conflicts of interest that were too messy, tax-wise, for most people who might consider investing. The second approach was something the foundation had been doing successfully, albeit on a smaller scale, for a couple of decades: the original $2.6 million investment in TOBI had led to a royalty payout of $19 million, which had been used, along with the Gates money, to launch the drug hunt with Aurora. It was compelling proof of how a royalty payment could be leveraged to further the foundation's mission. Now the $51 million they had already invested in Aurora (the $1 million for the pilot program, the $3.2 million for the prototype, and the $46.9 million for the discovery process) looked poised to yield some promising molecules—but to get a return on that investment, the foundation needed significantly more than this to move the effort forward.

Within a couple of hours, the six players had settled on the second option and began discussing potential donors—and the logistics of their efforts. It was a tricky endeavor. How would they lure new donors who didn't have a personal connection to the disease? How would they get them interested and excited? What tangibles could they offer? If they had been raising money for a hospital, they could offer naming rights to various wings. Could they convince donors to fund a certain molecule in exchange for attaching their name to the compound if it eventually came to market? After four hours, all the men were checking their watches and shifting restlessly in their chairs. They knew they wouldn't be able to resolve the issue of whom to tap for the money that day and were ready to leave. But they had made good progress.

Soon, the meeting drew to a close. But before Beall let the participants out the door, Mattingly raised the final issue: "I need a leader."

The table was quiet.

"I need someone who will lead this campaign. I don't have the connections or the clout," Mattingly said, looking each of them in the eye. They all remained silent.

"Well, I think you have your answer," said Hughes. "We'll have to leave it there for now."

Mattingly thanked everyone and watched helplessly as they filed out of the room and walked to the elevator. Niehaus, the last to leave, tapped Mattingly on the shoulder and pointed to Joe. "That's your leader. You know that, don't you?"

"I know," said Mattingly, as Joe disappeared into the elevator.

AFTER THE MEETING, MATTINGLY WAS EVEN MORE CONVINCED THAT Joe was the guy. Joe had a special way with people—he was smart, but not overpowering; confident, but not haughty. He was beloved— by Boston locals, his employees, Massachusetts governors, members of Congress, and even the sitting president, George W. Bush. During Joe's many visits to the foundation's headquarters in Washington, there had been many occasions where Mattingly had dropped Joe at the White House after the foundation business was complete. Mattingly had watched Joe work with both Dresing, when he was the CEO, and then Beall, when he took over. His efforts spanned generations and he could work with anyone, even figures as polarizing, stubborn, and different as Dresing and Beall.

After the summit, Mattingly brought Beall to Boston to see Joe. The two men didn't know each other well, but there was mutual respect between them, and the visit was proof that Beall also wanted Joe to take the role.

Around the same time, Niehaus called Mattingly to tell him that he wanted to donate: he liked the venture philanthropy approach that the foundation was embracing, and he wanted to give a million dollars to the effort. Mattingly was excited and grateful. He called Joe to tell him the good news and that he felt that this capital campaign would be successful. But to Mattingly's surprise, Joe told him, "Don't take that gift yet. I need to talk to him. He will give more." Joe was right.

It was at that point Mattingly knew that Joe would, despite his protests, become chair of the campaign. By December 2003, Joe had agreed.

In January 2004, the quiet phase of the fundraising began. A month later, the men of the New York summit met again to plan the campaign. How many top-, middle-, and base-tier donors would they need to target during this first stage of the campaign? Base-tier donors gave between $10,000 and $100,000; middle-tier donors gave between $101,000 and $1 million; the wealthiest donors gave a million dollars and up. In July, after some six months of all the participants plying their connections, Joe ended the quiet period of the capital campaign, dubbed Milestones to a Cure. Based on the donations so far, he set a goal of $175 million by 2010. The foundation would eventually give all of this to Vertex for more R&D for the doorman and two corrector drugs, and manufacturing and phase I and II clinical trials for all three.

CHAPTER 40

Molecular Architects
of Vertex West

2004–2005

In anything at all, perfection is finally attained
not when there is no longer anything to add, but
when there is no longer anything to take away.

—Antoine de Saint-Exupéry[1]

In early 2004, the Vertex CF team had an unexpected strategy they
wanted to pitch to the Cystic Fibrosis Foundation regarding which
drug, the corrector or the doorman, to create first. Fred Van Goor, who
was leading the biological arm of the CF project, and Paul Negulescu,
who was now in charge of Vertex San Diego, suggested shifting the balance of resources toward developing the doorman drug first. The suggestion was controversial. The doorman drug alone would work for only
4 percent of cystic fibrosis patients: those who had a particular mutation called G551D. That was a tiny market. But if it was effective, Vertex
argued, they would have half of the therapy needed to treat the majority
of patients with this disease.

The first to support the idea on the foundation side was the head of
medical affairs, Preston Campbell. He agreed with Vertex's perspective,
arguing that they should prioritize the doorman because it would provide sorely needed proof of concept that this wild idea of resurrecting a
dysfunctional protein could actually work. And if the doorman failed,
they could abort the corrector program and the whole CF effort entirely.
Once Negulescu assured Beall that Vertex would keep working on the
corrector if the doorman was successful, Beall endorsed the idea.

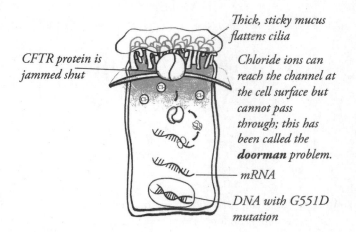

Thick, sticky mucus flattens cilia

CFTR protein is jammed shut

*Chloride ions can reach the channel at the cell surface but cannot pass through; this has been called the **doorman** problem.*

mRNA

DNA with G551D mutation

A sick airway cell with a doorman mutation like G551D. CFTR protein reaches the surface of the cell, but is then unable to open. The job of a doorman drug is to help the protein's channel, or "door," open so that chloride can pass through and help make the mucus liquid.

By the summer of 2004, Tom Knapp and Peter Grootenhuis's molecule screening efforts had yielded a hit. All the years of work they had done to date was to mark the starting line—to find a weak scaffold molecule they could build into something powerful enough to cure disease. The molecule they'd found was a mitten, and it was the job of medicinal chemists to transform it into a glove—a compound so effective that only a pill-sized quantity was required. In the molecule's current form, patients would need to take shovel-loads for an effect.

With a scaffold for the doorman finally in hand, the CF project's lead medicinal chemist, Sabine Hadida, needed chemists to work on both it and the eventual corrector, so she began hiring team members. One of the first was twenty-seven-year-old Jason McCartney. He had earned a master's in organic chemistry and then decided to go after a PhD—but three years into the program, burned out on school and academia, he dropped out to get a job instead. A hunch told him he was interested in medicinal chemistry, even though he wasn't completely sure what that was. McCartney was initially hired part-time for the corrector team, but Hadida soon saw that he was talented and brought him on full-time.

Around that time, Hadida hired Viji Arumugam, a Sri Lankan woman who had moved to Zambia with her family, then to the UK for her chemistry PhD, and finally to Tucson, Arizona. There she joined a company that was creating libraries of molecules, like the ones Knapp had screened to find a scaffold molecule. In fact, Aurora had been one of her former company's clients. Initially Arumugam joined McCartney on the corrector team, but Hadida soon moved her to Team Doorman, as the pressure to finalize that molecule rose.

In August, Vertex's head of chemistry hired Jinglan Zhou, originally from China, to join the San Diego chemists working on other projects. But when Hadida needed a senior medicinal chemist to supervise the doorman team, Zhou was happy to move into the CF group; she had worked next to Hadida at CombiChem and knew both her and Grootenhuis well. When she joined Hadida, Team Doorman was small: just two chemists based in China, charged not with making proprietary molecules, but rather chemical intermediates needed in large quantities, so that Vertex's chemists didn't have to make every molecule from scratch. Zhou's new job was to supervise them, a part-time chemist she brought with her from CombiChem, and soon Arumugam—a total of five chemists, including herself.

Hadida was careful when she hired people. She needed team players; there couldn't be any rivalry or competition around making certain molecules, or about working on Team Corrector versus Team Doorman. She needed people who took ownership of goals, not their own ideas or molecules or achievements. Such behavior would undermine the entire effort. Flexibility was also key, as the teams were fluid; chemists moved where they were needed most. And a sense of camaraderie was vital, as most of these scientists spent more time with their work family than their children and spouses at home.

As the medicinal chemists' work on the doorman launched in earnest, Hadida drew on her office whiteboard the structure of the hit that had now been promoted to lead molecule, using organic chemists' traditional shorthand notation. To the untrained eye, its black-and-white stick structures were cryptic: hexagons, with a sprinkling of oxygen, nitrogen, and hydrogen atoms, linked by single or double lines.

The scaffold molecule, quinolinone-3-carboxamide, was composed of just forty-eight atoms, and looked like a lobster. Two knitted hexagons made up the body; two single hexagons, attached to the body by a nitrogen atom, pointed in different directions, resembling claws.

The lobster-shaped "hit" molecule.

The first step in getting to know this lobster molecule was to strip it down to its essential components, in order to pinpoint which parts were critical to opening and closing the chloride channel. Hadida and her teammates chopped off one or two atoms at a time, then passed the edited molecule to Knapp's and Van Goor's people, who tested its effect on cells with the G551D mutation. Once they understood what parts were crucial to its doorman ability, Hadida and her team began altering and building upon these essential elements. Each chemist was performing artisanal chemistry, working on creating a single new molecule to test by adding and subtracting different atoms.

Knapp would take batches of newly created molecules—between ten and thirty, depending on the week—and test them at various concentrations in the cells carrying the doorman mutation. Once he had the results—graphs that revealed how chloride transport changed with low, medium, and high concentrations of each molecule—he shared them with Van Goor, and then the two of them would send the results and their interpretations to Hadida and her team. If, for example, Hadida had transformed a hexagon-shaped molecule into a pentagon, Knapp's team would test it on cells by, in essence, "asking" them which they preferred. If the cells' broken CFTR protein worked better with the hexagon, allowing more chloride to move and stimulating more fluorescence, then Knapp and Van Goor told Hadida's group to stick with that geometry for the next round of reactions. With each step, they asked the cell, *Does this work for you?* The brilliance of the light emanating from the cell lit the way toward molecules that were more and more effective at opening the chloride channel.

Knapp and Van Goor's results came every Thursday at 2 PM. After each member of Hadida's team had a chance to review the results alone, they brainstormed together in her office, arguing over the characteristics of the molecules they were creating, critically evaluating each one, and trying to figure out the next thing to try, to create a more effective molecule. With each new hypothesis, they made more atomic alterations and retested.

Some of their modifications connected the two lobster-claw hexagons, making them look like a pair of wire-rimmed spectacles. Other remodeling added a pentagon shape to a single molecule, creating a hexagon. Some of these changes improved the molecule's effect, but others had unintended consequences—like trying to make a hammer more powerful by making the head bigger, then finding that the larger head no longer fit into the space needed to pound a nail. Every time the chemists changed the structure of the molecule, they had to test its activity in the cell, then figure out the relationship between the two.

Part of the challenge was that these molecules were not flat, two-dimensional entities, as Hadida depicted them on the board. The chemists had to imagine them in three dimensions, seeing in their mind's eye how each molecule might bend and twist, anticipating which poses they might adopt as they helped open the broken chloride channel.

If Hadida and her team did their jobs well, all these little changes together would hopefully transform the weak doorman molecule into a powerful, muscular doorman drug that would open CFTR's chloride channel effortlessly and at low concentrations.

For Hadida, the job was a game: Had an alteration made the molecule a better fit—yes or no? Then she'd find the differences between each week's yeses and noes, and, together with her team, use those differences to conceive the next ten molecules. From these she'd choose the ones that seemed the most promising, and divvy them up between the chemists to make them.

Zhou, as senior chemist, was also charged with tracking the progress of each chemist on her team and helping each one troubleshoot. Sometimes a molecule looked simple on paper but then needed a complicated series of reactions to produce. Zhou usually assigned herself the most difficult molecule, thriving on the challenge and the creativity required. When she failed, Hadida was always supportive, telling her, Jinglan, if you cannot do it, nobody can.

Most of the time, for Zhou and the other chemists, frustration reigned. And 2 PM on Thursday was a particularly anxious time, as the chemists sat refreshing their browsers, waiting for Knapp's team to "publish" the weekly results. Each person was eager to see the data for the molecules they'd made the previous week. For Zhou, that data had the power to dictate her mood. Occasionally her response would be "Ohh, yeah!" More often it was "Oh God!"

Design, strategy, and planning were all needed to construct a molecule. What was the best order in which to link various atoms and molecules? Which type of chemical reactions would connect them? Certain atoms were drawn to each other, like iron to magnets; they wanted to bond. How could you keep them apart? Likewise, how could you join molecules that repelled each other, or keep them from connecting to the wrong atom?

There were multiple ways to build every molecule; the challenge lay in figuring out the most efficient approach—what reactions to force, which to block, and which to catalyze for creating the desired final molecule. Many reactions were fairly quick, requiring just a minute or so; others could take an hour or more. Sometimes the reactions were an extension of what the chemists learned in college and grad school. In other cases, the know-how to link certain atoms or molecules hadn't been discovered yet, requiring improvisation and invention. And sometimes no one, including Hadida, could figure out how to make certain molecules, so the chemists were constantly combing chemistry journals for ideas.

After each step, the chemists had to purify the intermediate molecule, dry it, and check that the substance they had just made was pure before the next step. A molecule with more than five steps could easily take several days. And each chemist was charged with making between five and ten each week, depending on the molecules' complexity. It was a packed routine, and every chemist had their way of doing things. As McCartney waited for molecules to dry or purify, he would move away from standing at his lab bench and sit at the desk reviewing data, planning the next experiments, ordering the chemicals for the next day's work, and brainstorming the next new molecules to build—always with music pounding in the background. He would run reactions that took more than an hour at the end of the day, so they would be complete by the time he returned in the morning. It was a never-ending cycle.

But as meticulous and organized as the chemists were in their designs for new molecules, the failure rate in the lab was sometimes as high as 90 percent. The chemistry simply wouldn't work: the molecules they produced were unexpectedly unstable and disintegrated, or erupted in flames, leaving behind a tiny pile of char.

Hadida guarded against this high failure rate with careful planning. She knew she would be disappointed when things didn't work, and unable to think clearly about next steps. So she preferred to do the thinking ahead of time. For each new molecule, she sketched out three possible routes on her whiteboard—like three alternate routes to a destination on Google Maps. This was science. There was no right or wrong way to build a molecule, just a whole bunch of possibilities. She invited anyone who walked into her office to contribute. It was a work in progress, a canvas that any team member could embellish.

Each week, Hadida gave Knapp's team a new batch of doorman candidates—vials containing less than a quarter-teaspoon of white or gray powder. And once all of the candidates had reached a certain level of potency, Knapp's team began testing every compound not only for its doorman potential, but for some fifteen other criteria—a list that had grown from Eric Olson's original spec list of six. At the top of the list was monogamy: it had to target CFTR only and could not cavort with other proteins in the body. It couldn't interfere with important enzymes in the liver that metabolize many common drugs, or trigger heart problems or other health issues. It had to be stable enough to remain active in the cell until a patient took their next dose. It had to be tough enough to survive a trip through the stomach and make it into the lungs and other organs damaged by the disease—but stay out of the brain, where the protein wasn't active in healthy people. Medicinal chemists had identified molecular traits over the years that lowered the chance of a potential drug sneaking through the blood-brain barrier, but Hadida's team wouldn't know for sure that their particular molecules wouldn't enter the brain until a drug candidate was tested in living animals. Finding a molecule that fit all the qualifications was like playing fifteen-dimensional Sudoku.

By fall 2004, the CF team was growing more and more optimistic that they would prevail, even though nearly everyone at Vertex Cambridge, and the industry, still thought that the CF project was outrageous and bound to fail. It wasn't uncommon for the medicinal chemists at Vertex Cambridge to laugh at the unconventional chemical structures Hadida's team was producing. The compounds didn't look the way drug-like molecules usually looked, and the chemistry used to make them wasn't traditional, either. Many of the molecules emerging from the doorman team violated the so-called Lipinski rule of five—a set of rules chemists used to evaluate whether a drug molecule was likely to be effective.[2] At a lot of pharmaceutical companies, chemists weren't allowed to make molecules that fell outside that box.

While the rules were not as restrictive at Vertex, Hadida got a lot of pressure from chemists within and outside of the company to restrict her thinking and follow those rules. She refused and followed her instincts. She, with the support of her team, was driven by the data. And the latest molecules they were creating made the cells fluoresce brighter, now meeting a key criterion known as the "positive dose response curve," meaning that as the concentration of the molecule rose, the fluorescence brightened in tandem. It showed that they were moving closer to creating a drug that could improve, extend, and maybe even save lives. Their exhilaration fueled the long hours, their fervor and expectations overriding their exhaustion.

CHAPTER 41

Rat to Man

2003–2004

Any discovery at all is thrilling. There is no
feeling more pleasant, no drug more addictive,
than setting foot on virgin soil.

—E. O. Wilson[1]

Back in early 2003, well before a hit molecule had been passed to
chemist Sabine Hadida's team, Vertex's leader of CF biology, Fred
Van Goor, had begun work on their next hurdle: obtaining the human
cells they would need to test the drug.

So far, they had been using rodent cells for their screens, but these
hadn't proved reliable enough. Mice engineered to carry the human
mutant CF protein didn't suffer lung disease; they only developed the
digestive problems. And when Tom Knapp's screening team tested hit
molecules that appeared promising in rat-cell tests on small quantities
of human airway cells provided by collaborators, just 10 percent of the
hits still worked. Clearly, there was a species difference that they didn't
fully understand.[2] Continuing to rely only on rat cells, and later living mice, could lead them down the wrong path, wasting tens of millions on developing a treatment that was great for rodents but failed in
humans. As far as Van Goor and Vertex's CF project leader Eric Olson
were concerned, they needed to confirm any hit identified by Knapp's
initial screens in *human* cells, from CF patients, *before* they could pass
it to Hadida's team to build the drug.[3]

Obtaining these cells was important for another reason. Once they
had a promising drug candidate, Olson and Van Goor were planning
an unorthodox pathway to clinical trials: testing it in human cells, then
moving straight to patients without testing in a CF animal first. That

wasn't how drug discovery was usually done. Drugs were almost always tested in animals modeling a specific disease—be it mice or monkeys or ferrets or pigs—before beginning clinical trials in humans. But Olson and Van Goor were adamant that, for this disease, that wasn't the best way forward. There *was* no animal that seemed to mirror the disease as it manifested in humans. And testing a drug in animals that didn't get the same disease wouldn't be able to provide evidence that drug was effective in the lungs. (Just because no animals were used to gauge efficacy, however, didn't mean no animals were used at all. After the drug had proved effective in rat and then human cells, it would still be given to animals to be sure that the molecule wasn't toxic and to measure its levels in the bloodstream—a step that was needed to figure out dosage for humans.)

This need for human CF cells had huge implications. Rat cells were easy to manage; they could be modified to express different CFTR mutations and then grown easily in the lab. But human cells were a limited, finite resource that came from autopsied lungs—*if* a family opted to donate the organs to research. Occasionally, a CF patient lucky enough to receive a transplant donated their diseased lungs, which were sent to one of several academic labs around the country that had mastered the art of collecting and nurturing these bronchial airway cells, growing them in the lab in the same single-layer manner the cells would grow in the airways of a living person. But human bronchial epithelial cells were precious, and their availability unpredictable.

At a 2003 quarterly meeting in San Diego, foundation president Bob Beall suggested that academic labs test the hits from Knapp's screens on human cells on behalf of Vertex. With investment from the foundation, labs in North Carolina, Iowa, and California had already mastered the art of growing these "airways in a dish." So, as far as Beall was concerned, Vertex didn't need to waste time and money reinventing the model. It was time many patients didn't have.

But while the science these labs used was reliable, it was university science—and couldn't be done with the speed and precision needed to support the rapid pace of molecule production in Hadida's team. Olson had collaborated with scientists in university labs before, and had disturbing visions of inexperienced graduate students or new lab personnel testing their precious molecules. He told Beall that because testing the drug candidates on patient cells was the most critical part of the research, the Vertex San Diego scientists themselves had to be the ones to do it. It wasn't negotiable.

Vertex San Diego chief Paul Negulescu, Olson, and Van Goor were all acutely aware of the urgency to move forward with testing in human lung cells. With chemist Peter Grootenhuis now in charge of choosing the chemical libraries that Knapp's team was screening in rat cells, more hits were emerging, and these hits needed to be confirmed in human cells. But the human lung cells they were receiving from their collaborators nationwide were both too few in quantity and poor in quality, unable to multiply robustly in the lab. Tim Neuberger, who now had the cell core facility producing dozens of cell types for all the different projects running in San Diego, suggested that lungs from a donor be sent directly to him in San Diego, and he could grow their cells in house. It was a bit of a stretch; he'd taken an anatomy course in college and worked with rodent organs, but had never touched a human one. Still, Neuberger was a modest man, not boastful or arrogant, so when he said he could handle human organs and harvest cells from them, Negulescu believed him.

However, neither Neuberger nor anyone else at Vertex San Diego knew how to dissect anything larger than a mouse. And it didn't make sense to send Neuberger somewhere else to learn: the cell core he led was responsible for supplying cells for all Vertex projects, not just CF, and he couldn't leave. So Negulescu tapped Van Goor to train with CF researchers around the country to learn how to collect the cells from donor lungs and cultivate them in the lab to create Vertex's own airways in a dish. From there, Van Goor and Neuberger together would refine the process and scale it up. So, beginning in 2003, Van Goor began regularly flying to Iowa to meet with former CF gene therapy researcher Mike Welsh, whose team taught him the art of cell cultivation. And he visited a lab at Stanford where CF researchers taught him how to dissect a lung.

By late summer 2004, Knapp had identified a few hits that he and Grootenhuis believed were worthy of promoting to lead molecule, and passed them to Hadida's team for chemical modification. And by late fall of that year, Hadida and her team had zeroed in on three doorman molecules, based on the initial scaffold, quinolinone-3-carboxamide, that were more powerful than anything they had ever seen.

The last two years of working on the project had been a grueling slog, and she had celebrated every small breakthrough, because they were very rare. But she also always braced for impact, because bad news was usually around the corner. This time, however, the news just kept getting better. The atomic tweaks that Hadida's team had made were adding up and the molecule was becoming increasingly powerful.

The best of the three molecules had a couple of features that had survived some seven hundred iterations from the original lead molecule. It had two hexagons on its left side, like a section of honeycomb. The right side of the molecule, however, had evolved dramatically. There was now a third hexagon that was sprouting two trimethyl groups; Negulescu thought they looked like two chicken feet. Those feet, it turned out, were the key: the "chicken" was ten times more potent, in Knapp's tests, than its most outstanding predecessors. The cells blazed brightly, like little atomic suns—a surefire sign that their chloride channels were open and working well.

The chicken-footed doorman, VX-770.

The chemist who had made the molecule was Viji Arumugam, and she felt as if she had won the jackpot. Anyone on Team Doorman could have made it, but Hadida and Jinglan Zhou had assigned it to her, and now it looked like a winner. Knapp's measurements suggested that the electrical activity created by the flow of chloride out of the cell matched the activity level seen in carriers, like the parents of CF patients, who had just one copy of the mutant gene. This level of activity seemed to enable carriers to have a normal, healthy life, suggesting that a drug that triggered that similar chloride movement might do likewise for those who were sick.

Now came the step Van Goor had been anticipating for weeks: testing whether this awkward-looking molecule, the chicken-footed doorman, was working. The cells he was using for this test were the same ones that Knapp had been using for his screens—rat cells with a human CFTR gene carrying the G551D mutation. The difference was that he was growing them as a single layer in a small petri dish, so that he could perform more precise measurements of the doorman's capabilities than Knapp had been able to do with the fluorescent dyes.

First, Van Goor placed the sheet of the rat cells into the Ussing chamber and measured the level of chloride ions moving through the mutant CFTR protein—practically nothing. Then he dissolved a few micrograms of the chicken-footed doorman, drizzled it on the rat cells, and placed the cells back in an incubator set to 98.6 degrees Fahrenheit—body temperature—for a few hours. When he removed the dishes and inserted them into the Ussing chamber again, he observed a dramatic surge in chloride ions moving through the channels.

Heart pounding, Van Goor immediately told Negulescu and the chemists what had happened, also calling Olson on the East Coast to share the stunning result. The whole team agreed that they had to replicate this several times before they told anyone outside the CF team. If this was somehow just a fluke, announcing it might jeopardize the whole project. After repeating the Ussing chamber measurement several times on different sheets of rat cells, the results were consistent. This doorman drug was definitely enabling the chloride to move out of the cell.

As Van Goor had been verifying his results, Knapp was consulting Olson's checklist, confirming that the molecule didn't have any undesirable qualities that harmed the cell. It didn't. The team's excitement was growing. The Ussing chamber provided *the* critical proof that the protein's chloride channel was now functional. All Vertex had to do now was transform it into a drug—figure out the right dose and mold it into an effective pill. A long, difficult path remained ahead, but Van Goor still felt a rush of excitement and optimism.

When Van Goor shared the news with Vertex founder and CEO Josh Boger, he asked Van Goor if there was a way to make the Ussing chamber data "more visual." It was, in a way, an odd request. Everything that Vertex did was about precision. Boger was a stickler for precisely quantifying the effect a drug had on a cell. And the biochemical data that Van Goor collected from the Ussing chamber was exactly that. But what Boger was asking for now was something qualitative that might provide a sense of the clinical impact. The two talked for a while, discussing what "visual" might entail.

Van Goor understood Boger's point of view. The data was compelling, but it didn't provide an idea of how the drug might alter the lung. Van Goor knew what he had to do. He needed to test whether the doorman drug that Hadida's team had made would change the appearance of cells from a person with CF. Such a visual demonstration might have the power to convince the doubting scientists and board members at Vertex Cambridge that what the San Diego team was doing might actually work.

The method that came to mind was using the airways in the dish he and Neuberger had been learning to grow over the last two years, since 2003. But in November 2004, they didn't yet have any airway cells from a patient with the G551D mutation—the one that only needed a doorman to open the CFTR protein's chloride gate. So instead Van Goor decided to create the demo Boger was requesting with airway cells from patients with the F508del mutation; because this mutation was more common, it was easier to secure small quantities of these cells.

Using techniques he had learned in Mike Welsh's lab in Iowa, he began by growing human airway cells with the F508del mutation in a nutritious broth for a few days. Then he transferred them into a rectangular plastic plate with six quarter-sized wells shaped like inverted top hats. Each hat contained a shelf made of fine mesh that created a false bottom, allowing the cells to be nourished with liquid from both above and below.

With his careful tending, the cells thrived, slowly forming a single cell layer. And as the cells grew and aged, they evolved, forming a community with tight junctions between them, just as they would inside the body. Four days in, Van Goor suctioned out some of the liquid, exposing the tops of the cells to air—just as would happen in the lungs of a baby after it was born. Air—the breath of life—was the critical trigger that pushed the airway cells to mature.

Over the next ten days the spherical cells morphed into tall, rectangular prisms. As their shapes changed, so did their identities. Some differentiated into "goblet" cells, responsible for secreting the river of mucus that trapped bacteria, viruses, and dirt before they could harm the lungs. Other cells sprouted what looked like blades of grass: the magical cilia, fingerlike projections critical for sweeping mucus, along with any trapped microbes, out of the lungs. Within a month, the carpet of cells in these little dishes had fully matured. Van Goor's airway in a dish was ready to use for creating Boger's demo.

This particular airway in a dish, cultivated with human airway cells from a patient with F508del, would typically need both a corrector and

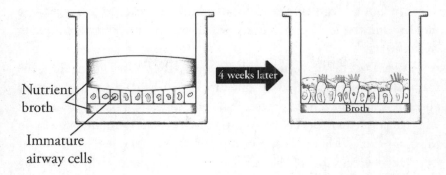

Immature human airway cells from the lungs of a patient with the F508del
mutation are grown in a nutrient broth for a few days. Then broth is removed
from the upper surface, and the cells are left to mature for the next month.

a doorman drug to make the protein work. But Van Goor used the temperature trick to move more CFTR protein to the cell's surface, eliminating the need for a corrector, so that a doorman drug was all the cells
needed.

Inside the dish, the surface of these mature airway cells was covered in a dry, hard mucus—as it would be in the lungs of children
with CF—that immobilized the cilia. Using the microscope camera,
Van Goor made a thirty-second black-and-white movie that showed
grainy dots—the airway cells—sitting perfectly still. Then he added the
chicken-footed doorman drug; dropped a couple of microscopic fluorescent latex beads, about the same size as the dust particles that can travel
deep into the lung, onto the surface of the mucus; and placed the cells
into the incubator.

Once the drug had been given time to work, Van Goor returned
to his microscope with the dish, placed it under the lens, and clicked
"record," hoping there would be a visible change. What he saw was
thrilling. The cells' surface was no longer dry, but had moistened, glim-
mering slightly; the mucus seemed to have liquefied. As he watched the
cells, mesmerized, the cilia swept the beads across the surface, like a
crowd passing a beach ball from one person to the next.

Although Van Goor wasn't using the cells that matched the first
patient group—those with the G551D mutation—this demo at least
provided preliminary evidence that the chicken-footed doorman drug
could open a broken CFTR protein and change physiology.

Van Goor, Negulescu, and Grootenhuis were excited to share these movies, and the ideal opportunity would present itself just a couple of weeks later.

Every December, Vertex held a science day in San Diego; all of the team leaders were expected to showcase their best data for one another and for Vertex's new head of research, Peter Mueller, a Bavarian scientist who had more experience[4] with drug development than anyone else in the company. Without his support, projects were scrapped.

At the time, most of the San Diego scientists were working on big diseases that were top priority for Boger and the rest of the Cambridge team; only a group of about twenty were focused on CF, and it was the last project on the day's agenda. The leaders for the anti-inflammatories group and the psoriasis, hepatitis, HIV/AIDS, and painkiller teams all took more than their share of time trying to impress Mueller. And as the hands on the clock inched closer to 6 PM, Van Goor saw that Mueller was getting antsy, checking his watch—ready to walk out and catch his flight back to Boston.

By the time Van Goor reached the podium and turned on his first slide, he had barely five minutes for his presentation on the lowest-ranking Vertex project. As Van Goor began to speak, Mueller was already rising from his seat and grabbing his bag. So with a quick glance at Grootenhuis and Negulescu, Van Goor fast-forwarded to the end of his presentation: the visual demonstration Boger had encouraged him to make showing the power of the in-development doorman drug.

Mueller glanced up and stopped. On the screen, the cilia beat back and forth rapidly, tossing around the fluorescent beads. Hypnotized by the image, Mueller put his bag back on the floor and sat down, leaning back in his chair. He recognized that this fluttering black-and-white clip wasn't just proof that the recent doorman molecules Hadida's group had designed fixed the chloride channel; it also linked a molecular change in the CFTR protein to a physical change in the airway cells. And it meant that the CF team might soon have a candidate drug ready to test in humans.

CHAPTER 42

A Christmas Gift

2004–2005

The scientific method is doing your
damnedest, no holds barred.

—Percy Bridgman[1]

Within a couple of weeks of CF biology leader Fred Van Goor's
December 2004 demonstration for Peter Mueller, the chicken-
footed doorman molecule had been "VXed"—company parlance for
promotion to an official drug candidate on its way to clinical trials.
Along with that came a new name, VX-770—VX for Vertex, and 770
for the molecule's number in the notebook of the chemist who made
it, Viji Arumugam. Once VXed, the molecule moved into the hands
of chemist Brian Bear, a new hire brought on in San Diego to scale
up the molecule's production. The quantity that Viji Arumugam had
made before was minute, less than a quarter teaspoon. This was plenty
for testing on cells, but only a small fraction of the approximately
100 grams the toxicology team in Cambridge would need to test the
molecule in animals. Those tests, where scientists monitored how the
drugs were processed in the animals' bodies, were vital for figuring out
the best human doses.

Making large batches of a compound required a different set of
skills from those needed to make the tiny quantities appropriate for
cells. For a start, everything on Bear's bench was supersized—all of the
glassware, including the flasks for brewing the reactions, was larger, as
was all the equipment. The heating plates were huge, and the mechanical stirrers used to mix the solutions looked like drill bits with propel-
lers attached to the ends. Each step took longer—especially any reaction
that had to be heated or cooled. And the process was also much more

379

dangerous; some steps used very strong acids, others released heat or gas, and the larger the quantity in production, the bigger the bang when that heat and gas were released. More heat meant increased risk of run-away reactions that could cause the compound to burst into flames or explode, destroying all the previous work. And unlike the medicinal chemists' version, the product that Bear was manufacturing had to be pure. Impurity risked causing harm to a patient that might be errone-ously blamed on the drug.

Once Bear had created the necessary quantity of VX-770 and sent it to Cambridge, the San Diego team would have no further role in devel-oping the molecule. They would only get progress reports along with the rest of the public when Vertex issued a press release—a safeguard against insider trading that was mandatory at pharmaceutical compa-nies. The transfer of their molecule wasn't unexpected or disappointing because, unlike when Hadida, Grootenhuis, and Zhou had designed molecules for past clients, this time they would be able to follow the story of their molecule to its end—whether that was into the body of a patient or the chemical scrap pile.

The development of the VX-770 doorman was now in the hands of Team Doorman's friend, colleague, and CF cheerleader Eric Olson, who would ensure that it began its journey through the animal tests. In the meantime, Hadida's San Diego team had the development of another CF drug to pour their creative energy into: a corrector. That work had been going well before the focus switched to the doorman; since then, it had been proceeding slowly, but still making solid prog-ress. Now they could focus all their attention on building a power-ful corrector molecule—the drug needed to correct the folding of the mutant protein in patients with the F508del mutation. If it passed mus-ter, it would later be added to the doorman drug for patients with the most severe form of the disease.

In 2003 and 2004, while Hadida's team had been developing the first drug, Van Goor had been learning more than how to grow airway cells in a dish. He had also learned how to dissect human lungs and collect the airway cells for cultivation. His teacher was Jeffrey Wine, a Stanford professor and one of the foundation's scientific advisors, who attended Bob Beall's quarterly meetings at Vertex San Diego. His moti-vation for involvement was personal: his daughter suffered from CF. So, beginning in 2003, whenever Wine's lab received a new pair of human lungs, he immediately called Van Goor, who jumped on the next flight to Stanford, where Wine and his postdocs taught him how to dissect the organs.

After three or four trips to Wine's lab, Van Goor had gained more confidence and began practicing on his own. He purchased pig lungs from a local butcher and dissected them in the lab as Tim Neuberger watched and learned. It was a messy, time-consuming process to separate the airways from the rest of the lungs, collect the cells, coax them to multiply in flasks, and then try to grow them into mock airways in special petri dishes. The entire process—from dissection to cell collecting to cell cultivation to a fully grown airway in a dish—took about five weeks.

Once the two men felt they could handle the entire process successfully, Chris Penland from the Cystic Fibrosis Foundation, a scientist and former student of Jeff Wine's, contacted the National Disease Research Interchange in Philadelphia, which worked with hospitals, clinics, and organ retrieval centers to procure organs and distribute them to approved researchers, and requested that both lungs of patients with CF and normal lungs (whose cells were needed as controls) be sent to Vertex.

The first one arrived unexpectedly late one evening when just Van Goor, Neuberger, and San Diego CF leader Peter Grootenhuis were in their labs. By that point, Van Goor was used to pig lungs, but when he opened the box containing ice-packed organs, he recoiled. The first thing he saw was a tongue, still attached to the trachea, which in turn then connected all the way down to the lungs. They had belonged to a woman in her thirties or forties who had been a heavy smoker—something that was immediately obvious when they cut into her lungs and saw balls of tar dotting the tissue. Energized by the delivery, Van Goor and Neuberger spent the whole night working on the dissection, practicing airway-cell collection and then beginning the difficult process of coaxing the finicky cells to grow and multiply. When Knapp and Hadida's team reached a point when they needed large quantities of human cells for testing, Neuberger hoped to be well practiced in the art of harvesting and growing them.

THE WEEK BEFORE CHRISTMAS 2004, AND JUST A WEEK OR SO AFTER Van Goor's presentation to Peter Mueller, Neuberger was walking around the University Town Center Mall in San Diego shopping for gifts when he received a call from the National Disease Research Interchange. They had just received a pair of lungs from a cystic fibrosis patient who carried the F508del mutation. Did Neuberger need these lungs?

Earlier in December, as Van Goor had been creating his doorman demonstration for Boger, Neuberger had taken over responsibility for lung dissection and cultivating airway cells from fresh organs to provide a plentiful supply for Knapp's team. This was critical because now that Hadida's group was refocusing on the corrector, they would need a deep supply of human cells from patients with the F508del mutation. Though he hadn't yet done a lung dissection all on his own, he'd been reading medical texts on harvesting cells and assisting Van Goor, and felt prepared. He'd ordered large quantities of all the chemicals he needed and bought the tools. Now here was the organ donation he'd been waiting for. He said yes.

The next day, Neuberger was attending evening Mass with his wife and son at St. Thérèse of Carmel Parish when the security guard at Vertex called. A courier had just delivered a package for him. He whispered a quick goodbye to his family and rushed from the church. When he arrived at about 11 PM, the building was deserted except for the guards. He picked up the Styrofoam box from the lobby and carried it to his lab, where he carefully lifted the lid and removed the ice packs. Inside, a set of lungs was soaking in clear liquid, wrapped within five plastic ziplock bags.

The lungs were healthier than he'd expected. They were also much smaller. Two healthy adult lungs together are about as big as a cantaloupe. These were less than half that size. A note inside the box said they were from a thirteen-year-old boy who had died from an adverse reaction to an antibiotic meant to fight a lung infection. Neuberger trembled slightly, realizing that the lungs in his hands were from a teenager exactly the same age as his son, the same age as a friend's sick daughter. *This could have been her,* he thought. He couldn't imagine what it must have been like for this kid's parents to sign the consent forms to donate his organs. They had given what was to them a sacred gift, hoping it might save other families from suffering. It didn't matter that this was Neuberger's first solo dissection; he couldn't screw it up.

Neuberger pulled up his checklist and protocols and began by preparing solutions to treat the lungs. Then he donned surgical gloves, a face mask, and a lab coat, and sterilized his instruments—forceps, scissors—like a surgeon prepping for an operation. By 1 AM, he was ready to begin. But the silence in the building was creepy, almost distracting, so he popped a disc into his CD player—the Beatles' *One* album—and hit the repeat button. It was going to be a long night.

He placed a tray of ice inside the biosafety cabinet—a sterile, glass-enclosed laboratory workbench—then put a second plastic bowl containing the lungs on top of the ice bath to keep the organs cool. He sat down on a stool and began rinsing the lungs—and that's when the ravages of disease became obvious. Though the lungs appeared pink and healthy on the outside, they didn't fill with air and float when submerged like healthy lungs. Instead, they sank, a dense package of bacteria, pus, and almost-solid plugs of hard, sticky mucus. Neuberger imagined how hard this child must have struggled to take a breath. Then he picked up the scalpel and started the dissection.

The cells he was after lined the large airways—the bronchi—which split off from the trachea and led into the lungs like two branches from a tree trunk. Neuberger held the lungs in place with the forceps as he slowly and carefully forced the scalpel along the airways, trying to separate the stiff, white, cartilaginous tubes from hard, gristly tissue. Where bacteria had destroyed healthy lung, the body had replaced it with fibrous scar tissue, and in some regions the tough material blunted his scalpel after just a few cuts, forcing him to replace the blade frequently.

For hours he wrestled with the organs, sawing tissue away until he was left with just two white-branched pipes, one from each lung. He then cut them into small inch-long sections and placed them in conical tubes filled with chemicals to dissolve the mucus. Every half hour, Neuberger replaced the rinse and by 5 AM, he was ready for the next step.

The airway cells were like a sheet of stamps glued to the inner surface of the tubes. To collect them, he treated the airways with an enzyme that sliced through the tissue matrix connecting the cells to the inner surface of the airway, and placed the tubes in a cold room—kept at 39 degrees Fahrenheit—where they would remain for the next thirty-six hours while the enzyme worked.

Now, finally, he could go home, shower, and sleep.

Neuberger returned the next evening, collected the airway tubes, and took them to his biosafety bench. There, he snipped them lengthwise, flattened them out, and then gently shot fine streams of water at their surface to dislodge the filmy sheets of cells, using just enough pressure to slide the ethereal, gauzy layer into a petri dish. It was exciting and nerve-racking—he would now be able to see whether the cells he'd harvested had survived. He placed the dish under the microscope, where a tiny lamp lit them up. There was no movement at all. Nothing. He had rinsed the mucus off these cells to release the cilia, so they should have been swaying back and forth, but these lay immobile.

He squinted through the eyepiece. Were they still glued down with mucus? Were they dead? Had he killed the cells? Distraught and breaking into a cold sweat, he stepped away from the microscope and went to grab a glass of water and think.

He returned a half hour later, ready to grab the cells and toss them out. But when he peered through the eyepiece one more time, to his surprise, relief, and delight, the cilia on the microscopic quilt of cells were wiggling, just a little. Just out of the fridge, the cells must have been too cold to move. The lamp on the microscope had warmed them up, and now he could see the cells were alive. He hadn't screwed up. With renewed confidence, Neuberger began the next step: separating this sheet into individual cells. He added an enzyme to the petri dish that destroyed the connections *between* the cells.

It took another day, but by Wednesday night, each airway cell floated freely, and Neuberger was ready to start growing them by pouring them into large flasks where they could multiply in a nutritious broth.

He and Van Goor had been struggling to grow cells—ones from their own dissections and ones they received from other labs—so Neuberger had spent a lot of time thinking about what conditions would make these airway cells happier and encourage them to multiply outside the lung. Imagining life as an airway cell, he'd remembered that in the lungs these cells sat on a layer of cells called fibroblasts. Perhaps if he grew some fibroblasts in the flask first, it would give the flask a homey, lung-like feel and make it easier for these airway cells to reproduce. So before adding the airway cells to the flasks of broth, he'd used the same flasks to grow fibroblast cells. He then removed the fibroblasts, but retained the liquid full of proteins and chemicals the fibroblasts had released. Next, he emptied all of the airway cells he had collected into the flasks, each of which contained a different combination of nutrients to figure out which mix of ingredients helped the airway cells grow best. Last, he put all of the flat, rectangular flasks into room-sized incubators temperature-controlled at 37 degrees.

Every individual's cells are unique, and there was no way to pinpoint the speed at which this teen boy's would multiply. Neuberger sampled the cells each day, checking numbers. Each flask was like its own tiny city where, if the population got too large, the environment would become toxic. As the cells grew and multiplied, they secreted various proteins that, at certain concentrations, would alter the cells' fates, changing them into other types of cells with different identities and functions. Patient airway cells, as long as they were growing comfortably in the flasks, were round. But if the broth wasn't changed at the right time, the cells transformed from spherical to pancake-like—flat, with a bubbly appearance—and this change was irreversible. So the cells required constant monitoring. When the population reached the verge of overcrowding, Neuberger divided each original flask into three, adding two additional new flasks of broth to allow the cells room to triple. The cells grew well, and by Friday evening, Christmas Eve, each set of three flasks were ready to be split again, into nine.

Neuberger worked through the night, throughout Christmas Day, and arrived back at home at 8 PM to find his wife and son waiting for him. Neuberger's wife was used to him working long hours, but this was different; he'd left her alone on a major holiday and she was visibly displeased. When he sat down on the sofa next to the Christmas tree, he looked at her. "At least I can come home and hug my son and open presents today," he told her, exhausted. "There's a family out there who is sitting around the Christmas tree without their son. So let's count our blessings."

Those words put his family's sacrifice into perspective.

After Christmas dinner, Neuberger catnapped and headed straight back to the lab to monitor the cells, a constant vigil. Within days, the sets of nine flasks had reached their population limits. Rather than risk splitting them again, he decided to be conservative; he froze the precious cells in hundreds of little tubes, which he stored in the cell bank. That way, when Knapp or Van Goor's team was ready to run tests on the evolving corrector molecule, they could defrost just a single tube.

Vertex now had plenty of F508del cells to work with, but these weren't mature cells with the same characteristics of those inside living lungs. The free-floating cells that Neuberger had frozen were "adolescents"; they were all identical, unlike "adult" airway cells, which formed a community—a collection of at least five cell types, each with their own identity and function. Some had cilia, some produced mucus, and some were stem cells that produced more airway cells. To test whether the correctors that Hadida's team made fixed the F508del mutation, Neuberger needed to simulate the structure and function of the living tissue that lined the human bronchi: he had to take these immature cells and create the same mini human airways in petri dishes he'd cultivated with Van Goor, but now on an industrial scale.

The process was never simple or predictable. While the cells from the Christmas lungs worked well, results with the next four organs they received weren't as good. Sometimes the harvest was plentiful, other times not, but Neuberger, with the help of his Vertex cell core partner Angela Kemnitzer, continued to hone the process over the course of 2005, generating plenty of model airways ready for drug testing. This was especially important because, by the fall of 2005, Hadida's chemists were developing increasingly potent correctors that needed to be tested in human cells.

CHAPTER 43

The Lucky Four

2005–2007

We look for medicine to be an orderly field of
knowledge and procedure. But it is not. It is an
imperfect science, an enterprise of constantly
changing knowledge, uncertain information, fallible
individuals, and at the same time lives on the line.

—Atul Gawande

Eric Olson, the Vertex Cambridge CF project director, knew Sabine
Hadida's team in San Diego was getting close to a corrector molecule that could also be VXed and sent to the East Coast for further
development. He also knew there was no time to waste, either on that
molecule, or on the doorman VX-770, already under his supervision in
Cambridge. There was a way they could speed up the development of
the drug so that it could be tested sooner in clinical trials. But it would
require the foundation to pitch in more money.

The reason was not that Vertex was short of money. In 2005 they had
more than $500 million in the bank. But, although Vertex was still run
by founder and CEO Josh Boger, it was a public company, and Boger was
accountable to his board of directors, to whom he had to present a budget every year. Any expense beyond $1 million, whether a piece of equip-
ment or a new research program, had to have the consent of the board.
When it came to CF, most of the board members were apathetic, some
disdainful. Until now, Vertex had invested very little of its own money,
so the board hadn't had any say in whether the project continued. That
was up to Boger. And Wall Street investors had absolutely no interest in
the program because the market was so small. Boger believed that the CF
project fulfilled what he considered Vertex's corporate mission, making

important medicines. But the project could only continue if all funding needs continued to be met by the foundation, at least until there was clinical evidence that this was a drug worth supporting.

Olson was understandably nervous when he arrived at the Cystic Fibrosis Foundation's office in Bethesda—along with Phil Tinmouth, who led business development from the Vertex Cambridge headquarters—to meet with Beall, Preston Campbell, Melissa Ashlock, and Chris Penland and share his idea for speeding up the clinical development of VX-770. Olson had conceived the idea several months earlier when he'd realized that there were a couple of stages of drug development that could run in parallel rather than one stage after the other. The animal testing required before VX-770 could be given to humans in a phase I trial would use the batch of VX-770 powder that Brian Bear had made, which could be dissolved and given orally as a liquid. This liquid form would also be given to healthy volunteers in the phase I trial and to patients in phase II. But before moving on to phase III, an entirely different set of chemists in Cambridge would still need to figure out how to regularly manufacture large batches of VX-770 and then package it into pill form—the form patients wanted. And this step was where they had an opportunity to speed up the process.

At the time, most pharmaceutical companies waited until the end of phase II before investing more time and money in determining how to manufacture the drug on a large scale. That was because this manufacturing step was expensive and time consuming, taking a year or more, and it didn't make sense to waste resources figuring out how to make a worthless drug. But this way of doing things meant there was often a significant lag between the second and third phases of clinical trials. If Vertex could complete these manufacturing steps in parallel with the preclinical animal studies, Olson reasoned, then the trials could proceed one after the other without delay. It was a gamble, because there was also the risk of the drug failing in phase II, but if VX-770 worked, it might save them a full year or more. He just needed an extra $20 million to do it.

It wasn't an ask Olson was anxious to make, but the request for additional funds was unavoidable. And while Olson was nervous—Beall provoked that reaction in everybody—he was also excited. After all, they now had a molecule to develop, and thanks to his efforts creating fertile ground in Cambridge, they might be able to save more lives by getting it into the hands of patients sooner.

Olson had alerted Ashlock to their impending visit so she could break the news to Beall that they wanted more funding, and when they walked into the fishbowl conference room on the foundation's second floor, Beall was furious. His face was bright pink, a color that was accentuated by his ring of white hair, and he was pacing next to the conference table like an aggravated rhinoceros. The foundation's original model, Beall's model, was that they would fund the discovery phase, creating a zero-risk project for Vertex. But then, once a promising molecule was discovered, Vertex was supposed to pay for its development: large-scale manufacturing, drug formulation, animal testing and toxicology, and the phase I, II, and III clinical trials. After meeting with Vicki Sato and Josh Boger in late 2001, the foundation had agreed to revise the agreement to share the cost of early clinical development— money Joe O'Donnell was in the midst of raising, having launched the public phase of the Milestones to a Cure campaign in 2004. Now here was Olson asking for more?

Beall felt the request was a shakedown, a violation of their understanding, and tore into Olson about trust and betrayal. We've lowered the risk enough, he told Olson. You should be able to pay for all of this now. Olson had worked with Beall for more than four years; he'd seen him angry and had been mentally prepared for him to go off the rails. Tinmouth, however, had never met Beall before, and blanched as the man unleashed a torrent of frustration, not just with the request but the timing. Beall had just finalized the 2006 budget with the foundation's board, never an easy task.

Olson quickly realized his approach had been wrong, and that now Beall felt backed into a corner. He needed to reframe the discussion, emphasize the opportunity rather than the cost, and make Beall feel like he was making a choice. He explained that Vertex could proceed the normal way—do all the animal studies in rats and dogs and the initial phase I and II human trials, wait on the outcomes, and then try to figure out how to make large batches of the drug. But that would take a couple of years. If instead, they ran the animal tests in parallel with the chemists starting work on the large-scale manufacturing, then they could save a year or more—and who knows how many lives. Olson suggested Beall hire consultants from the pharmaceutical industry and ask them to vet the plan. He didn't mind. In fact, he expected Beall would do his due diligence to make sure this wasn't just a ruse to squeeze more funds from the foundation.

To his surprise, after a couple of hours of sitting and talking through the idea with the rest of his team, Beall—either satisfied or worn out—flipped open his laptop and motioned Olson to pull up a chair, saying, "Okay, let's make a slide I can show the board tomorrow." He knew Olson and trusted him. He trusted his judgment that running the animal testing and manufacturing in parallel was the most direct path to a drug.

As Olson and Beall wrapped up the budget slide and the discussion, caterers brought in sandwich platters. Beall looked at the food, then at Olson. "I've got to get you guys $20 million and I have to feed you, too?"

Olson had won Beall over, but his job was not done. When Olson returned to Vertex, he also had to sell the plan to Boger and Sato and the heads of the divisions who would run the animal studies and the large-scale manufacturing. He hadn't told any of his colleagues at Vertex about his plan to run the work in parallel yet because he knew that once he already had the foundation's money, it would be harder for Vertex to refuse.

Olson focused on the upside as he met with his still-skeptical colleagues in Cambridge: "We were going to do the animal tests and the manufacturing, but now that the foundation is paying half the costs, Vertex can do them in parallel for half the price." Who didn't love a bargain? Olson's psychological play paid off; Vertex agreed.

The only thing that motivated Olson was getting a drug to sick kids sooner. Living in Cambridge, he had become connected to many families in the Boston area who had critically ill children walking the line between life and death every day.

Meeting the families and children had also become a powerful motivator for the scientists in the Cambridge labs, to whom CF was just a new disease with a weird molecule that had only been tested in cells. And now that the molecule would be in his Cambridge development crew's hands, an excited Olson wanted to lead a Vertex team in a Cambridge-Boston Great Strides walk, a fundraiser the San Diego team participated in yearly. But he was nervous about the company's commitment to the disease. If the doorman drug proved toxic in animals, Vertex might drop the molecule before ever testing it in humans. Then Vertex's participation in the walk the following year would be a glaring reminder of their failure. So he decided to walk in 2006 with just a few friends. If the drug was safe and passed the phase I trials, he would make sure that a large team of Vertexians, dressed in the company's royal-purple T-shirts, joined the fundraiser the following year. In the meantime, Olson's scientific team had to figure out how to make a form of VX-770 work not just in cells but living people. Unfortunately, that was proving more difficult than anticipated.

ONE OF THE PROBLEMS WITH VX-770 WAS THAT THE WHITE POWDER Brian Bear had created was a very stable crystalline form that didn't dissolve in water: if you gave the powder to a human being, it wouldn't dissolve in their stomach juices and enter their blood. And the San Diego team knew it. To test the molecule on the cells in San Diego, Knapp's team had first dissolved the powder in a clear liquid called polyethylene glycol, then added the solution to the cells. Polyethylene glycol was already used in a range of medical products and was safe for humans, but it had its share of side effects—diarrhea, nausea, stomach cramps, and flatulence—that made long-term use undesirable.

Still, the molecule was so powerful, so good at opening the CFTR door, that they chose to ignore the solubility issue and just focus on increasing its doorman potential. After all, there were lots of tricks chemists could use to help a molecule dissolve. They could figure out one with fewer side effects once they knew the drug worked in people. In the meantime, the polyethylene glycol–doorman solution was fine for animal testing. Vertex could easily feed it to healthy rats to learn whether the molecule made it into the bloodstream, and in what quantities; how long it lasted; and whether it caused any nasty side effects.

Once the drug was proved safe in rats, in May 2006 Vertex announced they would begin the first human trials.[1] The company planned to test the drug in some twenty individuals, most of them magnanimous healthy volunteers. The plan was to give these first recipients a liquid dose of VX-770 just as they had done for the rats, from once to multiple times a day, for fourteen days. Vertex sent the medicine to care centers participating in the phase I trial, where physicians running the study administered the drug at low, medium, and high doses to track how well each dose was tolerated and how long it lasted in the body. Did the drug cause an adverse reaction not foreshadowed in the rats? How well did healthy people tolerate single versus escalating doses of the drug? How did they feel on low versus high doses? How much VX-770 made it into the blood?

Once the healthy volunteers had completed the initial phase I trials in late 2006, and no one had suffered any side effects, trial-site physicians gave a single dose of the drug to a handful of relatively healthy CF patients with the G551D mutation, identified through the registry, to test whether the drug was safe for its intended patients. This was to be sure that their bodies processed the drug in the same way as healthy people's did. Once no significant reactions to the drug were found in those patients, either, the company began planning phase II.

While the clinical team was conducting the phase I trials with the doorman–polyethylene glycol mix, the chemists were still battling the insoluble nature of the molecule. It was a common problem in the pharmaceutical industry, and most companies chose to abandon such troublesome molecules and just start the discovery process again, hoping that the chemists would discover a more cooperative compound. But it had taken the company more than eight years to develop this first candidate drug and Vertex had no intention of starting over. VX-770 was very effective, and also very stable, another appealing quality; it would last on pharmacy shelves. They just needed to figure out how to get it to dissolve in patients' stomachs.

At the atomic level, the VX-770 powder that Brian Bear had made in San Diego was arranged in a regular lattice, creating a stable crystal. However, the chemists knew that VX-770 could be arranged in other forms, ones that lacked the rigidity of a crystal while remaining stable enough to sit on the pharmacy shelf for long periods without decomposing or unexpectedly transforming. The chemists worried that the wrong form of VX-770 would transform into an impotent crystalline state that wouldn't help the patient. They were right to be concerned; some forms of VX-770 appeared like powders when first made, but when the chemists mixed them with fluid simulating stomach acid to test how the drug fared in the gut, the powder immediately crystallized, refusing to dissolve, which rendered it useless.

Over the next year, the chemists created twenty-five different versions of VX-770. Each form had its own distinct appearance and properties. Some powders were gray, others white; some were stickier than Bear's original. Some were classic crystals; others, amorphous; still others were salts, made from a combination of VX-770 and another molecule latched together in a regular lattice. Making these forms was a little bit science and a little bit mystery and magic; some seemed stable but then, for unknown reasons, couldn't be replicated.

The purpose of this exhaustive effort was to find the version of the molecule that would work best as a tablet for patients. So Vertex's materials discovery group made as many forms of VX-770—crystalline and amorphous—as possible.

By the fall of 2006, they decided to embed VX-770 in a polymer, which would keep the molecule in an amorphous powdery form. When compressed into a pill, the VX-770-polymer combination passed all the tests: it was stable during manufacturing, on the shelf, and also in simulated stomach acid.

The drug was now ready for use in time for phase II trials—earlier than expected, thanks to Eric Olson's planning, and in the form that patients desired.

THERE WAS, HOWEVER, STILL ONE PROBLEM THAT MADE EVERYONE, especially Olson, nervous. So far, VX-770 had only been tested in rat cells carrying the human CFTR protein with a G551D mutation—never in *human* airway cells from a patient with the G551D mutation. With no animal model available, it was important that they at least test the drug in human airway cells before beginning the phase II trials and giving the drug to patients. Cystic fibrosis was a rare disease, and finding cells from CF patients with an even rarer mutation was tricky. But the foundation's head of medical affairs, Preston Campbell, had a hunch where the Vertex San Diego team might be able to secure some. He suggested Van Goor reach out to Joseph Pilewski, a pulmonologist at the University of Pittsburgh.

Pilewski had mastered the art of harvesting and growing airway cells from the diseased lungs of transplant recipients, work that the foundation had funded since the mid-1990s. Prior to 1995, these ravaged lungs had been tossed in the trash; when the foundation realized their value, they began paying for these precious organs to be sent to labs, like Pilewski's, where the mutant cells could be collected, grown, and sent to CF researchers worldwide. Pilewski's lab freezer contained several vials of airway cells that came from a CF patient who carried the G551D mutation. He packed the frozen tubes in a Styrofoam box filled with dry ice and mailed them to Van Goor, who defrosted them upon receipt and grew them in flasks of nutritious broth for a week to increase their numbers before transferring them into the wells of a rectangular plastic plate for maturation, just as he had done when creating the demo he had showed Peter Mueller in that head-turning meeting in 2004.

As before, the cultured layer of airway cells looked, to the naked eye, shiny and dry—almost crusty. Even though the cells secreted mucus, just as they would have in a living person's airway, that mucus was dehydrated and sticky. The cilia that should have been swaying back and forth in a sea of watery mucus were instead matted down by it, glued in place. But once Van Goor added a bit of VX-770 to the cells and put them in the incubator, the result was even more electrifying than what he'd seen previously with the F508del cells. The surface was wet, and the little cilia were now swaying back and forth as if someone had awoken them from a trance.

As he watched the cells, mesmerized, one thought filled his head: *The molecule works.* It was all very well to know the science—that if you fixed the misbehaving protein, allowing chloride to flow, in theory the cell should function normally—and to prove it with Ussing data. But it was something else entirely to see a cell with the same mutation as the patients in the imminent phase II trial physically responding to the drug.

Van Goor found Paul Negulescu and brought him in to take a look. Negulescu had experienced that high of discovery before, but what he felt now was much stronger. This was a flash—a jolt—of amazement and emotion at watching these sick cells become healthy and function normally in the span of minutes, all because of a molecule that this small team of scientists had designed and made with their own hands. It was the feeling of intimately connecting with nature.

Van Goor immediately sent a movie of the quivering airway cells to Olson, who began tearing up from joy and relief the moment he saw it. Here, finally, was some undeniable good news. Just months before they were scheduled to begin giving this drug to sick children in a clinical trial, they had clear proof that the chicken-footed doorman could fix the mutation.

This was a window into the lung. And their bridge to the clinical trials.

CHAPTER 44

The Doorman Launches the Era of Genetic Medicine

2007–2008

Two molecules that differ only by a few atoms may
have very different effects in the body. One may be
a lifesaving drug while the other is a poison.

—Mark Murcko[1]

The first CF patient to receive a dose of the VX-770 tablet in the phase II trial was a young man named Bill Elder Jr.—one of the 4 percent of patients with the G551D mutation who only needed a doorman drug to make the CFTR protein function.

Bill was born in Omaha, Nebraska, in 1987, and lived there until his family moved to Genesee, Colorado, a small town of 3,700 nestled in the foothills of the Rocky Mountains. He could see the mountains from his house, and herds of elk would come by regularly and block the driveway when he was on his way to school. He spent his childhood running and playing outside in the thin mountain air, the exercise keeping him fit, and so his lungs were stronger than they might have been had he lived somewhere else. He had no health issues until around age eight, when he started suffering from intense abdominal cramps that forced him to collapse on a couch or bed or pile of pillows to quell the pain. He began suffering from severe constipation and generally felt sick.

His parents took him to a range of doctors who did tests for anything that might trigger stomach cramps: lactose intolerance, Crohn's disease, celiac disease. The trek to various doctors continued for months until one physician noticed his clubbed fingernails—nails that grew almost dome-shaped from their nail beds. It was a sign of low oxygen levels that was linked to chronic lung disease. A scheduled sweat test almost immediately diagnosed cystic fibrosis.

Bill's family was referred to Dr. Frank Accurso, who ran the CF center at Children's Hospital in Denver. There, they met a team of experts who walked them through what it meant to have a child with this disease. Bill could see that his parents were sad and overwhelmed. He had a brother, Andy, five years younger; he would need to be tested, too. But Bill was happy that someone had figured out the cause of his bellyaches and that there was a pill that would make the pain go away. Because the disease had blocked his pancreas with mucus, he didn't have the enzymes he needed to digest his food. The pills were a substitute. He just had to take one every time he ate. No big deal.

When Accurso walked into the room, Bill had no idea that this man would become a lifelong mentor and the physician on whom Bill would model himself a couple of decades later. Accurso warmly addressed eight-year-old Bill first, and then the rest of the family. Accurso explained to Bill that the best way for him to remain healthy was to take charge of his disease. Rather than allowing his parents to seize the reins, he needed to administer his own medications and commit to daily exercise. If Bill was the one responsible for taking his pills (knowing what each one did), exercising, and doing his therapy, then he had a much better chance of living a long, healthy life.

Far from filling the young boy with dread about this disease, the knowledge and connection with Accurso empowered him. Back home, the whole family participated in his therapy. A new device, a vibrating vest designed to loosen lung mucus, was now available, and Bill used it to supplement the chest therapy that his dad practiced. His dad always made sure that it was fun, combining it with reading, movies, or hanging out.

A year after Bill's diagnosis, his mother was admitted to the graduate school of business at Stanford University, and the family moved to Atherton, a town near Palo Alto, flying back to Colorado for visits with Accurso. From then on, the family moved every few years as Bill's mother was relocated. After California, it was back to Colorado for a while, then Florida, then Southern California, and back to Colorado. At various times Bill was hospitalized when he caught a virus or bacterial infection, and was forced to spend time hooked up to an IV for a dose of antibiotics. Even with visitors, it was horrible and isolating. By the time he began college at Stanford University, his biceps were scarred from the PICC lines used to deliver long-term intravenous antibiotics.

Bill's first year at Stanford was rough, because he was determined to get the full college experience. He studied hard, socialized hard, and slept less than he used to—a schedule that left little time for therapy. He got sick more frequently in the dorms than he had at his parents' home and lost about a half liter of lung function just during freshman year. By the end of the year he was very sick and required outpatient IV antibiotics, though he escaped a stay in the hospital. Recognizing he had messed up, he spent the summer getting healthy. He was determined to make amends for neglecting his lungs all year and began exercising vigorously, with intense sprinting, interval running, and lots of therapy. He moved to his own apartment to avoid exposure to the infectious agents that were fixtures in the dorms. Exercise helped him bounce back from illness and he entered his sophomore year in good health.

While he was sitting in a biology class in the spring of 2006, he received an urgent text from Dr. Accurso: "Bill, I need to talk to you as soon as possible, 'cause we've got something amazingly groundbreaking. It's a new trial; please call me whenever you can."

Accurso had been studying cystic fibrosis since 1978, when he began his fellowship in pulmonary medicine, and in 1987 had taken the helm of the CF center in Colorado. There, he worked on newborn screening with Phil Farrell and studied inflammation in the lungs of CF infants.[2] In 2000, Preston Campbell had asked him to work with Cystic Fibrosis Foundation Therapeutics, Inc.; the new nonprofit drug discovery and development arm of the Cystic Fibrosis Foundation needed physician-scientists like Accurso, steeped in the physiology and science of the disease, to review industry proposals. Campbell set up a joint steering committee with Accurso and another handpicked expert, a couple of industry-chosen partners, and a fifth individual appointed jointly. In 2001, from his position on the committee, Accurso had a bird's-eye view as Aurora became Vertex, and he had followed the evolution of VX-770 from the start. He was the only practicing physician on the steering committee and attended meetings several times a year, contributing by critiquing Vertex's presentations, assessing their progress, and sharing his expertise about what else was possible and had been done in other studies. When it came to choosing a lead investigator for phase II of the doorman trial, Accurso was the natural choice, and to him, Bill Elder was a perfect patient to try the new medicine.

Though it was just a text, Bill could sense Accurso's excitement. So he called him back the moment class was done. There was a new medication, VX-770, that looked like it could correct the basic defect in CF, Accurso told him, but only for people with Bill's mutation. The phase II clinical trials were starting soon; Accurso said, "Drop your summer plans and come join the trial."

THE PHASE I TRIAL FOR VX-770 HAD PROCEEDED QUIETLY AND CAU-tiously, with most Vertex employees unaware of the drug's progress. Without fanfare, it marked the beginning of a new era in medicine. Not only was this the first drug that targeted the very root of a genetic disease—for cystic fibrosis, the malfunctioning CFTR protein—but, for the first time in the history of medicine, too, a drug had been designed for patients with a specific genetic mutation. If it worked at all, then by itself it would only benefit a select few with the right mutation—patients like Bill. But the hope was that if the drug was effective, it could be combined with a corrector to create a medicine that would work for almost all—nearly 90 percent—of CF patients.

For phase II, finding the right group of patients was absolutely critical. If they gave the drug to a patient with a mutation that didn't match VX-770, then it would have no effect and the trial and the molecule would fail. Beall had promised Sato and Boger a couple of years earlier that the foundation would facilitate the clinical trials by identifying the right patients. But that was only possible if the foundation knew which mutations the patients carried. In 2006, the registry that Warren Warwick had established in the 1960s had logged the mutation types of almost 87 percent[3] of the CF patients who visited the care centers. This was how the foundation knew that close to 90 percent of patients in the US carried one or two copies of the F508del mutation. But VX-770 was not designed for them, and to ensure the remaining trials would have enough subjects, Beall urged physicians to determine the mutations of the remaining 13 percent of patients before Vertex was ready to begin phase II.

After the 1989 discovery of the F508del mutation in Tsui's Toronto lab, many geneticists had begun examining the DNA of patients who clearly had CF but lacked the F508del mutation. One of those research-ers was Garry Cutting, a physician-scientist at Johns Hopkins School of Medicine in Baltimore. He was looking not only for other mutations, but also for modifier genes that somehow counteracted the CF muta-tions and enabled patients with two copies of one of the most severe mutations, F508del, to live longer with relatively mild disease. Francis

Collins once remarked that Doris Tulcin's daughter Ann must have had powerful modifiers; despite having two copies of the F508del mutations, she was still in relatively good health, swimming every day in her late sixties. After the discovery of CFTR was published in *Science*, Cutting began working closely with Lap-Chee Tsui's lab in Toronto, and together they discovered a handful of additional mutations that caused CF, including G551D.[4] Sixteen years later, the foundation needed to find as many people as possible who carried this super-rare mutation to keep the trials well stocked with patients.

Phase II was a much larger, more complicated endeavor than phase I, and included statisticians, clinical pharmacologists, safety specialists, and the clinical operations team from Bonnie Ramsey's Therapeutics Development Network, who were responsible for selecting test sites. TDN physicians advised Vertex on how to design these trials—a challenging feat for a rare disease, which Vertex had never worked on before. The first issue was recruiting enough subjects. Of the 30,000 cystic fibrosis patients in the US—already a small number—only 1,200 carried the G551D mutation. Of these, an even smaller number were the right age and in good enough health to participate. To enter the trial, a volunteer had to have cystic fibrosis with at least one copy of the G551D mutation, and be at least eighteen years old. The participants also had to have a lung function greater than 40 percent of that predicted for someone of their age, sex, and height without the disease.

As the trial leaders applied each criterion, the number of possible trial participants shrank. Phase III, if the drug got that far, posed an even larger problem. The phase III trial of Pulmozyme, the mucus liquifier, involved close to 1,000 patients; a trial of that size wouldn't be possible with this drug and population. And 1,000 was still on the low side for phase III clinical trials; some included 3,000 or more patients.

Recruiting patients wasn't the only obstacle Vertex faced. This trial's "endpoints"—the factors used to assess whether the drug was working—had never been used before in any clinical trial. With a cancer drug, imaging methods could be used to see whether tumors had shrunk or spread. The impact of an HIV/AIDS drug could be gauged by measuring whether virus levels in the blood rose or fell. Whether a diabetes drug was effective could be calculated from participants' blood glucose levels.

The first endpoint for VX-770 was the sweat chloride level, obtained through the classic sweat test used to diagnose CF patients since its discovery by Paul di Sant'Agnese in the late 1950s. Foundation drug development leader Melissa Ashlock had been thinking about clinical endpoints for research on cystic fibrosis since her gene therapy research days, and she knew that this first endpoint came with a problem: every center did the sweat test a little bit differently. Some, like Boston Children's, had their own homegrown method. Other centers used a commercial device to collect the sweat—but the readings were often inconsistent. But she and Frank Accurso, the trial's lead investigator, insisted that the sweat chloride levels be measured despite these challenges. It was the gold standard diagnostic test for the disease. If the drug worked, he reasoned, the patients' sweat might be less salty.

It was a controversial idea—after all, what if the drug worked but didn't alter the saltiness of the sweat? The curative powers of VX-770 had only been observed in cells in a dish. And there was no good in vitro model for sweat testing. But regardless of these concerns, Ashlock and Accurso developed a single sweat-collection protocol to use across all phase II trial sites and figured out how to train staff so all tests were done the same way, before being sent to Accurso's lab for analysis.[5]

The second endpoint, the nasal potential difference (NPD), was a research tool that Mike Knowles and Rick Boucher had first used to measure the electrical activity in the nose back in the early 1980s. Ashlock and Ron Crystal had used the same approach during their gene therapy trial, so she knew the procedure, as well as the challenges it presented. Taking the measurement was finicky, and required instrumentation that took three to four hours to set up. The test itself lasted about thirty minutes. And just like the sweat test, each center did it differently. To address this, Ashlock worked with a team led by J. P. Clancy, a physician-researcher at the University of Alabama and a leader on several of the foundation's committees, to visit each center to train and certify the staff in this technique. The NPD readout was a squiggly line on a strip chart recorder, and during the phase II trial, all test readouts were mailed to Clancy's team in Alabama for analysis.[6]

The third and last endpoint, the FEV1—the volume of air that an individual could exhale in one second—was the least novel, and most straightforward. This measure of lung function was used for many pulmonary diseases, and the test was done consistently across trial sites. It was sent to a data capture system at Vertex, later to be supplemented with analyzed data from the other two endpoints.

The sweat chloride and the NPD were direct measures of whether VX-770 fixed the CFTR protein. The FEV1, by contrast, was a reflection of how *well* CFTR was fixed. It was a measurement directly correlated with the patient's health and the state of their lungs. Patients barely noticed their abnormal sweat and NDP, but they all knew their FEV1. An FEV1 of less than 40 percent meant they should begin considering the possibility of a lung transplant; a patient with an FEV1 of 20 percent was skirting the line between life and death.

Ashlock was certain that, together, these three endpoints would reveal whether the drug was working. The sweat chloride test would show whether the CFTR in the sweat gland was working and making sweat less salty. The electrical activity in the nose would show whether the CFTR gate was opening, allowing chloride to flow in the nasal passages—and so, hopefully, also in the lungs. The FEV1 would reveal whether lung function was improving.

With the endpoints determined, the patients selected, and the nurses and physicians at the trial centers trained, the first part of phase II commenced in spring 2007.

JUST AS DR. ACCURSO HAD INSISTED, BILL DROPPED HIS PLANS FOR THE summer of 2007 and flew to Denver, where he entered VX-770's phase IIa trial at Children's Hospital—the first and smaller of two planned phase II trials, primarily designed to test for the most effective dose. The study was double blind, which meant neither the doctor nor participant knew whether the patient was receiving a placebo—a sugar pill designed to look like the real drug—or a low, medium, or high dose of the drug.

This trial lasted for fourteen days. Every day, Bill took a pill in the morning and a second in the evening, twelve hours later. On days seven and fourteen, he returned to the CF center so that the trained physicians there, responsible for charting the drug's safety and noting any adverse effects, could measure his lung function, test his sweat, and measure the voltage in his nose, which Bill considered brutally uncomfortable. The lung function measurements were sent directly to Vertex. The sweat collected in the sweat test was sent to Dr. Accurso in Colorado (who, despite recruiting Bill and others to the trial, had recused himself from seeing patients who were enrolled to avoid any actual or perceived bias as the trial's principal investigator). NPD readings were sent to Dr. Clancy in Alabama.

After the two-week study, Bill was disappointed. If he had received the drug and not the placebo, it didn't seem to be having any effect. The phase IIa trial took place across about a dozen sites, each one enrolling just one or two patients—too few to indicate to the patients or doctors involved whether the drug actually worked. Bill was the first of the twenty CF patients in the phase IIa trial and had no contact with other participants, so there was no way to know for sure whether others had felt an effect. All he knew for certain was that he didn't feel any different. Perhaps the drug was a bust.

The physicians who collected the sweat had no idea what their patients' chloride levels were after the treatment, because Accurso analyzed all the samples in Colorado. Likewise, they were unaware of the voltage-measurement results, because only the folks at the University of Alabama knew how to interpret the seismographic readout. The point was to keep everyone in the dark, to avoid jumping to conclusions about whether the drug was working and generating false expectations in the CF community.

Once all of the measurements had been analyzed (if needed) and then sent to Vertex, a team there "cleaned" the data, scanning for results that looked like typos or errors—salt levels or voltages that were biologically impossible. When the data looked reasonable, roughly four weeks later, this same team locked the dataset—no further changes could be made. Then a team of biostatisticians at Vertex accessed the clean data and crunched the numbers, a process that took another couple of weeks.

Robert Kauffman, as chief medical officer at Vertex, was first to see the phase IIa data. But Kauffman was an expert on viruses, brought on for the hepatitis C drug that Vertex was planning a phase III trial for in 2008; he wasn't familiar with chloride levels and voltage measures and didn't know what to make of the data. So he called Claudia Ordonez, formerly a pediatric pulmonologist at Boston Children's Hospital, who had recently joined Vertex as the medical director for Vertex's CF program, to help interpret—do you think there's anything here?

Ordonez looked at the numbers. She had no idea who received the drug and who got the placebo; she could only see the values for each unnamed patient's sweat chloride, FEV1, and NPD before the patient started treatment, during treatment, and after two weeks of treatment. But what she saw made her shake her head with disbelief and sit down to read through all of the data a second time.

A few days later, as Ordonez was in her office preparing the data and charts to present to a small group of Vertex leaders,[7] Olson walked past. She waved at him with a big smile. "Do you want to see the data?" He nodded eagerly and stepped in, pulling up a chair and reviewing the neatly assembled tables. At this first stage of phase II, all Olson wanted to see, all he expected to see, was proof of safety, not function—that the patients tolerated the drug and no one was harmed. He didn't anticipate therapeutic results after patients had been on the drug for just two weeks. These patients had been sick their whole lives; a drug couldn't reverse that in just a couple of weeks. Or so he thought.

When he looked at the sweat chloride, he was stunned. "It's remarkable," Olson whispered to Ordonez, as if speaking too loudly would change the results. As he ran his finger over the numbers slowly and carefully, he could barely believe what he was seeing. In just fourteen days, the drug's impact on patients was potent. Their salt/water balance had normalized and sweat chloride levels had fallen. All the participants had begun the trial with an average sweat chloride of about 100 millimolar, a sign that their CFTR was barely functioning. After just two weeks of taking the drug, chloride levels had dropped by between 30 and 50 points in all ten patients who received VX-770—regardless of dose—to between 45 and 75 millimolar.

Olson teared up. Could this molecule actually be doing what it was supposed to do? Salt levels above 60 were diagnostic of CF. Between 40 and 60 were inconclusive, and below 40 was normal. After treatment, some of these participants wouldn't have even tested positive in a screen for the disease. This was vital proof of concept that validated the leap of faith Beall and the team at Aurora had made ten years earlier, that a small, engineered molecule could fix this protein and treat the root of this vicious disease.

Most exciting was that those on a higher dose of the drug had a greater drop in salt. It was as if the more VX-770 was in these patients' bodies, the more malfunctioning CFTR channels it was able to fix. Olson had spent years dreaming of maybe triggering a 5- or 10-millimolar drop in sweat chloride. Now, sitting next to Ordonez, he was seeing results that were up to seven times better than that. The patients also had better lung function, drawing 12 percent more air into their lungs than before.

Ordonez presented her findings to CEO Josh Boger, president Vicki Sato, and head of research Peter Mueller. Olson sat in as well, and Fred Van Goor, who was visiting the Cambridge headquarters for meetings,

was also invited to join. The CF project still wasn't getting a lot of support at Vertex, even though they had started phase II trials. Van Goor's
life had been focused exclusively on the CFTR protein since he'd come
to Vertex seven years ago, but as far as most of the company was concerned, the project remained under the radar, quietly puttering along.
Boger had been instrumental in keeping it that way, to avoid drawing
too much attention from the board. Vertex chemists were still skeptical that, even if you could craft a molecule that would make CFTR
behave normally, the drug would actually stop the relentless march of
this deadly disease and help patients feel better. It had been a controversial decision to move resources from the corrector to the doorman, to
test whether fixing CFTR would lead to any improvements in health—
which is why the results of this clinical trial with VX-770 were so earth-
shaking. Now, much, much sooner than they hoped, it looked like the
drug was working.

Sitting in a corner room in the old Cambridge building, Ordonez
put up her first slide, showing the trial subjects' initial sweat chloride
levels. When she flicked to the next slide, the chloride levels dropped
like a rock. The slide after revealed another surprise—the restored electrical activity in the nose—generating rising murmurs through the
room. The next one showed how much more air patients could inhale
into their lungs.

When Van Goor saw the data, he knew that the theory they'd
pinned their hopes on was correct: they *could* resurrect a damaged protein, and make it work again in those struggling with the disease. He
knew Vertex had to keep going. They had to finish the still-in-process
phase IIb trials, complete a phase III, and get this medicine to patients.
This was the proof they needed to truly believe that the work they were
doing—not only with the doorman drug, but also with corrector molecules back in San Diego—could save lives.

After Ordonez's presentation, the attitude of Vertex's top executives
toward the CF project shifted. Finally, the board of directors had the
evidence they needed to change their minds about this project; the company was ready to invest their own money in clinical development. But
for Van Goor, the phase II data was more than just a sign of a promising medicine. This new drug would not only transform cystic fibrosis
treatment, Van Goor knew, but also disrupt the entire pharmaceutical
industry's strategy for developing new medicines. It was one of the most
important pieces of data in the history of cystic fibrosis. There was Paul

Quinton's discovery of the malfunctioning chloride channel. There was the discovery of the CF gene. And now there was this: the discovery that fixing CFTR's broken channel with a synthetic molecule could change a patient's entire physiology, from the saltiness of their skin to the volume of air in their lungs.

CHAPTER 45

Disruption

Fall 2008

Like all sciences, chemistry is marked by magic moments.
For someone fortunate enough to live such a moment,
it is an instant of intense emotion: an immense field
of investigation suddenly opens up before you.

—Yves Chauvin[1]

Seeing the results of the phase IIa data was one of the most exciting moments in CF project leader Eric Olson's career. He'd been in the pharmaceutical industry for twenty-five years and knew how rare it was to create a drug able to target the precise cause of a disease—in CF's case, a malfunctioning protein. But he also knew he had to calm down. Though early results were promising, Vertex was only halfway through the phase II trial.

Fifty percent of drugs fail in phase III, most commonly because they underperform and don't meet their endpoints.[2] But Olson wasn't worried about that. So far, VX-770 had met all three of its endpoints: lowering the saltiness of sweat, boosting the transit of chloride in and out of cells, and improving lung function. Olson was more concerned about unexpected, serious side effects. The first half of phase II had been small—just twenty patients. A bigger study could reveal an ugly side of the drug—an unanticipated or possibly deadly reaction. But at least Olson didn't need to fret about the final reason drugs failed: competition from a better product on the market. Vertex was well ahead of the few competitors (which the foundation was also funding) daring enough to try and make a drug for this disease.

Because the results from phase IIa were beyond anyone's expectations, Vertex was eager to publicize them; so far they'd been restricted to a select few. Such information would buoy investor confidence in the company and boost the price of Vertex stock. But as a courtesy, Olson knew that he had to share the results with Bob Beall, Preston Campbell, and Melissa Ashlock at the foundation before Vertex issued a press release.

An excited Olson called Ashlock, as the foundation's drug development leader, first. Rather than share the data immediately, he said only that her predictions of how the sweat chloride, the nasal potential difference, and the FEV1 would change were right on target—but that she still was not going to believe the results.

Olson didn't want to tell Beall or Campbell by phone. He wanted to show them the data in person and see their expressions as they combed through the numbers, so he asked Ashlock to keep mum about the results, and to arrange a meeting at the Hyatt in Bethesda and make sure that Bob Beall was there. Ashlock, in turn, told Campbell about the meeting, adding that she thought he would be happy with the data.

As it happened, Olson was scheduled to sit on a panel with Bob Beall in Washington, DC, just a couple of days after Vertex's CF program medical director Claudia Ordonez had presented the news to the Vertex executives. After the panel, thought Olson, would be an opportune moment to share the data.

When the panel was over, Ashlock and Campbell, who were sitting in the audience, crammed into a cab together with Olson and Beall and drove to the Hyatt. Beall knew nothing about the fact that Olson was planning to share the data of the phase II trial with him and had to be convinced to go to the Hyatt rather than right back to the Cystic Fibrosis Foundation. During the drive, Olson could barely sit still, like a child bursting with a secret. He anxiously made small talk, killing time as the car crawled through traffic along New York, then Connecticut, and finally Wisconsin Avenue.

In a suite on the second floor of the Hyatt, Ashlock, Campbell, and Beall sat around a dining table as Olson handed each of the foundation members a sealed manila envelope containing the summary. Silently, they tore the envelopes open. For thirty seconds, no one spoke, their eyes darting quickly across the charts and data on page after page. Campbell was the first to break the silence, asking about a p-value— a measure of likelihood that the results were due to chance versus the drug. But within five minutes, they were dissecting every data point. As the implications of the results became more tangible, their voices rose from a cautious whisper to full-throated speech. Because Ashlock had primed Campbell on the celebratory nature of the results, he had come prepared with a bottle of scotch, which he cracked open—even though it was only noon.

For all four of them, the moment was joyous, a turning point in the history of this fatal disease, and they were infused with optimism. As a physician, Campbell had watched dozens of children suffer and die. This data proved that the hypothesis of fixing a broken protein was correct. And, even if, for some unforeseeable reason, VX-770 failed in phase III, Campbell now had faith that scientists could build another molecule that *would* succeed.

This approach was going to work.

Vertex made the news public on March 27, 2008, revealing the first positive results for VX-770 after patients had taken the oral medication for just fourteen days.[3] In the press release, Accurso noted that the early results were unprecedented and remarkable because they had such a profound effect on so many symptoms of the disease. It was an important proof of concept.

That summer, as sweltering heat gripped DC and phase IIb was underway, Vertex submitted the compelling results from the earlier phase IIa trial to the FDA, in an effort to fast-track approval for phase III. Typically, in drug trials, companies had to wait after phase II while the FDA reviewed the data, before being given the go-ahead for phase III. But because the results were so overwhelmingly positive, because there had been no adverse health issues, and because there was no other treatment available to these patients, the Food and Drug Administration greenlit phase III trials for VX-770 in advance. The drug's efficacy was already so indisputable, the FDA saw no reason Vertex shouldn't start planning the final phase of the trial. Phase II would be complete in 2008. As long as phase IIb results were as incontrovertible as those from phase IIa, Vertex could begin phase III in 2009.

Vertex's readiness to begin the phase III trials immediately after phase II was the payoff for the plan Olson had pitched Beall a couple of years earlier. Rather than waiting for the end of phase II to figure out the complexities of manufacturing VX-770, the additional investment that Olson had pushed, plus the fast-tracking of phase II, meant that Vertex was now a couple of years ahead of their original timeline—and that, if phase III went well, the drug would reach patients sooner.

As satisfying as the results from the first half of phase II were, Olson knew that, for the families of children with CF, life remained grim. He'd first met the parents of one of those families, Maureen and Chuck Palermo, when they came to Vertex during the company's weekly Friday happy hour to meet Olson and his growing team. The couple had three children—Dan, Mark, and Lisa, all of whom had CF—but the family never all came together, because someone was always sick. The visiting children would explain that the brother or sister who wasn't present was having a hard time and was in the hospital for antibiotics.

In August 2007, Maureen and Chuck lost Dan just two months after his high school graduation and a family vacation in Savannah. Dan's condition had begun worsening in May, and while he had been able to leave the hospital for graduation, he'd returned almost immediately and stayed until the end of July, when he seemed well enough for a family vacation. He died three weeks before starting college, while VX-770 was in phase IIa clinical trials. When Olson passed the casket at the wake, he saw Chuck, Maureen, and their two remaining children, Mark and Lisa. Maureen, weary and eyes puffy, looked at Olson and said, "Can you guys hurry up while I still have two kids?"

The following spring, several months after Dan died and in the midst of phase IIb, Olson and a team from Vertex joined the 2008 Great Strides fundraiser, buoyed by the knowledge that VX-770 was on the right path. There Olson ran into the Palermos: Chuck, Maureen, and Mark. Now Lisa was in the hospital; her health had started to waver.

On October 20, 2008, Vertex shared the results of the second half of phase II—a twenty-eight-day trial, twice as long as the first, with nineteen patients.[4] Its strategy was similar to the previous phase II trials in which Bill Elder had participated, but the drug doses were higher. Its results underscored the earlier ones. At the highest doses, VX-770 was well tolerated and there were no serious adverse events. Lung function improved, sweat chloride dropped, and the electrical activity in the nose signaled that chloride was moving. And Vertex was already preparing to begin phase III trials—the final trials—early in 2009.

Three days later, the phase IIb results were announced at the 22nd Annual North American Cystic Fibrosis Conference in Orlando, to huge applause. While this was a tantalizing breakthrough for families with the G551D mutation, this group made up just 4 percent of the total CF population. Most parents in the audience had worked hard for the foundation and would need to keep doing so, having faith that eventually there would be a treatment for their children. But no one at the foundation or Vertex knew how long it would be before there was something for the nearly 90 percent who shared the most common mutation, discovered some twenty years ago.

The past two years had been an emotional roller coaster for Olson, swinging from the lows of Dan's death to the highs of a successful phase II trial, then to another crash when the Palermos' youngest child, Lisa, then fifteen, died on October 26, 2008—just after Olson returned to Cambridge from the conference. The news devastated Olson. Once again, he was too late. And he knew there were many other families around the country suffering similar excruciating losses, and who, for the time being, would continue to endure and keep fighting for their children.

CHAPTER 46

Tasting Like Average People

2010—2012

Now this is not the end. It is not even the beginning of
the end. But it is, perhaps, the end of the beginning.

—Winston Churchill[1]

From the moment Kim and Rob Cheevers's daughters Laura and Cate
were born, in 1998 and 2001, the family was surrounded by a circle
of supportive friends, including Joe and Kathy O'Donnell. They knew
the Cheevers because the girls' grandfather Gerry Cheevers was a childhood friend of Joe's.

The O'Donnells had also been the ones to recommend Laura and
Cate's physician, Dr. Allen Lapey, who was still passionately committed to treating cystic fibrosis fourteen years after Joey's death. He'd
told Kim and Rob from the beginning that chest physical therapy
had to be done twice every day—once in the morning and once at
night. Treat it just like how you would teach your children to brush
their teeth, he instructed. You do it every day no matter what. There
are no exceptions. You don't miss it because of a birthday, you don't
miss it because of Christmas. He prescribed some thirty pills—a mix
of enzymes, vitamins, and supplements—and recommended the girls
embrace an active lifestyle, because exercise naturally dislodged mucus
from the lungs, reducing the chance of bacterial infections. The girls'
natural passions made following Lapey's advice easy: Laura loved
dance and Cate, soccer.

Lapey gave the girls quarterly checkups at Mass General, monitoring their lung and liver function, digestion, and weight. The girls
loved Lapey. He was an old-fashioned Mister Rogers type who had an

all-consuming dedication to his patients. Laura and Cate thought he was quirky and kind. He had a love of birds and frequently described his birding escapades, which they found fascinating and hilarious.

Although Cate carried the same mutations as her older sister, Laura, the genetic lottery had given Cate a less severe form of the disease; Cate had inherited a different combination of genes from her parents that, together, tempered her illness. No matter how careful Kim and Rob were, Laura was always sicker, and she periodically developed serious lung infections that landed her in the hospital. Even with all her medical training and experience treating children with CF in the ICU, Kim found this terrifying. Doctors would anesthetize Laura to remove mucus from her lungs and then infuse her body with antibiotics through a central IV line for two weeks. Both procedures raised the risk of additional bacterial infections, blood clots, and liver damage. By the age of thirteen, Laura had been hospitalized five times. But Kim stayed positive. She knew the foundation had a drug in the works, something that would target the root of the disease. Although that was all she knew, she believed it would yield a treatment for her girls.

With their daughters' health always in jeopardy, the family closely tracked the latest CF research and devoted time to raising money for the foundation. Even with a busy schedule of daily physical therapy, doctor appointments, therapist visits, soccer, dance, school, and homework, Rob and Kim still ran four large fundraisers every year, and the Cheeverses were always among the top fundraising teams in Massachusetts, giving them close ties to the foundation. In 2009, the family raised nearly $200,000 through local wine tastings, billiards tournaments, and walkathons. Like Joe and Kathy with Joey, Kim and Rob were honest with their daughters. They explained to them when they were in first grade—when other children were only worried about their crayons and learning to read—that everyone had challenges, and theirs was managing this disease. The girls grew up strong willed, never complaining about the therapy, the dozens of pills, or all the additional work and exercise they had to do. Kim and Rob made it routine and the girls thrived. Cate and Laura sometimes needed a moment to catch their breath or rest when playing soccer or dancing, and as they got older, their close friends envied the sisters' ability to eat large quantities of high-calorie pastries and ice creams and salty snacks without any concern of getting fat. Other than that, their disease was often invisible.

While planning their 2009 fundraisers, Kim had heard rumors about phase III clinical trials for a treatment that was almost a cure. But she didn't know the details until Bob Beall called, brusque as ever.

"Kim, what mutations do your girls carry?" he barked.

"They have one G551D and one F508del. Why?"

Beall explained that Vertex would soon be enrolling patients for phase III trials that were expected to begin in early 2010 for a drug called VX-770. It was specifically designed for people with the G551D mutation. "I think Laura may be old enough to participate in the trial. I want you to discuss it with their physician."

Kim was stunned. Only 4 percent of the CF population carried this rare mutation. Now there was a possible treatment? For her girls?

The prospect of a clinical trial excited and terrified Kim. She knew that the side effects of any drug were unpredictable. The same drug that worked miracles in one person could prove toxic, even deadly, in someone else. But there was no foolproof way to predict how a patient would react.

When Kim saw Dr. Lapey the next day at the hospital, she asked his opinion.

"This is the one to try," he said enthusiastically. He'd been planning to contact them about the trial himself; Beall just beat him to it. "We can enroll Laura now." Cate could be enrolled in a trial for younger kids a few months later, in August, when that study began.[2] He explained that this was a double-blind trial: neither he nor they would know whether Laura was receiving the placebo or the drug while enrolled. Patients would receive the drug for twenty-four weeks—almost six months—and then be monitored for another twenty-four weeks after stopping the drug to determine the long-term impact. That meant the trial would take almost a full year to collect all the data.

Vertex announced the phase III trials in May 2009.[3] There would be three: one for patients twelve and older who had one or two copies of G551D; one for patients between six and eleven years old with one or two copies of G551D; and one for patients twelve and older with the more common F508del mutation. According to the work done at Vertex San Diego, the doorman drug shouldn't have any clinical impact on those patients with F508del because they had almost no CFTR protein on the cell surface for the doorman to open. But just in case there was an unanticipated therapeutic benefit not foreseen in experiments with airways in a dish, Vertex decided it was worth testing in a trial to be sure. All trials were scheduled to begin sometime in 2010, as Beall had told Kim.

Phase III clinical trials were rarely quick. The planning took time. Recruiting patients took time, even in cases like this where candidates with the right mutations were known. In addition, this was an international trial—there were patients outside of the US who were also participating. And while the full trial length planned was a year, the phase III trial period would extend beyond that, as patients could enter the trial, even at the same location, months apart.

"You understand?" Dr. Lapey asked.

"Yes, I get it," said Kim, surprised by Dr. Lapey's uncharacteristic excitement. Lapey had treated both of the girls from birth, and Kim thought of him as old-fashioned and maybe a little crotchety. As long as she'd known him, he'd never shown enthusiasm for any new treatment.

In fact, Dr. Lapey hadn't been excited about any drug that had come through the pipeline during the last three decades, because the impact of these treatments on his patients had been marginal at best. More often than not, he'd been disappointed. But this new drug from Vertex was different. It seemed to act fast and dramatically.

After Kim told Rob what Dr. Lapey had shared, they had questions, especially about the possible dangers. Lapey warned them of side effects like liver damage, but he veered away from sharing the possible benefits. He didn't want to raise their hopes or bias their observations.

With Lapey's endorsement, Laura began the trial in 2010. Other than taking the study pill twice a day, everything else about her treatment remained the same: twenty to thirty pills for food digestion, nebulizer treatments of Pulmozyme to break up mucus, and up to thirty minutes of physical therapy daily. Three days each week, a physical therapist provided extra support to keep Laura free of infection, and three days a week Laura danced—a more pleasurable route to keeping healthy.[4]

Weeks passed with Laura in the trial, but her health didn't improve. She remained weak, failed to gain weight, and continued to cough through the night, every night. At one point during the trial, Kim took her to Mass General, where she was hospitalized for two weeks to clean out her lungs and given massive doses of antibiotics. Was this the normal course of disease for Laura? Would she have been this sick anyway, or was the trial drug making it worse? Kim heard stories through the grapevine of the drug's positive impact on the disease, and she and Rob were convinced that she felt the same, or worse, because she had received a placebo. But they couldn't rid themselves of the worry that this might instead be a side effect of the medicine.

A few months later, Cate joined the VX-770 trial for younger children. Her reaction to the drug was very different. Within days her face had more color, and she kept telling her parents she "felt different inside." Kim and Rob suspected this was just talk until she went for her next checkup and they discovered she had gained a few pounds. Within three weeks, Rob and Kim were noticing more changes, and Cate was adamant that she was getting the real drug. As the trial progressed, Cate barely coughed, slept soundly through the night, and ate like a horse; her energy became bounceless. Laura remained sickly, and her nightly coughing left her drained and exhausted.

The stark difference between the sisters pained not just Kim and Rob but their grandfather Gerry Cheevers, who had been a goalie for the Boston Bruins from 1965 to 1972 and again from 1976 to 1980.[5] He knew Joe O'Donnell from growing up in Everett, and believed that if anyone could arrange for Laura to receive the real drug, it would be Joe. Unlike the girls' parents, who knew it wasn't possible to meddle with a clinical trial, Gerry didn't understand how clinical trials worked. He came to see Joe and told him about his granddaughters, describing how Laura was suffering as Cate thrived. Could Joe get Laura the drug?

Joe was used to helping people—making a phone call and fixing a problem. He enjoyed it. But in this case he was helpless. Joe explained that the trial was double blind, so no one knew who was on the drug and who had received the placebo—not the people who sent it out, the doctors, or the patients. And meddling with the trial could jeopardize the results and the chance of the drug getting approved for everyone. Gerry could see, and Kim and Rob could see, that one of his granddaughters had gained weight, and the other hadn't. One no longer coughed all the time. And one wasn't getting sick. But this was a problem that Joe couldn't fix. "I can't get her the drug," Joe told him.

Gerry began to sob. "What happens if she dies before she can get the drug, or something happens and it's too late?" Joe sat quietly with Gerry, knowing that there was nothing he could do to help.

When Joe told Kathy about how Gerry had come to him begging for the drug for Laura, they sat in silence. They remembered clearly, as if it were yesterday, how it felt to be so helpless, knowing that what happened to Joey was out of their hands. They had been in Gerry's shoes. They understood the desperation and fear. If they had two children, one thriving and one not, they would have done the same thing Gerry had. *Anybody would,* thought Kathy. She and Joe would have done anything to get a drug that might have saved their son. But they also understood how important it was that the trial be allowed to run its course.

During the second six months of the trial, after she was no longer taking the pills, Laura just became sicker; a full year later, the health of the two sisters differed dramatically. Cate had gained about seven pounds, her appetite had swelled, her lung function had jumped by more than 10 percent, and the saltiness of her sweat had fallen. More impressive, the changes had lasted some six months after she stopped taking the pills.

A couple of months after the end of the phase III trial, on February 23, 2011,[6] Vertex released the results for two trials: the one for patients older than twelve with the G551D mutation and the one for patients with F508del. Though Cate was in the third, ongoing trial, her response seemed typical among participants with the G551D mutation who had received the drug. The average weight gain was seven pounds compared to those in the placebo group, who like Laura had barely gained a pound over the entire year. Their average sweat chloride level had dropped by 45 millimolar, whereas the placebo group's chloride level still hovered at 100. In the treatment group, 67 percent had no need for antibiotics during the trial or the six months after, compared to just 41 percent in the placebo group.

For patients with two copies of F508del, there were no significant changes while on VX-770—more evidence that this group needed a corrector *and* VX-770 to improve their health.[7]

Even with these two studies completed, Vertex was not quite ready to file a New Drug Application with the FDA because they had one last phase III trial still in the works, the one with children between six and eleven. If these children could start a drug like VX-770 *before* they suffered major lung damage—like Cate—then maybe they could escape the horrors of this disease and even prevent many of the symptoms altogether. The final results of that second trial were still being analyzed and wouldn't be ready until after July 2011.

As the patients and their families waited, a surprising bit of news arrived: the FDA made VX-770 available, immediately, to anyone who had participated in phase III—and also to anyone too sick to wait for the FDA to approve it. That meant that Cate could resume taking VX-770, and Laura could start.

Once she did, her health improved quickly, just like Cate's had. Her face became rosier, she grew and put on weight, and her cough waned. Now both girls slept through the night and woke up refreshed. After a year on the drug, Laura had gained almost twelve pounds. Her lung function improved, though not as spectacularly as Cate's. Perhaps that was because she had been sicker, with more infections, which had racked up more damage and scar tissue in her lungs—changes that might not be reversible. But the transformation was still dramatic and she could now breathe deeply.

Then there was the salt.

Kim and Rob were never privy to the results of their daughters' sweat tests after the girls were on the treatment, but they didn't need to see the numbers. They could tell by just kissing them on the forehead and then licking their lips. Before, the girls had tasted salty, like little potato chips. Now, after the drug, they tasted normal. Or as Cate put it, "I taste like an average person."

THE ENTIRE CYSTIC FIBROSIS COMMUNITY COULD FEEL THE QUICKEN-ing drumbeat of progress. In June 2011, Vertex made public the results from the second phase III trial, the one in six- to eleven-year-old children. They echoed the earlier results. With the additional safety trials wrapped up, it was time to submit all the data to the FDA. This was no small task: every measurement taken from each trial participant during phase III was given to a team at the FDA to check every data point and redo every calculation to make sure that the results that Vertex was claiming were accurate. They would also determine whether Vertex could reliably manufacture identical batches of VX-770, which had been a tricky molecule to prepare in large quantities.

On October 19, 2011, Vertex submitted an application to the FDA to approve VX-770, which would soon be named Kalydeco.[8] Review of a first-of-its-kind medication like Kalydeco would typically take close to a year, depending on the results. But this was a special case. The data was indisputable, the disease severe, and there was no other drug on the market to treat it. Patients were desperate; many were dying. That included, on December 8, 2011, Chuck and Maureen Palermo's firstborn and last remaining child, twenty-four-year-old Mark Palermo, who died while waiting for a transplant.

Two months later, on December 15, 2011,[9] the FDA granted priority review for the drug, capping their time for weighing all of the drug data to six months. The FDA assigned priority review only for medicines that were major breakthroughs and for medicines for diseases for which no treatment existed. Both were true for VX-770. This meant the foundation and all the patients would learn whether the drug was approved sometime in April.

It was clear this drug produced significant therapeutic benefit as long as a patient had at least one doorman mutation, and most important, that it did so without causing harm.[10] But it wasn't yet clear whether the drug would stop the damage CF caused to the body from getting worse. Figuring that out would take years. And there was also a question that Vertex scientists like San Diego leader Paul Negulescu and lead biologist Fred Van Goor were already starting to address in the lab: What if patients with CF had a different type of mutation—not G551D, like the Cheeverses, but another—that allowed the CFTR protein to reach the surface of the cell but not open and close properly? Would VX-770 also work for those patients? Could this drug treat more than just the 4 percent of the CF population with G551D, expanding the market for this new medicine?

Joe O'Donnell knew this medicine was transformative. As chairman of the Milestones to a Cure campaign, he received a weekly report, along with Bob Beall, Preston Campbell, and Rich Mattingly, marking the latest progress. Joe had been watching the progress of VX-770 for several years and everything had been remarkably positive. He expected the drug to sail through the approval. But the moment he actually felt the impact that this drug would have on the community happened unexpectedly at his annual billiard-tournament fundraiser in late 2011.[11] Joe was leaning over the green felt of the billiard table, taking a shot, when Gerry Cheevers walked up to him. This was a different man from the one who had come to him months ago, desperate and fragile, frightened his granddaughter would die. No longer sleep deprived and deflated, his shoulders were back, his face animated, and his speech rapid, as if he couldn't share the news fast enough.

"Joe, you wouldn't believe what that drug is doing for Laura and Cate," Cheevers said. "It's amazing. Laura was on the placebo in the beginning—that's why she wasn't gaining weight."

"They're both on the drug now?" Joe asked.

"Yeah, it's phenomenal. Just two little pills—one in the morning, one at night. So simple."

Joe put his arm around Gerry, and for a few minutes, neither of them spoke. The drug targeted only about 1,200 people in the US, and maybe twice as many worldwide. But that was still a few thousand children and young adults who now had the chance to live longer, healthier lives.

When Joe arrived home from the tournament that evening, he shared the news with Kathy. Both of them sat quietly, tearing up in the living room of their Belmont home, where Joey had lived a quarter century earlier. This was a sweet victory. Sweeter than the discovery of the gene, with Joey newly gone and the couple believing he had just missed out on gene therapy and a cure. Now, the O'Donnells had a healthy family. And, rather than hearing about the progress from Rich Mattingly or Bob Beall at the foundation, Joe and Kathy heard it from a friend, the grandfather of children who were taking the drug and thriving. They knew the Cheeverses—Gerry, Kim and Rob, and the two girls—personally. And thanks to VX-770, Cate's and Laura's futures were a lot brighter.

Joe didn't just feel great joy for his friends; he also felt relief. All that had ever mattered to him was that he always did the best he could, for everyone and anything. He and Kathy knew they had done so for Joey. They never looked back on his life and second-guessed their decisions; there was no regret that they could have done this or that, which would have extended his life or prevented his death. They knew they had done everything they could. They had given their all. Now Joe felt a similar burden lifting. He had given all his energy to the foundation's Milestones campaign to help it raise the $175 million in R&D money needed for the doorman and the corrector drug, still in development—and now the first drug, the doorman drug, was real, and a potential lifesaver for Laura and Cate and thousands of others.

In mid-January 2012, Joe and Kathy flew to Washington for an annual fundraiser, where hundreds of parents and volunteers, many of whom the O'Donnells had known for at least a decade, gathered for cocktails. It was a community about which Joe and Kathy cared deeply: the families with a child hanging on the cusp, the single mothers who had just lost a child, the volunteers with ties through family or friends, and other generous souls who were just devoted to the cause. Many had been trudging along for years toward a goal that would not bring back their departed loved ones—a goal most people feared would never be realized as, year after year, little had changed and nothing transformative had happened. Yet these families had continued to fight. They came each year, each facing their own personal battles, but also thinking of how they could work together toward their common goal of curing this disease.

Each year, Bob Beall would give a progress report, delivering scraps of hope. Some years were good, like the one following the discovery of the gene, and the next, when parents were still hopeful about gene therapy. But the decade following had been brutal for the CF community. Since VX-770's clinical trials had begun, however, expectations had been rising cautiously. And now, after almost five decades, it felt like something momentous was happening. Tonight was altogether different from years past.

As Joe and Kathy walked into the ballroom, they were overwhelmed, tears welling in their eyes. The atmosphere was skin-tinglingly electric, the emotions of six hundred people synchronized—everyone sighing, crying, laughing at the same time. It wasn't a raucous celebration, like winning the World Series, but a spiritual one. It was a gentle exhale, a grateful release of a long-held breath—a brief, soulful pause to celebrate and appreciate what they had done together. It was an extraordinary evening, like nothing Joe had ever experienced.

What neither Joe, anyone at the foundation, nor any of the Vertex scientists knew during that event was that the FDA had been able to accelerate their review further, and was already on the precipice of announcing VX-770's approval. The data was so conclusive, the results so unarguably clear and positive, that the FDA completed its assessment in just slightly more than three months, granting approval on January 31, 2012.[12] It was one of the fastest drug approvals in FDA history.[13] The drug was deemed suitable for anyone older than six years with the G551D mutation.

Kathy first saw the news of early approval in an online story, and then heard it on the radio. Joe also saw it online. When he got home from work that evening, the two broke down, crying and holding each other. When they recovered, Joe poured them drinks and they celebrated quietly.

This was the first drug ever developed to treat patients with a specific genetic mutation, and its approval marked the beginning of a new era of medicines developed to match an individual's genes. It was the beginning of the personalized medicine revolution. The gene that caused CF had been among the first handful of genes discovered in the early days of gene mapping. CF had been one of the first diseases on which gene therapy was tested. Now it was the first to have a medicine capable of targeting the root of the disease.

But for the patients who carried this rare mutation, this string of firsts was irrelevant. All that mattered was that they now had a chance at good health and a long life.

Two weeks after the FDA's approval, on February 15, 2012, Bill Elder's mother could barely contain her excitement as she waited, staring at the sparkling snow blanketing the mountains and foothills surrounding the family's Colorado home, for the FedEx driver to deliver their first box. Bill had graduated from Stanford in spring 2011 and had spent the summer and fall applying to medical schools; he was now home waiting to hear. He had received the placebo in the phase II trial, and unlike the Cheeverses, who had seen the improvement in Cate's health during the trial, he had no idea what to expect—though he had been reading all the medical literature on the new drug.

When the truck pulled into the driveway, the driver passed a FedEx box to Bill's mother while Bill hurried into the kitchen to cook up a large plate of bacon and pour himself a tall glass of milk. He had to eat tons and tons of calories to stay well anyway—and while a plate of bacon wasn't the healthiest dinner choice, he wanted to give this new medicine the best chance of working. He ate a slice of bacon, took the pill with a large gulp of milk, and a couple of hours after finishing the food, went to sleep.

At 2 AM, Bill woke up and sat on the edge of his bed. Something was different, though he was too groggy at first to figure out what. He continued to sit, slowly breathing in and out through his nose, when he realized: he had never been able to breathe easily through his nose before. He grabbed a notebook from his desk and started jotting down everything he was feeling, the new sensations that were invading his body. Eventually he went running down the hallway and into his parents' room, yelling, "Kalydeco's working! Kalydeco's working!" His parents and his younger brother all woke up, startled at first, and then incredulous that the medicine could have worked that fast on a genetic disease that Bill had had his whole life.

But Bill was sure. He could smell something in the air of his home that he had never noticed before, thanks to spending his whole life essentially smell-blind from accumulated mucus: the scent of a candle laced with the piney aroma of a Christmas tree. Although he still couldn't verbalize what exactly was different, he knew the drug was working.

That was just the beginning. During the weeks that followed, when he went running in the mornings, he was able to go farther than he ever had before. The sweat on his arms, which before had left lines of white salt crystals as it dried, was now clear. As he ran past restaurants in the center of town, he discovered that each one had a distinct smell, depending on the food inside. He had no idea that it was possible to smell food from outside a restaurant. He could smell the trees as he walked through the forest; even lakes had their own scent. In fact, everything had its own distinct smell. The world had a whole layer of complexity that he had never noticed before.

WHEN THE US FOOD AND DRUG ADMINISTRATION APPROVED Kalydeco for patients with the G551D mutation in 2012, it was a huge victory not just for patients and their families, but also for Doris Tulcin and Bob Dresing—both still alive and following the drug trials—who had centralized the foundation, made it solvent, and set out the organization's mission; for Bob Beall, who had launched university research programs with hefty grants to lure the most talented scientists to this neglected orphan disease and taken a risk investing in a biotech company to develop a drug; for the devoted team of scientists at Aurora, and then Vertex, who had created a new type of drug that could fix a broken protein and in the process launched a new era of medicine; and for Joe O'Donnell and his Milestones campaign team, which raised the money that made that new drug possible. It also validated the foundation's funding model, venture philanthropy, and proved that its targeted approach to drug discovery was possible.

But Beall couldn't rest on his laurels. This was just the first step. Kalydeco worked for only 4 percent of CF patients. There were tens of thousands of people—close to 90 percent of CF patients—who carried the common F508del mutation. They needed a drug, too.

WHILE THE SUCCESS OF THIS NEW DRUG WAS THE PRIDE OF THE FOUN-dation, Beall was less happy about the price that Vertex had slapped on it: $294,000 a year.[14] It was one of the most expensive drugs in the world, priced in the same range as only a few others, also for extremely rare diseases.

Vertex hadn't developed the drug alone. The patient registry and the clinical trial network had been essential tools for getting companies interested in CF, and without them, Beall believed, there might never have been a drug. And Vertex's success wasn't just based on the money that the foundation had invested in the project. For decades, the foundation and the taxpayer-funded National Institutes of Health had supported the science that laid the groundwork for understanding the CFTR protein and the physiology of cystic fibrosis that made Kalydeco's development possible.

Many, including CF researcher and patient Paul Quinton, who had figured out the chloride problem in the early 1980s, felt Vertex was profiteering, and that the drug price violated the social contract Vertex had made with the foundation. After all, the foundation's contribution came from philanthropic donations given to Joe O'Donnell's Milestones campaign. Critics felt the company was exploiting the goodwill of the donors and gouging the patients. The foundation wasn't blameless in this, Quinton believed. By earning royalties from the sales of the Vertex drug, they had created a conflict of interest: they were making a profit from a drug that helped the people they were dedicated to assisting.

On July 9, 2012, a group of twenty-four researchers, physicians, and directors of CF care centers drafted a letter[15] expressing their concerns with the drug price to the newly appointed Vertex CEO, Dr. Jeffrey Leiden. For the next few months Leiden engaged in a dialogue with the signatories. But the discussions petered out after a few back-and-forths, and some six months later the issue seemed to lie abandoned, simmering under the surface, wrapped into a broader discussion of drug pricing.

It wasn't an easy situation for the foundation. No pharmaceutical company would have agreed to work with the foundation on a drug for CF and also relinquish control over pricing. The only way to lower the price of the medicine was to have more companies developing drugs and giving Vertex competition. But for the moment, Vertex had a monopoly.

The other issue was that the $19 million in royalty payments from the antibiotic drug TOBI, which had come from an arrangement similar to the one the foundation had with Vertex, was the seed, together with the Gates' money, that was used to hook Aurora in the first place. Now Beall and Campbell hoped that royalties from Kalydeco sales would provide more money that they could invest in the development of more and better drugs for the community. If the price went down, their royalties would, too. And, as Campbell knew well, the CF community derived a perverse benefit from this high drug price. Dozens of drug and biotech companies were now knocking on the foundation's door, eager to learn how they, too, could get into the CF space. When Beall hired Campbell in 1998, the two of them pounded the pavement for companies who would work on cystic fibrosis. From experience, they knew they had been lucky back then to find any company interested in the disease. If the price dropped significantly and the opportunity for huge profits fell, would anyone still want to develop drugs for this disease? At least now, in addition to investing in Vertex, they could stimulate some competition by providing grants to a broad collection of companies.

Then there was Vertex's perspective.

For a start, patients didn't pay $294,000 for their drug. For those with insurance, companies paid the cost and Vertex took care of every patient's copay. Patients who lacked insurance received the drug for free. That meant that the cost to the patient, when things worked as Vertex intended, was zero.

The pricing dilemma was also complicated by Vertex's larger financial situation. Every year since Boger had founded Vertex in 1989, the company had lost money. When he retired in 2009, a calculated move designed to provide Matthew Emmens,[16] Leiden's predecessor as CEO, with a big success early in his tenure, the company was on the verge of releasing a new hepatitis C drug, Incivek. It was the company's first successful drug that it would produce entirely on its own after twenty-two years in the drug discovery business.[17] After $4 billion in losses over those twenty-two years, 2011 was the first year Vertex turned a profit. The company prospered for a few more quarters, bringing in $1.6 billion over the next twelve months.[18] Then, less than a year later, on January 31, 2012, the FDA approved Kalydeco. For a short, glorious window, profits were soaring.

But in December 2013, Vertex's luck shifted. The FDA approved Gilead Science's Sovaldi, a better hepatitis C drug than Incivek, which crashed the market for Vertex's drug almost instantly.[19] As sales of Incivek plummeted, the company laid off 15 percent of their workforce—350 people—and in 2014, Vertex stopped manufacturing it.

Kalydeco was now the company's only source of revenue. It was an ironic twist: the same drug that many at Vertex had mocked for so long was now key to the company's survival.

CHAPTER 47

Milestones to a Cure

2004–2015

[Fundraising] isn't a simple process of begging—it's a process of transferring the importance of the project to the donor.

—Henry A. Rosso

When Rich Mattingly, the Cystic Fibrosis Foundation's chief operating officer, went to Joe O'Donnell to chair the foundation's capital campaign, he was far from a fundraising novice himself. Since Bob Dresing had hired Mattingly in 1980, Rich had spent more than twenty years building a strong network of chapters throughout the country that were fine-tuned fundraising machines, responsible for bringing in most of the foundation's money—money that was essential for supporting the CF centers. The executive director of each chapter was charged with leading a series of special events throughout the year and cultivating a Rolodex of committed donors, both big and small.

The directors had organized walkathons, galas, wine tastings, golf outings, and other special events throughout the country, and the more those events had grown in popularity, the more money was raised. But in 2004, when Bob Beall calculated what he would need to fund the work at Vertex, and Mattingly figured out roughly how much the chapters would bring in over the next six or so years, the difference was $175 million. That amount became the fundraising goal for the Milestones to a Cure campaign.

When Milestones kicked off in mid-2004, Mattingly was hoping that chapter executive directors would help him and Joe identify big local donors who supported their events so that he could test the waters and see whether they might make a larger donation. But initially that was a touchy issue. If Mattingly was diverting funds from the chapters' top donors, it might deflate the chapters' own events and decrease the amount that they were able to raise from local supporters. That, in turn, would reflect poorly on the chapters' fundraising teams. In response, Mattingly made it clear that if chapter directors helped to bring in a donor, they and their chapter would get the credit.

One event that Mattingly had launched almost ten years earlier, in 1995, was the Ultimate Golf Experience fundraiser, an ultra-high-end event at luxury golf courses around the country that attracted donors who could pay around $6,000 and up for a few days of golf. The first one was held at Pinehurst in North Carolina, and Mattingly had been able to attract eighty-eight people to the event—including Joe. It was an all-inclusive event in which guests enjoyed golf, the spa, relaxing, and being wined and dined, with high-level entertainment. The event was popular and became an annual affair, frequently selling out for the following year within a few months of the last. And its attendees, Mattingly knew, were people with the disposable income to donate major amounts to the foundation.

Mattingly had been running the event for about a decade and had cultivated many people there over the years whom he now hoped would be among the top-tier donors for the new fund drive.

Milestones was an odd campaign, as Joe told both Mattingly and everyone Joe asked for money. And the consultant whom Mattingly had hired at the start to provide some fundraising guidance echoed the sentiment. The foundation wanted money but had nothing to give in return. People who gave multimillion-dollar gifts typically did so in exchange for getting their name on a wing of a hospital, a building or school at a university, or an endowed professorship. The consultant told Mattingly that the foundation didn't have deep enough resources to draw $5 million, $10 million, and $25 million donors. You might get one $5 million donation, he told Mattingly, but that was probably it. You are never going to be able to raise $175 million.

Those words lit a fire under Mattingly and he took it as a challenge. He knew he could prove the guy wrong.

He began by hiring researchers whose job it was to examine the history of the foundation's previous events, particularly the upscale ones, and see who the big donors were. Using only public records, they would determine the backgrounds of major donors: What did they do for a living? How much income did they earn, and how much and what types of stock did they own? How much had they given to political campaigns? What issues were important to them? All this research was to figure out if they might be inclined to give more and how best to motivate them to do so. Such researchers were widely used for major gift initiatives at big institutions, but the foundation had never taken this approach.

As the research poured in, Mattingly and Joe organized the information into a pyramid that illustrated the top, middle, and base tiers of giving and how many donors they needed in each category to hit their $175 million goal. At the top of the pyramid were donors who gave a million or more. Based on other giving campaigns, Joe expected that between 80 percent and 95 percent of the money would come from this group, though it would include the fewest number of donors. Joe and Mattingly were hoping for more than ten such donors. Then there was the middle tier, those giving between $100,000 and $999,000. They needed about one hundred donors at this level. Finally, there was the base tier, those contributing gifts ranging from $10,000 to $100,000. They would need hundreds or even up to a thousand gifts in this category, on top of meeting their goals for the middle and top tiers, to reach the campaign's target.

The first segment of the campaign was the silent one, during which Joe and Mattingly raised close to half of the $175 million they needed. It was a promising sign they would be able to approach, if not reach, their goal. Then, once they had figured out the campaign's strategy and structure and took the campaign public, Mattingly hired a professional fundraiser from Children's Hospital named John Lehr to support him and Joe as they met with donors.

With Lehr working as Mattingly's right-hand man, identifying donors, providing research, and setting up the logistics of meetings, Mattingly and Joe began going on the road at least five days a month to meet donors. Mattingly was mindful of Joe's time and didn't invite him to walkathons or small events, where the volunteers were collecting lesser sums. While this group was critical for the foundation's fundraising, Joe's talents were unneeded there, as participants were mostly collecting money from sponsors and pledges. These were not prospects for big giving.

More often, Mattingly would bring Joe to galas in New York or Los Angeles or to big annual chapter fundraisers, gatherings where he could talk to a packed ballroom of guests, and later connect with particular ones whom the local chapter had identified as potentially interested in a larger financial commitment. People listened because Joe gave the campaign credibility and trustworthiness—and, equally as important, a human side. He was a dad, and he had lost his son, but he wasn't going to let anyone else lose their kid; he was fighting for them. And he made that clear. Guests, whether or not they were directly related to someone with CF, could sense his authenticity.

When the evening was over, Joe didn't run to catch a plane. He would stand around, have a drink with people, put his arm around them, and tell them, "I'm with you." He let them know they weren't alone. And though he obviously wasn't getting paid for anything he did for the foundation, he always stayed late and thanked people for coming. He'd follow up with a call within a day or two or send them a personal note. He knew how to take care of people and show his appreciation.

Sometimes Mattingly would take Joe to meet with a smaller group, usually connected personally or professionally to one of the foundation's major donors. Every now and then, one of Joe's friends, like Mitt Romney—who had always been very generous to the CF cause and had known Joey personally—would convene a group of their own friends at their home so that Joe could talk about the foundation's needs and goals in a more intimate setting.

Over the course of the campaign, Joe and Kathy also hosted some fifteen to twenty events in their Belmont home for locals to share their stories and for the O'Donnells to answer questions potential donors had about CF, why the foundation needed money, and how close they were to treatments.

In most cases, people at these gatherings were generous, and Joe could raise a lot of money for the Milestones campaign. But on one occasion he and Mattingly struck out completely.

The two were in Los Angeles talking with a potential top-tier donor who had the means to give a gift in the $25 million range. The donor took Joe and Mattingly to a series of bars where he regaled them with stories of Sinatra and other celebrities with whom he had signed deals. It had been a long evening; Joe still didn't have a good sense of whether this guy was going to deliver, and it was irritating him. So on the sidewalk between bars, Joe pulled the man aside and told him, "I need $5 million for my campaign—are you in?"

The guy wavered for a moment, then shook Joe's hand. "All right, I'm in for $5 million." But first the man wanted to take Joe and Mattingly to his lavish home in Beverly Hills just off Sunset Boulevard.

Joe had seen a lot of nice homes and wasn't interested in seeing another—until the man told the pair that he wanted to show them his game room. Then Joe perked up. His competitive spirit kicked in as he envisioned the room he had built for Joey filled with video games and pinball machines. Perhaps this guy wanted to challenge him; he could definitely beat him. He had played a lot with Joey, and though Joey always beat him, Joe was good. The pair entered the man's home, and he served both Joe and Mattingly more drinks before leading them to the game room.

It wasn't quite what Joe had expected. They sat down on leather couches in what looked like a lodge from the 1920s. Mounted on the walls were the heads of animals that the guy had shot at game reserves around the world. There was a bear, an elk, an antelope, and at least a half-dozen more animals staring down at him with their sad glass eyes. By the time they left the man's house, Joe had only a couple of hours before his flight. On the way to the airport, Joe called Mattingly, expressing his doubts that the guy was really "in." His hunch was right; the guy never gave a cent.

Fortunately, cases like that were rare.

In a more successful instance late in 2010, as Milestones was closing in on its goal, Joe only needed another $5 million to hit their $175 million target. He was a persistent fundraiser and he had a hunch who might give him the money. So when he saw Bob Niehaus, who had been present at the original New York brainstorming summit in May 2003, and whose nephew had CF, at an Ultimate Golf Experience event, he asked if Niehaus could donate a little more to fill the remaining gap. Niehaus was a generous philanthropist and had already given many large gifts to the foundation. So despite some hesitance on Niehaus's part, Joe kept pushing.

At one dinner following a day of golfing, Niehaus excused himself from the dinner table to go to the restroom. Feeling a similar urge, and with no sense of an appropriate time and place when it came to cystic fibrosis, Joe got up and followed him in. Niehaus was both incredulous and amused by Joe's nerve as he parked himself at an adjacent stall and kept the conversation going.

Joe asked, Well, are you in?

And finally, Niehaus responded: Yes. But I'm not going to shake your hand.

By 2011, THE MILESTONES CAMPAIGN HIT $176.3 MILLION, OVERSHOOT-ing their goal. It was an incredible achievement, and especially well timed, as Vertex had just sent all the data from the doorman drug's phase III trials to the FDA and was waiting for approval. Still, Mattingly and Beall were acutely aware of the cost of continuing the research that kept the corrector drug moving through the pipeline. The $175 million had been used up almost as fast as it came in: $20 million here, $40 million there, and so on, helping push both molecules to a stage of clinical development when Vertex would take over. If they were going to get the corrector to patients—and especially if they wanted to use the same fast track Olson recommended for the doorman—the foundation needed more money.

Typically, after a major giving campaign, an organization would finish the campaign with a big bang, a *Hallelujah, we did it, we were successful*, and then go dormant for a while. The last thing an organization would do was go back to its base and ask for more money. But the foundation had already run out of money, and needed to raise more as soon as possible.

Mattingly told Joe, who by that point had done more than a couple of hundred fundraising trips with him, that they had to keep going. Joe immediately told him no and that he was crazy. But the Cystic Fibrosis Foundation wasn't just any organization. Urgency was the motivation for everything it did. And as long as kids were dying from the disease, the foundation couldn't afford to take a dormant period just to be polite.

Joe realized Mattingly was right. Going dormant would send the wrong message to the donors and the CF community: that they had conquered CF. Nothing could be further from the truth. Ninety-six percent of patients still didn't have a treatment.

Joe agreed to chair the Milestones to a Cure II campaign, too. The target was $75 million.

CHAPTER 48

Tackling the Common Mutation

2004–2013

> We need enthusiasm, imagination, and the ability
> to face facts, even unpleasant ones, bravely.
>
> —Franklin D. Roosevelt[1]

Two of the cystic fibrosis patients waiting for Vertex's next drug were Paul and Sue Flessner's sons, Jonathan and Andy, both of whom carried two copies of the F508del mutation. Paul Flessner had been critical in facilitating the Cystic Fibrosis Foundation's connection to the Gates Foundation and securing the $20 million that helped fuel the CF project at Aurora before Vertex took over. He'd been a thoughtful and diligent counsel for the foundation, and continued to keep close track of the drug development. And as thrilled as he'd been for all the families who benefited from Kalydeco, he also felt a blow of disappointment knowing that first drug would not help his boys. They would have to wait.

While the doorman drug had been greedily hogging the limelight, the corrector molecule's development had continued, if slowly, back in San Diego. But once Vertex had seen signs from the early clinical trials for VX-770 that rehabbing the mutant CFTR protein was possible, safe, and might actually help patients, Vertex San Diego began ramping up the corrector's development.

From the in vitro experiments that Van Goor had done, it was clear that VX-770, the doorman, could not help the majority of cystic fibrosis patients because it only opened CFTR's damaged chloride channel at the cell's surface; it did nothing to rescue CFTR protein molecules so damaged that they were scrapped by the cell before they could reach the top of the cell, as occurred with the F508del mutation. Vertex's trial giving VX-770 to patients with the more common mutation had driven that point home: a doorman couldn't open a door on the cell surface if there was no door to open. Patients with two copies of this mutation (as opposed to patients with one copy of F508del and one copy of another, CF-causing mutation that allowed the CFTR protein to reach the surface) had one of the most severe forms of cystic fibrosis: their lungs were full of mucus, their pancreases were essentially destroyed, and they had sweat chloride levels greater than 100. For VX-770 to help these patients, it needed a partner drug: a corrector that refolded the mutant CFTR, enabling it to bypass the cell's quality control and travel to the cell's surface, where it still needed a doorman to open it up.

Back in late 2004, as the doorman molecule was progressing quickly in the hands of medicinal chemist Sabine Hadida and her team, high-throughput facility head Tom Knapp, the San Diego CF project leader Peter Grootenhuis, and head biologist Fred Van Goor were continuing to run high-throughput screening with rat cells in search of potential corrector molecules. By late 2005, after testing 164,000 molecules, they[2] found several hits that looked like good prospects and sent them to the medicinal chemists for development. The most promising looked like a caterpillar with two identical, hexagonal heads. This was the starting point, the lead molecule, the scaffold that Hadida's team would transform into a potent molecule, by trial and error, just as they did to produce VX-770.

Throughout 2006 and the first half of 2007, Grootenhuis, Van Goor, and Hadida and her chemists had been devoting their time entirely to building potential correctors, adding and subtracting atoms from the caterpillar molecule and testing them in cells. After the team had designed and tested more than 3,000 structural variations on that original molecule, chemist Jason McCartney was the one to first synthesize the winner.

The final corrector molecule bore little resemblance to its caterpillar-shaped ancestor. Over the course of the molecule's development, Hadida's team had elongated it, adding a couple more geometric shapes, a triangle and a pentagon, and a couple of fluorine atoms. The resulting shape helped more of the mutant protein fold properly and move to the surface of the cell, where the doorman could then open the channel and allow chloride to move.

The corrector molecule, VX-809.

Using cells that he had harvested from the lungs of the thirteen-year-old boy with two F503del mutations and from lungs from six other patients, cell facility manager Tim Neuberger created fresh airways in a dish on which Van Goor could test the corrector in combination with the doorman, VX-770. Seen through the microscope, these untreated airway cells with the F508del mutation were slathered in thick, dry mucus that had matted down the cilia, preventing all movement. But once treated with both VX-770 and the corrector, and then incubated at body temperature for forty-eight hours, those same cells moistened and the cilia began moving. It was just as they'd hoped—although the airway cells' surface wasnt quite as wet as when VX-770 was added to the airways in a dish grown with G551D cells.

Using an Ussing chamber, Van Goor saw that the electrical activity rose to 14 percent of that seen in healthy airway cells[3]—ones from patients without CF. That paled in comparison to what they had observed in the G551D cells with VX-770, where chloride transport reached close to 50 percent of that seen in healthy people—the same amount of chloride transport seen in CF carriers, who had one copy of the mutation and one normal copy.

The frustrating thing about developing the corrector was that none of the molecules that Hadida's team made ever raised the cell's activity above 14 percent of healthy protein function.[4] This ceiling was perplexing; now that the scientists had seen what was possible with VX-770, they were sure there was a way they could do better. Still, 14 percent was better than nothing, and by October 2007 the corrector was christened VX-809 and sent to the East Coast for animal testing, and to be formulated into a drug.

They were right to have persisted. In 2008, Hadida's team and Van Goor made a breakthrough. In their hunt for a corrector they had discovered several candidates, and found out that they didn't all behave the same way. Some seemed to be interacting with the CFTR protein at a different stage of the folding process than others. Experiments done by both Vertex and academic researchers[5] supported this idea. But what was most exciting was that sometimes, when two of Vertex's correctors were added to a cell at the same time, more CFTR reached the surface of the cell, and there was a larger jump in chloride transport than with either corrector alone.[6]

Later, in 2012, two teams, one based at the University of Western Ontario and another at the University of Texas Southwestern Medical Center, would figure out that the F508del mutation caused two discrete misfolds in the CFTR protein.[7] That was why it was impossible for one corrector alone to fix both problems, and why a single corrector couldn't raise chloride transport higher than 14 percent. The findings supported what Vertex had discovered four years earlier: that each corrector seemed to be targeting and correcting a different fold in the protein. It was as if the correctors were tailors that worked in tandem, one fixing a dress's misfolded pleat and a second fixing the sleeve before the dress could leave the store. The caterpillar-shaped corrector VX-809 seemed important for the first fold, but another corrector was still needed to help more of the CFTR protein fold well enough to reach the surface of the cell.[8]

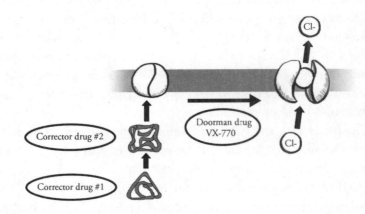

Adding together two types of correctors could boost
dramatically more CFTR to the surface than either one alone.
When combined with the doorman, the trio of molecules
promised a transformative therapeutic effect.

These findings implied that the ultimate drug for people with F508del would include two correctors and a doorman. Van Goor presented this finding and the data to Vertex's scientific advisory board in 2009, but the company wasn't in a position to fund such an ambitious drug. The doorman was about to begin phase III trials, and before they felt comfortable investing in developing a combination drug with three molecules—a doorman and two correctors—Vertex needed to see evidence that a drug combining the doorman with just one corrector would succeed.

The doorman, VX-770, was the firstborn overachiever that surpassed all expectations. The best of the correctors, VX-809, was like a younger child suffering from comparison to its older sibling. But it was an unfair comparison. The new molecule had a different function, after all, and met all of the original criteria that CF project leader Eric Olson and Van Goor had established early, including the goal of improving chloride transport to 10 percent of normal or better—which earlier studies suggested would lead to milder symptoms. VX-809 did that. It also boosted chloride transport to 14 percent of normal in cells in the lab. And while those lab results foreshadowed that VX-809 might not produce the kind of dramatic clinical benefits VX-770 was already demonstrating in phase IIa clinical trials, as long as it could improve the health of CF patients with the F508del mutation (the 45 percent of CF patients with two copies and/or the 41 percent with one), even just a little, then the drug had the potential to save lives while Vertex's scientists hunted for something better.

For the moment, VX-809 would do; in October 2007 the San Diego team sent the molecule off to Olson's team at Vertex Cambridge to ready it for clinical trials. In the meantime, Hadida's team continued work on other correctors in the same family as VX-809, in case the molecule wasn't effective or had harmful side effects. The hunt for a second corrector would have to wait until later.

IN DECEMBER 2008, AS VX-770 WAS FINISHING UP PHASE II TRIALS, VX-809—after proving safe in animal testing—was quietly given to healthy volunteers in a phase I trial to test whether it was safe in humans, too. It was. Phase II trials began in March 2009. Eighty-nine[9] patients older than eighteen, each with two copies of the F508del mutation, were given varying doses of either VX-809 only or a placebo. The trial's goal was to determine if the corrector alone had any therapeutic value, before adding it to VX-770. Although no one expected it to do much on its own, that hypothesis still had to be tested.

On February 3, 2010, Vertex announced that, on its own, VX-809 was safe for these patients and confirmed their hypothesis that the drug did not by itself change the saltiness of patients' skin, help them breathe better, or provide any other benefits.[10] And by October 18, 2010, Vertex was ready to begin phase IIa for the combination doorman *and* corrector.[11]

The results of this approximately monthlong trial would reveal whether Negulescu and Olson's vision of the corrector refolding the mutant protein and helping it reach the surface, and then the doorman opening the chloride channel, was correct—whether CFTR could in fact be fixed with a one-two punch.

When Vertex released the highly anticipated results of the first trial of the doorman-corrector combination eight months later on June 9, 2011,[12] they showed that, on average, sweat chloride dropped by 13 millimoles per liter. It wasn't the dramatic, fall-off-a-cliff drop they had observed with VX-770, when sweat chloride plummeted roughly 48 millimoles[13] into the zone occupied by people who didn't have CF. But for patients with the most common mutation, that dip in sweat chloride levels[14] was still large enough to provide hope the combo drug would improve their health and make them feel better.

Unfortunately, the trial also revealed something discouraging. Unlike VX-770, which had been free of side effects, in half the patients taking the doorman-corrector combo, the drug caused a tightening sensation in their chest—a feeling that the air wasn't able to flow freely. While it wasn't life threatening, it was unpleasant and unnerving. VX-809 also had a bizarre and unanticipated impact on the levels of its partner, VX-770: it increased levels of an enzyme in the liver that broke down VX-770. That meant that the combo drug required a higher dose of VX-770 than had been used in Kalydeco.

For these reasons, Vertex and the foundation had decided, a year previous, when the results about VX-809 first started coming in, to expand their collaboration by pushing another of the correctors Hadida's team had been working on into clinical trials. By June 2010, that second corrector, VX-661, entered phase I trials. And in April 2011, the foundation expanded its collaboration with Vertex to develop a second generation of corrector molecules beyond VX-661—an endeavor that would require tens of millions more dollars from Joe O'Donnell's

Milestones II campaign.[1] The foundation vice president, Preston Campbell, referred to the VX-661 molecule, made by senior chemist Jinglan Zhou, as a "kissing cousin" of VX-809 because the two were so similar when it came to increasing chloride transport. But what Campbell and others hoped was that VX-661 might not have the same tightening side effect. They wouldn't know this, however, until they tested it in people.

Meanwhile, trials on VX-809 continued; its side effects were not life threatening, and the drug might save lives. The trial plan for phase IIb, which began in October 2011, was to include patients from twenty-one clinical trial sites across the US and in Europe, New Zealand, and Australia. Qualified patients had to be sick but not too sick, the right age, and have two copies of the mutated F508del gene. But because this trial had so many more volunteers than the trials for VX-770 had—it was focusing on a more common mutation—physicians were also able to consider other criteria in selecting trial participants. Were these patients compliant? Did they take their meds on time, exercise, and do all the other things that CF patients are supposed to do to keep healthy? Had they participated in other clinical trials—a sign that they were perhaps more altruistic than average and willing to make a sacrifice for the rest of their community? The trial site physicians chose individuals likely to follow the rules and generate the most trustworthy data.

While Vertex had anticipated a lag in enrolling patients, the phase IIb trial filled up quickly and launched in October 2011, after Vertex had already filed for approval of VX-770. Patients had heard about the stunning results of the doorman and witnessed VX-770 already transforming the lives of a small sliver of their community who participated in the trials, so they had high hopes for this new drug combination.

Foundation president Bob Beall and VP Preston Campbell were excited, too. With phase III trials for VX-770 complete and the New Drug Application under consideration at the FDA, and phase II trials in progress for the VX-770/VX-809 combo drug, there were treatments in the works for almost 50 percent of all CF patients—potentially close to 90 percent, if the combo drug also worked for patients with just one copy of F508del. But although Beall and Campbell were optimistic, their expectations for the combo had been tempered by earlier reports from the San Diego team during their quarterly meetings that this molecule wasn't as powerful as VX-770; it didn't trigger the same dramatic jump in electrical activity in the cell tests and was unlikely to be as successful as the doorman.

At the end of June 2012—just five months after the FDA had approved VX-770, now known by its commercial name, Kalydeco—the CF community was expecting great things from this next medicine. And the news was good. Vertex announced that in phase II trials, the doorman VX-770 and corrector VX-809 together significantly improved the function of the patients' lungs. One-third of patients with two copies of F508del saw a 5 percent boost; another 19 percent saw a 10 percent improvement.

Had Kalydeco—a drug without side effects, which dropped sweat chloride dramatically—not set expectations so high, Beall and Campbell would have been elated with those numbers. The combo drug couldn't compare. Nevertheless, it was far from worthless—a 5 to 10 percent hike in lung function could save patients on the brink of death and improve the lives of others. This was still a victory, albeit a more modest one.

PHASE III FOR THE COMBO DRUG WAS SCHEDULED TO BEGIN IN MARCH 2013. The physician leading these combo trials was Michael Boyle, an associate professor of medicine at Johns Hopkins University Hospital and director of the hospital's Adult Cystic Fibrosis Program, which he had launched in 1999 when he was just thirty-five.

Boyle was a local; he hadn't left Johns Hopkins since entering medical school in 1986, completing his residency and a fellowship in pulmonary medicine, and then joining the university's faculty in 1999. He had originally planned a career in pulmonary and critical care, but after spending a couple of weeks in the hospital's pediatric cystic fibrosis clinic, where patients ranged in age from infants to young adults, he changed his mind. He was particularly struck by these young people in their late teens and early twenties—not much younger than himself—who were battling this disease, trying to figure out how to live and navigate what remained of their short lives. They had to spend hours every day taking medicines and enduring physical therapy just to stay alive. And yet, faced with this adversity, they hadn't given up on life. They were figuring out ways to go to college, get jobs, and even start families.

In 1999, the expected life span for these patients was around thirty-two years. The CF population was at a turning point, Boyle recognized, and with the encouragement of Hopkins's CF Center director, he launched the Adult CF Program at Hopkins. He hired a nurse and an administrator, and grabbed a phone and a stool. Then he collected the red medical charts of all forty patients older than eighteen, loaded them onto a wheelchair, and moved to his new office, several blocks away from the pediatric patients.

It was a good time to launch the program; in 1996, the foundation had issued a mandate to the care centers it funded that if they had more than fifty adult patients, their parent institution had to create a separate center dedicated to caring for them. Boyle's wasn't the first adult CF program; there were several around the country, including one in Philadelphia, where the CF Foundation had its inaugural meeting back in 1955. But Boyle's was the only one serving Washington, DC, Maryland, Virginia, and the Delaware region. As news of the center spread, many young CF adults who had slipped through the cracks after pediatric care returned to seek treatment.

The patient population was a good match for Boyle, who was calm and soft spoken, with an easygoing, unpretentious manner that put patients at ease. He made a lot of eye contact, listened more than he spoke, and recognized that balancing a patient's life and goals with their treatment was a negotiation. There was more to these patients than their disease, and he wanted them to get the most out of their shortened lives—a very different approach from that of older, more paternalistic and authoritarian physicians.

As Boyle quickly learned, treating adults was very different from treating children. In many cases, when a child was young, the doctor's focus was largely on the parents, explaining to them what had to be done for their child and how to do it. Now Boyle focused all his attention on the patient, while the parent, if they were there at all, remained in the waiting room. These patients were often very sick, with tough-to-treat infections now resistant to antibiotics that had worked when they were children. Some were teetering on the brink of death, waiting for lung transplants; some were concerned with their fertility; others were wrangling issues like losing their parents' insurance or claiming disability.

When Vertex initiated phase III trials for the VX-770/VX-809 combo drug in February 2013, they had no trouble finding volunteers, and Boyle was hopeful that some of his patients would be able to participate. Phase III was the most desirable of all the trials; it was the longest and the safest, and participants went in knowing the drug was effective (though not to what degree). Those who received the combo were eligible to keep taking it after the six-month trial, and those who got the placebo were guaranteed immediate access to the drug once the trial was over. The phase III trial was the most extensive CF drug planned trial to date: it would involve some 1,000 patients, participating from two hundred clinical trial sites across North America, Europe, and Australia.[16] The endpoints were to drop patients' salt chloride levels, improve their lung function, and help them gain weight.

By the time the phase III combo trial launched, Boyle had already supervised patients in several clinical trials, including Kalydeco's, and he had witnessed their extraordinary progress. Now he was keen to see what this new drug could do.

One of the CF patients in the trial, on the other side of the country from Boyle, was a now-grown Jonathan Flessner. He had attended Colorado College between 2007 and 2011, focusing on Romance languages and remaining fairly healthy, except for a flu during his senior year that temporarily dropped his lung function to 40 percent—a terrifying ordeal. After college, he'd returned home to Seattle, where he completed a computer science degree and made his health a priority, determined to avoid another experience like the flu if he could. Jonathan had participated in many clinical trials for CF treatments by his early twenties; both he and his brother, Andy, had volunteered to participate in the phase II combo trials, but because the locations were restricted and the trial size so small, neither had been accepted. Phase III was larger, and this time Jonathan was able to join.

His expectations weren't high. Like his father, he knew not to let his hopes rise too high or allow his life to revolve around the trial. And news of the combo's underwhelming results had already filtered through the CF community grapevine, so he knew that even if he got the drug, rather than the placebo, the effect wouldn't be transformative. And it wasn't. Though Jonathan didn't suffer any of the chest tightening that many participants reported as a side effect, he barely felt any difference during the trial—although he did find that he spent less time in the hospital. He didn't know whether he had received the combo drug or the placebo, but when the six months were up and he became eligible to take the drug, he did. Again, his body didn't feel any different. But given that he wasn't experiencing any side effects, he decided to remain on the medication until he knew more about the phase III results—or until there was another drug combo that was better.

CHAPTER 49

What Mutation Are You?

2013–2014

This is what personalized medicine is really going
to look like. It [will be] ten genes [causing] a
disease rather than one. But this is the model. This
is the model of how it's going to play out.

—Josh Boger

By early 2013, Vertex had multiple clinical trials in progress for CF treatments. Phase III trials for the VX-770/VX-809 combo drug were beginning. After being cleared for use in patients age six and older, Kalydeco alone was now being tested in two- to five-year-olds with the G551D mutation to see whether they could tolerate the drug. Finally, the kissing-cousin corrector—VX-661[1]—had just completed phase IIa trials. This new corrector, when given with Kalydeco, improved lung function by 9 percent, similar to VX-809. But the real advantage was it didn't trigger the frightening chest pains and breathing problems that VX-809 did.

Back in San Diego, biologist Fred Van Goor and chemist Sabine Hadida's teams were testing Kalydeco on human airway cells with other mutations that, like G551D,[2] affected the chloride channel. They wanted to know whether the CFTR protein these other mutations produced could also be convinced to work with the help of a doorman. These mutations included ones that produced CFTR proteins with "residual function," meaning that they worked a little, but not enough to keep a person healthy. Perhaps these patients, who made up between 5 and 10 percent of the CF population—between 1,500 and 3,000 people—would also benefit from the drug. To figure out whether Kalydeco would work for each new mutation, Van Goor's biology team first created individual

441

airways in a dish using cells from a patient with that mutation, and then tested whether the drug could liquefy the mucus and get the cilia swaying again. If the results from the airway in a dish were positive, Vertex would launch a small phase III trial for people with that mutation. Then, if the phase III trial was successful, the FDA would "expand" Kalydeco's "label," a phrase that meant adding a new group of patients to those eligible to take the drug.

Such precise clinical trials were only possible because of the foundation's registry, and how the foundation's leadership, in particular a physician named Bruce Marshall, had been pushing to find patients' genotypes—the two versions of the gene each patient carried—and include that data in the registry.

Initially, this was intended to help geneticists dissect the disease and identify modifier genes and environmental factors that made the disease better or worse. Figuring out which mutations were linked to more severe forms of the disease was also critical for anticipating the problems a patient might face, especially as newborn screening for a growing panel of mutations in the CF gene began gaining a foothold. The results enabled a physician to anticipate the course of the disease and take preventive measures, rather than having to wait and treat symptoms as they surfaced. Some mutations, for example, were linked to a dysfunctional pancreas, which meant that the child required enzymes to digest food and prevent malnutrition. Those children were also at higher risk for developing diabetes.

This desire to understand the consequences of all the mutations led to an international research project called the Clinical and Functional Translation of CFTR (CFTR2).[3] With funding from the Cystic Fibrosis Foundation, scientists at Johns Hopkins University, the Hospital for Sick Children in Toronto, and the Cystic Fibrosis Center in Verona, Italy, had compiled genetic and clinical data on 90,000 cystic fibrosis patients worldwide and, in the process, discovered more than 2,000 mutations.[4] Not all of these caused severe disease, but many did. As part of the CFTR2 project, the 374 most common mutations, representing some 97 percent of the global CF population, were described in detail: how they affected sweat chloride, lung function, the pancreas, and more.[5] Some, like F508del, affected a significant number of patients worldwide; others were ultra-rare, carried by less than a handful.

By 1997, eight years after the discovery of the gene, the foundation had logged the mutations for just over 64 percent of patients listed in their registry. That number rose steadily, by a couple of percent each year, to 87 percent by the time it became clear VX-770 would reach clinical trials. And by 2013, as Vertex was planning phase III trials for the VX-809 corrector, the genotypes of 97.5 percent of US patients with CF were known. With such a trove of genetic data in the registry, the foundation knew which mutations almost every member of the population carried and could quickly identify patients for any type of clinical trial. No other foundation for a disease, rare or common, had such a resource.

The work Vertex was doing to expand Kalydeco's label wasn't just a benefit to patients. Every new group eligible to take the drug added another slice of the CF market to Vertex's plate. By 2013, the price of Kalydeco had risen by more than $10,000 a year, to $307,000 per patient, so even a handful more patients made a difference to profits.[6] More than 2,000 people in North America, Europe, and Australia had at least one copy of the G551D mutation, but based on the laboratory groundwork the San Diego team had done, there were another 5,000 people with other types of chloride channel mutations who might also benefit from Kalydeco.

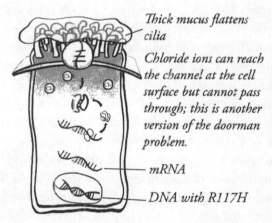

Thick mucus flattens cilia

Chloride ions can reach the channel at the cell surface but cannot pass through; this is another version of the doorman problem.

mRNA

DNA with R117H

A sick airway cell with the R117H mutation. There are many mutations like this one that produce a CFTR protein that rises to the surface but do not allow chloride to move through the channel, or don't allow much chloride to move through. Many of these proteins will work when exposed to the doorman drug.

The results of Vertex's label expansion work were first publicized in February 2014, when the FDA approved Kalydeco for CF patients with any of eight more mutations, in addition to G551D: G178R, S549N, S549R, G551S, G1244E, S1251N, S1255P, and G1349D. And from then on it seemed that every few months Vertex announced a new mutation that responded to Kalydeco. In late December 2014, the FDA approved the use of Kalydeco for patients older than six years with a so-called residual function mutation, R117H—some five hundred patients in the US. [7] One of those five hundred was Paul Quinton, the scientist who had made the critical breakthrough about how the CFTR protein functioned using sweat ducts gouged from his own arms and legs.[8] At seventy years old, he finally had a treatment. And with that latest approval, the total patients eligible to take Kalydeco, in addition to those with the G551D mutation, totaled 2,600 in the United States alone.[9]

As Vertex was testing mutation after mutation to figure out who else might benefit from the therapy, Kalydeco's unprecedented precision medicine was causing a stir in the wider biomedical and pharmaceutical community. Because of the collaboration between the Cystic Fibrosis Foundation and Vertex, cystic fibrosis had gone from an invisible, rare disease to one synonymous with personalized medicine and therapeutic breakthroughs. Vertex's San Diego team had transitioned from being the focus of internal and industry-wide derision to being leaders in cystic fibrosis and pioneers of an entirely new type of medicine. And some of the patients Vertex's drugs helped were becoming exemplars of how lives could be transformed as a result.

One such patient was Bill Elder Jr. Bill had been taking Kalydeco for about a year and a half and had completed a year of medical school when he was invited to testify in Washington at a congressional briefing on behalf of a local nonprofit and his Colorado senator Michael Bennet.[10] The focus of the briefing was the value of the FDA's "breakthrough therapy"[11] designation for new drugs, which had been used to expedite Kalydeco's approval. On July 24, 2013,[12] Bill shared with the panel how Kalydeco had completely transformed his life. Though many CF patients did advocacy work, this was Bill's first foray into it, and he hoped to convince his Washington audience that the FDA needed ways, like the breakthrough therapy designation, to allow more efficient drug evaluations when lives were at stake and patients had no options.

But Bill had little idea of the impact of his testimony until a year and a half later when, on a late December evening, he received a call from the White House. His phone rang just as he stepped into his apartment after an exhausting shift in family medicine, and the caller explained she was contacting him on behalf of First Lady Michelle Obama. "Michelle Obama would like to invite you as one of her guests to the State of the Union," she said. Sleep deprived and not processing the call properly, Bill told her he was in the middle of his family medicine rotation and had a lot of things scheduled; he had to clear the request with his boss at the hospital.

Bill was being invited to illustrate the value of the new Precision Medicine Initiative that President Obama would be announcing during the address. As the beneficiary of personalized gene-based medicine, and an example of how such medicine could change a person's life, he was a perfect choice.

Halfway through the January 20, 2015, State of the Union, the President described his new initiative plan and gave a shout-out to Bill: "I want the country that eliminated polio and mapped the human genome to lead a new era of medicine—one that delivers the right treatment at the right time. In some patients with cystic fibrosis, this approach has reversed a disease once thought unstoppable. Tonight, I'm launching a new Precision Medicine Initiative to bring us closer to curing diseases like cancer and diabetes—and to give all of us access to the personalized information we need to keep ourselves and our families healthier."[13]

After the speech, together with the rest of the First Lady's guests, Bill and his dad were ushered from the balcony, through a maze of secret staircases and passages, and guided to a tiny little room, where they met and shook hands with President Obama and had their picture taken.

A few days later, back in Ohio and still giddy from the State of the Union, Elder received another call from the White House office of science and technology. "This is very important: we would like you to come back to Washington, DC. We can't tell you what it's about, but this is a request directly from the president." This time Elder responded immediately: "I'll be there."

On January 30, 2015 Bill sat in the fourth row in the East Room of the White House as President Obama, flanked by the stars and stripes of the American flag and the President's flag, launched his Precision

Medicine Initiative.[14] Packed into the room were a list of the who's who of the new era of biomedical science, including NIH director Francis Collins, who had collaborated with Lap-Chee Tsui to discover the CF gene, led the effort to sequence the human genome, and who was now spearheading the new initiative. It was an inspiring day for Elder, who admitted that it was a thrill to meet not only President Obama for a second time, but also a cadre of scientists whom he considered his "nerdy scientific heroes."

Vertex's treatments for cystic fibrosis were leading the precision medicine movement, and the patients who benefited were the poster children. But as promising as things were for many CF patients, Bob Beall and Preston Campbell weren't satisfied. VX-809 had only been tested on patients with two copies of F508del, or 45 percent of the CF population; they still had half of all patients to worry about. And a small sliver of these wouldn't benefit from any of Vertex's therapies, because they had mutations that produced no protein, or proteins that were incomplete and couldn't be rehabbed with a doorman or corrector. These patients needed entirely new therapies that hadn't yet been conceived—and probably wouldn't be, unless someone was willing to invest not hundreds of millions but possibly billions in doing so. But after spending the last decade on two Milestones campaigns to pull together $250 million, how could Beall and Campbell even contemplate funding science that would cost billions?

CHAPTER 50

The Mother of All Deals

2014

> Bond didn't defend the practice. He simply
> maintained that the more effort and ingenuity you
> put into gambling the more you took out.
>
> —Ian Fleming, *Casino Royale*[1]

From the moment Cystic Fibrosis Foundation CEO Bob Beall had heard about Kalydeco's high price tag, back when the drug was first approved on January 31, 2012, he had been thinking about the contract that he, drug development leader Melissa Ashlock, lawyer Ken Schaner, and Aurora's chief knowledge officer John Mendlein finalized in May 2000 to pay Aurora $95 million over six to eight years to discover molecules that could fix CFTR. Although that agreement had evolved over the years, what remained the same was that, in the unlikely event that one or more drugs eventually resulted from that initial investment, the company, now Vertex, would repay the foundation with up to 12 percent[2] royalties on the drug's annual sales.

Beall was always anxious about the appearance of profiting from these drugs and wanted to reinvest those royalties as soon as possible into making better, more powerful medicines. But a trickle of royalties every quarter for the next twenty years wasn't enough. He and Preston Campbell wanted to sell the rights to those future royalties now, as they'd done with TOBI, so they could immediately plow that larger sum into even more ambitious research. The royalties from TOBI had brought in $19 million, which went straight into the pot to fund Aurora.

As soon as Kalydeco was approved in 2012, Beall and Campbell, with the go-ahead of the foundation's board of trustees, decided to sell a small slice of the drug's royalties. After all, even if the corrector VX-809,

447

with all its lackluster performance, was approved, they still needed to find a drug for the rest of the CF population. And Beall and Campbell had much more audacious plans beyond that: they wanted a cure, so a patient could walk into a doctor's office and leave without the disease. That was the reason that Campbell had left his position at Vanderbilt and come to the foundation. But despite recent technologies like the gene-editing tool CRISPR, which could correct errors in DNA like an editor could correct a typo in a book, a cure remained a lofty goal, and Beall and Campbell's immediate concern was getting treatments to all patients with CF. They couldn't stop at just 54 percent—patients with G551D (4 percent), with nine other similar mutations (4 percent), and with two copies of F508del (46 percent).

With Beall and Campbell's urging, in 2012 and 2013 the foundation sold first and second batches of royalties from Kalydeco to a Canadian pension fund for $156 million and $257 million, respectively.[3] The need for money had become even more urgent, because the foundation and Vertex were just completing phase II trials for VX-661, the less problematic corrector. By the beginning of 2015, Joe O'Donnell's Milestones II campaign had successfully reached its target of $75 million. The money was welcome; the agreement to accelerate VX-661 would cost another $20 million. And while the VX-770/VX-809 combo, the first treatment for people with two copies of the common mutation, was a helpful first step, Vertex could make, as CF biology leader Fred Van Goor and chemistry leader Sabine Hadida had discovered in 2009, an even better triple combo: a drug with one doorman and two correctors. Beall and Campbell wanted the royalty cash to support that vision.

And the triple-combo drug wasn't the only thing Beall and Campbell wanted to finance right away. They wanted to take bigger gambles on emerging treatments, and they wanted to do so fast. They wanted to launch programs that might not pay off for a decade: treatments using stem cells to regenerate parts of the lung and other organs that had been decimated by the disease; gene editing to fix everything from stem cells in the lung to CF-causing mutations in embryos still in the mother's womb; gene therapy strategies for patients with mutations that aborted the CFTR protein before it was complete or lacked the protein entirely. Both men knew that, while finding and approving drugs one by one would yield a stream of royalties for what they were planning—tens to a hundred or so million each year—they'd need a lot more money, a lot more quickly. Even the $400-plus million the foundation had received for the two sales of Kalydeco royalties wouldn't be enough.

By mid-2014, Vertex had just completed phase III trials of the VX-770/VX-809 combo, and both Beall and Campbell were fairly sure that the FDA was going to approve the drug for patients with two copies of F508del. If this drug, which worked for almost half of the CF population, had a similar price tag to Kalydeco, then the annual royalties would be substantial—in the hundreds of millions of dollars. With those potential royalties now on the horizon, they both thought it was good timing to cash out.

The two men reached out to Morgan Stanley to find a buyer who would pay them up front for the royalties that Kalydeco, plus potentially other drugs—like the combo that had resulted from the initial investment in Aurora and Vertex—would bring in for the foreseeable future. A consulting firm put together a projection of what Vertex would earn from Kalydeco over fifteen years, and the subsequent royalties, along with what a second drug might bring in over the same period if the FDA approved the combo in the next few months. The buyer that stepped up was Royalty Pharma, a private investment firm with a focus on healthcare royalties. The company was excited about Kalydeco and future combination drugs that included it, recognizing that the drugs Vertex had in its pipeline had the potential to treat more than 90 percent of the CF population.

Based on the estimate that Morgan Stanley had issued for the future earnings of these drugs, in November 2014 Royalty Pharma offered to pay $3.3 billion for revenue from Kalydeco and any other drug Vertex developed under the foundation's original 2000 agreement. It was a more than twentyfold return on their fifteen-year investment of $150 million. The upside for the foundation, other than receiving that lump sum, was that the money was theirs regardless of whether VX-770/VX-809 and other correctors in development were approved or not. And Royalty Pharma stood to do well over the long term if other drugs then in the pipeline were also approved.

The deal made the foundation the largest disease-focused charity in the United States and the fifth most financially powerful health nonprofit in the world, just a few pegs below their past benefactor, the Bill and Melinda Gates Foundation, and landed the CF Foundation on the cover of every major newspaper in the country.[4] Many articles raised concerns. Some critics believed that the $3.3 billion would discourage contributions from future supporters.[5] Others suggested that the amount was evidence the foundation was benefitting from Kalydeco's sky-high drug price and hadn't worked hard enough to convince Vertex to lower the price.[6] Perhaps, one article suggested, the foundation should be using that money to pay for patients' drugs.

But that wasn't going to happen. That money was solely for research; the royalties would be used to discover a cure and expand programs and services for people with CF.[7] It was the foundation's mission to invest in such things. It was why they had Kalydeco to begin with.

As it turned out, the sale of those royalties in November 2014 was well timed, because the future of the VX-770/VX-809 combo, still awaiting FDA approval after being submitted earlier that month, on November 5, was beginning to look a lot less certain.

THE PHASE III DATA SHOWED THAT THE VX-770/VX-809 COMBO DRUG worked. After just six months, patients had fewer lung infections and fewer hospitalizations, and had begun gaining weight.[8] For CF patients, every ounce was a victory, and fewer infections meant hope for a longer life. The announcement was the most anticipated news in the biotech industry, and it sent Vertex stock soaring 40 percent, landing at $93.57 a share.[9] It was further proof that drugs for orphan diseases like cystic fibrosis could be wildly lucrative—and thus financially feasible to pursue.

But reviewers at the FDA were having doubts about this second treatment, because the combo drug only increased lung function 2 to 4 percent, compared to Kalydeco's 12 percent. Maybe it wasn't worth approving, especially if its price was similarly stratospheric. So Anthony Durmowicz, a pediatric pulmonologist by training who was chosen as the cross-disciplinary team leader for the group reviewing the VX-770/VX-809 data at the FDA, decided to convene an advisory committee of outside experts to help determine whether the evidence supported the drug's approval.

At 8 AM on May 12, 2015, all the parties gathered at a Hilton in Gaithersburg, Maryland, in a large ballroom. It had been arranged like a courtroom. A panel of thirteen physicians, the judges, sat in front, facing the audience. To one side of the panel was a microphone for the speaker; on the other was a screen for presenting data and other evidence. The audience was a collection of patients, families, and physicians who had participated in and led the trials; foundation leaders Bob Beall and Preston Campbell; and representatives from other pharmaceutical companies who were intrigued by the drama. In many ways, this was like a court case, with the combo drug on trial.

For three hours, scientists, including Fred Van Goor, and physicians from Vertex presented data, beginning with the lab tests done in San Diego all the way through to the trial results, while panelists grilled them on the details. Physicians whose patients had participated in the phase III trial shared their opinions. After a lunch break, more than a dozen patients and their families gave powerful testimony. No matter what the numbers said, this drug, they felt, was a game changer.

As a young woman named Kate Marshall, a CF patient from York, Maine, told the committee, "Not losing 1 to 2 percent of my lung function each year is remarkable data . . . To not approve it for so many that have suffered will not only be confusing but cruel."

Six-year-old Ariana Masters, whose father had CF, told the committee, "Please vote for this drug to be approved so my daddy can stay healthy and play with me."

Jillian McNulty, who had flown in from Ireland, shared how, before the combo drug, she had spent between twenty-four and thirty weeks in the hospital every year of her life. But in the last year, she'd spent just eleven weeks in the hospital. She could now fight colds and flus without antibiotics.

After the public comment portion of the meeting was over, the panelists voted. The FDA was not obligated to follow the committee's recommendation, but they often did. That day, the committee voted twelve to one to approve the combo drug for patients with two copies of the F508del mutation.

As everyone filed out of the room, Van Goor felt his body sag with relief that the ordeal was now over. Stepping out of the ballroom and into the hall, he heard an eruption of voices. The families were laughing, crying, and taking pictures, and before he knew what was happening, people were descending upon him and hugging him, thanking him and Vertex for making the drug that would now be available to their loved ones.

A couple of months later, on July 2, 2015, the FDA granted their approval as well. Vertex christened their second successful drug for cystic fibrosis patients Orkambi. Treatments were now available for more than half the patients with this disease. It was a hard-won victory. But with the drug approved, and $3.3 billion in royalties and another $500 million or so in the bank, the foundation could now focus aggressively on investing in new treatments that would ensure no patient was forgotten.

CHAPTER 51

Very Personal Clinical Trials

2012–2017

Only those who will risk going too far can
possibly find out how far one can go.

—T.S. Eliot

Aerospace engineer Juliet Page had always suffered from lung issues. She'd never had a simple two- to three-day cold—it was always three to four weeks of sickness, battling pneumonia or bronchitis, and tossing back antibiotics to kill lung infections. Doctors had considered cystic fibrosis at various points in her life, and she'd even had several sweat tests that registered high salt. But her physicians never acted on it. She just kept coughing.

While the lung infections were irritating, they never really stopped Juliet from doing what she wanted. But in 2007, when she was forty-two, living in Annapolis, Maryland, she developed a dangerous bacterial infection called mycobacterium avium complex that she couldn't shake. Over the next five years, she remained on three different high-dose oral and inhaled antibiotics that made her permanently tired, nauseated, and depressed. Then there was the almost constant diarrhea. The side effects were so severe that she considered taking leave from her job and going on disability. Her physician was frustrated that he wasn't able to help Juliet and suggested she visit the CF clinic at Johns Hopkins University in nearby Baltimore. He didn't know much about CF, but he knew they had a lot of drugs for lung infections. Perhaps one would work for her.

At Johns Hopkins, Juliet met with a young pulmonologist named Patrick Sosnay. He had trained there in Baltimore and had decided to focus on cystic fibrosis after talking with Michael Boyle, the doctor who had launched Johns Hopkins's adult CF clinic. He'd admired the compassionate way that Boyle cared for his patients and wanted to practice medicine the same way.

A scientist to her core, Juliet had kept meticulous records of her own disease. She told both Dr. Sosnay and Dr. Boyle, who was supervising Sosnay, that she thought she had cystic fibrosis, then methodically laid out the evidence. A patient coming in suspecting they were suffering from CF wasn't a rare occurrence for Dr. Sosnay; others arrived from all over the world with similar claims. But Juliet was different. She'd had problems as a child, and the lung infections had gotten steadily worse over her life span. She was concerned about her sweat being salty. In short, there were many reasons for her to be tested. But even as Dr. Sosnay heard the evidence she laid out, as others had under similar circumstances, he wrote down in his notes based on his clinical exam that Juliet most likely did not have cystic fibrosis.

He was wrong. The results from the new sweat test he gave Juliet were enough to diagnose her. However, Sosnay also wanted to know which genetic mutations she carried—because she wasn't the typical CF patient. Her pancreas worked well and she didn't experience any of the digestive maladies that affected most patients. By the time of Juliet's diagnosis, in January 2008, Vertex was in the midst of phase II trials for VX-770, and the Cystic Fibrosis Foundation was urging physicians at all the care centers to identify every CF patient's mutations ahead of the drug becoming available. Sosnay took a blood sample from Juliet immediately, and by March he had an answer. After both copies of her CF gene were sequenced, letter by letter, they discovered one copy of the gene had the F508del mutation that most CF patients carried. But the other copy of her gene had two mutations. The first was G178E, a mutation carried by only two other known people in the world—two children in Italy. No one had ever seen the second mutation before.

Frustrated that there was no treatment yet in the works for either the common mutation or either of the other two she carried, Juliet started her own research, poking around online for more information. She came upon the database Lap-Chee Tsui had launched back in 1989 just after discovering the CF gene to catalog all its variants. By 2008 there were more than 2,000 known mutations.[1]

Studying a diagram of the CFTR protein, Juliet found that her G178E mutation was nested in a critical region responsible for forming the cell's chloride channel. Like G551D, the mutation VX-770 targeted, this mutation altered the shape of the channel, making it malfunction.

Juliet knew Vertex was in the midst of phase II trials for VX-770, and she asked Drs. Boyle and Sosnay when the company would begin clinical trials for people with rare mutations. That's when Dr. Boyle told her: there wouldn't be a clinical trial for her. There were too few people with her mutation for a pharmaceutical company like Vertex to do a clinical trial. Without a minimum number of patients, it was difficult for a study to prove that a drug significantly improved the symptoms of a disease or cured it. And without that evidence, the FDA wouldn't approve it.

But Juliet's question had raised an issue that had been vexing Dr. Boyle for some time. It wasn't feasible to make a drug to match every mutation, particularly when there were more than 1,700 CF mutations carried by fewer than fifty people worldwide—1,000 of them carried by no more than five people. What could be done for these patients with ultra-rare mutations?

At forty-seven, Juliet's FEV1—her lung function—was now at 55 percent. Every year it dropped a percent or two. And now she was getting sicker, with more infections each year. With no other options at the time, Dr. Sosnay prescribed nebulizer treatments, Pulmozyme, TOBI, and physical therapy to loosen the junk in her lungs, hoping that they could keep Juliet healthy enough until something better came along.

In 2012, Kalydeco was approved for patients, like Laura and Cate Cheevers, with the G551D mutation. When Juliet heard the news, she returned to Dr. Boyle (Dr. Sosnay was in Malaysia for a two-year appointment to help launch an arm of the Johns Hopkins Medical School) and asked if there was a way she could try the drug, even though she had a different mutation. To Juliet, the fact that her mutation was in approximately the same region of CFTR as G551D suggested that the doorman drug might be able to fix her mutation, too. But if you didn't have the G551D mutation, you couldn't get the drug—or at least, you couldn't get your insurance company to pay for it. Dr. Boyle could write Juliet a prescription, but she'd need proof that her mutation matched the approved one on the Kalydeco label before insurance covered it. And without insurance, the cost was steep—$28,000 per month.

Still, the only way to figure out whether the drug would work for Juliet was to try it. So Dr. Boyle began brainstorming how they could get the drug to test it for off-label use. Eventually, he decided he'd try to convince her insurance company to support a drug trial with just one patient: Juliet. He called and explained to them how the drug worked, telling them about Juliet's mutation and why he believed this experiment was worth doing. In April 2012, three months after Kalydeco's approval, and after multiple conversations with Dr. Boyle, the insurance company finally agreed to provide Juliet with a month's supply of Kalydeco—fifty-six pills, to be taken in two separate fourteen-day trial periods.

In the first week of May, Juliet met with Dr. Boyle at Johns Hopkins's CF adult care center in downtown Baltimore. He wanted to measure her pretrial sweat chloride levels and run a couple of other tests to get a baseline portrait of her health. He knew from the Kalydeco trials in patients with the G551D mutation that, if the drug worked, the salt levels in her sweat should plummet. That day, Juliet had a sweat chloride level of 70, per Paul di Sant'Agnese's sweat test—still the gold standard fifty years after its invention. The number was not super high, like that of most CF patients; only ten points higher than the level needed for a definitive CF diagnosis. Her lung function was 55 percent.

The next morning she took the first little blue pill—and then one every twelve hours for the next fourteen days.

Within hours of that first dose Juliet felt different. She began to cough and hack, her sinuses seemed to broaden and expand, and her body began purging hard, rubbery globs of mucus—material that had been lodged in her lungs and sinuses for what she figured must have been years. She called Dr. Boyle. "It's like I'm hacking up fur balls," she told him.

It was a promising sign, but not yet proof that Kalydeco was working. Juliet continued taking her pills for the next two weeks. On day fifteen, she drove from her home in Annapolis to the Johns Hopkins clinic, where the nurses sat her down for tests. Her lung function had increased by ten points, to 65 percent, after just two weeks on the drug—a significant change, if not as dramatic as the one seen in some users. But her sweat chloride had dropped to a staggeringly low 38. Below 40, she would no longer test positive for CF. "That's completely normal," Boyle told her, somewhat incredulous at the dramatic result.

Now, for the next stage of the study Boyle had designed for her, Juliet had to stop taking the drug for two weeks. After just twenty-four hours she could feel the difference. Her lungs felt more rigid, heavier; her breathing was slightly more labored, and she wasn't coughing up as much gunk.

Juliet was miserable and cranky at the end of this stoppage, arriving early for her appointment for sweat chloride measurement and breathing tests so she could start taking the drug again as soon as possible. After two weeks without the drug, her sweat chloride was 74—even higher than her pretrial benchmark, and back in the diagnostic zone for CF.

She barely waited until she had left the clinic before pulling the pill out of her purse and taking the long-awaited dose. And for the second two weeks on Kalydeco, she saw the same effects she had the first time: her lungs loosened with lubrication, she expelled more noxious, bacteria-filled mucus, and she felt healthy again. Indeed, when she returned to the clinic after taking the last pill, her sweat chloride was only 21.

Elated by Juliet's response to the drug, Dr. Boyle immediately wrote to the insurance company explaining that, with Kalydeco, her sweat chloride levels were normal—a sign the drug had worked for her. The insurance agreed to cover a one-year supply of the drug. But they wanted more follow-up before they approved it beyond that year: Did the drug improve her long-term health?

The answer was yes. That next year, Juliet didn't get sick. For the first time in her life she recovered from a seasonal cold in a few days, without antibiotics. She stopped using her nebulizers and breathing treatments, and her lung function remained stable.

With the success of Juliet's case, Dr. Boyle used the same strategy to get Kalydeco for two other patients with rare mutations. Other doctors around the US began following his lead, performing single-patient off-label case studies to figure out whether Kalydeco was a good match for their patients. In about forty cases, patients were able to get the drug that way.[2] When Boyle looked these patients up in the foundation's patient registry (the same one Warren Warwick had set up some forty years earlier), he saw that six months after patients began these make-shift Kalydeco trials, they'd all gained weight and their lung function had jumped, on average, some six percentage points.

IN FEBRUARY 2014, AFTER VERTEX PERFORMED A COMPLICATED, thirty-nine patient clinical trial, the FDA expanded the list of patients eligible for the drug to include eight more mutations—together, about

150 more CF patients—all of which affected how well the channel opened and closed.[3]

Juliet's mutation was not included in this new list. But a similar one, G178R, was. This DNA mutation changed the amino acid at position 178 of the protein from glycine to arginine, whereas Juliet's mutation changed the same amino acid from glycine to glutamic acid. Yet the drug still worked for her. Since being on Kalydeco she had managed to eradicate the bacterial infection in her lungs, hadn't had bronchitis or pneumonia or needed to take antibiotics, and had enjoyed stable lung function.

Dr. Boyle's experience with Juliet and several other patients had made him realize the FDA needed to change how they judged drugs, drug companies needed to revise how they tested them, and researchers needed to decide what to do about rare mutations. In 2016, now working for the Cystic Fibrosis Foundation as their senior VP of therapeutics development, he holed up for four days, developing an argument to present to the FDA—and Juliet's case was the perfect example to share. For the FDA, a clinical trial provided the hallowed gold-standard evidence for approving a drug for a particular set of patients—or denying it. Boyle's argument was that the FDA ought to accept other evidence, too, because there were patients with unique mutations who would never be eligible for a clinical trial. People like Juliet couldn't be ignored when there were approved therapies that might work.

Vertex had also been trying to figure out what to do about patients with rare mutations. But they were going about it from a different perspective. After Kalydeco's approval, Fred Van Goor's team at Vertex had begun testing other known rare CF mutations where the channel failed to open to see if Kalydeco would work on those, too. It was, after all, in their interest for the drug to work for as many patients as possible.

Originally, to expand Kalydeco's approval, they had studied additional mutations the same way they had the G551D mutation for which it first had been approved: testing them in rat cells engineered to carry a mutated human CFTR gene with a specific mutation; then in airways in a dish with human cells, where they could see the cilia sway back and forth if the drug worked; then in actual humans through clinical trials. That's what they did for R117H, the mutation carried by Paul Quinton and five hundred other CF patients.[4] But to follow this procedure for every known mutation, each one increasingly rare, was expensive and slow. And for many mutations there weren't enough patients to run a trial.

But whenever a drug worked in rat cells for those first ten mutations, it had also proved successful in human trials. Perhaps, Van Goor thought, as they had now proven that the drug was safe in ten clinical trials for ten different mutations, they could just use the rat cells to test whether Kalydeco was likely to work for a rare mutation? Then, if that test worked, they could skip the clinical trial and give it directly to patients.

Since 2010, Fred Van Goor's team of three San Diego biologists had engineered one hundred different lines of rat cells,[5] each carrying a human CFTR protein with a different rare CF mutation. Now, they tested all of these cells to see how much of the protein actually made it to the cell's surface, as well as how much chloride was able to move through the protein. Only if the protein actually made it to the surface was there a chance that the doorman drug would work. Then they added Kalydeco to each mutant cell and measured changes.

For cells with certain mutations, the results were dramatic; chloride activity spiked. For others, Kalydeco did almost nothing. In their testing they had included the R117H mutation, which responded to the drug, along with the common F508del mutation, for which clinical trials had already revealed that this drug alone had no effect. The results for both further demonstrated that it was possible to anticipate the impact—or lack thereof—of Kalydeco in patients based on rat cell testing.[6] At least for Kalydeco, which clinical trials had already proven to be safe and effective, these in vitro tests seemed an accurate proxy for a rare-mutation clinical trial. Vertex had a new type of test that could give patients with a one-of-a-kind mutation, like Juliet, confidence that the drug would work for them without testing it on them first. And if they could convince the FDA that this was safe, more patients could have their insurance pay for the drug.

When Vertex submitted these in vitro tests to the FDA, asking them to approve the drug for dozens more patients with rare mutations based on a few laboratory tests, the FDA had to conduct an inspection. Previously, tests in engineered rat cells had been used to decide which mutations should be included in human clinical trials. Now Vertex was asking the FDA to approve drugs for these patients based solely on these rat cell tests. Were the rat cells really a reliable proxy for human cells? The request was unprecedented, and the FDA wanted to ensure that the research was rigorous and robust.

Early in 2017, head of Vertex San Diego Paul Negulescu received a call from the reception desk that two FDA agents were waiting in the lobby to meet with him. He knew that the FDA was permitted to visit their research facility, and there was a protocol in place for just such an occasion. And while Negulescu was surprised to receive visitors that day, the visit itself was not a surprise. In fact, he had been expecting it.

Negulescu knew that approving a drug for additional mutations based on in vitro data and not a fresh clinical trial would set a new precedent for the FDA. And he knew that before they took such a step, they would want to see the original notebooks, review the calculations, examine the equipment used for the experiments, and confirm that all the data packed into the couple of summary tables in Vertex's application was real and accurate.

Negulescu escorted the FDA agents to a room, where they asked Van Goor to show them how the Vertex scientists ran all the tests with the cells, the instruments that were used, and the software that converted the data and performed the analysis. While Vertex's technology and science were proprietary—developed in house and not commercially available—nothing was off limits to the FDA.

So, Van Goor took them to a lab where one of the scientists, who had performed the original analyses eight years earlier that were cited in the drug application, was running similar experiments that day. He showed the agents each step: here is the plate, this is where I put the cells, this is where the electrodes go, here is the data coming out on the screen, this is the quality control, and then here is how it is entered into the notebook and "published."

The agents took notes. Stacks of laboratory notebooks were retrieved from the archives and from off-site, and over the course of two full days, the agents examined all the records.

By the end of the two days, when the agents told Negulescu and Van Goor they had completed their work, Negulescu couldn't help asking, "So what do you think?" While the agents didn't tell him Vertex was in the clear, they did say that what they had seen looked good.

That opinion was reinforced on May 17, 2017, when the FDA announced[7] that it had accepted Vertex's in vitro tests and approved Kalydeco for twenty-three additional so-called residual function mutations. As their name suggested, in these mutations, the chloride channel worked a little, but not well—like a door that would only open with a good shove. What was revolutionary about this drug approval was that it was the very first to be based on tests done in cells—not living humans.[8] It marked a new era of personalized, precision medicine: testing cells in the lab to figure out which individuals might respond to a therapy, a process called theratyping.

This new strategy for approving drugs was a victory for both Vertex and patients with super-rare mutations. It enabled a few more CF patients to benefit from Kalydeco's lifesaving properties, but its effects reached even further. In a blog describing the broader implications of this decision for the pharmaceutical industry, the FDA noted that for specific mutations, in vitro assay data could be used instead of small clinical trials when companies sought to expand the use of an approved drug with a clean safety profile to other members of the patient population.[9]

In the meantime, Boyle and Preston Campbell, who had become president and CEO of the CF Foundation when Bob Beall finally retired in 2015, were already focused on a new drug Vertex had in the works that might be effective for people with one *or* two copies of F508del—or nine out of ten cystic fibrosis patients.

CHAPTER 52

The Triple

2015–2017

Simply and clearly put: without the Cystic Fibrosis
Foundation funding, Vertex could not be in CF.
Without their innovative funding model, we might
be enthusiastic and passionate about transforming the
treatment paradigm for patients with this genetic disease.
But we wouldn't be able to act on that passion.

—Joshua Boger, Vertex founder and CEO

Watching Laura and Cate Cheevers go through the clinical trials for what became Kalydeco had been one of the most fascinating periods of Dr. Allan Lapey's career in cystic fibrosis. To see these two children side by side, one on the drug and one on the placebo, and then see their trajectories deviate within a couple of weeks, was striking. The way their coughs behaved, the way they gained weight, Dr. Lapey knew the medicine was working on a fundamental level that he had never seen before with any other treatment, including antibiotics. Yet even once both were taking the drug, Laura's and Cate's health differed from early on. Cate stopped coughing entirely with the treatment. But while Laura also felt better, gained weight, and was rarely ill, she continued to cough—and to Lapey, that meant there was room for improvement.

In 2015, Laura Cheevers was in her junior year of high school and doing well, thinking about where she might apply to college. She had been taking Kalydeco for more than three years when Dr. Lapey called Kim Cheevers to suggest that Laura enroll in a phase III trial of a new drug Vertex was testing, a combination of VX-770 and VX-661. It was similar to the recently approved combination drug, Orkambi, in that both contained the doorman VX-770. But, Lapey explained, the corrector molecule in Orkambi (VX-809, which had the chest-tightening side effects) had been replaced with a new, better corrector that was supposed to enhance VX-770's effects.

Molecularly speaking, the new corrector, VX-661, was very similar to VX-809. It had evolved from the same high-throughput screens as the VX-809 corrector and had been kept on a back burner in case VX-809 either didn't pan out, or had an unpleasant or dangerous side effect that only became apparent during clinical trials. In fact, as early as 2010, Eric Olson had suggested to then–foundation CEO Bob Beall and Campbell that they accelerate the development of VX-661, just as they had for the doorman drug. Vertex couldn't justify the expense at the time; they were pushing the doorman into clinical trials and working on the corrector, VX-809, and didn't have the resources to focus on a third molecule, especially when they still had no idea whether the doorman or the corrector would work in people. Having been through the same argument twice before with Olson, Beall and Campbell had agreed to provide an additional $20 million to shave two years off of this new molecule's development time by working on manufacturing practices and animal testing simultaneously.[1]

It was a prudent decision. As Vertex began testing VX-661 in phase I trials, phase II trials of VX-809 were already revealing problems, including the chest-tightening in some patients that felt like the constriction of asthma.[2] Compared to VX-809, VX-661 wasn't quite as effective in increasing the flow of chloride through the CFTR protein, but it was a superior drug. It produced the same therapeutic benefit as VX-809 but at a lower dose. It was distributed throughout the body more evenly than VX-809—which was important, as cystic fibrosis affected many organs and tissues. It also didn't interfere with breathing or boost the levels of liver enzymes that chewed up its partner drug, VX-770.

By May 2015, Vertex was facing an FDA advisory committee to discuss whether their first combo drug—VX-809 and VX-770, the doorman—was worth approving, and the company was busy preparing for that. But in the background, preparations kept chugging along for phase III trials on the new combo, VX-661 and VX-770, which had been scheduled to begin in early 2015 with four different sets of patients. Among those were patients who had two copies of the F508del mutation, and patients who had the F508del mutation in one copy of the gene and a different mutation in the other. If the new combo drug worked, it could provide two of the three molecules for a triple combination medicine that would reach close to 90 percent of CF patients.

When Dr. Lapey called Kim Cheevers about enrolling Laura in this new trial, she wasn't surprised. She had recently bumped into Jeffrey Leiden, Vertex's CEO, at a black-tie CF fundraiser in Boston, and he had told her that there were drugs coming through the pipeline that would make Laura feel even better.

Laura's sister, Cate, who had always been the healthier of the two, wasn't eligible for the trial because, with Kalydeco, her lung function was close to 100 percent; she was too healthy to qualify. She didn't need a nebulizer and had completely stopped physical therapy. Cate was also now an athlete in competitive sports, with a busy schedule that would have made it difficult to participate. Clinical trials were a big commitment, and during the first three months, the VX-661/VX-770 trial was intense. Every two weeks, beginning in the fall of Laura's junior year in 2016, Kim had to take Laura to the hospital, where her blood was drawn; her heart activity, breathing, and lung function measured; and sweat tests done.

Within a couple of weeks—even faster than when she had begun taking Kalydeco—Laura felt the new drug duet transforming her body chemistry, altering it in ways that Kalydeco alone had not. After beginning Kalydeco, the results had been so dramatic that she didn't think she could feel any better. With this new drug, she did. She used to feel jumpy and cough violently after her dance classes, but with VX-661/VX-770 the cough stopped completely, her breathing was easier, and she had even more stamina.

The new drug made Laura so much stronger and more robust that, in February of her senior year in high school, she traveled abroad for a two-week program in Spain. For Kim, the anticipation was horrible: Would Spanish authorities allow Laura to enter the country with an unapproved, unlabeled drug? Did she have all the other possible drugs she might need? She was continuing to take four other pills—digestive enzymes and supplements—because Kim didn't want her to stop while she was still growing. A couple of Laura's friends had learned how to do her physical therapy so they could help her keep up with it while in Spain. Kim also filled a notebook with instructions of what to do if Laura got sick. But when Laura connected with her mother using WhatsApp, Kim saw her daughter, as healthy as everyone else, having a fantastic time.

Vertex filed for approval of VX-661/VX-770 in July 2017, just as Laura was preparing to enter college at the University of Providence. The drug combo had changed her life. She was down to two physical therapy sessions a week and no longer needed a nebulizer. After the trial, Laura opted to remain on the drug combo while the FDA was assessing it. In February 2018, the new combo drug was approved and given the brand name Symdeko.[3] But this new drug was much more than a replacement for Orkambi. It contained what Vertex's Fred Van Goor had hypothesized, several years before, as two of the three components of a powerful new drug therapy, able to successfully treat any patient with at least one copy of F508del and also patients with other doorman mutations. Based on experiments that he and his team had run at Vertex, Van Goor anticipated that such a drug would need to contain three medicinal molecules: VX-661 (the corrector in Symdeko), VX-770 (the doorman in Kalydeco), and a new "second-generation" corrector that Hadida's team had spent the last couple of years sculpting from crude scaffolds.

A slew of studies from 2005 onward had provided evidence that when the amino acid phenylalanine was absent from position 508 in the CFTR protein—as it was in people with the F508del mutation—then the first critical fold didn't happen, and the cell's quality control system would destroy the misfolded, mangled protein. VX-809 (the first corrector developed) could rescue this folding step,[4] in the process boosting the chloride transport activity of the CFTR protein to 15 percent of what it would be in the lung cells of healthy people. But no matter how many different correctors Sabine Hadida's team of medicinal chemists made—and they had made some 25,000[5]—they couldn't get the activity higher than 15 percent. It was like a magic ceiling that the Vertex scientists couldn't smash. But Hadida's team had also observed that the correctors they were identifying seemed to fall in two groups—and that when they added one corrector from each group to a cell at the same time, there was a greater jump in the chloride transport than when they added each corrector alone. That suggested that each corrector was doing something distinct—and that using two corrector types together might make it possible to blast through that 15 percent ceiling.

Perhaps, Hadida thought, there were two critical stages of folding, and each one needed its own corrector. Findings from Van Goor's biologists supported this idea. When they played with the environmental conditions, incubating airway cells at 18 degrees Fahrenheit below body

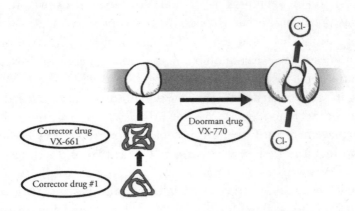

To discover a second corrector drug, the Vertex team
searched for molecules that, when added to cells with
the F508del mutation, would work with the doorman,
VX-770, and the corrector, VX-661, to dramatically boost
the amount of chloride moving out of the cell.

temperature—which slowed folding—that somehow fixed one step. Adding VX-809 then seemed to fix the other folding step. That was when they saw a significant jump in chloride transport.

In a presentation to Vertex's board in 2009, Van Goor told his audience that there seemed to be two critical locations on the F508del CFTR protein where correctors could bind that would stabilize the protein, help it fold properly, and therefore reach the surface in dramatically higher quantities. If they could find a second corrector to stabilize the CFTR protein, they might be able to boost the activity of the CFTR protein in patients with the common F508del mutation to more than 45 percent of normal—approximately what Kalydeco achieved for patients carrying the G551D mutation.

Confirmation that there were two critical folding steps also came from outside Vertex: a collaborative group of academics who were members of the CFTR Folding Consortium, ten labs dedicated to understanding why the mutated CFTR protein misfolded and finding molecules to correct it.[6] In 2012, a team led by Philip Thomas at the University of Texas Southwestern Medical Center, and another led by Gergely L. Lukacs[7] at McGill University, Montreal, showed that the F508del mutation caused two defects in the protein. A single corrector molecule, like VX-809, could only fix one, which was why the cells of patients who took it were only able to reach 15 percent of normal CFTR function.

Drug Combination	Resulting Chloride Transport
VX-770 (Kalydeco)	Approximately 50 percent[8] of the chloride transport of a healthy person with two normal CFTR genes (the same amount of chloride transport as healthy carriers who only have one normal CFTR gene)
VX-770/VX-809 (Orkambi)	About 25 percent[9] of the chloride transport of a healthy person with two normal CFTR genes
VX-770/VX-661	About 15–20 percent of the chloride transport of a healthy person with two normal CFTR genes
VX-770/VX-661/NEW CORRECTOR	More than 50 percent[10] of chloride transport of a healthy person with two normal CFTR genes (the same amount of chloride transport as healthy carriers who only have one normal CFTR gene)

Their experiments also revealed that two separate mutations in other regions of the protein could counteract the F508del mutation and restore its function to almost normal levels. That reinforced what Van Goor had suspected, based on Vertex's experiments years earlier with multiple correctors. Just as it took two mutations to counteract the F508del mutation, it would take two correctors, targeting two different regions of the protein, to "fix" it.

In October 2014, Sabine Hadida was promoted to lead chemist for all of Vertex San Diego. It had been a stressful year and she was behind schedule for producing the next-generation corrector—the third molecule for the triple. Her team had grown from three and a half people in 2002 to more than thirty by 2014, and they still hadn't created a second-generation corrector molecule worthy of testing in clinical trials. But they were making progress. The team of high-throughput screeners had discovered several promising hits that had been passed to the chemists in Hadida's team. And by November 2015, chemists Armando Urbina and Alina Silina had developed VX-440 and VX-152, correctors that each enhanced the effects of VX-661/VX-770 in human cells and were good candidates for becoming the third component of a triple combo drug. Both had met the criteria for clinical trials and were sent to Boston, where, if the molecules proved safe in animals, phase I trials would begin.

However, Van Goor, now leading Vertex's CF program—together with Peter Grooterhuis Paul Negulescu, Sabine Hadida, and Vertex CEO Jeffrey Leiden—were unwilling to risk these molecules not producing the "wow" effects that Van Goor and Hadida had predicted in 2009, and decided that they should also develop two more second-generation correctors as failure insurance. Just as VX-440 and VX-152 entered phase II trials in October 2016 after an uneventful phase I, another member of Hadida's team, Paul Krenitsky, synthesized another corrector, VX-659. Chemist Alex Abeda built the fourth contender, VX-445, which arrived in Boston in January 2017. To produce this final quartet of second-generation correctors had required a gargantuan amount of work: Hadida's chemists had made around 40,000 individual molecules.

By mid-2017, the first two second-gen correctors had completed phases I and II, and latecomers VX-659 and VX-445 were just entering phase II. But while each of the four correctors was being tested in phase I and II trials as its own triple combo with the Symdeko corrector and Kalydeco doorman, Vertex did not plan to go to phase III with all four triple combos—only the top two performers. Then the most effective combo drug, with the best safety record, would be submitted for FDA approval.

At the time, Vertex had at least half a dozen trials in progress—a couple involving the new triple-combination drugs, and several others aimed at expanding Kalydeco's and Orkambi's approval to younger and younger patients. The hypothesis was that if children began treatment almost from birth, patients might be spared both untimely death from early lung damage and the later horrors of CF that children who survived infancy faced. Dr. Lapey was certain that, if one of the triple combos worked, cystic fibrosis would no longer be much of a pediatric disease. The new drugs made it hard for him to envision much going wrong during the childhood of CF patients or even their early adulthood. No longer would he or any physician be able to pick a child with CF out from a crowd by their skinny, sickly appearance. For those on one of the Vertex drugs, being underweight was no longer an issue; ironically, for those used to eating constantly to keep up their body mass, weight gain was becoming a concern. Dr. Lapey was seeing his longtime patients on Kalydeco or Symdeko much less frequently, as they no longer needed to clean out their lungs with IV antibiotics over long hospital stays. For patients with the right mutations, cystic fibrosis might soon become a hidden disease.

ON JULY 18, 2017, FRED VAN GOOR HAD JUST SAT DOWN FOR DINNER
on the first night of a two-week cruise when he got a call from David
Altshuler, Vertex's chief scientific officer, in Boston. The midsized vessel was docked off the coast of Barcelona, Spain, and Van Goor left his
family in the dining room for a private section of the deck.

"Did you sign the document?" Altshuler asked him.

"No," Van Goor answered.

Altshuler told Van Goor to sign the nondisclosure agreement he'd
emailed him and then call him back.

Five minutes later, from the deck at the ship's bow, Van Goor did. It
was dark, the temperature a comfortable 80 degrees, and the deck was
mostly silent except for waves slapping the side of the ship.

Altshuler had just seen the results of the phase II trials for the two
new correctors—VX-440 and VX-152—in the triple drug combos. And
now that he had Van Goor's attention, he began spouting the data.
After the first few numbers, Van Goor pretty much stopped hearing
what Altshuler was saying, awestruck by what he'd heard so far.

Van Goor had already known that, together, these three compounds—the doorman and two correctors—could alter the lung cells
much more dramatically than either of the previous two-drug combos.
That knowledge was based on the science he had helped pioneer in the
lab. After all, the laboratory experiments with the little airways in a
dish, made with patients' lung cells, had predicted it. And they had yet
to be wrong.

The results that Altshuler was sharing were *exactly* as Van Goor had
expected. But actually hearing how the patients' lungs responded to this
new medicine, how dramatically it opened them up, made the news real
and overwhelming. He leaned on the ship's railing and wept, as other
passengers gave him a wide berth. Van Goor was unprepared for the
pent-up emotions that had been swirling in him for the past eighteen
years, and for how desperately he'd wanted these treatments to work.

Since coming to Vertex in 2001 to work on CF, there had only been
two other times when Van Goor had felt so deeply moved. The first was
when Kalydeco was approved. Van Goor had felt a rush of excitement
knowing that their first drug, for 4 percent of the CF population, was
going to work, but he'd also felt the weight of how much more there was
to learn and do. The second was the moment after the FDA advisory
meeting vote on Orkambi, when, after a harrowing day testifying about
the worth of the drug, he was greeted by a horde of parents and patients

who smothered him in hugs and tears, one after the other. Now there was this, the phase II data for the first two of the triple combinations, which seemed to finally promise the same kind of impact as Kalydeco, but for a much larger group of people.

The data showed that trial subjects' lung function had jumped by almost 10 percent for VX-152 and 12 percent for VX-440 in patients who had one copy of the F508del mutation and one copy of a minimal-function mutation, so named because this copy of the gene produced no CFTR protein.[11] Combined, these two mutations created a particularly severe and difficult-to-treat form of CF. For those with two copies of F508del, both triples improved patients' FEV1 and made their sweat less salty as well, if not quite as dramatically. Altshuler also shared good news about the phase I study for a third corrector, VX-659: it boosted lung function by almost 10 percent as well.

Van Goor's tears weren't quite joy; they were relief, maybe, or the realization that this was the beginning of the end of cystic fibrosis as they knew it: as a fatal, incurable disease.

CHAPTER 53

The Home Stretch

2018

If you have everything under control,
you're not moving fast enough.

—Mario Andretti

O ver the next year and a half, patients completed phase II trials
for the remaining triple combinations. By May 2018, Vertex had
pushed the two highest-performing correctors, VX-659 and VX-445,
into six-month phase III trials; VX-659's trial began in late February,
with VX-445's starting two months later. Each of these correctors were
combined with the Kalydeco doorman, VX-770, and the Symdeko corrector, VX-661, and given as a triple combination to patients who had
two copies of the F508del mutation—almost 50 percent of all patients
with CF. The triple combos were also tested in patients who had a single
copy of F508del plus one copy of the minimal-function mutation.

One of the patients who participated in this particular trial was a
young woman named Katy Farrell, born in 1985 in the Melrose Wake-
field Hospital where Joey had been born eleven years earlier. Katy developed pneumonia at two weeks old, and a second case before six months,
which raised red flags that prompted her doctor to order a sweat test.
The positive result led to a CF diagnosis and she ended up in the care of
Dr. Lapey, who'd treated Cate and Laura Cheevers and Joey.

Like the Cheevers sisters, Katy was athletic. She began Irish step-
dancing at age six before switching to soccer, softball, and, finally,
swimming—her passion. All these activities kept her in great physical
and mental health, and she didn't require hospitalization until she was
nine years old and in third grade. She then had another relatively clear

stretch of good health until her junior year of high school, when she began visiting the hospital twice a year for intravenous antibiotics.

Katy found the hospital stays depressing, so her mother, a nurse, administered her IV medications at home. And as soon as she was old enough, Katy learned to take care of this herself. But by 2011, when she was twenty-six, her veins had hardened with too much scar tissue, making it impossible to insert a needle. Dr. Lapey—still Katy's doctor, despite her adulthood, out of her desire not to tempt fate by switching from someone who'd kept her in such good health—instructed surgeons to insert a port into her chest through which drugs could be delivered.

Over the next year and a half, Katy's infections and hospital stays became more frequent and her lung function fell to 60 percent. In 2013, Katy began coughing up tablespoons of blood, a terrifying condition called hemoptysis that indicated bleeding in her infected lungs. Although it didn't happen that often, just every month or two, the first time Katy was certain she was dying and destined for a lung transplant—something she wasn't sure she wanted to endure. While antibiotics brought her infections and bleeding under control, her cough was relentless, and her lung function kept dropping, making her feel as if she were breathing through a straw. Her lungs hurt continuously, as if dozens of needles were poking her from the inside. And though the antibiotics had helped during high school, now she was allergic to several of them.

By 2015, when Katy's lungs were bleeding and filled with bacterial infections, she had started sleeping while sitting up. She was again thinking about the inevitability of a lung transplant when Dr. Lapey suggested that she enroll in the Symdeko trial—the combination of the VX-661 corrector and the VX-770 doorman. She carried two copies of F508del, so the trial was a good fit for her. She was lucky; she got the drug. Within a week, the combination had curbed, though not stopped, her coughing, and she could sleep again. The lung bleeds stopped and for the next two years, while remaining on Symdeko, she stayed out of the hospital.

The drug was good, but it wasn't perfect. Katy's passion was travel, and after her trips she was always run down and needed a course of antibiotics to kill off infections and get her health back on track. She had plans to travel to Ireland in early 2018 when Dr. Lapey suggested she postpone them and instead enroll in another trial: a phase III trial for the triple combo featuring VX-659. Katy was disappointed to abandon her plans, but she trusted Dr. Lapey; he said that this drug was even better than Symdeko. So she joined the trial.

Once again, she was lucky, getting the drug and not the placebo. She began taking the medicine on a Wednesday in February 2018, and by Saturday her cough had completely ceased. When she went hiking with her friends, they asked her if she was okay. Why wasn't she coughing?

On November 27, 2018, just before Thanksgiving and some sixteen months after Van Goor had first heard the phase II results while cruising off the coast of Spain, Vertex announced some early results for the triple combo that Katy had been taking, VX-659/VX-770/VX-661. Though the patients in this trial had only been on this particular triple combo for four weeks, the results were spectacular. The combination appeared to boost lung function by 14 percent in patients with just a single F508del mutation, and by an additional 10 percent in those, like Katy, with two copies of F508del who were already benefiting from VX-770/VX-661—results that overshadowed all the drugs to date, even VX-770. But as promising as these numbers were, they were preliminary—the Vertex CF team had to see how the drugs worked for the full six months of the phase III trial.

That same day, the company reported that the fourth and final corrector, VX-445, was about to begin phase III trials, both in patients who carried two copies of the common mutation and those who carried one copy of the common mutation plus a minimal-function one. Vertex hoped VX-445 would be an even better corrector than VX-659, and so planned to wait for those results to come through before submitting the most effective triple combination to the FDA.

After Katy had been taking the triple for almost a year, she finally flew to Ireland, enlivened by her newfound energy and lung function, both of which seemed to improve every day. She was so healthy, in fact, that she competed in the Metalman Swim Series in the open waters of Northwest Ireland in June and hiked a stretch of the famed five-hundred-mile Way of Saint James—or El Camino de Santiago—which pilgrims had followed for the past one thousand years on the Iberian Peninsula to reach the burial shrine of the Apostle James. For Katy, the hike was more of a spiritual trek than a religious pilgrimage—a rediscovery that her body was now strong and robust, and that there were many new roads that she could travel.

Typically when Katy returned from her travels, her body would be worn out and needing medicine to recover. But this time, when she returned after two and a half months of traveling in July 2019, she was in excellent health, with lung function at 92 percent—a number she hadn't seen since her college days. On this new drug she was free from the bondage of this disease and, for the first time at age thirty-three, thinking about how much she would love to have a family and live on a farm—both now possibilities.

By May 30, 2019, Vertex had all the data from both the phase III trials in hand. Over the past few years the company had tested four triple combos in phase II trials, then the VX-659 and VX-445 triple combos, each in their own phase III trials. VX-445 had emerged as the best corrector. The triple that included this talented molecule could even raise the FEV1 of patients with one F508del mutation and one minimal-function mutation—one of the most difficult gene pairs to treat—by a colossal 14.3 percent, and could drop sweat chloride by almost 42 points. For those with two copies of F508del, this triple improved lung function by more than 10 percent and dropped sweat chloride by 45 points.[1] Combined, these three molecules—VX-445, a second-generation corrector; VX-661, the first-generation corrector; and VX-770, the original doorman—resurrected the broken CFTR protein. The correctors tweaked the protein at two separate sensitive stages, refolding it and enabling a substantial quantity of it to reach the surface of the cell. There, the doorman helped the channel open, allowing chloride to flow in and out just as it would in a person without the disease, and restoring the balance of salt and water in the lung, gut, pancreas, and sweat glands.

Just shy of two months later, Vertex submitted an application to the FDA to approve the new triple. This drug combo would work for anyone who had at least one copy of the F508del mutation—almost 90 percent of all CF patients. The company requested a so-called priority review from the FDA, as they had for Kalydeco, to cut the drug review time from the standard twelve months to six.[2] For CF patients who were seriously ill and careening toward the lung transplant list or death, there was no time to waste.

FOR THE PAST DECADE, ROBERT COUGHLIN, FORMER STATE CONGRESSman and now president and CEO of the nonprofit Massachusetts Biotechnology Council (MassBio) in Cambridge, had been working to make the state the most conducive place in the world for biotech and pharmaceutical companies to innovate. Like Joe O'Donnell, Coughlin hadn't hit the snooze button in ten years because he believed that if he didn't work as hard as he could to help companies like Vertex reach their goals, his son Bobby—now fifteen years old and sick with CF-related end-stage liver disease—might die, as might other sick children. Still, though he hadn't admitted it to anyone, he never truly thought the day would come when Vertex might have something for his son. He didn't think it would happen—at least not in time for his Bobby.

His son had one copy of the F508del mutation, and one copy of a second, tricky mutation called 1717-1G>A. This second mutation was a minimal-function mutation[3]—meaning there was no protein to fix. His only hope was that Vertex would make a therapy that would work for the F508del mutation. And although Coughlin tracked all of Vertex's trials, in 2017 Bobby had already been too sick to participate in the phase II trials for any of the four triple cocktails—a tough blow, since this was the first treatment that held any possibility of treating his mutation.

For Bobby, the genetic lottery had given him really good lung function despite his CF, but a disease-prone liver. Coughlin was pretty sure that his son's healthy lungs were a result of his frenetic lifestyle. Bobby was, according to his dad, a hell of a hockey player. He was small but fast, furious and relentless—an above-average athlete overall, and always on the go. He had a short attention span and no interest in sitting around playing video games. He'd rather be moving. But at fifteen, his liver was swollen and unable to manufacture the proteins needed to make his blood clot. There was a risk that, if Bobby injured himself, he could bleed internally—so his parents had made him stop skating. He needed a liver transplant, and that was something not even these Vertex drugs could fix.

Coughlin had been trying hard not to follow this particular trial too carefully. He couldn't stomach the possible disappointment if the news was bad and help for his son was not forthcoming. But when Vertex got in touch to schedule a high-priority call on July 18, 2018, he knew it was good news. He felt it in his bones. And he was right: Vertex CEO Jeffrey Leiden was calling to tell him that the triple combo was working for patients like his son. Although he couldn't share the news with anybody until the market closed that evening, he spent the day on cloud nine.

When he finally told his family that night at dinner—his wife, Christine; his twenty-year-old daughter, Mary Kate; his eighteen-year-old son, Paul; and, of course, Bobby—there was a palpable sense of relief and joy. They were all keenly aware of how quickly cystic fibrosis could hurtle forward, progressing to a point of no return; Coughlin's children had seen their young cousins and other family members die from it. The pressure—the burden—of that fear had been ever present. But that day, July 18, Coughlin's whole extended family felt that weight lift.

Still, Coughlin knew that there would be at least another year of waiting before a drug could be approved. So he didn't dwell too much on the news, except as motivation for Bobby to continue to eat right, exercise—moderately—and avoid germs. It was more important than ever that Bobby's lungs not decline any further. And Bobby seemed to agree, starting to use his nebulizers first thing in the morning rather than waiting until the end of the day as he had been. He also seemed to have a little more bounce in his step, Coughlin thought—a good sign, because he knew how desperately Bobby wanted to forget about this disease and just be a normal kid.

Coughlin was grateful that Bobby was one of the nearly 90 percent of CF patients who were expected to benefit from the new triple-drug cocktail. But at the same time he couldn't stop thinking about the families whose children were in that last 10 percent. He knew that those kids couldn't be left behind. Back when the FDA approved Kalydeco for people with the G551D mutation and Orkambi for people with two copies of F508del, and he knew that his son wasn't among those who would benefit, it had made him feel, he said, like shit. He knew families who were still in that same position, and he knew how awful it was. Even though Vertex had overcome a big hurdle, they weren't at the finish line. And knowing that left Coughlin feeling uneasy.

CHAPTER 54

The Leftovers

2015–2020

> It turns out that even within the CFTR mutant kingdom, I
> am a rare breed. With two copies of a relatively rare W1282X
> mutation, the so-called "Ashkenazi Jewish mutation," my
> genetic composition places me in territory so uncharted,
> at least in the US, that there are no published statistics
> to indicate the size of my microcommunity. I am a rarity
> among the rare, an orphan of an orphan disease.
>
> —Emily Kramer-Golinkoff

After the triple-combo drug's approval, anticipated in early 2020, treatment would eventually be available for nine out of ten cystic fibrosis patients. But Emily Kramer-Golinkoff, born on January 9, 1985, wasn't one of them. She was diagnosed with CF at just six weeks old, and at the time, life expectancy for CF patients was their late twenties. The fact that Emily nevertheless had survived long enough to attend college at the University of Pennsylvania, and then gone on to get a master's degree in bioethics, was a testament to both her and her parents' commitment to hours of daily physical therapy and fistfuls of pills—and Emily's own unstoppable drive.

Emily's parents first learned what mutations she carried in the early 1990s, after a team at Johns Hopkins determined the specific sequence of her CFTR gene. It was the W1282X mutation—the most common mutation among the Ashkenazi Jewish population, which Batsheva Kerem had discovered after moving back to Israel from Lap-Chee Tsui's Toronto lab. At the time, this information was just another esoteric detail added to Emily's medical record, with no impact on her therapies or clinical course. But that changed in 2011, when she and the rest of the CF community realized that the therapies Vertex was developing were mutation specific. This was good news for CF patients with the mutations Vertex was working on—G551D and F508del—and other mutations that produced a CFTR protein that could be manipulated with one of Vertex's molecules. But in 2011, as far as twenty-six-year-old Emily and her parents could see, there was nothing in the pipeline for her.

In 2017, just a little more than sixty years after Wynne Sharples had launched the Cystic Fibrosis Foundation, there was hope that the following decade would increase the number of patients who had treatments to approximately 95 percent of people with CF. Emily and her parents were thrilled for the other families, but envious, too. For Emily it was like standing on a sinking ship and seeing lifeboats come for more and more of her fellow passengers . . . but there being no more lifeboats left for her.

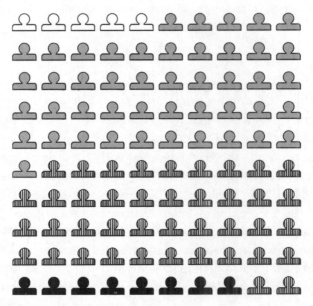

G551D and similar mutations: 8 percent
Two copies of F508del: 46 percent
One copy of F508del: 41 percent
Rare mutations with no drug: 5 percent

Emily had the unfortunate distinction of being what she felt was one of the leftovers, a CF patient whose mutation, W1282X, was rare and, as a nonsense mutation, couldn't be fixed by Vertex's protein modulator therapies. As for all proteins, the gene for her CFTR protein began with a start codon—three DNA bases that told the machinery in the cell to start reading the DNA code at a specific location and make a protein. A stop codon there marked where to stop transcribing the DNA, signaling that the protein was complete. But in Emily's mutation the stop codon appeared in the wrong location, like a period inserted in the middle of a sentence.[1] It terminated construction of her CFTR protein before it was completed, creating a useless strand of amino acids that was quickly scooped up by the cell's garbage collectors and sent to the trash compartment. That meant that Emily didn't even have a protein in her cells that could be corrected. She needed an entirely different type of therapy.

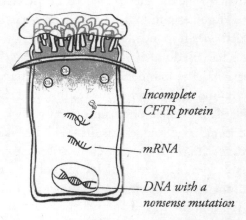

Incomplete CFTR protein

mRNA

DNA with a nonsense mutation

A sick airway cell from a patient with a nonsense mutation. The CFTR protein folds so poorly that it is terminated before it is complete and is destroyed by the cell. No CFTR reaches the cell surface.

But Emily hadn't been forgotten. When Bob Beall retired after thirty-five years at the foundation due to health problems after back surgery, he left the foundation in the expert hands of Preston Campbell. Campbell had guided the drug development with Beall for the past seventeen years, and when he took charge in 2015, he felt a strong moral obligation to the foundation's mission: developing therapies for *every* CF patient, not just those whose broken CFTR channels were amenable to the manipulations of a couple of correctors and/or a doorman. Once Campbell and Beall had witnessed the approval of Kalydeco for that first 4 percent of patients, they agreed that no patient would be left behind. It was a commitment to keep funding research that to the layperson sounded totally delusional, to develop medicines and therapies that seemed improbable, and to fight for healthcare legislation that would enable access to the pricey drugs that had been developed for every CF patient who could benefit. The foundation would continue to fund research "until it's done," which became their new tagline.

As of 2017, more than 2,000 mutations had been discovered in CF patients worldwide. If the triple-drug combo was approved, there would be treatments for people with about thirty-eight mutations total. Of all known mutations, about 650 produced a complete CFTR protein, meaning patients with those mutations might respond to one of Vertex's therapies. As of May 2017, the FDA allowed Vertex to provide drugs to patients with rare mutations if laboratory experiments with engineered cells predicted that the therapy would be effective. So, the foundation decided to engineer hundreds of cells with these 650 specific mutations to see which ones responded to Vertex's drugs—a strategy, theratyping, that Vertex had used previously in place of clinical trials for people with rare mutations. The foundation intended to share the results with Vertex, let Van Goor's team verify the results,[2] and then submit an application to the FDA to allow physicians to prescribe drugs for patients with these additional rare mutations. Dr. William Skach, whom Campbell had hired as senior VP of research in 2014, estimated that theratyping might enable Vertex to get medicine to another 2 or 3 percent of patients whose numbers were too small to be included in a clinical trial.

The foundation was able to theratype cells with rare mutations because, in September 2016, flush with funds from the $3.3 billion payment they'd received for Kalydeco, Orkambi, and the other drugs Vertex had in development, it had launched its own $6 million laboratory, in Lexington, Massachusetts, less than twenty miles from Vertex headquarters. The purpose of the CFF Therapeutics Laboratory was multifaceted. Researchers at universities still needed permission from their institutions' lawyers before they could share their resources, even those developed with foundation research grants, and wrangling intellectual property agreements with various universities was time consuming and frustrating. So Campbell had pitched Beall the idea that they should set up their own lab to supply tools and patient cells, and train scientists interested in pursuing CF research. If the foundation developed such resources, they could share freely with no strings attached.

But the Lexington lab wasn't solely intended to support others. It was also screening chemical libraries to discover hit molecules that could be developed into drugs. Skach wanted to pioneer drug discovery for molecules that could fix nonsense and other types of mutations that Vertex's drugs couldn't treat. His hope was that scientists at the Lexington lab would discover molecules that could mask the misplaced period that stopped mRNA production, creating an incomplete protein. If that mutation could be masked, the protein-making machines in the cells would just skip over or read through the mutation without stopping and complete the entire protein. Such molecules were called "read-through" agents, and by 2017, the CFFT laboratory scientists had screened 80,000 molecules and began identifying some that, at least initially, looked like promising candidates for the job. If read-through agents could be developed, it might take care of another 3 to 4 percent of the remaining CF patients. Plus, nonsense mutations were not specific to cystic fibrosis; for example, cases of beta thalassemia, Duchenne muscular dystrophy, Hurler syndrome, and Rett syndrome were caused by nonsense mutations. Finding a molecule that addressed these CF mutations could have implications well beyond cystic fibrosis.

To accelerate the development of treatments for nonsense mutations, in January 2019 Skach brought in twenty experts in the field of protein manufacturing to brainstorm and learn about these mutations and work on how they might make a read-through drug for CF. Skach and Lexington, Massachusetts, lab leader Martin Mense also began hiring lab staff to bolster expertise in gene editing and stem cells. Skach wasn't just looking for a treatment. As the leader of all the foundation's basic and translational research, biotechnology efforts, and collaborations with pharmaceutical companies, his goal—and the foundation's new goal—was much more ambitious: curing the disease.

It was something no one would have considered possible a decade earlier.

Vertex's treatments, while a huge step in effective therapy, were not cures. CF patients needed to take the medication every day for the rest of their lives. Skach was interested in finding a way to rewrite the genetic code, some kind of onetime treatment after which patients would be able to say, "I used to have CF." He wanted to use new technologies like CRISPR, which enables researchers to edit DNA the way a writer edits a sentence—clipping a word, erasing a typo, restoring missing words or phrases—to fix genetic mutations in the bodies of living CF patients. He wanted to fund teams that were exploring the most unorthodox, highest-risk, and potentially most rewarding approaches to a cure.

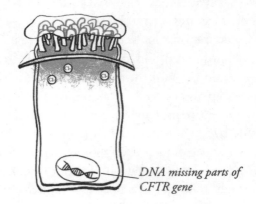

DNA missing parts of CFTR gene

A sick airway cell that is missing parts of the CFTR gene. The cell does not make any CFTR protein at all, and so cannot benefit from corrector or doorman drugs. Mutations like these require a completely novel approach, like gene therapy, in which a healthy copy of the CFTR gene is delivered to the cell.

Skach was also driving new research in gene therapy, the attempt to use viruses to deliver healthy DNA into lung cells. Gene therapy had failed twenty years earlier. But now, two decades later, many of the initial problems had been overcome and the strategy was enjoying a resurgence. Some of the original pioneers were leading these gene therapy efforts for CF, and in the previous few years, gene therapy had been successfully used to treat sickle cell disease,[3] cancer,[4] and one type of inherited vision loss.[5] Some companies were exploring how to remove samples of patients' airway stem cells (the cells that generated all the other cell types populating the lungs' airways), correct the mutation in them, and then somehow return the healthy, repaired cells to the patient's lungs. No one knew how to do this for the lung, but twenty years before, no one had known how to do what Vertex had done, either. Audacious ideas, the foundation leaders knew, could yield incredible results.

At Yale University in 2019, for example, a team of researchers succeeded in using a type of gene editing to cure cystic fibrosis in a baby mouse while it was still in the womb. At day fifteen of gestation the Yale team delivered biodegradable nanoparticles carrying the editing instructions into the amniotic fluid surrounding the developing fetus. As the fluid moved in and out of the fetus's lungs, it carried the nanoparticles with it. The natural DNA repair systems in the fetal mouse's lung cells then used the instructions in those nanoparticles to edit and correct the mutation. While not every cell in the entire fetal mouse was corrected, when the mice were born their lungs appeared healthier and the editing seemed to have yielded some therapeutic benefit. Although the approach was years away from testing in humans, it was proof that such molecular surgery—in a living animal while in the womb—was possible.[6]

Just as the CF foundation was committed to leaving no patient behind, Vertex Pharmaceuticals had made the same pledge. They were also looking to the next generation of therapies that might work for these patients. Already they had collaborations with the leading gene-editing companies and were also looking into technologies to deliver updated forms of gene therapy. Fred Van Goor and Paul Negulescu could see that creating medicines for the last 5 percent of CF patients was going to be just as hard as it was for the first 95 percent, and involve just as many problems.

And so the two scientists were, in a way, back where Paul Negulescu had begun in 1998, when he first connected with Bob Beall—with a series of new, wild strategies about how to create novel medicines to cure this disease, betting on ideas that wouldn't come to fruition for maybe ten to fifteen years.

Emily didn't have that much time to wait, however. In 2012, when Kalydeco was approved, Emily was twenty-seven years old, and time was running out for her. She'd spent most of seventh and eighth grade in the hospital with lung infections and digestive complications, and after graduating from college in 2007, she developed cystic fibrosis–related diabetes, a common complication affecting about 40 percent of adult CF patients.[7] Once CF destroyed the pancreas, a patient's body could no longer produce insulin to control blood sugar levels. As Emily's blood sugar rocketed up, the high blood sugar levels, along with other factors, fueled the bacterial infections in her lungs, destroying large swaths of tissue. By the time the doctors were able to get her blood sugar and the infections under control, Emily had lost 15 percent of her lung function, bringing it down to 37 percent.

Emily and her family had the heartbreaking realization that, despite Kalydeco and the other drugs in Vertex's pipeline, nothing in the works would treat her nonsense mutations. Yet Emily wasn't inclined to stand by while others invested in promising technology that might yield clinical trials over a decade away. She and her family launched an independent nonprofit called Emily's Entourage to fund CF research. After her team of friends and family easily raised $8,000 in 2010 from a simple stair-climb fundraiser, Emily realized they could aim higher and be more ambitious. So she and her mother, Liza Kramer, decided to focus their CF fundraising on a single target: nonsense mutations. Emily and Liza had seen what the Cystic Fibrosis Foundation had managed to do and thought they could do something similar, albeit on a smaller scale, by raising money to fund early-stage research and push risky concepts and technology.

To identify promising science and serve as an advisor, Emily enlisted the help of Kevin Foskett,[8] a friend's father and a well-known CF researcher. Foskett, an original member of the Cystic Fibrosis Foundation's scientific advisory team that traveled to Aurora in 1998, began assembling a board of scientists to guide Emily's Entourage toward the most exciting, leading-edge research that was still stuck in university laboratories without the resources to push it forward. With a panel of ten distinguished researchers who had devoted their careers to this disease—including Batsheva Kerem; nephrologist Alan Verkman, with whom Bob Beall had an early conversation about high-throughput screening; and Garry Cutting, a geneticist who had spent the previous twenty years focused on cataloging all of the CFTR mutations and correlating each one with disease severity—Emily threw herself into fundraising and advocacy. As she raised money through galas, marathons, grassroots social media, and community events, she also sought to identify the barriers and gaps that were preventing researchers from studying nonsense mutations.

One personal contribution Emily made was her cells. Because she had two identical nonsense mutations, her cells were particularly valuable to researchers wanting to test the impact of nonsense mutation drugs. She traveled around the country donating intestinal cells, nasal cells, and even stem cells from her lungs. She created a registry for other people who carried nonsense mutations and were interested in being contacted directly for clinical trials. This differed from the CFF's registry, which required physicians at the care centers to identify appropriate patients from the registry and pair them up with the right clinical trial. Her tactic drastically cut down the time needed to recruit patients for trials. Emily's registry also empowered the patient, giving them more control over their ability to participate—without a physician intermediary.

While earning her master's in bioethics, Emily had worked in Penn Medicine's strategic marketing department. After graduating, however, she left her job to immerse herself in the biology of CF, ongoing CF research, and particularly all research focused on nonsense mutations, alongside advocating for patients and raising money. In many cases, nonsense mutations had flown under the radar for CF researchers, as well as those working on other diseases, and there was little recognition among biotech and pharmaceutical companies of how unmet this need was. So Emily began traveling around the country and speaking to the scientists at these companies, including Vertex. She built relationships, putting a human face on the group of patients that wouldn't benefit from the current protein-modulator therapies. And she kept reminding them that helping 95 percent of the CF patients wasn't good enough. After a couple of years, these scientists often told Emily that they were working on the descendants of cells she herself had donated.

Emily was keen to fund a broad range of approaches, as long as patients with nonsense mutations would benefit.[9] By early 2020, her foundation had raised more than $6 million and had provided eighteen grants to researchers at universities all over the country. They had pro-vided funding to academic researchers who were using viruses called bacteriophages—Latin for *bacteria eaters*—to kill microbes that were resistant to antibiotic drugs.[10] This treatment could help any patient with antibiotic-resistant infections, not just those with CF. Emily's Entourage had also provided funding to a researcher at Georgia Tech who was experimenting with using nanoparticles to deliver molecules like DNA or the more transient mRNA (the intermediate molecule that was translated into proteins) directly to lung cells. If they could suc-cessfully deliver a gene or a sliver of mRNA that encoded the normal version of the CFTR protein to a patient's lung cells, the patient might begin manufacturing a healthy CFTR protein—an approach that could be used regardless of what mutation a person carried.[11] Another grant went to a researcher in the Netherlands who was using cells like Emi-ly's to grow organoids—small clumps of lung tissue—in a petri dish to test whether the Vertex drugs, or other new CF drugs, would help treat nonsense mutations. Through an unusual arrangement with the Univer-sity of Iowa and Philadelphia life-sciences funder Militia Hill Ventures, Emily's Entourage used venture philanthropy to provide seed funding for a new gene therapy company called Spirovant Sciences that was test-ing several types of viruses to deliver genes into the lungs.

The CF Foundation was funding similar ventures, and more of them. But Emily felt that her organization—because it was small and scrappy, as she described it—was more nimble, able to respond faster to new discoveries and push research toward application earlier than other venture capital operations. And if one of these therapies became ready for clinical trials, Emily would be able to volunteer immediately.

Emily's Entourage was going strong, but Emily herself was in a delicate situation. Since Kalydeco's 2012 approval, Emily's lung function had been declining in fits and spurts. With such low function, she was suffering from lung bleeds, and in May 2018, one of her lungs partially collapsed. A transplant team was closely following her health, and Emily was at the point of having to weigh the costs and benefits of a double lung transplant—something she had hoped to avoid for as long as possible. By the fall of 2019, she had become resistant to most antibiotics, which meant that if she became sick with new lung infections, there might not be any effective drugs to treat her. Her lung function was at only 30 percent. Yet the drastic state of her lungs wasn't obvious by looking at her. She was slim but still striking, with shoulder-length dark hair and a pale complexion, and she still spoke energetically and passionately about all the things she was working on: the collaborations, the research, and patient empowerment.

Emily was optimistic. She understood the high probability that the work she was doing wouldn't benefit her. But she and the rest of Emily's Entourage were working as fast as they could, with a pace and sense of purpose she found therapeutic. As a means of self-preservation, she actively denied the statistics and chose to believe that she might defy the odds.

CHAPTER 55

A New Generation

2017–2020

If you're in the luckiest 1 percent of humanity, you owe it to
the rest of humanity to think about the other 99 percent.

—Warren Buffett

O f the two O'Donnell girls, the younger, Casey, was the first to get
married. The ceremony took place on a spectacular summer day
at the family's Cape Cod home on August 22, 2015. The lucky man
was Michael Buckley, who had crossed paths with the O'Donnell family almost a decade earlier, in the summer of 2006, when he was randomly assigned to be Joe's caddy one day at the golf course. Joe made an
immediate connection with Mike and hired him to caddy for the rest of
that summer, and the following summer as well. That's when Mike met
Casey, and his group of friends merged with hers and Kate's. At the time
he was a college senior and she, a high school senior. The two kept in
touch over the years, when Mike moved to Chicago after college. But it
was only in 2011, in New York, after Casey had graduated from Harvard
and moved there to work for a startup, and Mike had relocated there
to work at the New York Stock Exchange, that the two began dating.

Casey describes herself as Type A and is a curious blend of her parents' personalities. Always the more serious of the sisters and often mis-
taken as the older of the two, she was introverted, with a reserved Irish
sensibility, like her mother, Kathy. She was also focused, thoughtful,
and analytical like her dad when assessing situations and making decisions. Casey had always been decisive and independent, and when it
came to having a baby, she was no different. Her parents were older and
she wanted to have kids young, while they were still in good health and
could enjoy their grandchildren.

Always practical, Casey had raised the necessity for genetic testing with Mike before they were married. She'd known of her carrier status since the age of ten, when Joe had taught both her and her sister, Kate, how CF genes were passed from one generation to the next, and the implications for children who received two of them. In sixth grade, when Casey learned basic genetics in school, she had already heard the lesson. For her own children, Casey wasn't taking any chances. Although Mike had no family history of CF that he knew about, he was of Irish and Italian ancestry, which meant there was a one-in-twenty-nine chance that he might be a carrier. And she herself was evidence that even if he were a carrier, there were medical procedures they could use—like in vitro fertilization—to be sure they would have a healthy child. To her, the cost would be worthwhile.

Fortunately, Mike's tests came back negative. But when Casey was later tested, too, at twelve weeks into her pregnancy—not just for CF mutations, but for mutations associated with eighty-six other diseases—her tests were less reassuring.

Casey knew she carried one bad CFTR gene because of the in utero genetic testing that Joe and Kathy had done in 1989. But back then, before the discovery of the gene, tests had looked for genetic markers near the gene rather than the gene itself—so Casey didn't know what mutation she carried. The only information that Kathy had been given was that Casey was a carrier and wouldn't have CF—and she hadn't been screened for anything else. Now, she discovered, in addition to carrying a CF mutation, she carried three other mutations, for disorders with long technical names that she'd never heard of—but that, if two copies were inherited, caused severe disease.

When it came to CF, Casey learned she didn't have the common F508del mutation, for which Vertex now had a therapy. Instead, she had a severe minimal-function mutation called 1717-1G>A.[1] This copy of Casey's CFTR gene didn't produce any CFTR protein. It was a mutation carried by only 1 percent of CF patients, largely in the UK and Ireland.

Though the report freaked Casey out, she knew enough genetics to understand one critical fact: at worst, the baby might also become a carrier of these mutations—they wouldn't suffer from any of the diseases.

Had the results been different—had Mike also been a carrier—the couple would have had the fetus tested. And if those tests came back positive for CF or any of the other severe genetic diseases, Casey knew that she and Mike would have chosen to terminate the pregnancy. She described herself as unapologetically pro-choice; after all, her family knew firsthand the horrors of living with a genetic disease.

For Joe, the birth of Blair Buckley, his healthy granddaughter, on June 24, 2017, changed his world. Grandmother Kathy took one look at Blair and was immediately smitten. Though Joe claimed he didn't find newborns that interesting, it wasn't long before baby Blair had her grandfather wrapped around her finger. Joe was a big guy with a low, booming voice and it took a little while to win Blair over, but before long she was grabbing toys, bringing them over to him, and talking his ear off. Casey, Mike, and Blair lived a couple of floors below the O'Donnells in an apartment building in Boston, and Joe visited most mornings and came over to see Blair most evenings, too. Coming in, he would always say, "Where's the kid? Where's the kid?" with a big grin on his face.

Two years later, on July 7, 2019, Casey and Mike had a second child: a healthy baby boy named James Daniel Buckley.

Both Kate and Casey had gone to Harvard for undergraduate education; Kate continued in her father's footsteps and earned her MBA at Harvard Business School. Like her dad, Kate was extroverted and gregarious, befriending people easily. She enjoyed big gatherings and meeting people. She and Casey had spent so much of their childhood attending events and visiting places where they rubbed shoulders with older, high-powered businesspeople, greatly accomplished scholars, and even world leaders that Kate was more comfortable in a group of older CEOs than many people her own age. Both sisters, however, were very aware of their privileged upbringing. They were careful about mentioning such influential contacts to anyone other than close friends and family, wary of being judged as spoiled and not wanting to set themselves apart from other children.

A couple of years after getting her MBA, in 2017, Kate landed a job as assistant director of Student and Young Alumni Engagement at Harvard Business School. The job was essentially networking with recent alumni and connecting them to the university fundraising network—a natural fit for someone who had grown up with a master fundraiser and had participated in Joey Fund events from as early as she could remember.

Kate had always loved fundraising. It tickled her competitive spirit and allowed her to indulge her enthusiasm for event planning and bringing people together. Her early fundraising experiences had taught her that almost everybody wants to give, either their time or their money, and wants to feel connected to a cause; usually, they just don't know how. The most successful way to get people involved, Kate found, was to give them something personal to care about. When she spoke to people about the Joey Fund, telling them the story of what had happened to her family, they would connect with her and her parents. Often this led to a desire to get involved personally and become an advocate for CF— just because they knew the O'Donnells' story. Once people learned how the disease had impacted her family, they wanted to help.

One problem that she and Casey had recognized as they got older and continued to do more work for the Joey Fund was that many events tended to attract only an older crowd. When she was still in business school, Kate had challenged herself to figure out how to introduce a new generation to philanthropy. Part of that was updating the Joey Fund website, an internet relic dating back to the early 1990s. The World Wide Web had just been coming into its own when one of Joey's friends, Andrew Plunkett, had built the site for Joe and Kathy, but by 2013 the site looked clunky and barely functioned. So Kate contacted Andrew, who had always stayed connected to the family and still frequently helped out at Joey Fund events, to see if he would help her update the site and teach her how to run and maintain it. It was a savvy marketing move: with a rebuild of the aging Joey's Park in Belmont planned for 2013, people would be looking for information. She and Casey had also begun incorporating social media like Facebook and Instagram into their fundraising platform, and were thinking about new types of events that might attract younger supporters to the cause, like music festivals.

The annual movie premiere remained a hit with younger people. Kate had spoken about the Joey Fund at various business school events, and when it came time for the premiere, many of her peers were interested. But the event was held in a mall, a twenty-five-minute drive from where most of her classmates lived, in downtown Cambridge, and could only travel via public transit. To overcome that barrier, Kate hired a school bus to shuttle her friends to and from the event. She told them they could attend for free or give a donation, whichever they preferred.

For the first event, most of the fifty attendees made a donation instead of buying the $125 ticket—a steep price tag for students. But Kate understood it was difficult for someone who'd never attended such an event to understand why they were paying $125 for food, drinks, and a movie. The next year, after enjoying their first event, many of her friends attended the premiere again—increasing their donation or even paying full price.

When Kate started working at the business school a few years later, she extended the tradition to her new, young workmates—busing them to and from the event to get them involved. From that night on, many of them became regular attendees, bringing friends and spouses and spreading the tradition further.

Around the same time, Kate suggested two other modern twists to the movie premiere to appeal to younger attendees: changing the swag from shirts, clocks, and picture frames to more utilitarian items, like Yeti-style wine tumblers and reusable bags; and adding a photo booth. She also began planning events at places that would attract a younger crowd, like Night Shift Brewing and Oath Pizza. Like Joe, Kate didn't see the impending milestone of having a treatment for 95 percent of patients as a reason to stop fundraising for the Joey Fund. She knew that there were plenty of patients, from those in their early teens to people in their forties, who had already sustained significant lung damage that couldn't be fixed with the therapies Vertex had developed. These people would continue to need treatments, and who knew what other conditions and needs might arise as patients finally had the opportunity to age. Then there were the 5 percent of patients with other mutations—like the one Casey carried—who still needed a treatment. They couldn't be left behind.

In 2020, Kate began moving into a leadership role at the Cystic Fibrosis Foundation, becoming vice chair of Tomorrow's Leaders with the foundation's Massachusetts/Rhode Island chapter. Through that new role, alongside her work at the Joey Fund, Kate, and Casey with her, was committed to keeping the fight alive as long as people like Joey continued to suffer from this disease.

AFTER HAVING RAISED A COMBINED $250 MILLION FOR THE FIRST AND second Milestones campaigns for the Cystic Fibrosis Foundation and an additional $125 million through the Joey Fund, Joe was still raising money. Although the foundation was now flush with funds, with $4 billion in their bank account after selling its royalty stream for the Vertex drugs, Joe knew developing a new type of therapy for that 5 percent of patients who didn't benefit from Vertex's doorman or correctors would be expensive.

Over the past twenty years, the foundation had invested more than $225 million into Aurora and Vertex. And Vertex had invested some $7 billion of its own money to develop Kalydeco, Orkambi, Symdeko, and the new triple combo—roughly $1.8 billion per drug. Designing other pathbreaking therapies to treat people with nonsense mutations—such as using viruses or nanoparticles to deliver genes into the lungs, or reengineering patients' stem cells through gene editing and returning them to the patient—would require billions. In 2015 the foundation spent $152 million investing in cutting-edge research; in 2016, $172 million; in 2017, $177 million; and in 2018, $196 million.

Then, in 2019, at the annual North American Cystic Fibrosis Conference, William Skach, the foundation's chief scientific officer, announced that the foundation was kicking off a new era. Their new goal: cure cystic fibrosis for all patients, regardless of which mutation they carried. They were going to invest $500 million in stem cells, gene editing, the new generation of gene therapy, and research that would likely revolutionize medicine and drug development—just as they had done for the past twenty years.

Raising this money would be even more difficult than it had been ten years ago. The perception among donors was that the foundation had more money than it needed, after the $3.3 billion it had received for the rights to future Vertex drug royalties. And with 95 percent of CF patients cured, donors thought, there couldn't be much more to do—and they were therefore less willing to give. But Joe knew that at $1.8 billion per drug, $3.3 billion wouldn't go far, especially given the foundation's strategy of aggressive investment in new technologies.

So Joe, in addition to soliciting immediate donations, had begun using a different approach on his standing army of aging, wealthy donors: legacy giving. Rather than asking people to give now, he asked that they consider leaving a portion of their estate to the foundation. Leave us $10,000, $100,000, or a million in your will, he would suggest. Leaving it all to your children might erode their productivity and their desire to contribute to society, so give it to a major disease charity instead. Joe thought it was a beautiful idea, and something that he planned to do himself.

When it came to his own estate planning, Joe would ensure his family was comfortable and secure after he was gone. But he intended to shift the majority of his wealth into a new foundation, perhaps called something along the lines of the O'Donnell Family Foundation, which Kate and Casey would run jointly as copresidents. His reasons were both selfish and humanitarian. Running the foundation would keep his daughters involved in philanthropy and endow them with influence—that was the selfish part. The humanitarian part was that, with this large endowment, his daughters would have the opportunity to do good things and make a difference in the lives of others.

As of early 2020, Joe wasn't sure what was next for the Joey Fund. But for the time being, he and Kathy, but increasingly Kate and Casey (to whom their parents had officially passed the reins in 2013), planned to keep running events—perhaps funneling less of the money raised to the CF foundation, and holding back significantly more than the 5 percent they usually reserved for struggling local families, especially after seeing the impact of the coronavirus pandemic on the CF community. Still, funding new research felt more important than ever to Joe. Advances in stem cell therapies, gene therapies, and gene editing would benefit not just CF patients still without a treatment, but all people with genetic diseases, rare and common.

For more than a decade, other health nonprofits looking to replicate the success of the CF Foundation and the Joey Fund had met with Joe, Bob Beall, and Preston Campbell for guidance. They approached Joe the way early cystic fibrosis activist Doris Tulcin and her father, George Frankel, had Basil O'Connor, to hear about how he'd built a fundraising operation to support polio patients and vaccine research with the National Foundation for Infantile Paralysis. What was the secret to curing genetic diseases? How had Joe raised a quarter-billion dollars to fund research? How could these other small health nonprofits become billion-dollar, agenda-setting philanthropies that pioneered new medical research? How could they, too, craft royalty deals that would bring a windfall of profits down the road?

At least once a month since 2012, when Kalydeco was approved, someone from a nonprofit would make an appointment to talk with Joe about how he did what he did. And over the past decade his thoughts had coalesced. He'd learned five things over the past forty-five years since Joey was born that he believed a foundation needed in order to succeed.

First, you had to centralize your efforts around a disease. A foundation had to have a single mission, and couldn't be democratic. In his opinion—which he'd exercised in his own family-run business—you had to have just one decision maker. Otherwise, people tended to sit around and squabble, which would slow progress. Bob Dresing had insisted on such an autocratic setup for the CFF so he could whip it into a single, powerful, centralized unit. You also couldn't have ten independent foundations all dedicated to the same cause, whether breast cancer or Alzheimer's disease or lung cancer. Ultimately they would splinter the pool of resources and needlessly duplicate research efforts—a clear waste of money.

The second element was what Joe called the "Bob Beall factor." To succeed, a foundation needs a strong leader with a solid grasp of the science who can guide the group's efforts toward the most promising research.

Third, you need someone to lead the volunteers—someone like Rich Mattingly, who could connect with both the chapters and the patients.

Fourth, you need talented volunteer fundraisers, like Doris Tulcin and Joe himself, who had credibility, personal experience with the disease, and a knack for connecting with people in an authentic way.

The last ingredient was patience.

EPILOGUE

2019–2020

The fact that people think it's impossible
doesn't bother me anymore.

—Preston Campbell

On October 21, 2019, barely four months after Vertex had submitted the drug for review, the FDA approved the triple—a medicine that the company named Trikafta. Preston Campbell, foundation CEO, was sitting on a couch in his office, strategizing and preparing for the upcoming North American Cystic Fibrosis Conference in Nashville, Tennessee, when he heard an excited banging on his door. Mike Boyle, who was preparing to step into Campbell's shoes as CEO when Campbell retired at the beginning of 2020, and two other employees burst into the room to share the news—the triple had been approved more than five months earlier than they or anyone else had expected.

Before long they were joined by Patty Burks, the Cystic Fibrosis Foundation's director of clinical trial affairs, whose son had died from CF in 1993 at just eleven years old. For Burks, as for so many parents, the approval of this triple was a victory and a catharsis. She had lost her child, but thanks to the efforts of thousands of parents, volunteers, and donors, others wouldn't. Over the next few hours the celebration grew, as others with personal links to the disease poured into the office, weeping with joy after having spent most of their lives battling this disease.

Patients who now have a drug
Patients still waiting for a treatment

For Fred Van Goor back in San Diego, news of the approval filled him with intensely personal joy. As treatments had been approved over the last seven years, beginning with Kalydeco in 2012, he had toyed with the idea of planting a rose garden to commemorate the percentage of patients who had effective treatments. The garden was an homage to the Cystic Fibrosis Foundation's "65 roses" event, a reference to a child in the early 1950s who misheard "cystic fibrosis" as "sixty-five roses." The rose was also the foundation's adopted symbol of hope. But it wasn't until a few years ago that he'd been able to execute his vision. After moving to a new home, he began filling the perimeter of his yard, surrounding the swimming pool, with different varieties of roses: a sweet-smelling Barbra Streisand rose; one with delicate white flowers; tough, viny varieties with scarlet blooms. With the triple combination, the vast majority of patients would have a drug—and he had planted fifty-eight rosebushes in recognition. But once the approval came through, he decided not to wait until Vertex figured out treatments for the remaining patients before he planted all sixty-five. He would plant them now—a promise that he and all his colleagues in the San Diego labs would continue working until there was a medicine for every patient.

Paul Negulescu, who had returned Bob Beall's call to Aurora more than twenty years before, had felt the impact of that seemingly small decision every time news of an FDA approval had rippled through the CF community: Kalydeco in 2012, Orkambi in 2015, and Symdeko in 2018. Now, with the approval of the triple Trikafta, most patients—including many he had met when they had visited Vertex San Diego over the past eighteen years—were now guaranteed better health. They could expect to live long lives and die the same mundane deaths as many others. Shortly after the approval, Lola and Ashton Ferguson, Sean Young, Bobby Coughlin, and Jonathan Flessner began taking the triple and thriving. Doris Tulcin's daughter Ann, whom Dorothy Andersen didn't expect to live a year, went on Trikafta, too, at sixty-seven years old. Doris, still an honorary trustee of the CFF Board at ninety-two years old, reports that the impact on Ann's health has been transformative.

Since Vertex acquired Aurora, Negulescu had watched the company grow from one focused on developing immunosuppressants for autoimmune diseases and medicines for HIV and hepatitis C, to one that had developed four approved drugs for CF patients and was now "All in for CF." Cystic fibrosis had made Vertex one of the country's most innovative pharmaceutical companies, and now the company had pledged to find a cure for every last patient with the disease.

Bob Beall got the news of Trikafta's approval from a friend with cystic fibrosis who was also an FDA employee. The two already had dinner plans for that evening, and she called to tell him that they would have a lot more to celebrate. Then came calls from Preston Campbell, Mike Boyle, and a string of friends. Under Beall's leadership, the foundation had worked with Vertex to develop drugs for 50 percent of CF patients—a milestone for the disease and for personalized medicine generally. Now the work he had begun had led to treatment for so many more. Beall was excited for the patients, many of whom would now have a normal life span. And he was thrilled that the drugs were so safe. But he was also reminded that venture philanthropy and drug discovery were not for the faint of heart. You had to be willing to take on a lot of risk—and he knew that the foundation had been incredibly lucky.

Joe and Kathy O'Donnell first got the news of the drug's approval around 10 PM, while at dinner in Athens, Greece, with close friends. Joe's phone began chiming; both he and Kathy began receiving emails and texts by the dozens. Preston Campbell, MassBio CEO Bob Coughlin, and Vertex founder Josh Boger all called to talk with Joe. And the calls just kept coming late into the night: from close friends, from families whom he and Kathy had counseled and helped over the years, from patients and physicians—everyone who had watched and supported the O'Donnells' struggle and commitment over the last forty-five years.

In mid-2019, after I learned the results of Casey's prenatal tests and discovering that she didn't carry the mutation most common among those of Irish descent, I asked Joe and Kathy if, for the sake of my curiosity and this book, they both would take genetic tests to see which mutations each of them carried. I wanted to know whether, after they'd spent forty-five years raising money for the foundation and helping drive the discovery of drugs like the triple combo, it would have worked for Joey.

Having worked with me for seven years on this book, Joe and Kathy were accustomed to my questions, but this request was still a surprise. Joe told me bluntly that it was "a pain in the ass" to get it done, but did it anyway. Kathy had an interest in genealogy, and had already used a 23andMe kit to dive into her ancestry, so she was curious to know for herself, too.

The results showed that Kathy carried the less common mutation, which she'd passed to Casey. Joe, on the other hand, carried the common F508del mutation. Both mutations were severe. But for children like Joey who carried this combination, there was now a powerful treatment. For Kathy, that knowledge brought some comfort. For Joe, it did not.

I'd thought that if Joe knew he'd helped to fund a drug that would have worked for Joey, it would make him happy. But the gene Joey carried was irrelevant to him. It didn't give him any satisfaction or joy to know that this medicine would have worked for his son. That hadn't been his motivation all these years.

What does make him happy is knowing that infants and teens and young and middle-aged adults who are suffering from this disease now have a good shot at a full and healthy life, and the opportunity to have a family and goals unhindered by this disease. And that his children have genetic testing to keep the disease at bay.

Today, it is probably unsurprising that Joe O'Donnell has agreed to chair a third Milestones to a Cure. This time the goal is $500 million—twice as much as the previous campaigns put together.

Patients with cystic fibrosis and other lung diseases need more help, the care centers need more funding, and developing a cure will almost definitely take more than what the foundation currently has in the bank. And for Joe, working on CF will always be a labor of love—the most important thing he has ever done, and will continue to do, in his life.

ACKNOWLEDGMENTS

Just as it took a small army to develop treatments for this disease, it took a village to help me tell the story of this quest.

The cystic fibrosis community is a tight-knit group, woven together by the need for unity among those battling a rare, debilitating disease and their single-minded desire to cure it. As I researched this book, I had the privilege to connect with many people—ultimately several hundred—who have played a pivotal role in caring for patients, designing drugs, raising money, and bringing the community together. While I was only able to include a fraction of them in this book, it wouldn't have been possible without their contributions.

I am forever indebted to Joe and Kathy O'Donnell for believing in the value of this story and in my ability to tell it. Joe and Kathy understood the ambitious nature of the saga: the human perspective told through their own eyes and those of other families, the riveting roller coaster of revolutionary science, the unusual growth and strategy of the Cystic Fibrosis Foundation, and the birth of a new era of medicine. And they allowed me, an enthusiastic reporter with no connection to cystic fibrosis, to step into their lives and their home, welcoming me with warmth and sincerity. They opened doors for me, made introductions, and, most of all, shared some of the most exquisite and excruciating moments of their time with Joey. They allowed me to dig and probe into memories that were painful to relive—and they did so with grace and generosity. Their daughters, Kate and Casey, freely recounted memories, recollections, and deeply personal moments that had shaped their and their parents' lives—and helped me track down their busy parents when I couldn't reach them.

I am also grateful to the other parents, many of whom I had never met before, who trusted me with heart-wrenching stories of their constant fight to keep their children alive, and to the many adult patients who detailed their medical odysseys to help illustrate the ravages of this disease. Thank you to the entire Cheevers family—it was amazing to see in person the impact of these drugs on Laura and Cate over five years and watch their transformation into healthy and strong young women. Thank you to Paul and Sue Flessner, Bob Coughlin, Jennifer Ferguson, and Katrina Young for sharing their stories of having children with CF in the new millennium. I'm grateful for your trust. I'm also appreciative of Danny Bessette, Bill Elder, Jonathan Flessner, Katy Farrell, and Emily Kramer-Golinkoff for providing deeply personal accounts of what this disease has done to their bodies, and how the disease progressed and changed their lives.

Dr. Allen Lapey was a key source in helping me tell the O'Donnells' story and those of several other families and patients included in this book. To help me fill in the history, he retrieved Joey's medical records from the archives—a task I know caused him tremendous emotional strain, given how close he was to Joey. The way in which he connects with his patients is truly inspiring, and I can understand why those who survive into adulthood don't want to leave his care.

There are not enough words to thank the Cystic Fibrosis Foundation, which embraced this project from the beginning, thanks to president and CEO Bob Beall and his successor, Preston Campbell. Both of them granted numerous interviews throughout the years it has taken me to write this book, and generously helped me reach out to physicians and patients who then shared their stories with me. Preston fielded an endless string of queries from me, ranging from the intricacies of lung biology to the earliest interactions with Aurora—and I'm thankful for his unfailing attention. Enormous gratitude also to Doris Tulcin, Robert Dresing, and Rich Mattingly, who gave me a rich portrait of the foundation during the 1950s, '60s, and '70s, and of the scientific pioneers working during that same time; to Bonnie Ramsey, who also provided colorful recollections of the foundation's early days and the work of other pioneers, including Harry Shwachman; to Mike Boyle, who spent many hours explaining the logistics and results of clinical trials; and to Bill Skach, who invited me to the CF Foundation's science retreats, which allowed me to learn about the most cutting-edge CF science from the researchers themselves. Thanks to Chris Penland for providing a broad-ranging history of the foundation's scientific pursuits; to Martin Mense for access to the Cystic Fibrosis Foundation Therapeutics Laboratory; to Bruce Marshall for providing details about the patient care centers; to Anthony Durmowicz for the behind-the-scenes details on the workings of the FDA and its decision-making process;

to Ken Schaner, the man behind the foundation's royalty agreements; to Mary Dwight, for sharing the art of crafting public policy; and to chief of staff Marybeth McMahon, for her expansive knowledge of the foundation and community—and for helping me getting me the access I needed to many sources. Thank you, too, to communications director Jessica Rowlands for helping track down foundation facts, statistics, and photos.

Dozens of scientists in academia and at Vertex helped me understand the nuances of cystic fibrosis, gene cloning, and gene therapy and the complexities of drug development. Thank you to Paul Quinton for meeting with me and demystifying the science behind CF's salty sweat. And thank you to Mike Knowles and Rick Boucher for spending hours unraveling for me the science of lungs with CF. It was a great pleasure to talk with Lap-Chee Tsui about his quest to discover the CF gene—and many thanks for bringing together the lab members who are still working at Toronto's Hospital for Sick Children for me to speak with. I'm grateful to him, Batsheva Kerem, Johanna Rommens, Jack Riordan, and Aravinda Chakravarti for sharing their roles in and experiences of the hunt for the CF gene and for fact-checking those sections of the book.

Many thanks to NIH Director Francis Collins and Mitch Drumm who, as the American half of the gene-hunting team, described the Michigan effort. I'm also grateful to Dr. Collins for carving out the time to fact-check the gene-hunting sections of the book just as the coronavirus pandemic was beginning to grip the United States.

Thank you to Jim Wilson and Ron Crystal for detailing the story of their gene therapy efforts to treat CF; and to Melissa Ashlock, who provided a bridge from the world of gene therapy into the drug discovery effort at the foundation.

I appreciate my time spent with the late Nobel laureate and Aurora Biosciences founder Roger Tsien, whose humility and passion for teaching science still move me to this day. Thank you to Josh Boger for his unedited descriptions of Vertex's early days and also for his ability to explain science in a historical context with wit and humor.

I couldn't have told the drug discovery story without the access Vertex granted me, admitting me to their labs and coordinating conversations with their scientists. Thank you to Fred Van Goor, Sabine Hadida, Jinglan Zhou, Viji Arumugam, Jason McCartney, Brian Bear, Tom Knapp, Tim Neuberger, Angela Kemnitzer, and the late Peter

Grootenhuis. Their individual contributions to developing drugs for this disease and their dedication to all patients was one of the most inspiring parts of writing this book. They are true heroes. A special shout-out goes to scientists Paul Negulescu and Eric Olson, who, in addition to their scientific contributions, fact-checked the book. Additional cheers to Eric for fielding a maddening stream of questions over the years, up until the day the manuscript was due to production. And Heather Nichols, thank you for enabling all of this and more.

My book benefited from a couple of rich sources that provided key primary information that couldn't be found anywhere else. *Cystic Fibrosis in the 20th Century* provided firsthand accounts from many of the field's scientific pioneers and early founders of the Cystic Fibrosis Foundation. Thanks to British physician James Littlewood for pointing me to the website cfmedicine.com, which provided a dense and comprehensive history of CF and its medical and scientific breakthroughs, and links to seminal papers in the field. Thanks to the Archives Program at Boston Children's Hospital and the Archives & Special Collections of the Augustus C. Long Health Sciences Library of Columbia University Irving Medical Center for providing access to the notes, letters, and photographs of Dorothy Andersen, and to the Office of NIH History and Stetten Museum of the National Institutes of Health.

Thank you to my brilliant editor Leah Wilson at BenBella Books, who guided me through writing my first book with patience, compassion, and thoughtful comments. She has a mind-blowing ability to recall every detail in this book and process complex science all while keeping the narrative moving. When the story exploded two years ago and the science breakthroughs at Vertex looked like they would soon yield a treatment for 90 percent of patients, Leah and the whole BenBella team embraced my vision for this book and gave me the time to follow this through and tell the whole story. And on top of all that, Leah made sure that this book went to production before she went to deliver her baby boy.

Thank you to James Fraleigh for your light touch and attention to detail—copyediting was painless because of you.

I would also like to thank my agent, Ethan Bassoff, for matching me with my publisher and supporting my vision for this ambitious book.

I was lucky to find Cecilia Wallace, a dazzling new graduate of Oberlin College, to work part time as my research assistant during the frantic final year of writing this book. Not only did she passionately dive into obscure medical and historical documents and Congressional records—fact-checking, seeking permissions, and more—but her unfailing drive, even as coronavirus brought everything to a standstill, was invaluable in completing this book. She will make a wonderful doctor someday.

Thanks go as well to Carlo Abulencia: without your careful transcriptions of hundreds of interviews for me over the years, it would have been impossible for this dyslexic writer to distill the mountains of information I collected.

A lifetime of thanks goes to my editor at *Discover* magazine, Pam Weintraub, who first gave me the opportunity to write about Kalydeco and unlock the story of this community and this disease.

Many thanks, also, to my boss Beth Daley and my fellow editors on the science and technology and the health and medicine desks at *The Conversation* for supporting my month of leave to work on the book just as coronavirus began to swamp the news cycle. Thank you so much for having my back during this time.

To my dear friend Sharon Guynup, who bullied me into writing my first book proposal; edited it dozens of times; and read drafts, rallied my flagging spirits, and provided great journalism tips for me as I approached the finish line.

To my dear friends Rob and Stephanie Marlin-Curiel, and their daughters, Emma and Annika (who gave up her bed at least a dozen times during my visits to Boston), thank you for true interest, friendship, generosity, and many dinners as I completed my reporting. Thank you to Selin Cherian-Rivers, who devoted her sharp mind to reading the last version of the book and making suggestions that made many parts better; and to Leila Hebshi, for her inspirational title suggestion.

Thank you to my parents, Bhavna and Pravin Trivedi, and sister, Mamta Trivedi, who, in addition to their love and encouragement, made astute observations and recommendations to countless versions of the book throughout the years and helped shape its tone and content. I'm grateful for their forgiveness when I missed holidays or cut vacations short to focus on this project—I've spent too little time with them.

Finally, I'm grateful to my husband Chad Cohen, who has loved and nurtured me—not just through his wildly creative meals but also by being the sole breadwinner during periods when I needed to focus on this book. I'm grateful to him for his friendship and emotional support, for talking through the stories and the science, and for his tough, candid feedback. I also appreciate the heartwarming cheerleading from Sonali and Dhruv Cohen, our wonderful children, who always had faith that I could complete this book—even though, as they commented one day, they couldn't remember a time in their short lives when I wasn't working on it. They bring me joy every day.

HISTORY OF THE CYSTIC FIBROSIS FOUNDATION AND CF SCIENCE AND DRUG DISCOVERY

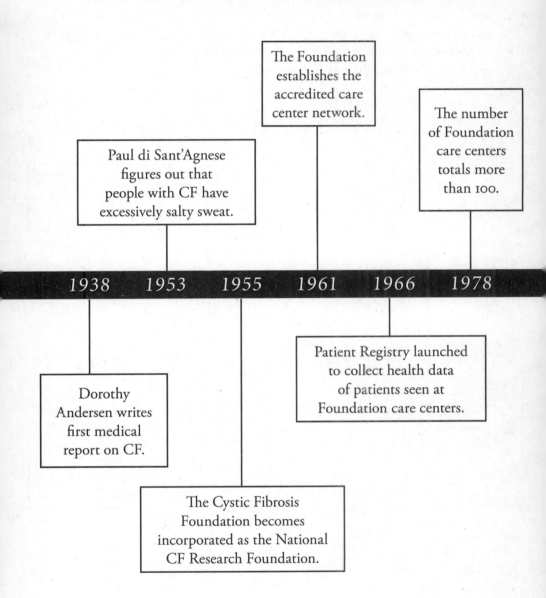

The Foundation establishes the accredited care center network.

The number of Foundation care centers totals more than 100.

Paul di Sant'Agnese figures out that people with CF have excessively salty sweat.

1938   1953   1955   1961   1966   1978

Dorothy Andersen writes first medical report on CF.

Patient Registry launched to collect health data of patients seen at Foundation care centers.

The Cystic Fibrosis Foundation becomes incorporated as the National CF Research Foundation.

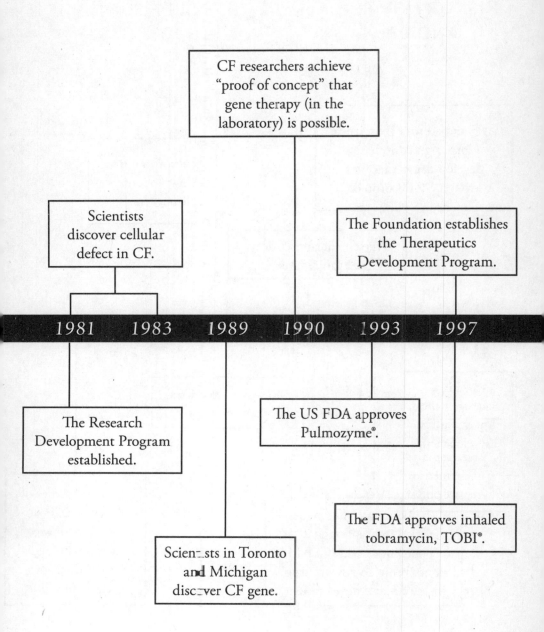

CF researchers achieve "proof of concept" that gene therapy (in the laboratory) is possible.

Scientists discover cellular defect in CF.

The Foundation establishes the Therapeutics Development Program.

1981   1983   1989   1990   1993   1997

The Research Development Program established.

The US FDA approves Pulmozyme®.

The FDA approves inhaled tobramycin, TOBI®.

Scientists in Toronto and Michigan discover CF gene.

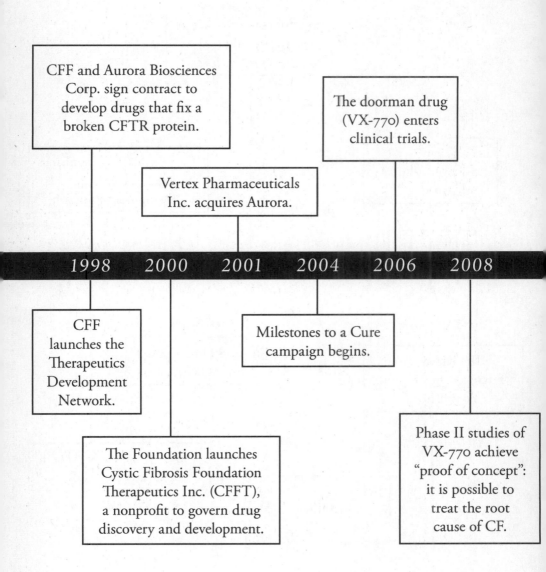

CFF and Aurora Biosciences Corp. sign contract to develop drugs that fix a broken CFTR protein.

The doorman drug (VX-770) enters clinical trials.

Vertex Pharmaceuticals Inc. acquires Aurora.

1998   2000   2001   2004   2006   2008

CFF launches the Therapeutics Development Network.

Milestones to a Cure campaign begins.

The Foundation launches Cystic Fibrosis Foundation Therapeutics Inc. (CFFT), a nonprofit to govern drug discovery and development.

Phase II studies of VX-770 achieve "proof of concept": it is possible to treat the root cause of CF.

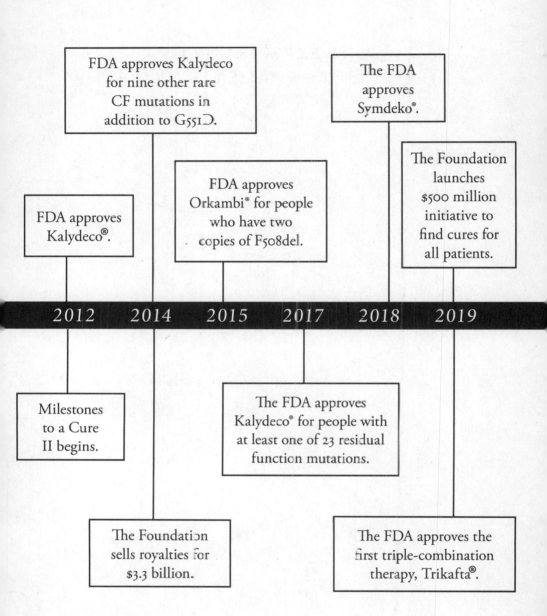

FDA approves Kalydeco for nine other rare CF mutations in addition to G551D.

The FDA approves Symdeko®.

The Foundation launches $500 million initiative to find cures for all patients.

FDA approves Kalydeco®.

FDA approves Orkambi® for people who have two copies of F508del.

2012   2014   2015   2017   2018   2019

Milestones to a Cure II begins.

The FDA approves Kalydeco® for people with at least one of 23 residual function mutations.

The Foundation sells royalties for $3.3 billion.

The FDA approves the first triple-combination therapy, Trikafta®.

LIFE EXPECTANCY OF
CYSTIC FIBROSIS PATIENTS, 1940–2017

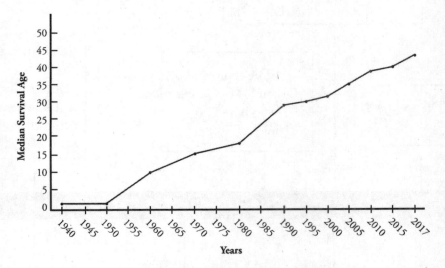

Source: Cystic Fibrosis Foundation's Patient Registry

ENDNOTES

1 In Latin and in modern paraphrase, ideas that survive in two ancient writers, Theocritus and Cicero.

CHAPTER 1

1 "A Narrative History of Mass General," Massachusetts General Hospital, accessed January 20, 2020, https://www.massgeneral.org/museum/history.
2 "Overview of the Research Institute," Massachusetts General Hospital, accessed January 20, 2020, https://www.massgeneral.org/research/about/overview-of-the-research-institute.
3 Steven Jacobs, "Surgeon, First to Replant Severed Limb, Dies at 70," *Harvard Crimson,* October 15, 2002.

CHAPTER 2

1 "Michael S. Brown, the Nobel Prize in Physiology or Medicine 1985: Facts," NobelPrize.org, accessed January 10, 2020, https://www.nobelprize.org/prizes/medicine/1985/brown/facts/.
2 "Harold E. Varmus, the Nobel Prize in Physiology or Medicine 1989: Biography," NobelPrize.org, accessed January 20, 2020, https://www.nobelprize.org/prizes/medicine/1989/varmus/biographical/.
3 "Robert J. Lefkowitz, the Nobel Prize in Chemistry 2012: Facts," NobelPrize.org, accessed January 10, 2020, https://www.nobelprize.org/prizes/chemistry/2012/lefkowitz/facts/.
4 Marvin J. Weinstein et al., "Gentamicin, a New Antibiotic Complex from Micromonospora," *Journal of Medicinal Chemistry* 6, no. 4 (July 1, 1963): 463–64, https://doi.org/10.1021/jm00340a034.
5 "Gentamicin was first approved for use in the United States in 1970 and remains in wide use." "Gentamicin," in *LiverTox: Clinical and Research Information on Drug-Induced Liver Injury* (Bethesda, MD: National Institute of Diabetes and Digestive and Kidney Diseases, 2012).
6 "Commonwealth of Massachusetts Election Statistics 1974," Office of the Secretary of the Commonwealth, http://archive.org/details/electionstatisti19741975mass.

CHAPTER 3

1 Science Service, "Viosterol Official Name for Irradiated Ergosterol," *Journal of Chemical Education* 7, no. 1 (January 1, 1930): 166, https://doi.org/10.1021/ed007p166.
2 B. Dowd and J. Walker-Smith, "Samuel Gee, Aretaeus, and the Coeliac Affection," *British Medical Journal* 2, no. 5909 (April 6, 1974): 45–47.
3 Samuel Gee, "On the Coeliac Affection," *Saint Bartholomew's Hospital Reports* 24 (1888): 17–20.
4 "Celiac Disease," National Institute of Diabetes and Digestive and Kidney Diseases, https://www.niddk.nih.gov/health-information/digestive-diseases/celiac-disease.

5 Stefano Guandalini, "A Brief History of Celiac Disease," *Impact: A Publication of the University of Chicago Celiac Disease Center* 7, no. 3 (2007); Dorothy H. Andersen, "Cystic Fibrosis of the Pancreas and Its Relation to Celiac Disease: A Clinical and Pathologic Study," *American Journal of Diseases of Children* 56, no. 2 (August 1, 1938): 344–99.

6 Ellen S. More, *Restoring the Balance: Women Physicians and the Profession of Medicine, 1850–1995* (Cambridge, MA: Harvard University Press, 1999), 98.

7 Ellen S. More and Marilyn Greer, "American Women Physicians in 2000: A History in Progress," *Journal of the American Medical Women's Association* 55, no. 1 (2000): 5–9.

8 Barbara Sicherman and Carol Hurd Green, "Dorothy Hansine Andersen," in *Notable American Women in the Modern Period: A Biographical Dictionary* (Harvard University Press, 1980), 18–19.

9 Stephanie Clague, "Dorothy Hansine Andersen," *Lancet Respiratory Medicine* 2, no. 3 (March 2014): 184–85.

10 Reem Gerais, "Dorothy Andersen (1901–1963)," in *The Embryo Project Encyclopedia* (Embryo Project, June 18, 2017), last modified July 4, 2018, https://embryo.asu.edu/pages /dorothy-andersen-1901-1963.

11 "The Florence R. Sabin Papers: Biographical Overview," US National Library of Medicine, accessed January 20, 2020, https://profiles.nlm.nih.gov/spotlight/rr/feature /biographical.

12 Reem Gerais, "Dorothy Andersen (1901–1963)," in *The Embryo Project Encyclopedia* (Embryo Project, June 18, 2017), last modified July 4, 2018, https://embryo.asu.edu/pages /dorothy-andersen-1901-1963.

13 "Department History," Columbia University Department of Pathology and Cell Biology, June 19, 2017, https://www.pathology.columbia.edu/about-us/department-history.

14 H. A. Carithers, "The First Use of an Antibiotic in America," *American Journal of Diseases of Children* 128, no. 2 (August 1974): 207–11.

15 R. W. Cantrell, R. A. Bell, and W. T. Morioka, "Acute Epiglottitis: Intubation versus Tracheostomy," *Laryngoscope* 88, no. 6 (June 1978): 994–1005.

16 H. A. Carithers, "The First Use of an Antibiotic in America," *American Journal of Diseases of Children* 128, no. 2 (August 1974): 207–11.

CHAPTER 4

1 Lysbeth Cohen, *Dr. Margaret Harper: Her Achievements and Place in the History of Australia* (Sydney, Australia: Wentworth Books, 1971).

2 Margaret Harper, *The Parents' Book* (Angus & Robertson, 1926); Lysbeth Cohen, *Dr. Margaret Harper: Her Achievements and Place in the History of Australia* (Wentworth Books, 1971), 38; Erwin H. Ackernecht, "Anticontagionism between 1821 and 1867," *Bulletin of the History of Medicine* 22, no. 5 (1948): 562–93; Yvonne E. Cossart, "The Rise and Fall of Infectious Diseases: Australian Perspectives, 1914–2014," *Medical Journal of Australia* 201, no. 1 (July 7, 2014): 11–14; "Vaccine History Timeline," Health.vic, https://www2.health.vic.gov .au:443/public-health/immunisation/immunisation-schedule-vaccine-eligibility-criteria /vaccine-history-timeline.

3 A. E. Garrod and W. H. Hurtley, "Congenital Family Steatorrhoea," *Quarterly Journal of Medicine* 6, no. 2 (January 1, 1913): 242–58, https://doi.org/10.1093/oxfordjournals .qjmed.a069344; Reginald Miller and Herbert Perkins, "Congenital Steatorrhoea," *Quarterly Journal of Medicine* 14, no. 53 (October 1, 1920): 1–9, https://doi.org/10.1093/qjmed /os-14.53.1.

4 Margaret Harper, "Two Cases of Congenital Pancreatic Steatorrhoea with Infantilism," *Medical Journal of Australia* 2 (November 15, 1930): 663–64.

5 Ibid.

6 P. Rothman, "Obituaries: Arthur Hawley Parmelee M. D. (1883–1961), Pioneer Pediatrician to the Newborn," *American Journal of Diseases of Children* 103, no. 2 (February 1962): 197–200.

7 A. H. Parmelee, "The Pathology of Steatorrhea," *American Journal of Diseases of Children* 50, no. 6 (December 1, 1935): 1418–28.

8 Ibid.

9 A. E. Garrod and W. H. Hurtley, "Congenital Family Steatorrhoea," *Quarterly Journal of Medicine* 6, no. 2 (January 1, 1913): 242–58; Cecil Clarke and Geoffrey Hadfield, "Congenital Pancreatic Disease with Infantilism," *Quarterly Journal of Medicine* 17, no. 68 (July 1, 1924): 358–64.

10 Julius H. Hess and Otto Saphir, "Celiac Disease (Chronic Intestinal Indigestion): A Report of Three Cases with Autopsy Findings," *Journal of Pediatrics* 6, no. 1 (January 1, 1935): 1–13.

11 A. H. Parmelee, "The Pathology of Steatorrhea," *American Journal of Diseases of Children* 50, no. 6 (December 1, 1935): 1418–28.

12 Kenneth D. Blackfan and S. Burt Wolbach, "Vitamin A Deficiency in Infants: A Clinical and Pathological Study," *Journal of Pediatrics* 3, no. 5 (November 1, 1933): 679–706.

13 Kenneth D. Blackfan and Charles D. May, "Inspissation of Secretion, Dilatation of the Ducts and Acini, Atrophy and Fibrosis of the Pancreas in Infants: A Clinical Note," *Journal of Pediatrics* 13, no. 5 (November 1, 1938): 627–34.

14 Dorothy H. Andersen, "Cystic Fibrosis of the Pancreas and Its Relation to Celiac Disease: A Clinical and Pathologic Study," *American Journal of Diseases of Children* 56, no. 2 (August 1, 1938): 344–99.

CHAPTER 5

1 Dorothy H. Andersen, "Cystic Fibrosis of the Pancreas and Its Relation to Celiac Disease: A Clinical and Pathologic Study," *American Journal of Diseases of Children* 56, no. 2 (August 1, 1938): 344–99.

2 Jacob Greenberg, "An Attempt to Reproduce Cœliac Disease Experimentally in Young Animals by Excluding the External Pancreatic Secretion from the Intestine," *Yale Journal of Biology and Medicine* 6, no. 2 (December 1933): 121–53.

3 Claude Bernard, "Mémoire sur le pancréas et sur le role du suc pancréatique dans les phénomènes digestifs, particulièrement dans la digestion des matières grasses neutres," *Supplemént aux comptes rendus hebdomaires des séances de l'Académie des sciences* 1 (1856): 379–563.

4 D. Wright Wilson, "Claude Bernard," *Popular Science Monthly* 84 (June 1914): 567–78.

5 *The Transactions of the American Medical Association* circa 1860 features an article on "a case of diarrhea adiposa," which includes Dr. de la Tremblaye's six-year-old patient with fatty diarrhea and diseased pancreas. The case is featured amid discussion of Claude Bernard's "Memoire sur le Pancréas." Both the article and "Memoire" refer to de la Tremblaye's case appearing in the 1852 edition of the "Recueil des travaux de la société médicale d'Indre et Loire." 2° serie, 3° chapter, 4° trimester.

6 Dorothy H. Andersen, "Cystic Fibrosis of the Pancreas and Its Relation to Celiac Disease: A Clinical and Pathologic Study," *American Journal of Diseases of Children* 56, no. 2 (August 1, 1938): 398.

7 Alfred F. Hess, "The Pancreatic Ferments in Infants," *American Journal of Diseases of Children* 4, no. 4 (October 1, 1912): 205.

8 Dorothy H. Andersen, "Pancreatic Enzymes in the Duodenal Juice in the Celiac Syndrome," *American Journal of Diseases of Children* 63, no. 4 (April 1, 1942): 643–58.

9 Archive at Columbia University. Letter from J.O to D.A.

CHAPTER 6

1 John Herbers, "The 37th President Is First to Quit Post," *New York Times,* August 9, 1974.

2 Bruce Gellerman, "It Was Like a War Zone: Busing in Boston," part 1, *WBUR,* September 4, 2014, https://www.wbur.org/news/2014/09/05/boston-busing-anniversary; "Forced Busing to Boston Schools Begins Today; Boycott Threatened," *Los Angeles Times,* December 10, 1974; Jack Tager, *Boston Riots: Three Centuries of Social Violence* (Northeastern University Press, 2001); Jack Porter, "What South Boston Wants Is 'Free Choice,'" *Los Angeles Times,* December 17, 1974.

3 Chris Woolston, "Moving Forward with Cystic Fibrosis," *Knowable Magazine,* May 29, 2018.

4 "About ABCD Nonprofit," Action for Boston Community Development, accessed January 11, 2020.

CHAPTER 7

1 Paul di Sant'Agnese, "Experiences of a Pioneer Researcher: Discovery of the Sweat Electrolyte Defect and the Early Medical History of Cystic Fibrosis," in *Cystic Fibrosis in the Twentieth Century: People, Events and Progress* (A. M. Publishing, 2001), 19.

2 Anthony di Sant'Agnese, email message to author, March 26, 2016.

3 D. A. Christie and E. M. Tansey, eds., *Cystic Fibrosis: The Transcript of a Witness Seminar Held by the Wellcome Trust Centre for the History of Medicine at UCL, London, on 11 June 2002* (London: Wellcome Trust Centre for the History of Medicine, 2004); H. A. Carithers, "The First Use of an Antibiotic in America," *American Journal of Diseases of Children* 128, no. 2 (August 1974): 207–11.

4 H. A. Carithers, "The First Use of an Antibiotic in America," *American Journal of Diseases of Children* 128, no. 2 (August 1974): 207–11.

5 Alexander Fleming, "On the Antibacterial Action of Cultures of a Penicillium, with Special Reference to Their Use in the Isolation of B. Influenzæ," *British Journal of Experimental Pathology* 10, no. 3 (June 1929): 226–36; "Alexander Fleming, The Nobel Prize in Physiology or Medicine 1945: Biographical," NobelPrize.org, https://www.nobelprize.org/prizes/medicine/1945/fleming/biographical/; Susan Aldridge, John Parascandola, and Jeffrey Sturchio, *Commemorative Booklet: The Discovery and Development of Penicillin 1928–1945* (American Chemical Society and the Royal Society of Chemistry, 1999); "Streptococcal Infections," US National Library of Medicine: MedlinePlus, October 21, 2016, https://medlineplus.gov/streptococcalinfections.html; American Chemical Society International Historic Chemical Landmarks, "Alexander Fleming's Discovery and Development of Penicillin," American Chemical Society, https://www.acs.org/content/acs/en/education/whatischemistry/landmarks/flemingpenicillin.html.

6 Paul di Sant'Agnese and Dorothy Andersen, "Celiac Syndrome; Chemotherapy in Infections of the Respiratory Tract Associated with Cystic Fibrosis of the Pancreas; Observations with Penicillin and Drugs of the Sulfonamide Group, with Special Reference to Penicillin Aerosol," *American Journal of Diseases of Children* 72 (July 1946): 17–61.

7   D. A. Christie and E. M. Tansey, eds., *Cystic Fibrosis: The Transcript of a Witness Seminar Held by the Wellcome Trust Centre for the History of Medicine at UCL, London, on 11 June 2002* (London: Wellcome Trust Centre for the History of Medicine, 2004).

8   "Acute Disseminated Encephalomyelitis Information Page," National Institute of Neurological Disorders and Stroke, March 27, 2019, https://www.ninds.nih.gov/Disorders /All-Disorders/Acute-Disseminated-Encephalomyelitis-Information-Page.

9   Anthony di Sant'Agnese, email message to author, March 26, 2016.

10   "City Swelters in 91.4 Degree Heat; Mid-West Industry Slowed; More Torrid Weather Due Here Today," *New York Times,* August 26, 1948.

11   Yas Kuno, *The Physiology of Human Perspiration* (London: J. and A. Churchill, 1934).

12   Paul di Sant'Agnese et al., "Abnormal Electrolyte Composition of Sweat in Cystic Fibrosis of the Pancreas; Clinical Significance and Relationship to the Disease," *Pediatrics* 12, no. 5 (November 1953): 549–63.

13   "News and Announcements," *Pediatrics* 12, no. 3 (September 1, 1953): 335.

14   Dorothy H. Andersen, "Pancreatic Enzymes in the Duodenal Juice in the Celiac Syndrome," *American Journal of Diseases of Children* 63, no. 4 (April 1, 1942): 643–58.

15   Joanna H. Fanos, "'We Kept Our Promises': An Oral History of Harry Shwachman, M.D.," *American Journal of Medical Genetics Part A* 146A, no. 3 (February 1, 2008), 287.

16   Ibid., 287.

17   Ibid., 288.

18   K. A. Misch and H. M. Holden, "Sweat Test for the Diagnosis of Fibrocystic Disease of the Pancreas; Report of a Fatality," *Archives of Disease in Childhood* 33, no. 168 (April 1958): 179–80.

19   Lewis E. Gibson, "A New Era in Diagnosis: The Sweat Test," in *Cystic Fibrosis in the Twentieth Century: People, Events and Progress* (A. M. Publishing, 2001). Some findings applied to the East Asian population. He learned that the Japanese, for example, lacked apocrine sweat glands in their armpits—which reduced body odor. The unfortunate Japanese women born with these glands were labeled outcasts and unmarriageable, and encouraged to endure a painful surgical procedure to correct the defect, leading Gibson to speculate that this may have been the origin for their interest in perspiration.

20   L. E. Gibson and R. E. Cooke, "A Test for Concentration of Electrolytes in Sweat in Cystic Fibrosis of the Pancreas Utilizing Pilocarpine by Iontophoresis," *Pediatrics* 23, no. 3 (March 1959): 545–49.

21   Stephanie Clague, "Dorothy Hansine Andersen," *Lancet Respiratory Medicine* 2, no. 3 (March 2014): 184–85.

22   Reem Gerais, "Dorothy Andersen (1901–1963)," in *The Embryo Project Encyclopedia* (Embryo Project, June 18, 2017), last modified July 4, 2018, https://embryo.asu.edu/pages /dorothy-andersen-1901-1963.

23   Paul di Sant'Agnese, "Bronchial Obstruction with Lobar Atelectasis and Emphysema in Cystic Fibrosis of the Pancreas," *Pediatrics* 12, no. 2 (August 1953): 178–90; Paul di Sant'Agnese, "The Pulmonary Manifestations of Fibrocystic Disease of the Pancreas," *Diseases of the Chest* 27, no. 6 (June 1, 1955): 654–67.

CHAPTER 8

1   Evelyn Graub and Milton Graub, "Origin and Development of the Cystic Fibrosis Foundation," in *Cystic Fibrosis in the Twentieth Century: People, Events and Progress* (A. M. Publishing, 2001), 149.

2   Jia-Qiang Huang et al., "The Selenium Deficiency Disease Exudative Diathesis in Chicks Is Associated with Downregulation of Seven Common Selenoprotein Genes in Liver and Muscle," *Journal of Nutrition* 141, no. 9 (September 2011): 1605–10.

3   Foundation Launch and Correspondence—Letter from WS to DA in 1955. 55–56 (footnoted in chapter 5), July 25, 1955.

4   "Wynne Sharples Ballinger, US Paediatrician," Science Photo Library, https://www.sciencephoto.com/media/876890/view/wynne-sharples-ballinger-us-paediatrician; "Wynne Sharples Ballinger (1923-2008)," Smithsonian Institution Archives, February 16, 2012, https://siarchives.si.edu/collections/siris_arc_306467.

5   "Wynne Sharples Nobleman's Bride," *New York Times,* April 8, 1951.

6   "History's Keepers: The Legacy of the National Society of The Colonial Dames of America," NSCDA, https://nscda.org/historys-keepers-the-legacy-of-the-nscda/.

7   Joanna H. Fanos, "'We Kept Our Promises': An Oral History of Harry Shwachman, M.D.," *American Journal of Medical Genetics Part A* 146A, no. 3 (February 1, 2008): 284–93.

8   The President's Commission on the Assassination of President Kennedy, "Warren Commission Hearings," vol. 9 (1964), 199.

9   Evelyn Graub and Milton Graub, "Origin and Development of the Cystic Fibrosis Foundation," in *Cystic Fibrosis in the Twentieth Century: People, Events and Progress* (A. M. Publishing, 2001).

10 Doris F. Tulcin, *Memoirs of a Monarch: A Chronicle of My Life* (iUniverse, 2008), 35.

11 "George Frankel, Oilman, 78, Dies," *New York Times,* April 8, 1971.

12 "New Disease," *Time,* March 1, 1954, 38.

13 James W. Barton, "Future Grave for 80 Per Cent of Cystic Fibrosis Children," *Bergen Evening Record,* April 12, 1957.

14 Howard A. Rusk, "War on Cystic Fibrosis," *New York Times,* October 13, 1957.

CHAPTER 9

1   Doris F. Tulcin, *Memoirs of a Monarch: A Chronicle of My Life* (iUniverse, 2008), 41.

2   Frank Deford, "When the Dread Disease Struck in Her Own Home, Doris Tulcin Declared War on Cystic Fibrosis," *People,* December 6, 1976.

3   David M. Oshinsky, *Polio: An American Story,* 1st ed. (Oxford University Press, 2006).

4   Ibid., 47.

5   Ibid., 48–50.

6   "Eddie Cantor and the Origin of the March of Dimes," March of Dimes, https://www.marchofdimes.org/mission/eddie-cantor-and-the-origin-of-the-march-of-dimes.aspx.

7   David M. Oshinsky, *Polio: An American Story,* 1st ed. (Oxford University Press, 2006), 113.

8   Ibid., 187.

9   Doris F. Tulcin, *Memoirs of a Monarch: A Chronicle of My Life* (iUniverse, 2008), 44.

10 Paul di Sant'Agnese, "Organizing the Foundation's Medical Program: Collaboration from the Beginning," in *Cystic Fibrosis in the Twentieth Century: People, Events and Progress* (A. M. Publishing, 2001).

11 Dorothy Hansine Andersen, letter of November 4, 1959, Columbia University archive of Dorothy Hansine Andersen's letters.

12 Wynne Sharples, letter of December 4, 1959, Columbia University archive of Dorothy Hansine Andersen's letters.

13 Ibid.

14 "Wynne Sharples Bellinger (1923–2008)," Smithsonian Institution Archives, February 16, 2012, https://siarchives.si.edu/collections/siris_arc_306467.

CHAPTER 10

1 "About the Cystic Fibrosis Foundation: Our History," Cystic Fibrosis Foundation, accessed January 11, 2020, https://www.cff.org/About-Us/About-the-Cystic-Fibrosis-Foundation /Our-History/.
2 Atul Gawande, "The Bell Curve," *New Yorker,* December 6, 2004.
3 Carl Doershuk, "The Matthews Comprehensive Treatment Program: A Ray of Hope," in *Cystic Fibrosis in the Twentieth Century: People, Events and Progress* (A. M. Publishing, 2001).
4 Ibid.
5 Ibid., 68.
6 Ibid., 71.
7 Warren Warwick, "Center Directors Remember: Minnesota," in *Cystic Fibrosis in the Twentieth Century: People, Events and Progress* (A. M. Publishing, 2001).
8 Cecil J. Nesbitt, "Cecil J. Nesbitt, 89: Mathematician Wrote Actuary Science Books," *Los Angeles Times,* October 28, 2001.
9 Carl Doershuk, "Growth of the Foundation's Medical Program," in *Cystic Fibrosis in the Twentieth Century: People, Events and Progress* (A. M. Publishing, 2001).
10 The raw data describing the natural history of the disease between 1950 and 1980 was lost in a trucking accident as the CFF was moving from Atlanta to Washington, DC.

CHAPTER 12

1 "Daniel Flood, 90, Who Quit Congress in Disgrace, Is Dead," *New York Times,* May 29, 1994; William C. Kashatus, *Dapper Dan Flood* (Pennsylvania State University Press, 2010), 285.
2 Doris Tulcin, "Doris Tulcin's Story: One Family's Special Mission," in *Cystic Fibrosis in the Twentieth Century: People, Events and Progress* (A. M. Publishing, 2001), 170.
3 Previously the National Institute of Arthritis and Metabolic Diseases; until June 23, 1981, the National Institute of Arthritis, Metabolism, and Digestive Diseases; and until April 8, 1986, the National Institute of Arthritis, Diabetes, and Digestive and Kidney Diseases, then finally National Institute of Diabetes and Digestive and Kidney Diseases (NIDDK).
4 National Cancer Act of 1971, Pub. L. No. 92-218 (1971).
5 National Diabetes Research and Education Act of 1974, Pub. L. No. 93-354 (1974).
6 "National Institute of Diabetes and Digestive and Kidney Diseases," The NIH Almanac, January 3, 2020, https://www.nih.gov/about-nih/what-we-do/nih-almanac /national-institute-diabetes-digestive-kidney-diseases-niddk.
   On July 23, 1974, the National Diabetes Mellitus Research and Education Act (P.L. 93-354) was signed into law. The National Commission on Diabetes, authorized by this act, was chartered on September 17, 1974. The act authorized diabetes research and training centers, and an intergovernmental diabetes coordinating committee that included representatives from the NIAMDD and six other NIH Institutes.
7 "National Diabetes Statistics Report, 2017" (Atlanta, GA: Centers for Disease Control and Prevention, 2017), https://www.cdc.gov/diabetes/data/statistics-report/index.html. At the time of this report, 30.3 million people in the US had diabetes (9.4 percent of the

population), and 84.1 million adults (aged eighteen years or older) had prediabetes (33.9 percent of the adult US population).

8   House Report 1219, "A Mandate from the Subcommittee on Appropriations for the Department of Labor and Department of Health, Education and Welfare of the Committee on Appropriations of the US House of Representatives," stated, "The committee is convinced that cystic fibrosis research deserves priority attention and therefore directs that a study be undertaken and completed within the next year on what is being done in research and patient care in cystic fibrosis and allied diseases."

9   Doris F. Tulcin, *Memoirs of a Monarch: A Chronicle of My Life* (iUniverse, 2008), 77.

10 *Cystic Fibrosis: State of the Art and Directions for Future Research Efforts* (Bethesda, MD: National Institutes of Health, 1978).

CHAPTER 13

1   September 23, 1980. Extension of remarks. P. 26963. Hon. Silvio O. Conte of Massachusetts. Proclamation 4859, September 17, 1981; Proclamation 4978 – September 24, 1982.

CHAPTER 14

1   R. C. Boucher, P. A. Bromberg, and J. T. Gatzy, "Airway Transepithelial Electric Potential in Vivo: Species and Regional Differences," *Journal of Applied Physiology: Respiratory, Environmental and Exercise Physiology* 48, no. 1 (January 1980): 169–76.

2   M. Knowles, J. Gatzy, and R. Boucher, "Increased Bioelectric Potential Difference Across Respiratory Epithelia in Cystic Fibrosis," *New England Journal of Medicine* 305, no. 25 (December 17, 1981): 1489–95.

3   M. Knowles, J. Gatzy, and R. Boucher, "Relative Ion Permeability of Normal and Cystic Fibrosis Nasal Epithelium," *Journal of Clinical Investigation* 71, no. 5 (May 1, 1983): 1410–17; M. R. Knowles, et al., "Abnormal Ion Permeation through Cystic Fibrosis Respiratory Epithelium," *Science* 221, no. 4615 (September 9, 1983): 1067–70.

CHAPTER 15

1   P. M. Quinton, "Dueling with Cystic Fibrosis: Finding the Chloride Defect," in *Cystic Fibrosis in the Twentieth Century: People, Events and Progress* (A. M. Publishing, 2001).

2   US Centers for Disease Control and Prevention, "Questions and Answers About Tuberculosis" (Atlanta: US Department of Health and Human Services, 2014), https://www.cdc.gov/tb/publications/faqs/pdfs/qa.pdf.

3   Texas Department of Health Tuberculosis Elimination Division, "Re: Tuberculosis Screening in Schools," February 8, 2002, www.dshs.texas.gov/idcu/investigation/forms/tbcp02-08-02.doc.

4   "School-based screening for TB infection among children was started in the 1950s when infection and disease rates were higher than at present." Allan B. Bloch, "Screening for Tuberculosis and Tuberculosis Infection in High-Risk Populations Recommendations of the Advisory Council for the Elimination of Tuberculosis," *Morbidity and Mortality Weekly Report* 44(RR-11) (September 8, 1995): 18–34, https://www.cdc.gov/mmwr/preview/mmwrhtml/00038873.htm.

5   "TIRR Memorial Hermann Celebrates 60 Years of Pushing the Envelope in Rehabilitation," *TIRR Memorial Hermann Journal*, Winter 2020, http://tirr.memorialhermann.org/journal/2020-winter/celebrating-60-years-of-rehabilitation/.

6 A. Spock et al., "Abnormal Serum Factor in Patients with Cystic Fibrosis of the Pancreas," *Pediatric Research* 1, no. 3 (May 1967): 173–77; B. H. Bowman, L. H. Lockhart, and M. L. McCombs, "Oyster Ciliary Inhibition by Cystic Fibrosis Factor," *Science* 164, no. 3877 (April 18, 1969): 325–26; J. A. Mangos and N. R. McSherry, "Sodium Transport: Inhibitory Factor in Sweat of Patients with Cystic Fibrosis," *Science* 158, no. 3797 (October 6, 1967): 135–36; J. A. Mangos and N. R. McSherry, "Studies on the Mechanism of Inhibition of Sodium Transport in Cystic Fibrosis of the Pancreas," *Pediatric Research* 2, no. 5 (September 1968): 378–84; J. A. Mangos, N. R. McSherry, and P. J. Benke, "A Sodium Transport Inhibitory Factor in the Saliva of Patients with Cystic Fibrosis of the Pancreas," *Pediatric Research* 1, no. 6 (November 1967): 436–42.

7 Wikipedia, s.v. "Sweat Gland," last modified January 15, 2020, 21:27, https://en.wikipedia.org/wiki/Sweat_gland.

8 M. Knowles, J. Gatzy, and R. Boucher, "Relative Ion Permeability of Normal and Cystic Fibrosis Nasal Epithelium," *Journal of Clinical Investigation* 71, no. 5 (May 1, 1983): 1410–17.

9 P. M. Quinton, "Chloride Impermeability in Cystic Fibrosis," *Nature* 301, no. 5899 (February 3, 1983): 421–22.

CHAPTER 17

1 "Mary Weiss, Trailblazer in the Fight Against CF, Dies at 77," Cystic Fibrosis Foundation, April 19, 2016, https://www.cff.org/News/News-Archive/2016/Mary-Weiss-Trailblazer-in-the-Fight-Against-CF-Dies-at-77/.

2 Doris F. Tulcin, *Memoirs of a Monarch: A Chronicle of My Life* (iUniverse, 2008), 82-84.

3 "CF Basic Research Centers," Cystic Fibrosis Foundation, 2020, https://www.cff.org/Research/Researcher-Resources/CF-Basic-Research-Centers/.

CHAPTER 18

1 "About SickKids," SickKids, accessed February 6, 2020, http://www.sickkids.ca/AboutSickKids/index.html.

2 Gabriel C. Lander et al., "Bacteriophage Lambda Stabilization by Auxiliary Protein GpD: Timing, Location, and Mechanism of Attachment Determined by CryoEM," *Structure* 16, no. 9 (September 10, 2008): 1399–1406.

3 Paloma Martínez, "Manuel Buchwald: una vida de viaje, de ciencia y de música" (Radio Canadá Internacional, October 14, 2015).

4 Tania N. Petruzziello-Pellegrini et al., "Cystic Fibrosis in Canada: A Historical Perspective," *Canadian Journal of Respiratory, Critical Care, and Sleep Medicine* (December 21, 2018): 1–10, https://doi.org/10.1080/24745332.2018.1470910.

5 "Directory: Mary Corey," SickKids, accessed January 20, 2020, http://www.sickkids.ca/AboutSickKids/Directory/People/C/Mary-Corey-Staff-profile.html.

6 L. Wijcik, M. Buchwald, and J. R. Riordan, "Induction of Alkaline Phosphatase in Cultured Human Fibroblasts. Comparison of Normal Cells and Those from Patients with Cystic Fibrosis," *Biochimica Et Biophysica Acta* 585, no. 3 (July 4, 1979): 374–82.

7 J. R. Riordan, N. Alon, and M. Buchwald, "Plasma Membrane Lipids of Human Diploid Fibroblasts from Normal Individuals and Patients with Cystic Fibrosis," *Biochimica Et Biophysica Acta* 574, no. 1 (July 27, 1979): 39–47.

8 From 1961 to 1966, Robert W. Holley, Har Gobind Khorana, Heinrich Matthaei, Marshall W. Nirenberg, and colleagues cracked the human genetic code. Marshall W. Nirenberg

and J. Heinrich Matthaei, "The Dependence of Cell-Free Protein Synthesis in E. Coli upon Naturally Occurring or Synthetic Polyribonucleotides," *Proceedings of the National Academy of Sciences of the United States of America* 47, no. 10 (October 1961): 1588–1602.

9   At the time the estimated number of human genes was more than 100,000. After the genome was sequenced, scientists discovered there were only between 20,000 and 25,000 human genes. "Understanding Our Genetic Inheritance: The Human Genome Project, FY 1991–1995" (United States Department of Health and Human Services, April 1990); International Human Genome Sequencing Consortium, "Finishing the Euchromatic Sequence of the Human Genome," *Nature* 431, no. 7011 (October 21, 2004): 931–45.

10 D. Botstein et al., "Construction of a Genetic Linkage Map in Man Using Restriction Fragment Length Polymorphisms," *American Journal of Human Genetics* 32, no. 3 (May 1980): 314–31.

11 "Andree Marie Dozy," *SFGate*, May 4, 2003, https://www.sfgate.com/news/article /DOZY-Andree-Marie-2618910.php.

12 Y. W. Kan and A. M. Dozy, "Polymorphism of DNA Sequence Adjacent to Human Beta-Globin Structural Gene: Relationship to Sickle Mutation," *Proceedings of the National Academy of Sciences of the United States of America* 75, no. 11 (November 1978): 5631–35.

13 Ibid.

14 Y. W. Kan and A. M. Dozy, "Antenatal Diagnosis of Sickle-Cell Anaemia by D.N.A. Analysis of Amniotic-Fluid Cells," *Lancet* 2, no. 8096 (October 28, 1978): 910–12.

15 D. Botstein et al., "Construction of a Genetic Linkage Map in Man Using Restriction Fragment Length Polymorphisms," *American Journal of Human Genetics* 32, no. 3 (May 1980): 314–31.

16 These are called "restriction fragment length polymorphisms" (RFLPs, pronounced "rif-flips").

17 "How Many Words Are There in the Bible?," Word Counter, December 8, 2015, https:// wordcounter.net/blog/2015/12/08/10975_how-many-words-bible.html. There are 3,116,480 characters in the King James Authorized Bible.

18 Adele Glimm, *Gene Hunter: The Story of Neuropsychologist Nancy Wexler* (Joseph Henry Press, 2006).

19 Lawrence K. Altman, "Researchers Report Genetic Test Detects Huntington Disease," *New York Times,* November 9, 1983.

20 J. F. Gusella, et al., "A Polymorphic DNA Marker Genetically Linked to Huntington's Disease," *Nature* 306, no. 5940 (November 17, 1983): 234–38.

CHAPTER 19

1   Elie Dolgin, "The Most Popular Genes in the Human Genome," *Nature* 551, no. 7681 (November 22, 2017): 427–31.

2   L. Pauling and H. A. Itano, "Sickle Cell Anemia a Molecular Disease," *Science* 110, no. 2865 (November 25, 1949): 543–48.

3   V. M. Ingram, "A Specific Chemical Difference Between the Globins of Normal Human and Sickle-Cell Anæmia Hæmoglobin," *Nature* 178, no. 4537 (October 13, 1956): 792–94.

4   M. F. Perutz et al., "Structure of Hæmoglobin: A Three-Dimensional Fourier Synthesis at 5.5-Å. Resolution, Obtained by X-Ray Analysis," *Nature* 185, no. 4711 (February 1960): 416–22.

5   "Alpha Thalassemia X-Linked Intellectual Disability Syndrome," Genetics Home Reference, March 17, 2020, https://ghr.nlm.nih.gov/condition/alpha-thalassemia-x-linked -intellectual-disability-syndrome.

6 S. Ottolenghi et al., "The Severe Form of Alpha Thalassaemia Is Caused by a Haemoglobin Gene Deletion," *Nature* 251, no. 5474 (October 4, 1974): 389–92.

7 J. M. Taylor et al., "Genetic Lesion in Homozygous Alpha Thalassaemia (Hydrops Fetalis)," *Nature* 251, no. 5474 (October 4, 1974): 392–93.

8 Alexander Fleming, "On the Antibacterial Action of Cultures of a Penicillium, with Special Reference to Their Use in the Isolation of B. Influenzæ," *British Journal of Experimental Pathology* 10, no. 3 (June 1929): 226–36; R. R. Porter, "The Hydrolysis of Rabbit Y-Globulin and Antibodies with Crystalline Papain," *Biochemical Journal* 73 (September 1959): 119–26.

9 Robert Johnson, "History of the Cystic Fibrosis Research Trust" (June 1984), http://www.cfmedicine.com/history/futureotherhistory.html#one.

10 "Friedreich Ataxia Fact Sheet," National Institute of Neurological Disorders and Stroke, https://www.ninds.nih.gov/Disorders/Patient-Caregiver-Education/Fact-Sheets/Friedreichs-Ataxia-Fact-Sheet.

11 "Carrier Testing for Cystic Fibrosis," Cystic Fibrosis Foundation, accessed January 23, 2020, https://www.cff.org/What-is-CF/Testing/Carrier-Testing-for-Cystic-Fibrosis/.

12 "Collaborative Research; International Research Team Finds Genetic Link to Cystic Fibrosis," Business Wire October 9, 1985. "We have made a significant investment in the RFLP project—several million dollars to date—and we will continue to invest in it because we believe it will result in diagnostic tests for many genetic diseases." "Science and Technology Report and Outlook, 1985–1988" (United States Office of Science and Technology Policy, 1989).

13 "LOD Score," National Human Genome Research Institute, https://www.genome.gov/genetics-glossary/LOD-Score.

14 L. Roberts, "The Race for the Cystic Fibrosis Gene," *Science* 240, no. 4849 (April 8, 1988): 141–44; L. Roberts, "Race for Cystic Fibrosis Gene Nears End," *Science* 240, no. 4850 (April 15, 1988): 282–85.

15 Robert Mullan Cook-Deegan, "Survey of Genome Science Corporations: Contract Report, Prepared for the Office of Technology Assessment, US Congress," March 1994, http://citeseerx.ist.psu.edu/viewdoc/download?doi=10.1.1.461.5746&rep=rep1&type=pdf. "The claims in the original patent application for the CF linkage were quite broad, claiming the ultimate gene and any markers discovered closer to it. That patent was abandoned when it became clear the cost of pursuing and defending it would not be balanced by financial benefits, especially after the CF gene itself was found, and a patent filed on behalf of the University of Michigan and its collaborators."

16 R. Saltus, "Biotech Firms Compete in Genetic Diagnosis," *Science* 234, no. 4782 (December 12, 1986): 1318–20.

17 R. G. Knowlton et al., "A Polymorphic DNA Marker Linked to Cystic Fibrosis Is Located on Chromosome 7," *Nature* 318, no. 6044 (December 28, 1985): 380–82.

18 R. White et al., "A Closely Linked Genetic Marker for Cystic Fibrosis," *Nature* 318, no. 6044 (December 28, 1985): 382–84.

19 B. J. Wainwright et al., "Localization of Cystic Fibrosis Locus to Human Chromosome 7cen-Q22," *Nature* 318, no. 6044 (December 28, 1985): 384–85.

20 P. Newmark, "Testing for Cystic Fibrosis," *Nature* 318, no. 6044 (December 28, 1985): 309.

21 L. C. Tsui et al., "Cystic Fibrosis Locus Defined by a Genetically Linked Polymorphic DNA Marker," *Science* 230, no. 4729 (November 29, 1985): 1054–57.

22 Martin Farrall et al., "First-Trimester Prenatal Diagnosis of Cystic Fibrosis with Linked DNA Probes," *Lancet* 327, no. 8495 (June 21, 1986): 1402–5.

CHAPTER 20

1   Sally Squires, "The Seeds of Disease; Researchers Have Developed New Techniques to Predict Who Will Inherit Genetic Diseases," *Washington Post,* July 23, 1986.
2   Sally Squires, "The Ethics of Genetic Counseling; Reading the Human Blueprint Before Nature Has Its Way," *Washington Post,* November 26, 1986.

CHAPTER 21

1   Hui-Chuan Lai et al., "Delayed Diagnosis of US Females with Cystic Fibrosis," *American Journal of Epidemiology* 156, no. 2 (July 15, 2002): 165–73.
2   *The Newborn Screening Story* (Association of Public Health Laboratories, 2013).
3   "If Phenylketonuria (PKU) Is Not Treated, What Problems Occur?," National Institute of Child Health and Human Development, December 1, 2016, https://www.nichd.nih.gov /health/topics/pku/conditioninfo/untreated.
4   Ibid.
5   "PKU Test for Phenylketonuria: Purpose, Procedure, Results," WebMD, https://www .webmd.com/children/pku-test#1.
6   Cynthia C. Chernecky and Barbara J. Berger, *Laboratory Tests and Diagnostic Procedures* (New York: Elsevier Health Sciences, 2012).
7   "Screening for Phenylketonuria (PKU): US Preventive Services Task Force Reaffirmation Recommendation," *Annals of Family Medicine* 6, no. 2 (March 2008): 166.
8   *The Newborn Screening Story* (Association of Public Health Laboratories, 2013).
9   Lynn M. Taussig et al., "Neonatal Screening for Cystic Fibrosis: Position Paper," *Pediatrics* 72, no. 5 (November 1, 1983): 741–45.
10 H. Shwachman, A. Redmond, and K. T. Khaw, "Studies in Cystic Fibrosis: Report of 130 Patients Diagnosed Under 3 Months of Age over a 20-Year Period," *Pediatrics* 46, no. 3 (September 1970): 335–43.
11 J. R. Crossley, R. B. Elliott, and P. A. Smith, "Dried-Blood Spot Screening for Cystic Fibrosis in the Newborn," *Lancet* 1, no. 8114 (March 3, 1979): 472–74.
12 "Newborn Screening for CF," Cystic Fibrosis Foundation, https://www.cff.org /What-is-CF/Testing/Newborn-Screening-for-CF/.
13 N. Fost and P. M. Farrell, "A Prospective Randomized Trial of Early Diagnosis and Treatment of Cystic Fibrosis: A Unique Ethical Dilemma," *Clinical Research* 37, no. 3 (September 1989): 495–500.
14 Beginning in 1991, the IRT test was confirmed with a DNA test.
15 After 1991 the IRT test was combined with a DNA test that scanned for the most common mutation in the CFTR gene: F508del.
16 P. M. Farrell et al., "Nutritional Benefits of Neonatal Screening for Cystic Fibrosis. Wisconsin Cystic Fibrosis Neonatal Screening Study Group," *New England Journal of Medicine* 337, no. 14 (1997): 963–69.

CHAPTER 22

1   Aravinda Chakravarti, "Ching Chun Li (1912–2003): A Personal Remembrance of a Hero of Genetics," *American Journal of Human Genetics* 74, no. 5 (May 2004): 789–92.
2   A. Chakravarti et al., "Nonuniform Recombination within the Human Beta-Globin Gene Cluster," *American Journal of Human Genetics* 36, no. 6 (November 1984): 1239–58.
3   Ushma S. Neill, "A Conversation with Francis Collins," *Journal of Clinical Investigation* 122, no. 11 (November 1, 2012): 3797–98.

4   Peter B. Moore, David Eisenberg, and Jason Kahn, *Biographical Memoirs: Donald M. Crothers* (National Academy of Sciences, 2018).

5   "Donald Crothers Obituary," *New Haven Register,* March 23, 2014.

6   "1968: First Restriction Enzymes Described," National Human Genome Research Institute, April 26, 2013, https://www.genome.gov/25520301/online-education-kit-1968-first-restriction-enzymes-described.

7   "1972: First Recombinant DNA," National Human Genome Research Institute, April 26, 2013, https://www.genome.gov/25520302/online-education-kit-1972-first-recombinant-dna.

8   Stanley N. Cohen et al., "Construction of Biologically Functional Bacterial Plasmids In Vitro," *Proceedings of the National Academy of Sciences of the United States of America* 70, no. 11 (November 1973): 3240–44.

9   "1976: First Genetic Engineering Company," National Human Genome Research Institute, April 26, 2013, https://www.genome.gov/25520305/online-education-kit-1976-first-genetic-engineering-company.

10   A. M. Maxam and W. Gilbert, "A New Method for Sequencing DNA," *Proceedings of the National Academy of Sciences* 74, no. 2 (February 1, 1977): 560–64; F. Sanger and A. R. Coulson, "A Rapid Method for Determining Sequences in DNA by Primed Synthesis with DNA Polymerase," *Journal of Molecular Biology* 94, no. 3 (May 25, 1975): 441–48; F. Sanger et al., "Nucleotide Sequence of Bacteriophage ΦX174 DNA," *Nature* 265, no. 5596 (February 1977): 687–95.

11   Alexandra Minna Stern, *Eugenic Nation: Faults and Frontiers of Better Breeding in Modern America* (University of California Press, 2016).

12   "Future and Past Annual Meetings," American Society of Human Genetics, https://www.ashg.org/meetings/future-past/.

13   Y. W. Kan and A. M. Dozy, "Polymorphism of DNA Sequence Adjacent to Human Beta-Globin Structural Gene: Relationship to Sickle Mutation," *Proceedings of the National Academy of Sciences of the United States of America* 75, no. 11 (November 1978): 5631–35.

14   Paul Berg et al., "Potential Biohazards of Recombinant DNA Molecules," *Science* 185, no. 4148 (July 26, 1974): 303.

15   V. B. Reddy et al., "The Genome of Simian Virus 40," *Science* 200, no. 4341 (1978): 494–502.

16   Edward J. Benz, "Bernard Gregoire Forget, MD," *Transactions of the American Clinical and Climatological Association* 127 (2016): lxxxviii–xci.

17   J. L. Slightom, A. E. Blechl, and O. Smithies, "Human Fetal G Gamma- and A Gamma-Globin Genes: Complete Nucleotide Sequences Suggest That DNA Can Be Exchanged between These Duplicated Genes," *Cell* 21, no. 3 (October 1980): 627–38.

18   F. S. Collins et al., "G Gamma Beta+ Hereditary Persistence of Fetal Hemoglobin: Cosmid Cloning and Identification of a Specific Mutation 5' to the G Gamma Gene," *Proceedings of the National Academy of Sciences of the United States of America* 81, no. 15 (August 1984): 4894–98.

19   F. S. Collins and S. M. Weissman, "Directional Cloning of DNA Fragments at a Large Distance from an Initial Probe: A Circularization Method," *Proceedings of the National Academy of Sciences of the United States of America* 81, no. 21 (November 1984): 6812–16.

20   "Genetics and Genomics," University of Michigan Medical School, January 6, 2017, https://medicine.umich.edu/medschool/education/phd-programs/about-pibs/graduate-programs/genetics-genomics.

21   Edward W. Holmes, "Of Rice and Men: Bill Kelley's Next Generation," *Journal of Clinical Investigation* 115, no. 10 (October 1, 2005): 2948–52.

22 F. S. Collins et al., "Construction of a General Human Chromosome Jumping Library, with Application to Cystic Fibrosis," *Science* 235, no. 4792 (February 27, 1987): 1046–49.

23 J. L. Marx, "Hopping Along the Chromosome," *Science* 228, no. 4703 (May 31, 1985): 1080.

CHAPTER 23

1 Mike Barnicle, "1,000 Celebrate the Life of a Courageous Fighter," *Boston Globe,* November 28, 1986.

CHAPTER 24

1 Batsheva Kerem et al., "In Situ Nick-Translation Distinguishes between Active and Inactive X Chromosomes," *Nature* 304, no. 5921 (July 1983): 88–90.

2 Batsheva Kerem et al., "Mapping of DNAase I Sensitive Regions on Mitotic Chromosomes," *Cell* 38, no. 2 (September 1, 1984): 493–99.

3 Xavier Estivill et al., "A Candidate for the Cystic Fibrosis Locus Isolated by Selection for Methylation-Free Islands," *Nature* 326, no. 6116 (April 1987): 840–45.

4 Andrew Veitch, "Victory 'Near' over Cystic Fibrosis," the *Guardian,* April 30, 1987; Fiona Harari, "Reported Gene Find Lifts Hope for Cure," the *Age,* May 2, 1987.

5 "Scientists Find Genetic Key to Cystic Fibrosis," *St. Petersburg Times,* April 30, 1987.

6 Peter Gorner and Jeff Lyon, "Deadly Gene Losing Its Mystery," *Chicago Tribune,* April 30, 1987.

7 Wen-Hsiung Li and Matthew A. Saunders, "News and Views: The Chimpanzee and Us," *Nature* 437, no. 7055 (September 1, 2005): 50–51.

8 F. Liang et al., "Gene Index Analysis of the Human Genome Estimates Approximately 120,000 Genes," *Nature Genetics* 25, no. 2 (June 2000): 239–40; International Human Genome Sequencing Consortium, "Initial Sequencing and Analysis of the Human Genome," *Nature* 409, no. 6822 (February 2001): 860–921; Robert Kanigel, "The Genome Project," *New York Times,* December 13, 1987.

9 J. R. Riordan et al., "Identification of the Cystic Fibrosis Gene: Cloning and Characterization of Complementary DNA," *Science* 245, no. 4922 (September 8, 1989), 1069.

10 Ibid., 1067.

CHAPTER 25

1 "Human Gene Mapping 10: Report of the 10th International Workshop on Human Gene Mapping," *Cytogenetics and Cell Genetics* 51, no. 1–4 (1989).

2 Landon Curt Noll, "English Names of the First 10000 Powers of 10," http://lcn2.github .io/mersenne-english-name/tenpower/tenpower.html.

3 Rebecca Kolberg, "Cystic Fibrosis Gene Found," *United Press International,* August 23, 1989.

4 Lydia Zajc, "Cystic Fibrosis Finding Should Improve Treatment, Testing," *United Press International,* August 24, 1989.

5 Frances Kelly, "Researchers Score Genetic First—Sick Children's Hospital Team Isolates Cause of Cystic Fibrosis," *Toronto Star,* August 24, 1989.

6 Thomas H. Maugh, "Researchers Isolate Gene That Causes Cystic Fibrosis," *Los Angeles Times,* August 25, 1989.

7 Jean Seligmann and Daniel Glick, "Cystic Fibrosis: Hunting Down a Killer Gene," *Newsweek,* September 4, 1989.

8   Sandra Blakeslee, "Scientists Develop New Technique to Track Down Defects in Genes," *New York Times,* September 12, 1989.

9   Lydia Zajc, "Cystic Fibrosis Finding Should Improve Treatment, Testing," *United Press International,* August 24, 1989.

10 Ibid.

11 Daniel E. Koshland, "The Cystic Fibrosis Gene Story," *Science* 245, no. 4922 (September 8, 1989): 1029.

12 T. Shoshani et al., "Association of a Nonsense Mutation (W1232X), the Most Common Mutation in the Ashkenazi Jewish Cystic Fibrosis Patients in Israel, with Presentation of Severe Disease," *American Journal of Human Genetics* 50, no. 1 (January 1992): 222–28.

13 Patrick Stafler et al., "The Impact of a National Population Carrier Screening Program on Cystic Fibrosis Birth Rate and Age at Diagnosis: Implications for Newborn Screening," *Journal of Cystic Fibrosis: Official Journal of the European Cystic Fibrosis Society* 15, no. 4 (2016): 460–66.

CHAPTER 26

1   "Defining Moments: Cloning Insulin," Genentech April 7, 2016, https://www.gene.com /stories/cloning-insulin.

2   D. V. Goeddel et al., "Expression in Escherichia Coli of Chemically Synthesized Genes for Human Insulin," *Proceedings of the National Academy of Sciences of the United States of America* 76, no. 1 (January 1979): 106–10.

3   "Defining Moments: The Approval," Genentech, May 12, 2016, https://www.gene.com /stories/the-approval.

4   Vageesh S. Ayyar, "History of Growth Hormone Therapy," *Indian Journal of Endocrinology and Metabolism* 15, no. S3 (September 2011): S152–65.

5   J. B. Armstrong and J. C. White, "Liquefaction of Viscous Purulent Exudates by Deoxyribonuclease," *Lancet* 2, no. 6641 (December 9, 1950): 739–42; W. S. Chernick, G. J. Barbero, and H. J. Eichel, "In-Vitro Evaluation of Effect of Enzymes on Tracheobronchial Secretions from Patients With Cystic Fibrosis," *Pediatrics* 27 (April 1961): 589–96.

6   J. Salnikow et al., "Bovine Pancreatic Deoxyribonuclease A. Isolation of Cyanogen Bromide Peptides; Complete Covalent Structure of the Polypeptide Chain," *Journal of Biological Chemistry* 248, no. 4 (February 25, 1973): 1480–88; T. H. Liao et al., "Bovine Pancreatic Deoxyribonuclease A. Isolation of Cyanogen Bromide Peptides; Complete Covalent Structure of the Polypeptide Chain," *Journal of Biological Chemistry* 248, no. 4 (February 25, 1973): 1489–95.

7   Orphan Drug Act of 1983, Pub. L. No. 97–414, 2049 (1983).

8   Isabel Stenzel Byrnes and Anabel Stenzel, *The Power of Two: A Twin Triumph over Cystic Fibrosis* (University of Missouri Press, 2007), 9.

9   Lawrence Fisher, "Rehabilitation of a Biotech Pioneer," *New York Times,* May 8, 1994.

CHAPTER 27

1   B. W. Ramsey, P. M. Farrell, and P. Pencharz, "Nutritional Assessment and Management in Cystic Fibrosis: A Consensus Report. The Consensus Committee," *American Journal of Clinical Nutrition* 55, no. 1 (January 1992): 108–16.

2   Lawrence Fisher, "Genentech's Drug to Treat Cystic Fibrosis Is Approved," *New York Times,* December 31, 1993.

3   Associated Press, "Inhalant Drug Approved for Cystic Fibrosis," *New York Times,* December 24, 1997.

4   Jack Levy et al., "Bioactivity of Gentamicin in Purulent Sputum from Patients with Cystic Fibrosis or Bronchiectasis: Comparison with Activity in Serum," *Journal of Infectious Diseases* 148, no. 6 (December 1, 1983): 1069–76.

5   P. M. Mendelman et al., "Aminoglycoside Penetration, Inactivation, and Efficacy in Cystic Fibrosis Sputum," *American Review of Respiratory Disease* 132, no. 4 (October 1985): 761–65.

6   B. W. Ramsey et al., "Efficacy of Aerosolized Tobramycin in Patients with Cystic Fibrosis," *New England Journal of Medicine* 328, no. 24 (June 17, 1993): 1740–46.

7   Ibid.

8   "NCATS Inxight: Drugs—Tobramycin," National Center for Advancing Translational Sciences, https://drugs.ncats.io/drug/VZ8RRZ51VK; Associated Press, "Inhalant Drug Approved for Cystic Fibrosis," *New York Times,* December 24, 1997.

9   "NCATS Inxight: Drugs—Tobramycin," National Center for Advancing Translational Sciences, https://drugs.ncats.io/drug/VZ8RRZ51VK.

10 "Chronic Obstructive Pulmonary Disease," Centers for Disease Control and Prevention, June 6, 2018, https://www.cdc.gov/copd/index.html.

11 Ana Swanson, "Big Pharmaceutical Companies Are Spending Far More on Marketing than Research," *Washington Post,* February 11, 2015.

12 A. Bruce Montgomery et al., "Aerosolised Pentamidine as Sole Therapy for *Pneumocystis carinii* Pneumonia in Patients with Acquired Immunodeficiency Syndrome," *Lancet* 330, no. 8557 (August 29, 1987): 480–83.

13 Robin Marantz Henig, "AIDS: A New Disease's Deadly Odyssey," *New York Times,* February 6, 1983.

14 "HIV/AIDS Historical Time Line 1981–1990," U.S. Food and Drug Administration, January 5, 2018, http://www.fda.gov/patients/hiv-timeline-and-history-approvals/hivaids-historical-time-line-1981-1990.

15 *Tufts Center for the Study of Drug Development Impact Report* 15, no. 1 (January 2013). The Tufts report says: "While nine out of 10 clinical trials worldwide meet their patient enrollment goals, reaching those targets typically means that drug developers need to nearly double their original timelines . . . 11% of sites in a given trial typically fail to enroll a single patient, 37% under-enroll, 39% meet their enrollment targets, and 13% exceed their targets."

16 Bonnie W. Ramsey et al., "Intermittent Administration of Inhaled Tobramycin in Patients with Cystic Fibrosis," *New England Journal of Medicine* 340, no. 1 (January 7, 1999): 23–30.

17 "Lung Capacity and Aging," American Lung Association, August 26, 2019, https://www.lung.org/lung-health-and-diseases/how-lungs-work/lung-capacity-and-aging.html.

18 Gary K. Chikami, "Approval Letter for New Drug Application 50-753" (Center for Drug Evaluation and Research, December 22, 1997), https://www.accessdata.fda.gov/drugsatfda_docs/nda/97/50753-mr.pdf.

CHAPTER 28

1   R. C. Mulligan and P. Berg, "Expression of a Bacterial Gene in Mammalian Cells," *Science* 209, no. 4463 (September 19, 1980): 1422–27.

2   Ibid.

3   James M. Wilson, "Genetic Diseases, Immunology, Viruses, and Gene Therapy," *Human Gene Therapy* 25, no. 4 (April 1, 2014): 257–61.

4   Wilson, "Genetic Diseases, Immunology, Viruses, and Gene Therapy"; J. M. Wilson et al., "Retrovirus-Mediated Transduction of Adult Hepatocytes," *Proceedings of the National Academy of Sciences of the United States of America* 85, no. 9 (May 1988): 3014–18.

5   Wilson et al., "Retrovirus-Mediated Transduction of Adult Hepatocyte."

6   J. M. Wilson et al., "Correction of the Genetic Defect in Hepatocytes from the Watanabe Heritable Hyperlipidemic Rabbit," *Proceedings of the National Academy of Sciences of the United States of America* 85, no. 12 (June 1988): 4421–25.

7   J. M. Wilson et al., "Temporary Amelioration of Hyperlipidemia in Low Density Lipoprotein Receptor-Deficient Rabbits Transplanted with Genetically Modified Hepatocytes," *Proceedings of the National Academy of Sciences of the United States of America* 87, no. 21 (November 1990): 8437–41.

8   J. R. Chowdhury et al., "Long-Term Improvement of Hypercholesterolemia After Ex Vivo Gene Therapy in LDLR-Deficient Rabbits," *Science* 254, no. 5039 (December 20, 1991): 1802–5.

9   Natalie Angier, "A New Gene Therapy to Fight Cholesterol Is Being Prepared," *New York Times,* October 29, 1991.

10   "Third Annual North American Cystic Fibrosis Conference, Tarpon Springs, Florida, October 11–14, 1989. Abstracts," *Pediatric Pulmonology* 4 (1989): 1–167.

11   Robin Marantz Henig, "Dr. Anderson's Gene Machine," *New York Times,* March 31, 1991.

12   R. M. Blaese et al., "T Lymphocyte-Directed Gene Therapy for ADA-SCID: Initial Trial Results After 4 Years," *Science* 270, no. 5235 (October 20, 1995): 475–80.

13   Natalie Angier, "Girl, 4, Becomes First Human to Receive Engineered Genes," *New York Times,* September 15, 1990.

14   The virus that W. French Anderson and R. Michael Blaese used in this clinical trial was a retrovirus, which inserts the therapeutic gene into the genome of the patient's cells. Adenovirus does not.

15   Robin Marantz Henig, "Dr. Anderson's Gene Machine," *New York Times,* March 31, 1991.

16   D. P. Rich et al., "Expression of Cystic Fibrosis Transmembrane Conductance Regulator Corrects Defective Chloride Channel Regulation in Cystic Fibrosis Airway Epithelial Cells," *Nature* 347, no. 6291 (September 27, 1990): 358–63.

17   Ibid.

18   M. L. Drumm et al., "Correction of the Cystic Fibrosis Defect in Vitro by Retrovirus-Mediated Gene Transfer," *Cell* 62, no. 6 (September 21, 1990): 1227–33.

19   Natalie Angier, "Team Cures Cells in Cystic Fibrosis by Gene Insertion," *New York Times,* September 21, 1990.

20   L. Roberts, "Cystic Fibrosis Corrected in Lab," *Science* 249, no. 4976 (September 28, 1990): 1503.

CHAPTER 29

1   Nicholas Wade, "Death Leads to Concerns for Future of Gene Therapy," *New York Times,* September 30, 1999.

2   Ruth SoRelle, "Survey Finds Americans Back Gene Therapy, Research," *Houston Chronicle,* September 29, 1992.

3   Manish Mohanka, Danai Khemasuwan, and James K. Stoller, "A Review of Augmentation Therapy for Alpha-1 Antitrypsin Deficiency," *Expert Opinion on Biological Therapy* 12, no. 6 (June 2012): 685–700.

4   Wallace P. Rowe et al., "Isolation of a Cytopathogenic Agent from Human Adenoids Undergoing Spontaneous Degeneration in Tissue Culture," *Proceedings of the Society for Experimental Biology and Medicine* 84, no. 3 (December 1, 1953): 570–73.

5   M. A. Rosenfeld et al., "Adenovirus-Mediated Transfer of a Recombinant Alpha 1-Antitrypsin Gene to the Lung Epithelium in Vivo," *Science* 252, no. 5004 (April 19, 1991): 431–34.

6   M. A. Rosenfeld et al., "In Vivo Transfer of the Human Cystic Fibrosis Transmembrane Conductance Regulator Gene to the Airway Epithelium," *Cell* 68, no. 1 (January 10, 1992): 143–55.

7   J. C. Olsen et al., "Correction of the Apical Membrane Chloride Permeability Defect in Polarized Cystic Fibrosis Airway Epithelia Following Retroviral-Mediated Gene Transfer," *Human Gene Therapy* 3, no. 3 (June 1992): 253–66.

8   Garry R. Cutting, "Two Steps Closer to Gene Therapy for Cystic Fibrosis," *Nature Genetics* 2, no. 1 (September 1992): 4–5.

9   J. A. Whitsett et al., "Human Cystic Fibrosis Transmembrane Conductance Regulator Directed to Respiratory Epithelial Cells of Transgenic Mice," *Nature Genetics* 2, no. 1 (September 1992): 13–20.

10 Stephen C. Hyde et al., "Correction of the Ion Transport Defect in Cystic Fibrosis Transgenic Mice by Gene Therapy," *Nature* 362, no. 6417 (March 1993): 250–55.

CHAPTER 30

1   John Hoffman, "Admission," *First in Human* (Discovery Channel, August 10, 2017).

2   Steven L. Brody et al., "Acute Responses of Non-Human Primates to Airway Delivery of an Adenovirus Vector Containing the Human Cystic Fibrosis Transmembrane Conductance Regulator cDNA," *Human Gene Therapy* 5, no. 7 (July 1, 1994): 821–36.

3   Ibid.

4   R. G. Crystal et al., "Administration of an Adenovirus Containing the Human CFTR cDNA to the Respiratory Tract of Individuals with Cystic Fibrosis," *Nature Genetics* 8, no. 1 (September 1994): 42–51.

5   Ibid.

6   J. Zabner et al., "Adenovirus-Mediated Gene Transfer Transiently Corrects the Chloride Transport Defect in Nasal Epithelia of Patients with Cystic Fibrosis," *Cell* 75, no. 2 (October 22, 1993): 207–16.

7   R. G. Crystal et al., "Administration of an Adenovirus Containing the Human CFTR CDNA to the Respiratory Tract of Individuals with Cystic Fibrosis," *Nature Genetics* 8, no. 1 (September 1994): 42–51.

8   Associated Press, "Big Victory in War on Cystic Fibrosis: Gene Therapy Corrects Key Defect," *Chicago Tribune,* October 15, 1993.

9   R. H. Simon et al., "Adenovirus-Mediated Transfer of the CFTR Gene to Lung of Nonhuman Primates: Toxicity Study," *Human Gene Therapy* 4, no. 6 (December 1993): 771–80.

10 J. F. Engelhardt et al., "Adenovirus-Mediated Transfer of the CFTR Gene to Lung of Nonhuman Primates: Biological Efficacy Study," *Human Gene Therapy* 4, no. 6 (December 1993): 759–69.

11 Warren King, "Gene Rides Virus to Fight Cystic Fibrosis," *Seattle Times,* November 27, 1995.

12 Terence R. Flotte et al., "Phase I Trial of Intranasal and Endobronchial Administration of a Recombinant Adeno-associated Virus Serotype 2 (RAAV2)-CFTR Vector in Adult

Cystic Fibrosis Patients: A Two-Part Clinical Study," *Human Gene Therapy* 14, no. 11 (July 20, 2003): 1079–88

13   Richard B. Moss et al., "Repeated Adeno-associated Virus Serotype 2 Aerosol-Mediated Cystic Fibrosis Transmembrane Regulator Gene Transfer to the Lungs of Patients with Cystic Fibrosis: A Multicenter, Double-Blind, Placebo-Controlled Trial," *Chest* 125, no. 2 (February 2004): 509–21.

14   Richard B. Moss et al., "Repeated Aerosolized AAV-CFTR for Treatment of Cystic Fibrosis: A Randomized Placebo-Controlled Phase 2B Trial," *Human Gene Therapy* 18, no. 8 (August 2007): 726–32.

15   "Using Fatty Droplets to Import Genes into the Nose, Breakthrough in the Search for Cystic Fibrosis Cure," *Daily Mail,* December 31, 1994.

CHAPTER 31

1   Sheryl Gay Stolberg, "The Biotech Death of Jesse Gelsinger," *New York Times,* November 28, 1999.

2   James M. Wilson, "Lessons Learned from the Gene Therapy Trial for Ornithine Transcarbamylase Deficiency," *Molecular Genetics and Metabolism* 96, no. 4 (April 1, 2009): 151–57.

3   "Gene-Therapy Trials Must Proceed with Caution," *Nature News* 534, no. 7609 (June 30, 2016): 590.

4   James M. Wilson, "Lessons Learned from the Gene Therapy Trial for Ornithine Transcarbamylase Deficiency."

5   Nicholas Wade, "Death Leads to Concerns for Future of Gene Therapy," *New York Times,* September 30, 1999.

6   Melinda Wenner, "Tribulations of a Trial," *Scientific American,* September 2009.

7   CDC Media Relations, "1996 HIV/AIDS Trends Provide Evidence of Success in HIV Prevention and Treatment" (Centers for Disease Control and Prevention, February 1996).

8   GlobeNewswire, "Tufts Center for the Study of Drug Development Assessment of Cost to Develop and Win Marketing Approval for a New Drug Now Published," March 10, 2016.

9   Oleg Larin, "Principle of Combinatorial Chemistry," *Combinatorial Chemistry Review,* August 1, 2011, http://www.combichemistry.com/principle.html.

10   Richard Saltus, "The Genome Project; a 'Superstar' Takes over the Helm; Renowned Gene Hunter Called Perfect Fit for the Job," *Boston Globe,* February 1, 1993.

11   G. M. Denning et al., "Processing of Mutant Cystic Fibrosis Transmembrane Conductance Regulator Is Temperature-Sensitive," *Nature* 358, no. 6389 (August 27, 1992): 761–64.

CHAPTER 32

1   "O'Donnells Donate $30 Million," *Harvard Gazette,* March 20, 2012, sec. Campus and Community.

CHAPTER 33

1   "89% of Trials Meet Enrollment, but Timelines Slip, Half of Sites Under-Enroll," *Tufts Center for the Study of Drug Development Impact Report* 15, no. 1 (January 2013). The Center reports, "While nine out of 10 clinical trials worldwide meet their patient enrollment goals, reaching those targets typically means that drug developers need to nearly double their original timelines . . . 11% of sites in a given trial typically fail to enroll a

single patient, 37% under-enroll, 39% meet their enrollment targets, and 13% exceed their targets."

2   Maura O'Leary et al., "Progress in Childhood Cancer: 50 Years of Research Collaboration, a Report from the Children's Oncology Group." *Seminars in Oncology* 35, no. 5 (October 2008): 484–93.

3   Ibid.

4   E. N. Pattishall, "Negative Clinical Trials in Cystic Fibrosis Research," *Pediatrics* 85, no. 3 (March 1990): 277–81; Katharine Cheng et al., "Randomized Controlled Trials in Cystic Fibrosis (1966–1997) Categorized by Time, Design, and Intervention," *Pediatric Pulmonology* 29, no. 1 (2000): 1–7.

5   "Dick Kronmal, UW SPH History Project," video, 55:15, University of Washington School of Public Health, March 5, 2013, https://www.youtube.com/watch?v=4DGwOAB7QEE.

CHAPTER 34

1   G. Grynkiewicz, M. Poenie, and R. Y. Tsien, "A New Generation of Ca2+ Indicators with Greatly Improved Fluorescence Properties," *Journal of Biological Chemistry* 260, no. 6 (March 25, 1985): 3440–50.

2   Anders Lennartson, "The Colours of Chromium," *Nature Chemistry* 6, no. 10 (October 2014): 942.

3   "Roger Y. Tsien, the Nobel Prize in Chemistry 2008: Biographical," NobelPrize.org, accessed February 19, 2020, https://www.nobelprize.org/prizes/chemistry/2008/tsien/biographical/.

4   Atsushi Miyawaki et al., "Fluorescent Indicators for Ca 2+ Based on Green Fluorescent Proteins and Calmodulin," *Nature* 388, no. 6645 (August 1997): 882–87.

5   Quyen T. Nguyen et al., "Surgery with Molecular Fluorescence Imaging Using Activatable Cell-Penetrating Peptides Decreases Residual Cancer and Improves Survival," *Proceedings of the National Academy of Sciences* 107, no. 9 (March 2, 2010): 4317–22.

6   Stephen J. Lippard, "Roger Y. Tsien (1952–2016)," *Science* 354, no. 6308 (October 7, 2016): 41.

7   Sacha Pfeiffer, "Five Things You Should Know about Kevin Kinsella," *Boston Globe*, August 28, 2015.

8   William A. Wells, "Shine a Light, Aurora Biosciences Corporation," *Chemistry & Biology* 4 (August 1, 1997): 537–38.

9   Jorge Cortese, "At the Speed of Light," *Scientist*, July 10, 2000.

10 Tapan K. Chaudhuri and Subhankar Paul, "Protein-Misfolding Diseases and Chaperone-Based Therapeutic Approaches," *Federation of European Biochemical Societies Journal* 273, no. 7 (April 2006): 1331–49.

11 "NCI Drug Dictionary: Genistein," National Cancer Institute, February 2, 2011, https://www.cancer.gov/publications/dictionaries/cancer-drug/def/genistein.

12 Gerene M. Denning et al., "Processing of Mutant Cystic Fibrosis Transmembrane Conductance Regulator Is Temperature-Sensitive," *Nature* 358, no. 6389 (August 1992): 761–64.

13 J. E. González and R. Y. Tsien, "Improved Indicators of Cell Membrane Potential That Use Fluorescence Resonance Energy Transfer," *Chemistry & Biology* 4, no. 4 (April 1997): 269–77.

CHAPTER 35

1   John Markoff, "Bill Gates's Brain Cells, Dressed Down for Action; Pressed to Innovate, Microsoft Relies Again on an Inner Circle," *New York Times,* March 25, 2001.
2   Paul Flessner, email message to Bonnie Ramsey, July 26, 1999.
3   "Leadership: Bill Gates," Bill & Melinda Gates Foundation, https://www.gatesfoundation .org/who-we-are/general-information/leadership/executive-leadership-team/bill-gates.
4   Paul Flessner, email message to Bill Gates, September 14, 1999.
5   Bill Gates, email message to Paul Flessner, September 14, 1999.
6   Reuters, "National News Briefs; Gateses Give $20 Million to Cystic Fibrosis Group," *New York Times,* October 24, 1999.
7   "$2.7 Million Raised in One Night for Cystic Fibrosis Research," PR Newswire, October 27, 1999.

CHAPTER 36

1   Frederick R. Adler et al , "Lung Transplantation for Cystic Fibrosis," *Proceedings of the American Thoracic Society* 6, no. 8 (December 15, 2009): 619–33.
2   Mandy Carranza, "Genetic Tests May Bring Hope, Inspire Fear," CNN, January 31, 2007: "Whatever the uncertainty over genetic discrimination, Americans have become concerned about the possible misuse of genetic information, especially in health insurance and employment. A June 2000 *Time* magazine/CNN poll found that 75 percent of those surveyed wouldn't want their insurance company to have information about their genetic codes. Some members of Congress have been working since 1996 to pass legislation that would prevent health insurers and employers from requesting or requiring genetic tests and bars insurers from raising premiums based on test results."
3   National Library of Medicine, "Turner Syndrome," Genetics Home Reference, October 2007.

CHAPTER 37

1   Lewis H. Sarett and Clyde Roche, "Max Tishler," *National Academy of Sciences Biographical Memoirs,* vol. 66 (Washington, DC: National Academies Press, 1995), 353–70.
2   Sarett and Roche, "Max Tishler," 358.
3   Joshua Boger et al., "Novel Renin Inhibitors Containing the Amino Acid Statine," *Nature* 303, no. 5912 (May 1933): 81–84.
4   "Remembering Pan Am Flight 103," *Federal Bureau of Investigation News,* December 14, 2018, https://www.fbi.gov/news/stories/remembering-pan-am-flight-103-30-years-later -121418.
5   Syracuse University Libraries, "Victim: Irving Stanley Sigal," Pam Am Flight Lockerbie Air Disaster Archives.
6   Vertex Pharmaceuticals, "Vertex Pharmaceuticals Researchers Report Three-Dimensional Structure of Hepatitis C Helicase Enzyme; Report in Structure," EurekAlert, January 15, 1998.
7   CenterWatch, "FDA Approved Drugs: Agenerase (Amprenavir)," CenterWatch: The Trusted Source for Clinical Trials Information.
8   Vertex Pharmaceuticals, "2000 Form 10-K" (United States Securities and Exchange Commission, March 26, 2001).

9   Andrew Pollack, "Vertex Buys Biotechnology Rival for $592 Million," *New York Times,* May 1, 2001; Vertex Pharmaceuticals, "Form 8-K" (United States Securities and Exchange Commission, April 30, 2001), https://investors.vrtx.com/node/13931/html.

10 Aurora Biosciences, "2000 Form 10-K" (United States Securities and Exchange Commission, March 15, 2001).

11 Robert F. Higgins, Sophie LaMontagne, and Brent Kazan, "Vertex Pharmaceuticals and the Cystic Fibrosis Foundation: Venture Philanthropy Funding for Biotech," Harvard Business School Case Collection, October 2007.

12 Ibid. Conversation adapted from this Harvard Business School case study.

CHAPTER 38

1   Antoine de Saint-Exupéry, *The Wisdom of the Sands,* trans. Stuart Gilbert (Harcourt, Brace & World, 1950).

2   Tinsley H. Davis, "Biography of E. P. Greenberg," *Proceedings of the National Academy of Sciences of the United States of America* 101, no. 45 (2004): 15830–32.

3   W. C. Fuqua, S. C. Winans, and E. P. Greenberg, "Quorum Sensing in Bacteria: The LuxR-LuxI Family of Cell Density-Responsive Transcriptional Regulators," *Journal of Bacteriology* 176, no. 2 (January 1994): 269–75.

4   Andrew Pollack, "Drug Makers Listen in While Bacteria Talk," *New York Times,* February 27, 2001.

5   Pfizer, "2000: Pfizer Joins Forces with Warner-Lambert," https://www.pfizer.com/about/history/pfizer_warner_lambert.

6   "Vertex Pharmaceuticals to Acquire Aurora Biosciences for $592 Million," *Boston Business Journal,* April 30, 2001.

7   Dennis P. Curran, Sabine Hadida, and Sun-Young Kim, "Tris(2-Perfluorohexylethyl)Tin Azide: A New Reagent for Preparation of 5-Substituted Tetrazoles from Nitriles with Purification by Fluorous/Organic Liquid-Liquid Extraction," *Tetrahedron* 55, no. 29 (July 16, 1999): 8997–9006.

CHAPTER 39

1   "Power 2005: The Old Guard," *Boston* magazine, March 22, 2012.

2   "Team: Robert H. Niehaus," GCP Capital Partners, http://www.gcpcapital.com/team/bio.aspx?team_Name=Robert_Niehaus.

CHAPTER 40

1   Antoine de Saint-Exupéry, *Wind, Sand and Stars*, trans. Lewis Galantiere (Harcourt Brace Jovanovich, 1967), 46.

2   Christopher A. Lipinski, Franco Lombardo, Beryl W. Dominy, and Paul J. Feeney, "Experimental and Computational Approaches to Estimate Solubility and Permeability in Drug Discovery and Development Settings," *Advanced Drug Delivery News* 46, no. 1–3 (2001): 3–26.

CHAPTER 41

1   E. O. Wilson, *Consilience: The Unity of Knowledge* (Knopf Doubleday, 2014).

2   Anne Harding, "More Compounds Failing Phase I," *Scientist,* September 12, 2004.

3 Personal communication with Preston Campbell, January 30, 2019.

4 Barry Werth, *The Antidote: Inside the World of New Pharma* (Simon & Schuster, 2014), 91.

CHAPTER 42

1 E. O. Wilson, *Consilience: The Unity of Knowledge* (Knopf Doubleday, 2014).

CHAPTER 43

1 Vertex Pharmaceuticals, "Vertex Pharmaceuticals Initiates Phase I Development for VX-770 in Cystic Fibrosis," Business Wire, May 17, 2006.

CHAPTER 44

1 Mark Murcko, "Discovering New Medicines Is as Much About Culture as Science," the *National*, April 12, 2014.

2 T. Z. Khan et al., "Early Pulmonary Inflammation in Infants with Cystic Fibrosis," *American Journal of Respiratory and Critical Care Medicine* 151, no. 4 (April 1995): 1075–82.

3 Personal communication with Preston Campbell, January 30, 2019.

4 Cystic Fibrosis Centre at the Hospital for Sick Children, "Mutation Details for c.1652G>A," Cystic Fibrosis Mutation Database, April 25, 2011; G. R. Cutting et al., "A Cluster of Cystic Fibrosis Mutations in the First Nucleotide-Binding Fold of the Cystic Fibrosis Conductance Regulator Protein," *Nature* 346, no. 6282 (July 26, 1990): 366–69.

5 Frank J. Accurso et al., "Sweat Chloride as A Biomarker of CFTR Activity: Proof of Concept and Ivacaftor Clinical Trial Data," *Journal of Cystic Fibrosis: Official Journal of the European Cystic Fibrosis Society* 13, no. 2 (March 2014): 139–47.

6 George M. Solomon et al., "An International Randomized Multicenter Comparison of Nasal Potential Difference Techniques," *CHEST* 138, no. 4 (October 1, 2010): 919–28.

7 Vertex Pharmaceuticals. "Vertex Announces Positive Results for VX-770, an Oral Investigational Agent That Targets a Defective Protein Responsible for Cystic Fibrosis," Business Wire, March 27 2008.

CHAPTER 45

1 Yves Chauvin, "Olefin Metathesis: The Early Days," Nobel Lecture, Stockholm University, December 8, 2005, https://www.nobelprize.org/prizes/chemistry/2005/chauvin/lecture/.

2 Alberto Grignolo and Sy Pretorius, "Phase III Trial Failures: Costly, but Preventable," *Applied Clinical Trials* 25, no. 8 (August 1, 2016).

3 Vertex Pharmaceuticals, "Vertex Announces Positive Results for VX-770, an Oral Investigational Agent That Targets a Defective Protein Responsible for Cystic Fibrosis," Business Wire March 27, 2008.

4 Vertex Pharmaceuticals, "Vertex Announces Positive 28-Day Results for VX-770, an Oral Investigational Agent That Targets a Defective Protein Responsible for Cystic Fibrosis," Business Wire, October 20, 2008.

CHAPTER 46

1 "From the Archive: Mr. Churchill on Our One Aim," the *Guardian*, November 10, 2009.

2   Vertex Pharmaceuticals, "Vertex Pharmaceuticals Reports Second Quarter 2009 Financial Results and Highlights Recent Clinical Progress and Business Development Activity," Business Wire, August 5, 2009.

3   Vertex Pharmaceuticals, "Vertex Pharmaceuticals Initiates Phase 3 Registration Program for VX-770, an Oral CFTR Potentiator Targeting the Defective Protein Responsible for Cystic Fibrosis," Business Wire, May 27, 2009.

4   Ibid.

5   "Gerry's Biography," Gerry Cheevers, http://www.gerrycheevers.com/cheevers-bio.php; The Goalies Archive, "#30 Gerry Cheevers," Boston Bruins Goaltending History, http://www.goaliesarchive.com/bruins/cheevers.html.

6   Vertex Pharmaceuticals, "Phase 3 Study of VX-770 Showed Profound and Sustained Improvements in Lung Function (FEV1) and Other Measures of Disease Among People with a Specific Type of Cystic Fibrosis," Business Wire, February 23, 2011.

7   Ibid.

8   Vertex Pharmaceuticals, "Vertex Submits Application for Priority Review and Approval of KALYDECO™ (VX-770, Ivacaftor) in the U.S. as First Potential Medicine to Target the Underlying Cause of Cystic Fibrosis," Business Wire, October 19, 2011.

9   Vertex Pharmaceuticals, "FDA Grants Priority Review for KALYDECO™ (Ivacaftor), the First Potential Medicine to Target the Underlying Cause of Cystic Fibrosis," Business Wire, December 15, 2011.

10 Pamela B. Davis, "Therapy for Cystic Fibrosis—The End of the Beginning?," *New England Journal of Medicine* 365, no. 18 (November 3, 2011): 1734–35.

11 Ibid.

12 Vertex Pharmaceuticals, "FDA Approves KALYDECO™ (Ivacaftor), the First Medicine to Treat the Underlying Cause of Cystic Fibrosis," press release, January 31, 2012, https://investors.vrtx.com/news-releases/news-release-details/fda-approves-kalydecotm-ivacaftor-first-medicine-treat.

13 Ibid.

14 Andrew Pollack, "F.D.A. Approves Cystic Fibrosis Drug," *New York Times*, February 1, 2012.

15 David M. Orenstein et al., "Concerned Physicians to Jeff Leiden," July 9, 2012, https://www.medpagetoday.com/upload/2013/5/17/CFletter.pdf.

16 Luke Timmerman, "Vertex CEO Josh Boger Retiring in May; Matthew Emmens to Fill Role," *Xconomy*, February 5, 2009.

17 Andrew Pollack, "Second Drug Wins Approval for Treatment of Hepatitis C," *New York Times*, May 23, 2011.

18 Matthew Herper, "For Vertex Pharmaceuticals, Can One Billion-Dollar Breakthrough Beget Another?," *Forbes*, September 5, 2017.

19 Drugs.com, "FDA Approves Sovaldi for Chronic Hepatitis C," December 6, 2013, https://www.drugs.com/newdrugs/fda-approves-sovaldi-chronic-hepatitis-c-3986.html.

CHAPTER 48

1   Franklin D. Roosevelt, "Oglethorpe University Address: The Country Needs, the Country Demands Bold, Persistent Experimentation," in *Public Papers and Addresses of Franklin D. Roosevelt,* vol. 1 (Random House, 1938), 639.

2   Fredrick Van Goor et al., "Correction of the F508del-CFTR Protein Processing Defect in Vitro by the Investigational Drug VX-809," *Proceedings of the National Academy of Sciences* 108, no. 46 (November 15, 2011): 18843–48.

3 Ibid.

4 Ibid.

5 Ying Wang et al., "Additive Effect of Multiple Pharmacological Chaperones on Maturation of CFTR Processing Mutants," *Biochemical Journal* 406 (August 13, 2007): 257–63.

6 Goor et al., "Correction of F508del-CFTR Protein Processing Defect by the VX-809."

7 Wael M. Rabeh et al., "Correction of Both NBD1 Energetics and Domain Interface Is
   Required to Restore ΔF508 CFTR Folding and Function," *Cell* 148, no. 1–2 (January 20,
   2012): 150–63; Juan L. Mendoza et al., "Requirements for Efficient Correction of ΔF508
   CFTR Revealed by Analyses of Evolved Sequences," *Cell* 148, no. 1–2 (January 20, 2012):
   164–74.

8 Gergely L. Lukacs and A. S. Verkman, "CFTR: Folding, Misfolding and Correcting the
   ΔF508 Conformational Defect," *Trends in Molecular Medicine* 18, no. 2 (February 2012):
   81–91; Hong Yu Ren et al., "VX-809 Corrects Folding Defects in Cystic Fibrosis Transmembrane Conductance Regulator Protein through Action on Membrane-Spanning
   Domain 1," *Molecular Biology of the Cell* 24, no. 19 (October 1, 2013): 3016–24.

9 Vertex Pharmaceuticals, "Vertex Announces Results from Phase 2a Trial of VX-809 Targeting the Defective Protein Responsible for Cystic Fibrosis," Business Wire, February 3,
   2010.

10 Ibid.

11 Vertex Pharmaceuticals, "Vertex Pharmaceuticals Initiates Clinical Trial to Evaluate
   Combination Regimens of VX-770 and VX-809 Targeting the Defective Protein Responsible for Cystic Fibrosis," Business Wire, October 18, 2010.

12 Vertex Pharmaceuticals, "Interim Phase 2 Data Showed a Combination of VX-770 and
   VX-809 Improved Function of the Defective Protein That Causes Cystic Fibrosis in People with the Most Common Form of the Disease," Business Wire, June 9, 2011.

13 Vertex Pharmaceuticals, "Phase 3 STRIVE Study of VX-770 Showed Durable Improvements in Lung Function (FEV1) and Other Measures of Disease Among People
   with a Specific Type of Cystic Fibrosis," Business Wire, June 10, 2011, https://investors.vrtx
   .com/news-releases/news-release-details/phase-3-strive-study-vx-770-showed-durable
   -improvements-lung.

14 Vertex Pharmaceuticals, "Interim Phase 2 Data Showed a Combination of VX-770 and
   VX-809 Improved Function in People With the Most Common Form of the Disease."

15 Vertex Pharmaceuticals, "Vertex Announces Presentation of New Data on VX-770 and
   VX-809 at North American Cystic Fibrosis Conference," Business Wire, October 3, 2011.

16 Vertex Pharmaceuticals, "Vertex Announces Initiation of Pivotal Phase 3 Program of
   VX-809 in Combination with Ivacaftor for the Treatment of People with Cystic Fibrosis
   Who Have Two Copies of the F508del Mutation," Business Wire, February 26, 2013.

CHAPTER 49

1 Vertex Pharmaceuticals, "Treatment with VX-661 and Ivacaftor in a Phase 2 Study
   Resulted in Statistically Significant Improvements in Lung Function in People with Cystic Fibrosis Who Have Two Copies of the F508del Mutation," Business Wire, April 18,
   2013.

2 Vertex Pharmaceuticals, "Data from Phase 2 Combination Study of VX-809 and Ivacaftor
   in People with Cystic Fibrosis Who Have the Most Common Genetic Mutation (F508del)
   Presented at North American Cystic Fibrosis Conference," Business Wire, October 11,
   2012.

3   Cystic Fibrosis Foundation, Johns Hopkins University, and Hospital for Sick Children, "The Clinical and Functional Translation of CFTR (CFTR2)," 2011, https://cftr2.org/.

4   Cystic Fibrosis Foundation, "Types of CFTR Mutations," https://www.cff.org/What-is-CF/Genetics/Types-of-CFTR-Mutations/.

5   Cystic Fibrosis Foundation, "Find Out More About Your Mutations," https://www.cff.org/What-is-CF/Genetics/Find-Out-More-About-Your-Mutations/.

6   Barry Werth, "A Tale of Two Drugs," *MIT Technology Review*, October 22, 2013.

7   Vertex Pharmaceuticals, "U.S. Food and Drug Administration Approves KALYDECO™ (Ivacaftor) for Use in Eight Additional Mutations That Cause Cystic Fibrosis," Business Wire, February 21, 2014, https://investors.vrtx.com/static-files/d2bc4c18-0a58-42f0-9a8c-eeb70b56bcd6.

8   Vertex Pharmaceuticals, "U.S. Food and Drug Administration Approves KALYDECO® (Ivacaftor) for Use in People with Cystic Fibrosis Ages 6 and Older Who Have the R117H Mutation," Business Wire, December 29, 2014.

9   Vertex Pharmaceuticals, "Vertex Announces Results of Phase 3 Study of Ivacaftor in People with CF Who Have the R117H Mutation," Business Wire, December 19, 2013.

10 William Elder, "Answering a Compelling Need: Expediting Life-Saving Treatments to Patients," Friends of Cancer Research Congressional Briefing, July 24, 2013, Washington DC, https://www.focr.org/events/answering-compelling-need-expediting-life-saving-treatments-patients.

11 Jonathan J. Darrow, Jerry Avorn, and Aaron S. Kesselheim, "The FDA Breakthrough-Drug Designation—Four Years of Experience," *New England Journal of Medicine* 378, no. 15 (April 12, 2018): 1444–53.

12 William Elder, "Answering a Compelling Need: Expediting Life-Saving Treatments to Patients," Friends of Cancer Research Congressional Briefing, July 24, 2013, Washington DC, https://www.focr.org/events/answering-compelling-need-expediting-life-saving-treatments-patients.

13 Barack Obama, "2015 State of the Union Address," January 20, 2015, https://www.cnn.com/2015/01/20/politics/state-of-the-union-2015-transcript-full-text/index.html; Vital Signs, "Medical Student William Elder Jr. Featured in President Obama's Speech at White House Press Conference," Boonsoft School of Medicine, Winter 2015, https://medicine.wright.edu/about/news-and-events/vital-signs/article/medical-student-william-elder-jr-featured-in-president-obamas-speech-at-white-house-press-conference.

14 Jocelyn Kaiser, "Obama Gives East Room Rollout to Precision Medicine Initiative," *Science,* January 30, 2015, https://www.sciencemag.org/news/2015/01/obama-gives-east-room-rollout-precision-medicine-initiative.

CHAPTER 50

1   Ian Fleming, *Casino Royale* (Jonathan Cape, 1953).

2   Jonathan D. Rockoff and Joseph Walker, "Cystic Fibrosis Foundation Sells Drug's Rights for $3.3 Billion," *Wall Street Journal,* November 19, 2014.

3   Esther Kim and Andrew W. Lo, *Venture Philanthropy: A Case Study of the Cystic Fibrosis Foundation* (Social Science Research Network, April 23, 2019).

4   John Tozzi, "This Medical Charity Made $3.3 Billion From a Single Pill," Bloomberg, July 7, 2015, https://www.bloomberg.com/news/features/2015-07-07/this-medical-charity-made-3-3-billion-from-a-single-pill.

5   Brady Dennis, "Are Risks Worth the Rewards When Nonprofits Act like Venture Capitalists?," *Washington Post,* July 2, 2015.

6 Andrew Pollack, "Deal by Cystic Fibrosis Foundation Raises Cash and Some Concern," *New York Times*, November 19, 2014.

7 Carolyn Beeler, "With Big Money Comes Big Questions for Cystic Fibrosis Foundation," *The Pulse* (National Public Radio, December 5, 2014).

8 Vertex Pharmaceuticals, "Two 24-Week Phase 3 Studies of Lumacaftor in Combination with Ivacaftor Met Primary Endpoint with Statistically Significant Improvements in Lung Function."

9 Andrew Pollack, "Vertex's 2-Drug Cystic Fibrosis Pill Shows Promise," *New York Times*, June 24, 2014, https://www.nytimes.com/2014/06/25/business/international/vertexs-two -drug-cystic-fibrosis-treatment-shows-promise-in-clinical-trials.html.

CHAPTER 51

1 *CFTR2: Clinical and Functional Translation of CFTR,* https://cftr2.org; Patrick R. Sosnay et al., "Defining the Disease Liability of Variants in the Cystic Fibrosis Transmembrane Conductance Regulator Gene," *Nature Genetics* 45, no. 10 (2013): 1160–67, doi:10.1038 /ng.2745; Gudio Veit et al. "From CFTR Biology Toward Combinatorial Pharmacotherapy: Expanded Classification of Cystic Fibrosis Mutations," *Molecular Biology of the Cell* 27, no. 3 (2016): 424–33, doi:10.1091/mbc.E14-04-0935.

2 Personal communication with Michal Boyle, June 28, 2017.

3 Vertex Pharmaceuticals, "U.S. Food and Drug Administration Approves KALYDECO™ (Ivacaftor) for Use in Eight Additional Mutations That Cause Cystic Fibrosis," Business Wire, February 21, 2014, https://investors.vrtx.com/static-files/ d2bc4c18-0a58-42f0-9a8c-eeb70b56bcd6.

4 Vertex Pharmaceuticals, "U.S. Food and Drug Administration Approves KALYDECO® (ivacaftor) for Use in People with Cystic Fibrosis Ages 6 and Older Who Have the R117H Mutation," December 29, 2014, https://investors.vrtx.com/news-releases /news-release-details/us-food-and-drug-administration-approves-kalydecor-ivacaftor-use.

5 Fred Van Goor, Haihui Yu, Bill Burton, and Beth J. Hoffman, "Effect of Ivacaftor on CFTR Forms with Missense Mutations Associated with Defects in Protein Processing or Function," *Journal of Cystic Fibrosis* 13 (2014): 29–36.

6 Ibid.

7 US Food & Drug Administration, "FDA Expands Approved Use of Kalydeco to Treat Additional Mutations of Cystic Fibrosis," May 17, 2017, https://www.fda .gov/news-events/press-announcements/fda-expands-approved-use-kalydeco-treat -additional-mutations-cystic-fibrosis.

8 Anthony G. Durmowicz, Robert Lim, Hobart Rogers, Curtis J. Rosebraugh, and Badrul A. Chowdhury, "The U.S. Food and Drug Administration's Experience with Ivacaftor in Cystic Fibrosis. Establishing Efficacy Using in Vitro Data in Lieu of a Clinical Trial," *Annals of the American Thoracic Society* 15, no. 1 (2017), https://doi.org/10.1513 /AnnalsATS.201708-668PS.

9 Tony Durmowicz and Mike Pacanowski, "Novel Approach Allows Expansion of Indication for Cystic Fibrosis Drug," US Food & Drug Administration, May 18, 2017, https://www.fda.gov/drugs/news-events-human-drugs /novel-approach-allows-expansion-indication-cystic-fibrosis-drug.

CHAPTER 52

1   Vertex Pharmaceuticals, "Vertex and Cystic Fibrosis Foundation Therapeutics to Collaborate on Discovery and Development of New Medicines to Treat the Underlying Cause of Cystic Fibrosis," April 7, 2011, https://investors.vrtx.com/news-releases/news-release-details/vertex-and-cystic-fibrosis-foundation-therapeutics-collaborate.

2   https://www.accessdata.fda.gov/drugsatfda_docs/label/2018/211358s000lbl.pdf.

3   Vertex Pharmaceuticals, "Two Phase 3 Studies of the Tezacaftor/Ivacaftor Combination Treatment Met Primary Endpoints with Statistically Significant Improvements in Lung Function (FEV1) in People with Cystic Fibrosis," March 28, 2017, https://investors.vrtx.com/static-files/f15217ac-4a8b-436a-9215-79144ec2e59b.

4   Fred Van Goor et al., "Correction of the F508del-CFTR Protein Processing Defect in Vitro by the Investigational Drug VX-809," *Proceedings of the National Academies of Sciences USA* 108 (2011): 18843–48.

5   Personal communication with Fred Van Goor, March 15, 2019.

6   J. L. Mendoza et al., "Requirements for Efficient Correction of ΔF508 CFTR Revealed by Analyses of Evolved Sequences," *Cell* 148, nos. 1–2 (2012): 164–74, doi: 10.1016/j.cell.2011.11.023.

7   Wael M. Rabeh et al., "Correction of Both NBD1 Energetics and Domain Interface Is Required to Restore ΔF508 CFTR Folding and Function," *Cell* 148, nos. 1–2 (2012): 150–63, doi: 10.1016/j.cell.2011.11.024.

8   Frederick Van Goor, et al., "Rescue of CF airway epithelial cell function in vitro by a CFTR potentiator, VX-770," *Proceedings of the National Academy of Sciences of the United States of America* 106, no. 44 (2009): 18825–30. doi:10.1073/pnas.0904709106.

9   Frederick Van Goor, et al. "Correction of the F508del-CFTR protein processing defect in vitro by the investigational drug VX-809," *Proceedings of the National Academy of Sciences of the United States of America* 108, no. 46 (2011): 18843–48. doi:10.1073/pnas.1105787108.

10 Dominic Keating, et al., "VX-445-Tezacaftor-Ivacaftor in Patients with Cystic Fibrosis and One or Two Phe508del Alleles," *New England Journal of Medicine* 379 (2018):1612–20. doi: 10.1056/NEJMoa1807120. Epub 2018 Oct 18.

11 Vertex Pharmaceuticals, "Vertex Announces Positive Phase 1 & Phase 2 Data from Three Different Triple Combination Regimens in People with Cystic Fibrosis Who Have One F508del Mutation and One Minimal Function Mutation (F508del/Min)," July 18, 2017, https://investors.vrtx.com/news-releases/news-release-details/vertex-announces-positive-phase-1-phase-2-data-three-different.

CHAPTER 53

1   Vertex Pharmaceuticals, "Vertex Selects Triple Combination Regimen of VX-445, Tezacaftor and Ivacaftor to Submit for Global Regulatory Approvals in Cystic Fibrosis," Business Wire, May 30, 2019.

2   https://www.fda.gov/patients/learn-about-drug-and-device-approvals/fast-track-breakthrough-therapy-accelerated-approval-priority-review.

3   Fernando Augusto Lima Marson, Carman Sílvia Bertuzzo, and José Burecu Ribeiro, "Classification of CFTR Mutation Classes," *Lancet Respiratory Medicine* 4, no. 8 (2016): PE37–38, http://dx.doi.org/10.1016/S2213-2600(16)30188-6.

CHAPTER 54

1   T. Shoshani, et al., "Association of a Nonsense Mutation (W1282X), the Most Common Mutation in the Ashkenazi Jewish Cystic Fibrosis Patients in Israel, with Presentation of Severe Disease," *American Journal of Human Genetics* 50, no. 1 (1992): 222–28.

2   Bijal P. Trivedi, "Cystic Fibrosis Foundation Opens Drug Discovery Lab," *Science* 353, no. 6305 (2016): 1194–95, DOI: 10.1126/science.353.6305.1194.

3   Jean-Antoine Ribeli, et al., "Gene Therapy in a Patient with Sickle Cell Disease," *New England Journal of Medicine* 376 (2017): 848–55, DOI: 10.1056/NEJMoa1609677.

4   US Food and Drug Administration, "FDA Approval Brings First Gene Therapy to the United States," August 30, 2017, https://www.fda.gov/news-events/press-announcements /fda-approval-brings-first-gene-therapy-united-states.

5   US Food and Drug Administration, "FDA Approves Novel Gene Therapy to Treat Patients with a Rare Form of Inherited Vision Loss," December 18, 2017, https://www .fda.gov/news-events/press-announcements/fda-approves-novel-gene-therapy-treat -patients-rare-form-inherited-vision-loss.

6   Emily Mullin, "Gene Editing Fixes Cystic Fibrosis Gene in Mice Before Birth," *Nature Medicine,* May 8, 2019, https://www.nature.com/articles/d41591-019-00012-x.

7   K. Kayani, et al, "Cystic Fibrosis-Related Diabetes," *Frontiers in Endocrinology* 9, no.20 (2018), DOI:10.3389/fendo.2018.00020.

8   "Scientific Advisory Board," Emily's Entourage, accessed April 11, 2020, https://www .emilysentourage.org/about-emilys-entourage/scientific-advisory-board/.

9   "Awarded Grants," Emily's Entourage, accessed April 13, 2020, https://www .emilysentourage.org/awarded-grants/.

10 Abby Ellin, "Cystic Fibrosis Patients Turn to Experimental Phage Therapy," *New York Times,* May 17, 2019, https://www.nytimes.com/2019/05/17/well/live/cystic-fibrosis -patients-turn-to-experimental-phage-therapy.html.

11 Piotr S. Kowalski, et al., "Delivering the Messenger: Advances in Technologies for Therapeutic mRNA Delivery," *Molecular Therapy* 27, no. 4 (2019): 710–28. doi: 10.1016/j .ymthe.2019.02.012.

CHAPTER 55

1   Fernando Marson, Carmen Bertuzzo, and Jose Ribeiro, "Classification of CFTR Mutation Classes, *Lancet,* August 1, 2016, https://www.thelancet.com/journals/lanres/article /PIIS2213-2600(16)30188-6/fulltext.

INSERT PHOTOGRAPHY CREDITS

Page 1: courtesy of Joe and Kathy O'Donnell.

Page 2: courtesy of Joe and Kathy O'Donnell.

Page 3: With permission, from I. M. Modlin and M. Kidd, eds., *The Paradox of the Pancreas* (Weesp, the Netherlands: 2003), 280 *(top)*; courtesy of the Archives & Special Collections, Columbia University Health Sciences Library *(bottom left and right)*.

Page 4: courtesy of the Office of NIH History and Stetten Museum *(top left)*; courtesy of the Boston Children's Hospital Archives, Boston, Massachusetts *(top right)*; courtesy of the author *(bottom)*.

Page 5: courtesy of Doris Tulcin *(top left)*; courtesy of the Cystic Fibrosis Foundation *(top right and bottom right)*; courtesy of Mike Boyle *(bottom left)*.

Page 6: courtesy of the author *(top left)*; courtesy of UW Medicine *(top right)*; courtesy of the Cystic Fibrosis Foundation *(bottom)*.

Page 7: public domain *(top)*; courtesy of SAGE Publications, from "Ultra-High Throughput Screening," *Laboratory Automation News* 2 (no. 4), September 1997 *(middle right and bottom)*; courtesy Vertex Pharmaceuticals *(middle left)*.

Page 8: courtesy of Vertex Pharmaceuticals.

INDEX

A

ABC transporters, 203
Abeda, Alex, 468
Accurso, Frank, 396–398, 400–402, 408
adult CF programs, 438–439
Agenerase, 333
alpha-I antitrypsin deficiency, 247–249
Altshuler, David, 469–470
amino acids, 141, 202–204
aminoglycosides, 13
amniocentesis, 163
amylase, 31
Andelman, Eddie, 126
Andersen, Dorothy Hansine
    background of, 18–20
    and Children's Exocrine Research
        Foundation, 57
    congenital heart defect study by, 53
    cystic fibrosis distinguished from celiac
        disease by, 26–27, 30–32
    digestive enzyme testing by, 31–32, 45
    di Sant'Agnese's relationship with, 45,
        46, 53–54
    education emphasized by, 62
    and fundraising, 56
    and Graub's diagnosis, 55–56
    Matthews' consulting with, 75
    on NCFRF medical education
        committee, 59, 69–70
    penicillin used by, 47
    relationships with patients' families,
        32–33
    and sweat test, 50
    and Tulcin, 59–61
    work to identify cystic fibrosis, 17–18,
        20–30
Andersen, Hans Peter, 19

Anderson, W. French, 243–244, 250
Andretti, Mario, 471
Angelou, Maya, 224
Angier, Natelie, 240
antibiotics
    aerosolizing, 229–230
    ciprifloxacin, 186
    in "cleanouts" for patients, 39
    combinations of, 75
    first-generation, 46
    for Flessners, 308, 310
    inhaled tobramycin, 229–235
    for lung infections, 13, 229
    in Matthews' treatment program, 76
    for Joey O'Donnell, 14, 101, 186, 187
    penicillin, 46–47
    technology in development of, 341
    for Woglom, 20
Aretaeus of Cappadocia, 18
Arumugam, Viji, 364–365, 374, 379
Ashlock, Melissa Rosenfeld. see also
        Rosenfeld, Melissa
    and Aurora CF research, 300, 304
    and Aurora's Gates Foundation funding,
        314, 315
    and clinical endpoints for research,
        400–401, 407
    and Vertex CF research, 351–352,
        407–408
    and Vertex funding negotiations, 337,
        388, 389
Asimov, Isaac, 236
Aurora Biosciences Corporation, 271, 290
    CF Foundation investment in, 492
    drug development project for CF, 327–
        328, 335–338, 342–351, 447

drug development team for CF drug,
    339–351
drug screening work of, 328, 333–335
founding of, 294–295
Gates Foundation funding for, 312–317
Negulescu's work at, 296–306, 314, 339
Quorum acquired by, 341
structure of, 296
Vertex's acquisition of, 328, 333–335,
    341–342 (see also Vertex
    Pharmaceuticals Inc.)
autonomic nervous system, 52

B

Ballmer, Connie, 314
Ballmer, Steve, 311–312, 314
banana diet, 18
Barbero, Giulio, 86–87
Barnett, Dick, 126
Barry, Amy, 355–358
Barry, Jamie, 355–356, 358
Barry, Peter, 355–356, 359
Bartlett, Frederic H., 30
Beall, Robert, 92
    and Aurora's research, 297–300, 302,
        304–306, 311–317, 337–338
    and care center funding, 132–133
    at CF Foundation, 96, 129, 264, 266
    and CFF Therapeutics Laboratory, 480
    and CF gene discovery, 213
    and Cheevers family, 413
    and Collins's research, 183–184, 197
    and corrector drug testing, 437, 438
    drug development focus of, 283–290
    and Flessner, 310–311
    and funding for research, 266–268, 282,
        356, 388–390, 420, 430, 446–449
    and fundraising for Vertex, 359–361
    and Gates Foundation funding, 311–317
    and Genentech's research, 214–215,
        217–218
    and gene therapy, 237, 243, 245, 249,
        264
    and high-throughput screening, 269–
        271
    and human cells for CF drug testing,
        372
    and inhaled tobramycin project, 231–
        234

and Kalydeco, 422–423
at NIH, 92–96
and Quorum research, 341
research centers established by, 133–135
retirement of, 379
and royalties from Kalydeco, 423, 424,
    447–448
and Trikafta approval, 497
and Vertex CF research, 351, 352, 363,
    407–408
and VX-661 development, 463
and VX-770/VX809 approval, 451
on Williamson's gene identification, 194
Bear, Brian, 379–380, 391, 392
Bennet, Michael, 444
Benzer, Seymour, 206
Bernard, Claude, 29–30
Bessette, Danny, 211–212
Bier, August, 3
Bill & Melinda Gates Foundation, 312,
    449–450
Blackfan, Kenneth, 25–26
Blaese, R. Michael, 243–244, 250
Blanc, William, 75
Blank, Steve, 327
Boat, Tom, 308
Boger, Joshua
    and Aurora Biosciences acquisition,
        333–335
    background of, 328–303
    and CF drug project, 335–338, 390
    on CF funding for Vertex, 462
    and funding for CF research, 387–388
    at Merck, 330–332
    on personalized medicine, 441
    and results of VX-770 trials, 403
    and Trikafta approval, 498
    and Vertex CF drug development, 375
    Vertex founded by, 332
    and Vertex's work, 332–334, 353
Boger, Ken, 337
Boucher, Richard, 104–111, 119, 122, 203,
    247, 400
Boyer, Herbert, 175, 215
Boyle, Michael, 438–440, 454–458, 461,
    495, 497
breakthrough therapy drugs, 444–445
Bridgman, Percy, 379
Buchwald, Manuel, 140–143, 152

Buckley, Blair, 489
Buckley, James Daniel, 489
Buckley, Michael, 487–489
Buffett, Warren, 487
Burks, Patty, 495
Bush, George W., 361

C

Campbell, Preston, 286–289
　Accurso recruited by, 397
　and Aurora's research, 302, 304
　background of, 285–287
　and CFF Therapeutics Laboratory, 480
　and corrector drug testing, 437, 438
　and funding for research, 446–449
　and fundraising for Vertex, 359–362
　on impossibilities, 495
　and new Vertex drug, 461
　and rare mutations therapies, 479
　retirement of, 495
　and royalties from Kalydeco, 423–424,
　　447–448
　and Trikafta approval, 497–498
　and Vertex CF drug development, 337,
　　363, 393, 407–408
　and Vertex funding negotiations, 388
　and VX-661 development, 463
　and VX-770/VX809 approval, 451
Canadian Cystic Fibrosis Foundation
　(CFF), 78, 147
cancer research, 93, 284
Cantor, Eddie, 67
carriers, xv–xvi, 8, 28, 213
Carter, Jimmy, 102–103
cause of cystic fibrosis, 56, 122
celiac disease, 18, 22, 24–32
CFTR Folding Consortium, 466
Chakravarti, Aravinda, 172–173, 207–309
Chauvin, Yves, 406
Cheevers, Cate, 319, 411–413, 415–418, 462,
　464
Cheevers, Gerry, 411, 415–416, 418–419
Cheevers, Kim, 318–320, 411–415, 417, 464
Cheevers, Laura, 319, 411–418, 462,
　464–465
Cheevers, Rob, 318–320, 411, 412, 414–415,
　417
chest physical therapy
　as children grow larger, 39

　for Joey, 14, 36, 39, 100
　Matthews' demonstration of, 76
　by Nadeau, 80–84
　Shwachman's teaching about, 75, 80–82
Children's Exocrine Research Foundation,
　57–60
Children's Oncology Group (COG), 284,
　287–288
Ching Chun Li (C.C.), 172–173
chorionic villus sampling, 163–164, 324
chromosome-jumping technique, 179–184,
　196
chromosomes, 141, 146, 153–154
Churchill, Winston, 411
ciprofloxacin, 186
Clancy, J. P., 400, 401
"cleanouts" for patients, 39
Cleveland CF clinic, 74–77, 133
Cleveland Health Fund, 89–90
Clinical and Functional Translation of
　CFTR (CFTR2), 442
cloning
　of genes, 215, 217
　public attention to, 266
Cohen, Stanley, 175
Cole, Jeff, 183
Collaborative Research, Inc., 152–158, 192
Collins, Francis, 173–174
　background of, 174–175
　and CF gene identification, 196–197,
　　206–207, 209–211, 240
　and gene therapy for CF, 240, 242, 249
　and Human Genome Project, 268–269,
　　300
　and Precision Medicine Initiative, 446
　research of, 175–184
　as Tsui's competitor, 192
　on Ann Tulcin, 399
　and Williamson's incorrect gene
　　identification, 194
Collinson, Stuart, 334–335
CombiChem, 348, 350
combinational chemistry, 268
Comprehensive Treatment Program,
　74–78, 133, 168
Cooke, Robert E., 53
Cooper, Lee, 160
Copp, Father Rodney, 188
Corey, Mary, 141

Coughlin, Bobby, 321, 322, 475, 476
Coughlin, Christine, 320–321, 476
Coughlin, Mary Kate, 476
Coughlin, Robert (Bob), 319–322, 474–
     476, 497–498
Cousin's Club, 74
Crandall, J. Taylor, 359–362
Crothers, Donald, 174
Crystal, Ronald, 247–251, 253–259, 262,
     266, 400
Cutting, Garry, 398–399, 484
cystic fibrosis (CF). *see also specific topics
     and researchers*
  carriers of, xv–xvi, 8, 28
  cause of, 56, 122
  described, xv
  distinguishing celiac disease from, 24–32
  divorce rates associated with, 308
  genetic test for, 162, 163
  health of patients with, 39
  health problems caused by, 106
  identification of, 17–18, 20–30
  impacts on families' lives, 278–279
  Joey's early signs of, 3
  life expectancy for patients with, 9, 74,
     438, 477, 501
  medical community's knowledge of, 54,
     61–62, 73
  misdiagnosis of, 3–5, 7, 11, 325
  NCFRF research and medical education
     centers for, 72–79
  public perception of, 133
  research on origin of, 85–86
  support groups for, 38–39
  symptom idiosyncrasies with, 73
  timeline for, 501–505
  transmission of, 8–9, 28
Cystic Fibrosis Canada, 140
Cystic Fibrosis (CF) Foundation, xvi–xviii
  accomplishments of, 267
  care centers of, 116, 132–133, 227–228,
     286–287
  and cure research, 481, 492
  Dresing's presidency of, 129–132
  drug development network of,
     283–290 (*see also* Aurora
     Biosciences Corporation; Vertex
     Pharmaceuticals Inc.)
  and drugs for rare mutations, 480

  and funding for research, 184, 197, 210,
     266–268, 277, 289, 306, 315–
     317, 335–338, 341, 354–362,
     387–390, 446–450, 462, 486
  fundraising by, 126–135, 309, 320,
     322–323, 354–362, 419–420,
     422, 425–430, 491–494, 498–499
  Genentech's assistance from, 217–218,
     222
  and gene therapy, 236, 237, 243, 249,
     482
  headquarters relocation for, 91–92
  and inhaled tobramycin project, 231,
     233–235
  Joey Fund money for, 282
  and Kalydeco price, 422–424
  NCFRF renamed as, 84, 91
  nonprofit drug discovery and
     development arm of, 397
  Kate O'Donnell at, 491
  Joe O'Donnell on board of, 126
  patient registry of, 398, 442, 443
  public service announcements by,
     102–103
  research centers of, 129, 133–135
  scientific meeting hosted by, 93–94
  size and power of, 449–450
  Tulcin as president of, 91–92
Cystic Fibrosis Foundation Therapeutics,
     Inc., 397
Cystic Fibrosis Foundation (CFF)
     Therapeutics Laboratory, 480–482
Cystic Fibrosis Research Foundation Trust
     (CF Trust), 150–151
Cystic Fibrosis Research Inc., 220
Cystic Fibrosis Research Institute of
     Pennsylvania, 70
Cystic Fibrosis Transmembrane
     Conductance Regulator (CFTR),
     209, 343–344. *see also individual gene
     mutations*
  combination drugs for, 436–441, 449–
     452, 462–474
  corrector drugs for, 298, 431–440, 463,
     465–468, 474
  doorman drug for, 363–386, 416–422,
     431–434, 457–461
  and gene therapy, 236 (*see also* gene
     therapy)

and ion channel research, 335, 339
  misfolding of, 434, 465, 479
  rare mutations therapies, 478–482
  repairing, 269–271, 299–305, 327,
      346–349, 351
  understanding role of, 291
cytomegalovirus, 6

D

Daschle, Tom, 94
Davies, Kay, 151–152
Deford, Frank, 185
de la Tremblaye, Dr., 30
Denton, Robert, 69, 70, 75
diabetes, 93
diagnosis of cystic fibrosis
  Andersen's test for digestive enzymes,
      31–32
  chorionic villus sampling, 163–164, 324
  genetic testing, 158–159, 162–164,
      318–320, 322, 324
  importance of early diagnosis, 31, 47,
      75, 168–169, 171
  with iontophoresis, 52–53, 75–76
  IRT test for, 170–171
  and misdiagnoses, 3–5, 7, 11, 325
  in newborns, 75
  newborn screening for, 166–171
  at one year old, 166–167
  sweat test for, 48–53, 75–76
digestive enzymes, 31–32
di Sant'Agnese, Paul
  acute disseminated encephalomyelitis in,
      47–48
  Andersen's relationship with, 45, 46,
      53–54
  background of, 45–46
  in CF Foundation's treatment consensus
      group, 228
  early treatments by, 45–48
  education emphasized by, 62
  Farrell's work with, 168–169
  and fundraising, 56
  and lack of knowledge about cystic
      fibrosis, 61–62
  Lapey's work with, 13
  Matthews' consulting with, 75
  on NCFRF medical education
      committee, 55, 69

research by, 13, 53–54
  and Shwachman, 168–169
  sweat test developed by, 48–52
  at Tulcin's fundraiser, 64–65
DNA, 141–147, 173–178, 216. see also
    genetics
DNA circles, 178–179, 182
DNase, 216–222, 233
Docter, Jack, 226, 227
Doershuk, Carl, 89, 227, 228, 308
Donis-Keller, Helen, 152–153, 156–157, 192
Donovan, Anne, 129
Dozy, Andree M., 146
Dresing, Robert (Bob)
  and Boat, 308
  as CF Foundation president, 129–132
  as CF Foundation vice president, 91–92,
      95
  on CF gene discovery, 211
  and Collins's research, 184
  Flessner's meeting with, 309
  fundraising by, 129, 134–135, 494
  and gene therapy, 237, 245, 264
  and inhaled tobramycin project, 231
  lobbying by, 94
  and research funding, 89–91
  and research network creation, 96
  and Tulcin, 39, 91, 129
  on Williamson's gene identification, 194
Drive-In Concessions of Mass., Inc.,
    42–44, 355
drug therapy(-ies), 13, 46–47, 375–376. see
    also individual drugs and companies
  breakthrough therapy drugs, 444–445
  categories of, 298
  CF Foundation's drug development
      network, 283–290 (see also Aurora
      Biosciences Corporation)
  CF Foundation's royalties from, 315,
      360, 423–424, 447–450, 452
  and combinational chemistry, 268
  combination drugs, 436–441, 449–452,
      462–477, 495–497
  computers in drug design, 330–333
  corrector drugs, 298, 302–303, 352,
      380, 386 (see also under Vertex
      Pharmaceuticals Inc.)
  cost of Kalydeco, 422–424, 443, 447,
      450

costs of developing, 266–267, 336, 492
doorman drug, 352, 364–370, 374–380
    (*see also* VX-770 [doorman drug])
Flessner's dissatisfaction with, 310
high-throughput screening for, 269–271,
    342
and Human Genome Project, 268–269
human trials for CF drug, 391–392,
    395–410, 413–417
inhaled tobramycin, 229–235
preferred form of, 336
for rare mutations, 478–481
structure-based drug design, 332–333
testing process for, 295–297, 388
Drumm, Mitchell, 181–183, 196–198, 201,
    202
Durmowicz, Anthony, 450
dyes for experiments, 292–294, 297, 334

E

Early, Joe, 94
Earnshaw, Katherine, 64
Elder, Andy, 396
Elder, Bill, Jr., 395–398, 401–402, 421–422,
    444–446
Eli Lilly, 215
Eliot, T.S., 453
Emerson, Ralph Waldo, 139
Emily's Entourage, 484–486
Emmens, Matthew, 424

F

F508del mutation, 206
    airway cells from patients with, 382–386
    as cause of CF, 244
    drug development related to, 300–303,
        352, 380 (*see also individual drugs*)
    gene therapy for, 244
    in Israel, 212
    missing protein in, 347
    prevalence of, 352
familial hypercholesterolemia (FH),
    239–240
Farber, Sidney, 51, 59
Farrall, Martin, 206–207
Farrell, Katy, 471–474
Farrell, Philip, 166–171
Ferguson, Ashton, 322–324, 497
Ferguson, Jennifer, 322–325

Ferguson, Lola, 324–325, 497
Ferguson, Matt, 322–324
Fischer, Martin H., 11
Fleming, Alexander, 47
Fleming, Ian, 447
Flessner, Andrew, 307–310, 313, 431, 440
Flessner, Jonathan, 307–310, 313, 431, 440,
    497
Flessner, Paul, 307–316, 431
Flessner, Sue, 307–310, 316
Flood, Daniel, 86–87, 94
Food and Drug Administration (FDA)
    and breakthrough therapy drugs,
        444–445
    DNase approvals, 218–219
    and drugs for rare mutations, 455,
        457–461, 480
    and gene therapy, 250
    and gene therapy human trials, 254
    Pulmozyme approval, 222
    synthetic human insulin approval, 215
    TOBI approval, 235
    VX-659/VX-770/VX-661 combo drug
        (Trikafta) approval, 474, 495
    VX-770 (Kalydeco) approvals, 44, 416–
        418, 420, 422, 424
    and VX-770 trials, 408
    VX-770/VX-661 combo drug approval,
        465
    VX-770/VX-809 combo drug approval,
        449–452, 463
Forget, Bernie, 177
Foskett, Kevin, 484
Frankel, George, 59
    connections of, 60
    death of, 89
    fundraising work by, 60–61
    and March of Dimes, 68
    and move of NCFRF to New York, 63,
        72
    and O'Connor, 66
    research funding from, 85, 86
    and Sharples, 61, 70
Fredrickson, Don, 196
Frizzell, Ray, 270–271, 290
fundraising, 56–71
    birth of national foundation for, 58–59
    by the Cheevers, 412, 413
    for cure research, 492–494

by Cystic Fibrosis Foundation, 126–135,
320, 322–323, 354–362, 425–430,
491–494
by Emily's Entourage, 484–486
by Flessner, 311–314
by the Graubs, 56–58
by Jenny Lesnick Fund, 56
Joey Fund, 277–282, 489–491
with legacy giving, 492–493
and Mattingly, 354–361
by NCFRF, 38, 59, 62–64, 68–71,
85–90
NIH study of, 94–96
by O'Connor, 66–68
by the O'Donnell family, 126–128,
354–362, 419–420, 425–430,
489–494, 498–499
planning for, 357–358, 427, 494
public service announcements for,
102–103
by small groups, 56–58, 60–61, 64–65
by Tulcin, 61, 64–66, 68, 69, 88–90,
94–96

G

G178E mutation, 454, 455
G178R mutation, 458
G551D mutation
drug development related to, 347, 352,
363, 374–375, 393–394, 399,
413, 416, 420 (*see also individual
drugs*)
prevalence of, 399, 409
Ganz, Wilbur, 232–233
GAP Conferences, 86
Gates, Bill, 64, 311–314
Gates, William H., Sr., 312, 315, 316
Gatzy, John, 119
Gawande, Atul, 387
Gee, Samuel, 18
Gehlerter, Thomas, 180
Gelsinger, Jesse, 264–266
gene cloning, 215, 217
gene editing, 481–483, 492, 493
Genentech, 232
drugs developed by, 215–216
gene therapy by, 259
Pulmozyme development by, 214, 216–
223, 229, 233

gene therapy, 180, 235–266, 482–483
alpha-I antitrypsin deficiency, 247–249
Anderson and Blaese's research, 243–
244, 250
CF Foundation focus on, 214
CF Foundation funding for, 236, 237,
492, 493
Collins' work on, 211
Crystal's human trials, 253–259
Crystal's research, 247–252, 262
delivery of genes by viruses, 236–237,
241–243, 251–252
familial hypercholesterolemia, 239–240
first CF trials in humans, 253–263
Flessner's dissatisfaction with, 310
Lesch–Nyhan syndrome, 237–239
personalized medicine, 420–421
by Targeted Genetics, 262
Welsh's human trials, 253, 254, 259–260
Welsh's research, 244–245
Williamson's research, 252, 262–263
Wilson's human trials, 253, 254, 260–
261, 264–266
Wilson's research, 237–243, 245, 252,
262
genetic counseling, 318–319
genetics
chromosome-jumping technique, 179–
184, 196
for cystic fibrosis, 162–164
gene cloning, 215, 217
gene mutations, 199–200, 202–205,
220, 442, 478–482 (*see also
individual mutations*)
genes, 143–144, 152–156, 172, 198,
200
genetic markers, 147, 148, 152–158,
172–174, 183, 198
genotypes of CF patients, 442, 443
identification of CF gene mutations,
191–213, 220
medical genetics, 9–10, 175–184
polymorphisms, 145–147, 152, 153
recombinant DNA technology, 174–176
research in, 139–159
Williamson's incorrect gene
identification, 193–197
genetic testing, 158–159, 162–164, 318–320,
322, 324, 488, 498

genome mapping, 145–147, 150–153, 200
gentamicin, 13
Genzyme, 259
Georgia Warm Springs Foundation, 66
Gibson, Lewis, 52–53
Gillick, Pat, 97
Glaxo Wellcome, 333
Graub, Evelyn, 55–58
Graub, Lee, 87
Graub, Milton, 55–59, 72, 85–88
Green, Chris, 308–309
Greenberg, E. Peter, 340, 341
Greenberg, Jacob "Jack," 29, 30
green fluorescent protein (GFP), 293
Grootenhuis, Peter
    background of, 348
    and corrector drug development, 432
    and human cells for CF drug testing,
        373
    and Vertex CF drug development, 348–
        349, 351, 353, 364, 378, 381, 467
growth hormone deficiency, 215
Gusella, James, 148
Guthrie, Robert, 167

H
Haas, Sidney, 18
Hadida, Sabine, 349–351
    corrector drugs development, 380, 382,
        432–435, 465, 467–468
    doorman drug development, 364–370,
        373–374
Hanna-Attisha, Mon, 214
Harkin, Tom, 94
Harper, Margaret Hilda, 23–24
Harriman, Averill, 69
Harrison, Gunyon, 78, 115–117
Harvard Medical School, 5
health policy, 92–96, 171, 321–322
Helfand, Lynne, 321
hemoglobin research, 149–150, 177–178
hemoptysis, 472
high-throughput screening, 269–271, 342
Hospital for Sick Children ("Sick Kids"),
        139, 140, 147, 210, 211
Howard Hughes Medical Institute
        (HHMI), 180, 210–211
Huertas, Esther, 318

Hughes, Thomas, 359–362
human genetics, 178, 182
Human Genome Project, 200, 268–269,
        300
Huntington's disease, 148, 162, 163
Huxley, Thomas, 253

I
immunoreactive trypsinogen (IRT) test,
        170–171
Incivek, 424
Ingram, Vernon, 149–150
inhaled tobramycin, 229–235
insulin, 215
ion channels, 299, 335, 339, 343
iontophoresis, 52–53, 75–76

J
Jackson, Jesse, 45
James, Forrest Hood "Fob," 134
Jenny Lesnick Fund, 56
Joey Fund, xvii, 277–282, 489–491, 493
Joey's Park, 276, 490
Johnson, Nancy, 80

K
Kalydeco, xvi, xviii, 417, 421–424, 467,
        497. see also VX-770
    in combo drugs, 436–441, 468, 471
    cost to patient for, 424
    expanding approval for younger patients,
        468
    impact on patients' lives, 444, 464
    Page's individual trial of, 454–458
    price of, 422–424, 443, 447, 450
    sale of royalties for, 447–450
    testing with mutations other than
        G551D, 441–444, 457–461
Kan, Yuet Wai, 146, 150, 176
Kauffman, Robert, 402
Keefer, Chester, 47
Kelley, William, 180–181, 237, 238
Kelliher, Albert, 36, 37, 39, 101, 102
Kelliher, Dick, 36, 37
Kelliher, Joan, 36, 37
Kelliher, Margaret, 36, 37, 39, 101
Kemnitzer, Angela, 346, 386

Kennedy, John F., 116
Kerem, Batsheva, 191–192, 195, 197–202, 204, 206, 208–213, 484
Kerem, Eitan, 191–192
Kinsella, Kevin, 294–296, 331, 332
Kirkman, Henry Neil, 175
Knapp, Tom
  and corrector drug development, 432
  fundraiser participation by, 336
  and high-throughput screening, 299–300, 305, 327–328
  and human cells for CF drug testing, 371, 372, 373
  and Vertex CF drug development, 339, 343, 346, 348–349, 364, 366–369, 391
Knowles, Michael, 107–111, 119, 247, 400
Kornberg, Arthur, 86
Koshland, Daniel, 212
Kramer, Liza, 484
Kramer-Golinkoff, Emily, 477–479, 483–486
Krenitsky, Paul, 468
Kronmal, Dick, 285
Kuno, Yas, 49, 50

L

Landauer, Kenneth, 72–74, 78, 85
Lapey, Allen, 83
  background of, 12–14
  as Cheevers daughters' doctor, 319, 411–414
  and combination drugs, 462–464, 468, 472–473
  as Farrell's doctor, 471–473
  and Joey's final illness and death, 186–188
  and Joey's frequent infections, 37
  Joey's treatment by, 14–16, 124
  Kathy's encouragement from, 34
  at Mass General CF unit, 13–14
  O'Donnells' choice of, 10–12
  and O'Donnells' genetic test, 163, 164
  O'Donnells' visit with, 161–162
Lasker, Mary, 69
Lawson, David, 151
Lederberg, Joshua, 86
Lehr, John, 427

Leiden, Jeffrey, 180, 423, 464, 467, 475
Lesch–Nyhan syndrome, 237–239
Lesnick, Dr., 56
Levinson, Art, 217
life expectancy with cystic fibrosis, 9, 74, 438, 477, 501
lipase, 31
Lovel, Mrs. Percy, 151
Lukacs, Gergely L., 466
lung research, 104–111, 380–386
lungs, 9
  with cystic fibrosis, 9, 21, 27, 106
  dislodging mucus from (see chest physical therapy)
  DNase thinning of mucus in, 216–222
  functioning of, 105–106
  and gene therapy, 242–243, 245
  in healthy people, 105–106
  in Joey, 12
  structure of, 105
  testing health of, 76
  and vitamin A deficiency, 26
lung transplants, 319

M

March of Dimes, 68
Marcuse, Edgar K., 226
Marshall, Bruce, 442
Marshall, Kate, 451
Martin, Denise "Bubs," 39, 97–98, 102, 124
Martin, Florence, 37, 39
Mason, Mary Louise, 19
Massachusetts General Hospital, 5, 13–14, 100–101
Masters, Ariana, 451
Matthews, LeRoy, 74–78, 86–89, 168, 216, 228–230
Mattingly, Richard, 131–132, 267, 354–361, 425–430
May, Charles D., 26, 51, 59
McCartney, Jason, 364, 365, 368–369, 432
McClintock, Barbara, 17
McCreery, Robert, 88, 91
McElvaney, Gerald, 255–259
McNulty, Jillian, 451
Mead, Margaret, 55
medical expenses, 40

Mendlein, John, 314–315
Menino, Tom, 280
Mense, Martin, 481
Merck, 85–86, 330–332
messenger RNA, 141, 144, 149–150
Microsoft, 309–314, 316
Miller, Dusty A., 240
"mist tents," 44, 75
Mohrenschildt, George de, 58
molecular biology, 174
molecular genetics, 180
Montgomery, Bruce, 233–234
Moore, Stanford, 217
Morgan, Wayne, 285
Mueller, Peter, 378, 403
Mulcahy, Jerry, 82–83, 101
Mulligan, Richard, 239
Murcko, Mark, 342, 353, 395

N

Nabel, Gary, 180
Nadeau, John, 80–84, 100, 101
Natal, Robert, 72, 78, 85
National Cystic Fibrosis Research
    Foundation (NCFRF), 38
  allocation of funding for research by,
      70, 90
  education emphasis of, 62–63
  founding of, 59
  Frankel as trustee for, 61
  fundraising by, 38, 59, 62, 64, 68–69,
      85–90
  GAP Conferences of, 86
  headquarters of, 72, 91–92
  local chapters of, 59, 61, 64, 70–71
  and March of Dimes, 68
  and Matthews' Comprehensive
      Treatment Program, 77–78
  move to Atlanta for, 88–89
  move to New York for, 63, 72
  O'Donnell's help from, 38–39
  origin of, 58–59
  patient registry of, 77–79
  regional and national trustees of, 90
  renaming of, 84, 91 (see also Cystic
      Fibrosis [CF] Foundation)
  research and medical education centers
      of, 72–79, 85

Sharples' departure from, 69–70
National Cystic Fibrosis Week, 102, 133
National Foundation for Infantile
    Paralysis (NFIP), 60, 65–68
National Heart, Lung, and Blood Institute
    advisory council, 95
National Institute of Arthritis and
    Metabolic Diseases (NIAMD), 54,
    87, 92–93
National Institutes of Health (NIH)
  cystic fibrosis recognized by, 54
  cystic fibrosis study by, 94
  di Sant'Agnese's work at, 54, 168
  Farrell's work at, 168–169
  funding from, 86–87, 94–96, 133, 194,
      196, 284
  and gene therapy, 251, 253–255, 262,
      265
  Human Genome Project at, 268–269
  Lapey's work at, 12–13
Negulescu, Paul, 290, 467
  background of, 291–294
  and Boger, 334–335
  CF drug development team of, 339–351
  and CF foundation, 336
  and cure for CF, 483
  drug discovery research at Aurora, 296–
      306, 314–316
  and FDA inspection of facility, 460
  and Ferguson's speech, 323
  and Gates Foundation funding, 314–316
  and growth of Vertex, 497
  and human cells for CF drug testing,
      373
  and treatment approvals, 497
  and treatment for rare mutations, 483
  and Vertex acquisition of Aurora,
      334–335
  and Vertex CF drug development, 374,
      375, 378, 394
  and Vertex CF research, 351, 363
  vision of, 327
Nesbitt, Cecil J., 77–78
Neuberger, Tim, 344–345, 373, 381–386,
    433
newborn screening, 166–171
Newmark, Peter, 158
Niehaus, Bob, 355–356, 359–362, 429–430

Niehaus, Kate, 355–356
night coughing, 161–162
nonsense mutations, 213, 478–479, 481,
    483–486

O

Obama, Barack, 445–446
Obama, Michelle, 445
O'Connor, Basil, 60, 65–68
O'Donnell, Casey, 275–276, 281–282, 487–
    491, 493, 498
O'Donnell, Dennis, 35
O'Donnell, Denny, 35, 37, 101
O'Donnell, Joe, xvi–xviii, 487
    business ventures of, 40–44, 125–126,
        279, 355
    and Casey's marriage, 487, 488
    and CF public service announcements,
        102–103
    and Cheevers family, 411, 415–416,
        418–419
    contacts and partnerships of, 279–280
    Coughlin's advice from, 321
    and early illness in Joey, 3–8
    and family foundation funding, 493
    and family's remembrance of Joey,
        275–276, 358
    fundraising by, 126–128, 277–282, 280,
        354–362, 425–430, 491–494,
        498–499
    genetic testing for, 498
    and granddaughter Blair, 489
    and Joey Fund, 277–282
    Joey's at-home care by, 36, 37, 43
    and Joey's diagnosis, 8–9
    and Joey's early elementary years, 97–
        101, 123
    and Joey's final illness and death,
        185–190
    and Joey's first Christmas, 35–36
    and Joey's medical expenses, 40, 43, 44
    and Joey's treatment, 14–16
    and Kate's birth, 190
    and Kathy's pregnancy with daughter,
        162–165
    and Massachusetts NCFRF support
        group, 39
    and Nadeau, 80

stress management by, 37–38
support network for, 39
and Trikafta approval, 497–498
visit at Lapey's cabin, 161–162
and VX-770 approval, 418–420
work at Harvard, 40, 42
O'Donnell, Joe (Poppy), 35, 39, 101
O'Donnell, Joey, xvi, xvii, 498
    at-home care for, 16, 36–37, 39, 83–84,
        100, 123
    continuing infections for, 37–38
    cystic fibrosis diagnosis for, 7–9
    cytomegalovirus of, 6
    discharge from hospital for at-home
        treatment, 34–35
    early hospital admission of, 5–7
    elementary school years of, 97–102,
        123–125
    family's memories of, 275–276
    feeding tube for, 16, 36
    final illness and death of, 185–188
    frequent infections in, 37, 100–101,
        125, 164
    funeral for, 188–189
    homecoming from early treatments, 16,
        34–35
    and Joey Fund, 277–282
    Joey's Park, 276, 490
    and Kathy's pregnancy, 163–165
    Lapey as doctor for, 10
    medical expenses for, 40, 43, 44
    misdiagnosed illness in, 3–5, 7, 11
    Nadeau's physical therapy for, 83–84
    and public service announcements, 103
    reporters' interviews with, 123–124
    treatment plan for, 11–12, 14–16, 44,
        100–101, 123–124, 186, 187
    visit at Lapey's cabin, 161–162
    visit with friends to morgue, 160–161
O'Donnell, Kate, 162–165, 185, 190, 275–
    276, 281–282, 489–491, 493
O'Donnell, Kathy, xvi–xviii
    career before Joey's birth, 40
    and Casey's marriage, 487
    and Cheevers family, 411, 415–416, 419
    and early illness in Joey, 3–8
    and family's remembrance of Joey,
        275–276

and friends' visits to hospital, 160
and fundraising, 428, 493
genetic testing for, 488, 498
and granddaughter Blair, 489
and Joey Fund, 277–282
Joey's at-home care by, 36–37, 39, 83, 123
and Joey's diagnosis, 7–9
and Joey's discharge from hospital for at-home treatment, 34–35
and Joey's early elementary years, 97–101, 123–125
and Joey's final illness and death, 185–190
and Joey's first Christmas, 35–36
and Joey's treatment, 14–16
and Kate's birth, 190
Lapey chosen by, 11
and Massachusetts NCFRF support group, 39
pregnancy with daughter, 162–165
stress management by, 38
support network for, 39
talk with Joey's friends after his death, 189–190
and Trikafta approval, 497
visit at Lapey's cabin, 161–162
and VX-770 approval, 419, 420
O'Donnell, Mary, 35, 37, 101
O'Donnell, Neil, 35
O'Donnell, Ruth, 35, 37
O'Donnell, Ted, 35
O'Donnell, Teresa (Nonni), 35–37, 39, 101
O'Farrell, Maggie, 34
Olson, Eric, 337, 347
    Aurora's job offer to, 341–343
    background of, 339–341
    and corrector drug criteria, 435
    and funding for research, 387–391
    and human cells for CF drug testing, 371–373
    and Vertex CF drug development, 375
    and Vertex CF research, 351, 352–353
    and VX-661 development, 463
    and VX-770 testing, 379, 393, 394, 403–404, 406–410
Ordonez, Claudia, 402–404

Orkambi, 452, 462–463, 467, 468, 497. see also VX-770/VX-809 combo drug
ornithine transcarbamylase deficiency (OTC), 264–265
Orphan Drug Act, 218, 232, 233
Osler, William, 218

P

Page, Juliet, 453–458
Palermo, Chuck, 409, 417
Palermo, Dan, 409
Palermo, Lisa, 409, 410
Palermo, Mark, 409, 417
Palermo, Maureen, 409, 417
pancreas, 21–22
    with celiac disease, 22
    with cystic fibrosis, 24–31
    and digestion, 29–30
pancreatic enzyme test, 31–32
Paracelsus, 191
Parcells, Bill, 280
Parker, H. Stewart, 310
Parmelee, Arthur, 24–25
PathoGenesis, 232–235, 267, 285, 289
patient advocacy, xviii
patient registry
    in Canada, 141
    for nonsense mutations, 484
    in United States, 77–79, 398, 442, 443
Patterson, Paul, 87
Pauling, Linus, 149
penicillin, 46–47
Penland, Christopher, 300, 315, 381, 388
personalized medicine, 420–421, 444
Perutz, Max, 150
Pfizer, 341
pharmaceutical industry. see also drug therapy(-ies)
    and CF Foundation's drug development network, 283, 288–289 (see also individual companies)
    and costs of drug development, 266–267
    drug testing in, 295
    and gene therapy revolution, 268
phenylalanine, 202
phenylketonuria (PKU) testing, 167
physical therapy, 169. see also chest physical therapy

The Physiology of Human Perspiration
(Juno), 49
Pilewski, Joseph, 393–394
Plunkett, Andrew, 490
Pogue, Richard, 77
polio research, 65–68
polio treatment, 65–66
polio vaccine, 60, 67–68
polymorphisms, 145–147, 152, 153
precision medicine, 444–446
Precision Medicine Initiative, 445–446
prenatal genetic testing, 162–164
Prontosil, 20
proteins, 142. *see also specific proteins*
in CF patients and healthy persons, 141,
209
families of, 203–204
ion channels, 299, 335, 339, 343
manufacturing of, 215
and structure-based drug design,
332–333
Protropin, 215
Proust, Marcel, 23
Pulmozyme development, 214–223, 229,
233, 267

Q

Quinton, Paul, 112–122, 142–143, 203, 423,
444
Quorum Sciences, 340, 341

R

R117H mutation, 443, 444, 458, 459
Ramsey, Bonnie
and antibiotic treatment for lungs, 229
background of, 224–227
at Boston Children's Hospital, 224–225
and CF Foundation's drug development
network, 283–289
in CF Foundation's treatment consensus
group, 228
as Flessner's physician, 309, 310
and inhaled tobramycin, 230–235
and Microsoft funding, 311, 312
and Pulmozyme development, 218, 219,
229, 233
at Seattle Children's, 226–227

read-through agents, 481
Reagan, Ronald, 218
recombinant DNA technology, 174–176,
214
recombination (crossovers), 154
relationships of doctors and patients/
families, 32–33, 76, 84
renin inhibitors, 330
research, 53–54. *see also individual
researchers, companies, and conditions*
allocation of funds for, 89
birth of national foundation to promote,
58–59
CF Foundation centers, 129, 133–135
and CF Foundation drug development
network, 283–290, 335–338 (*see
also* drug therapy(-ies); *individual
companies*)
CF gene identification, 191–213, 220
on chromosome jumping, 179–184
for curing CF, 481, 483, 492
to distinguish CF from celiac disease,
31–32
dyes for, 292–294, 297, 334
early discoveries of new disease, 17–18,
20–30
early small groups' fundraising for,
56–58, 61, 64–65
Frankel's promotion of, 60–61
funding for, 70, 73, 85–87, 89–91,
94–96, 116, 131, 133, 147, 184,
194, 196, 197, 210, 266–268,
277, 284, 289, 306, 311–317,
335–338, 341, 354–362, 387–391,
422–423, 430, 446–450, 462,
484–486, 492, 493
fundraising for (*see* fundraising)
genetic (*see* gene therapy; genetics)
on hemoglobin, 149–150, 177–178
on lungs, 104–111
by National Cystic Fibrosis Research
Foundation, 38
at National Institutes of Health, 12–13
by NCFRF centers, 73
on nonsense mutations, 481, 484–486
on polio, 65–68
on salt–water balance, 107–111
stem cell, 481, 482, 492, 493

on sweat glands, 115–122, 142–144, 200

Tulcin's promotion of, 59–61

Ribicoff, Abraham, 60, 64, 65, 95

Rink, Tim, 334

Riordan, John "Jack," 141–144, 199, 200, 202–204, 209, 210

Rommens, Johanna, 191–198, 200–203, 206, 209, 210, 212

Romney, Mitt, 280, 428

Roosevelt, Franklin D., 60, 65–66, 67, 431

Rosenfeld, M. Geoffrey, 294

Rosenfeld, Melissa, 248–252, 255–256, 258. *see also* Ashlock, Melissa Rosenfeld

Rossi, Paul Del, 98, 127–128, 189

Rosso, Henry A., 425

royalty payments, 315, 360, 423–424, 447–450, 452

Royalty Pharma, 449

Rozmahel, Richard, 201–202

S

Sabin, Florence Rena, 19

Sabin, Gary, 359–362

Sagan, Carl, 104

Saint-Exupéry, Antoine de, 339, 363

salt transport, 116–122, 270

salt–water balance, 104, 107–111, 403

Satcher, David, 166

Sato, Vicki, 335–338, 341, 390, 403

Schaner, Ken, 314, 315

Schering, 13

Seattle Children's, 226–227, 233–234

Selden, Robert, 286

1717-1G>A mutation, 475, 488

Shak, Steven, 214–219, 229, 233

Shaltz, Sy, 57

Sharples, Philip, 58, 59

Sharples, Wynne "Didi," 57–63, 69–70, 75

Shaw, George Bernard, 264

Sherlock Holmes, 72

Shwachman, Harry (Henry), 10, 14, 224
    and aerosolizing of antibiotics, 229–230
    background of, 50–51
    in CF Foundation's treatment consensus group, 228
    chest physical therapy taught by, 75, 80–82

and di Sant'Agnese, 168–169
    education emphasized by, 62
    humanity and devotion of, 225
    as leading CF physician in country, 168
    Matthews taught by, 75
    on NCFRF medical education committee, 59, 69
    Sharples' funding for, 58
    and sweat test, 50–52
    Tulcin's checkups by, 69

Shwachman, Irene, 50, 51

sickle cell disease, 176, 177

Sigal, Irving, 332

Silina, Alina, 467

Skach, William, 480–482, 492

Smith, Alan, 259, 260

Smith, Arnold, 218, 229–231, 233, 235

Smith, G.T., 354

Soprano, Debbie, 128, 277, 278

Sosnay, Patrick, 453–455

Spirovant Sciences, 485

St. Mary's Hospital Medical School, 150

Steffan, Michael, 325–326

Stein, William H., 217

stem cell research, 481, 482, 492, 493

Stenzel, Anabel, 220–223

Stenzel, Isabel, 220–223

Stepek, Marjorie, 77

Stockton, William, 149

Summerhayes, Donna, 140

Summerhayes, Doug, 140

support groups, 38–39

sweat glands research, 116–122, 142–144, 200

sweat test(s), 75
    di Sant'Agnese's development of, 48–50
    Gibson and Cooke's iontophoresis test, 53, 75–76
    for Joey, 4, 7
    for newborns, 75
    pros and cons of, 169
    Shwachman's use of, 50–52

Symdeko, 465, 468, 471, 497. *see also* VX-770/VX-661 combo drug

T

Targeted Genetics, 262, 288, 289, 309–310

Teresa, Mother, 123

testing. *see* diagnosis of cystic fibrosis; genetic testing

thalassemia, 149–150

Therapeutics Development Network (TDN), 284–290, 399

theratyping, 480

Thomas, Philip, 466

Thompson, Craig, 180

Thoreau, Henry David, 85

Tinmouth, Phil, 388, 389

Tishler, Max, 329–331

tissue plasminogen activator (TPA), 216

tobramycin, inhaled, 229–235

Tobramycin Inhalation Solution (TOBI), 235, 267, 289, 315, 399, 423, 447

treatments for cystic fibrosis, xvii–xviii, 496. *see also* drug therapy(-ies); gene therapy

antibiotics, 13, 14, 20, 39, 46, 75, 76, 101, 186, 187, 308

chest physical therapy, 14, 36, 39, 75, 76, 80–84, 100

di Sant'Agnese's work on, 45–48

with early diagnosis, 31

for Flessners, 308–310

gentamicin, 13

inhaled tobramycin, 229–235

Joey's treatment plan, 11–12, 14–16, 44, 100–101, 123–124, 186

lack of medical consensus on, 73–74

lack of standard protocols for, 72

lung transplants, 319

Matthews' Comprehensive Treatment Program, 74–78

"mist tents," 44, 75

national CF Foundation's improvement of, 228

nonsense mutation therapies, 478–479, 481, 483–486

nutrition, 75, 76, 228

penicillin, 46–47

periodic "cleanouts" for patients, 39

Pulmozyme (DNase), 214–223, 229

for rare mutations, 478–482

testing of, 284

unproven, 168

Van Goor's garden commemorating, 496

Trikafta, 492, 495–497. *see also* VX-659/ VX-770/VX-661 combo drug

Truman, Harry, 60

trypsin, 31, 32

Tsien, Roger, 271, 290–298, 334

Tsui, Lap-Chee

background of, 139–140

and CF gene identification, 191–203, 206–211, 240

and Chakravarti, 173

Collins' work with, 196–197

presenting at Royal Society, 192–193

refinement of approach by, 172

research by, 144–148, 152–158, 191– 203, 399

and Williamson's incorrect DNA sequence, 193–196

tuberculosis (TB), 113–114

Tucker, Ron, 150, 151

Tulcin, Ann, 59–61, 69, 88, 356, 399, 497

Tulcin, Bob, 60

Tulcin, Doris

and Ann's disease, 59–60, 69, 88, 497

as CF Foundation president, 127

and Dresing, 89, 91, 129

fundraising by, 61, 64–66, 68, 69, 88– 90, 94–96, 129, 134–135

and Genentech's research, 214–215, 217–218

and Kalydeco approval, 422

on National Heart, Lung, and Blood Institute advisory council, 95

as NCFRF president, 91

and O'Connor, 66

O'Donnell's visit from, 127

and Orphan Drug Act, 218

and research network creation, 96

and Sharples' resignation, 70

Twain, Mark, 172

U

Urbina, Armando, 467

V

Van Goor, Fred, 469

airway cells grown by, 380–381, 383– 394

background of, 343–344

on CF drug development team, 339, 343–344, 363

and combination drugs, 451, 465, 469

and corrector drugs, 352, 432–435, 467, 469–470

and doorman drug, 366, 367, 374–378, 403–404

and effective treatments garden, 496

and FDA inspection of facility, 460

and human cells for CF drug testing, 371–373

and treatment for rare mutations, 458–459, 483

Verkman, Alan, 270, 290, 484

Vertex Pharmaceuticals Inc., xvi–xviii, 357

acquisition of Aurora Biosciences, 328, 333–335, 341–342

CF Foundation funding for, 337, 354–362, 387–390, 422–423, 430, 462, 492

clinical trials for CF drugs, 391–392, 395–410, 413–417, 435–442, 455, 464, 469–473, 475

combination drugs, 436–441, 449–452, 462–466, 469–476

corrector drugs development, 431–441, 463, 465–468, 471, 474

doorman drug development, 363–386, 371–374, 379–380, 387–394, 397–410, 413–422, 431–434, 436, 437, 471

drug development project for CF, 335–338, 342–353, 477–478

drug development team, 339–351, 364–365

drug production by, 379–380, 389–393

and drugs for rare mutations, 480

drug testing work of, 333–334

founding of, 332

gene therapy technologies research, 483

growth of, 497

and Incivek, 424

Kalydeco testing with mutations other than G551D, 441–444, 457–461

and precision medicine, 444, 446

and price of Kalydeco, 422–424, 443, 447

and sale of drug royalties, 449

structure-based drug design at, 332–333

Vetter, Joseph, 244

vitamin A deficiency, 25–26, 29, 30

VX-152, 467, 468, 469–470

VX-440, 467, 469–470

VX-445, 468, 471, 473, 474

VX-659, 468, 470, 471–474

VX-659/VX-770/VX-661 combo drug, 473–476. see also Trikafta

VX-661 (corrector drug), 436–437, 441, 463, 471. see also Symdeko

VX-770 (doorman drug), 379–380, 391–394, 397–410, 413–421, 431–434, 436, 437, 471. see also Kalydeco

VX-770/VX-661 combo drug, 462–465, 467. see also Symdeko

VX-770/VX-809 combo drug, 436–441, 449–452, 463

VX-809 (corrector drug), 431–440, 463, 465–466. see also Orkambi

W

W1282X mutation, 213, 477, 478–479

Wallace, William, 74

Walton, David, 7

Warner Lambert, 340, 341

"War on Cystic Fibrosis," 62–63

Warwick, Warren, 77–79

Washington, Denzel, 275

Watson, James, 86, 268

Weaver, Harry, 67–68

Weicker, Lowell, 94

Weinberg, Amy, 356

Weinberg, John, 356, 359–362

Weiss, Mary, 134

Weissman, Sherman, 174, 177–179

Welsh, Michael, 244–245, 253, 254, 259–260, 269–270, 301–302, 373

Wexler, Nancy, 148

Whedon, Donald, 93, 94

White, Raymond, 157, 158, 192

The William H. Gates Foundation, 311–317

Williams, Judy, 285, 289

Williamson, Robert, 149–152, 157, 158, 192–197, 206, 207, 252, 262–263

Wilson, E. O., 291, 371

Wilson, James, 180, 237–243, 245, 252–254, 260–262, 264–266

Wine, Jeffrey, 380–381

Woglom, Katherine, 20, 46

Wohl, Mary Ellen, 227
Wolbach, Simeon Burt, 25–26

Y

Young, Katrina, 325–326
Young, Robert, 325–326
Young, Sean, 325–326, 457

Z

Zhou, Jinglan, 365, 367–368, 374
zoo blot, 198–199
Zucker, Charles, 294

ABOUT THE AUTHOR

Photo by Chad Cohen
and Dhruv Cohen

BIJAL TRIVEDI'S award-winning writing has been featured in the *Best American Science and Nature Writing 2012*, *National Geographic*, *Scientific American*, *Wired*, *Science*, *Nature*, *Economist*, *Discover*, and *New Scientist*. Her work has taken her from the Mexico–Guatemala border where she covered the use of genetically modified mosquitoes for fighting the dengue virus to behind the scenes at Massachusetts General Hospital where she watched trauma surgeons induce profound hypothermia to save pigs with life-threatening injuries to Moscow's Star City where she blasted off with space tourism entrepreneurs on the "Vomit Comet" for astronaut training.

Bijal's undergraduate fascination with biochemistry and molecular biology at Oberlin College compelled her to pursue a master's degree in molecular/cell/developmental biology at UCLA. Her love of writing drew her to journalism rather than to a lab bench—and to a second master's degree in science journalism from New York University. She has edited the National Institutes of Health Director's Blog and helped launch the National Geographic News Service in partnership with the New York Times Syndicate, which she wrote for and edited. She currently works as a science and technology editor for the *Conversation*.

Bijal lives with her husband Chad and her two children, Dhruv and Sonali, in Washington, DC.